Spatial Representation

Spatial Representation

Problems in Philosophy and Psychology

Edited by

Naomi Eilan, Rosaleen McCarthy
and Bill Brewer

BLACKWELL
Oxford UK & Cambridge USA

First published 1993

Blackwell Publishers
108 Cowley Road
Oxford OX4 1JF
UK

238 Main Street, Suite 501
Cambridge, Massachusetts 02142
USA

British Library Cataloguing in Publication Data

A CIP catalogue record for this book is available from the British Library.

Library of Congress Cataloging-in-Publication Data

Spatial representation / edited by Naomi Eilan, Rosaleen McCarthy, and
Bill Brewer.
p. cm.
Includes bibliographical references and index.
ISBN 0–631–18355–8 (alk. paper)
1. Space perception. 2. Mental representation. I. Eilan, Naomi.
II. McCarthy, Rosaleen A. III. Brewer, Bill.
BF469.S675 1993
153.7′52—dc20 92–43144
CIP

Typeset in 9½ on 11pt Erhardt
by Graphicraft Typesetters Ltd, Hong Kong
Printed in Great Britain by TJ Press Ltd, Padstow, Cornwall

This book is printed on acid-free paper

Contents

Contributors

Janette Atkinson is a member of the Medical Research Council's Senior Scientific Staff, and co-founder of the Visual Development Unit at the University of Cambridge. She has worked widely in human neuroscience, especially on the normal and abnormal development of the brain mechanisms of vision in infancy and early childhood.

Bernard Balleine graduated in psychology from the University of Sydney and is now a Research Fellow at Jesus College, Cambridge. Current research interests include the behavioural and neurophysiological basis of goal-directed action, particularly with regard to the influence of motivation processes.

Oliver Braddick is Reader in Vision in the Department of Experimental Psychology at the University of Cambridge, and co-founder of the Visual Development Unit. He has worked widely in the areas of spatial and pattern vision, motion perception, stereo vision and the development of these aspects of visual processing in infancy.

Bill Brewer is a Fellow in Philosophy at St Catherine's College, Oxford; he was formerly a Research Fellow on the Spatial Representation Project at King's College, Cambridge. He works and has published on issues in the philosophy of mind and action, metaphysics and epistemology. He is married with a three-year-old daughter.

John Campbell has been Fellow and Tutor in Philosophy at New College, Oxford since 1986; before that he was Research Lecturer at Christ Church, Oxford. He has held visiting appointments at UCLA and the Australian National University and has been visiting Fellow at the King's College Research Centre. He has published on metaphysics and the philosophy of mind and language.

Lynn Cooper is Professor and Chair of the Department of Psychology at Columbia University. She received her BA degree from the University of Michigan and her PhD from Stanford University. She is a member of the American Academy of Arts and Sciences.

Anthony Dickinson is a University Lecturer in the Department of Experimental Psychology at the University of Cambridge. His interest in basic learning and motivational processes is reflected in *Mechanism of Learning and Motivation* (1979, edited with

R. A. Boakes) and *Contemporary Animal Learning Theory* (1980) and he is currently the editor of the comparative and physiological psychology section of *The Quarterly Journal of Experimental Psychology*.

Naomi Eilan is a Senior Research Fellow in Philosohy at the King's College Research Centre, Cambridge, where she works on problems in consciousness, the self and spatial thought. Since joining the Spatial Representation Project at King's she has focused on drawing on work in developmental, cognitive and neuropsychology in approaching traditional philosophical puzzles in this area.

Roberta Klatzky is a Professor of Psychology at the University of California, Santa Barbara. She received a BSc in mathematics from the University of Michigan and a PhD in psychology from Stanford University. Her research has concerned the ways that people represent and think about aspects of the world that are not easily verbalized. She has had a long-term collaboration with Susan Lederman on object representation and categorization through the modality of touch.

Susan Lederman is a Professor of Psychology and Computing and Information Sciences at Queen's University in Kingston, Canada. Her research interests generally involve how people sense, think about and manipulate objects with their hands with and without vision. With Roberta Klatzky, she has investigated the nature of haptic object representation and classification, and addressed how knowledge o biological tactile systems might be applied to the design of sensor-based autonomous and teleoperated robotic systems.

Rosaleen McCarthy is a Senior Research Fellow on the Spatial Representation Project at the King's College Research Centre and University Lecturer in the Department of Experimental Psychology in Cambridge. She has worked as a clinician and has published widely as a researcher in cognitive neuropsychology. Together with Professor Elizabeth Warrington she is co-author of *Cognitive Neuropsychology: A clinical introduction*.

Michael Martin is a Lecturer in Philosophy at University College, London; he was formerly a Junior Research Fellow at Christ Church, Oxford. He has written several articles on the philosophy of perception.

Andrew Meltzoff is a Professor of Psychology at the University of Washington and head of the Development Psychology Unit. He received his BA from Harvard and his DPhil from Oxford. He is a Fellow of the American Academy for the Advancement of Science, The American Psychological Association and the American Psychological Society. Professor Meltzoff serves on the Editorial Board of the *British Journal of Developmental Psychology* and *Infant Behavior and Development*. He was recently given the James Mckeen Cattell Award.

Ruth Millikan is Professor of Philosophy at the University of Connecticut. She graduated from Oberlin College and received her PhD from Yale University. She has published several books: *Language, Thought and Other Biological Categories* (1984), *White Queen Psychology and other Essays for Alice* (1993) and has written numerous articles on philosophy of psychology, philosophy of biology and philosophy of language of *Mind*, *Journal of Philosophy* and *Philosophical Review*.

Margaret Munger received her BA degree from the University of Chicago and her MA and MPhil from Columbia University. She is currently a doctoral candidate at Columbia University.

John O'Keefe has been Professor of Cognitive Neuroscience in the Department of Anatomy and Developmental Biology at University College, London since 1987. He studied engineering at New York University and psychology and philosophy at City University of New York, where he gained a BA in psychology in 1963. He received an MA in physiological psychology from McGill University in 1964 and a PhD in 1967. In 1992 he was elected Fellow of the Royal Society in recognition of his contribution to the understanding of the spatial and memory functions of the brain.

Christopher Peacocke is Waynflete Professor of Metaphysical Philosophy in the University of Oxford and a Fellow of Magdalen College. He has written on the philosophy of mind, the philosophy of language and logic, and the philosophy of psychology. His books include *Sense and Content* (1983), *Thoughts: An essay on content* (1986) and *A Study of Concepts* (1992). He was formerly the Susan Stebbing Professor of Philosophy at King's College, London, and has held visiting appointments at Berkeley, UCLA and Michigan. He was a Fellow of the Center for Advanced Study in the Behavioral Sciences at Stanford in 1983–4, and was elected a Fellow of the British Academy in 1990.

Julian Pears is a Research Fellow at the King's College Research Centre, Cambridge.

Herbert Pick Jr is a Professor of Child Psychology at the University of Minnesota and Director of Graduate Studies of the Cognitive Science Program. He received his PhD from Cornell University in 1960. His research interests are in spatially co-ordinated behaviour broadly construed. The particular problems range from how people use perceptual input to guide their locomotion to how wayfinders navigate through the wilderness using topographic maps. In between these poles he and his students have investigated how children maintain spatial orientation in their environments.

Elizabeth Spelke is a Professor at Cornell University in the fields of psychology, human development and cognitive studies. Her research has focused on the early development of perception and cognition of objects, space and number.

Michael Tye is Professor of Philosophy at Temple University and King's College, London. He is the author of two books: *The Metaphysics of Mind* (1989) and *The Imagery Debate* (1991). He is also the author of articles in *Philosophical Review*. *Journal of Philosophy*, *Mind* and other leading philosophical journals.

Gretchen Van de Walle is a graduate student at Cornell University in the fields of psychology and cognitive studies. Her research has focused on the early development of perception of motion, space and objects.

Preface

The contributions to this volume originated as presentations to an interdisciplinary workshop organized by the Spatial Representation Project at the King's College Research Centre, Cambridge, in April 1991. This brought together a dynamic group of philosophers and psychologists of many persuasions, actively working on a broad range of issues connected with spatial representation. It involved a number of people whose contribution is not explicitly represented by a chapter in this book. We are sure we speak for all our authors in thanking everyone who attended the workshop for the lively and constructive discussions which arose both in and out of the scheduled sessions. These clearly made a significant impact on the revised versions of presented material which appear in this volume.

The workshop marked the half-way point of a four-year interdisciplinary research project on spatial representation funded by a grant to the Research Centre at King's College, Cambridge, from the Leverhulme Trust. The project was set up in the belief that there is a great deal of overlap in many of the most important and central questions being asked in distinct areas of research on spatial perception, cognition and action, and that our understanding of these pivotal problems will benefit greatly from bringing together the insights and methods of several distinct areas in psychology and philosophy. We have since discovered both how difficult and yet also how rewarding it is working to make good this abstract hunch. The workshop put real flesh on these rewards, as the individual contributions, and the opportunities for cross-disciplinary exchange which they provided, will, we believe, make evident.

The project consists of a permanent group of four full-time researchers from a variety of branches of psychology and philosophy, along with a number of visitors, two of whom have been intimately associated with the project for most or all of the time since its inception in September 1989. We, the editors, would like first of all to express our deep gratitude to these very close colleagues: Julian Pears, who is the fourth permanent member of the project; and John Campbell and Tony Marcel, the two long-term visitors, both of whom have contributed massively and formatively to our joint research. All three have made invaluable contributions both to the seminar series we have been running throughout the project and to the production of this book, from the earliest ideas about the workshop right through to the end of the editorial process.

We have been very fortunate also to have had a number of other visitors, who have stayed with us for various lengths of time, and who have made very stimulating contributions to our work. We would like in particular to thank John O'Keefe and Christopher Peacocke, both of whom have been very supportive of the project from the outset, and whose work has provided stimulating input to our joint research. Christopher Peacocke's comments on previous drafts of our various editorial introductions were extremely useful. Martin Davies was also very helpful in improving these sections of the book. Other visitors who have been instrumental in shaping our thought in the area include Kathleen Akins, Renée Baillargeon, Jonathan Cole, Mike Martin, Brian O'Shaughnessy, Jacques Paillard, Wolfgang Prinz and Michael Turvey.

Without the support of the Leverhulme Trust, there would have been no project. We hope this collection goes some way to justify and express our thanks for their funding and commitment to our research. The current academic climate makes innovative adventures across disciplinary boundaries even more difficult than they have always been. As well as creating the possibility for a book of this kind, and perhaps further follow-up collections and other related publications, the Leverhulme Trust has enabled us to go some way towards establishing an interdisciplinary, philosophical–psychological approach to issues in spatial representation, issues that lie at the heart of long-standing and fundamental problems concerning the relation between mind and world.

King's College Research Centre has always had a commitment to ventures of this kind. We are very grateful to its Convenor, Martin Hyland, and the other managers for their continual support, advice and encouragement throughout the project. The whole book would also be in a far worse state without Martin's patient comments. Many thanks also to the Provost and Fellows of King's College, Cambridge, for granting us Research Fellowships for the duration of the project, which has enabled our work to benefit from a quite unique intellectual environment.

It has been a pleasure to work with Blackwell Publishers in putting this volume together. We would like in particular to thank Stephan Chambers and Steve Smith for their encouragement and advice, and Jack Messenger for excellent copy-editing of the manuscript, which did much to improve it.

Finally, we would like to express our gratitude to Rosemarie Baines for transforming an unruly collection of papers in every conceivable form, into a clean and uniform typescript.

Naomi Eilan
Rosaleen McCarthy
Bill Brewer

General Introduction

Naomi Eilan, Rosaleen McCarthy and Bill Brewer

The aim of these introductory remarks is to explain something of the rationale for holding a joint philosophy and psychology workshop on spatial representation, most of the papers delivered at which are collected in this volume. In Part I we give a very general and somewhat philosophically oriented description of why we believe issues of spatial representation are so central to an understanding of the way the mind represents the external world, by describing four basic but abstract ingredients in our conception of an external spatial world which we believe the chapters in this volume help to articulate. In section II we give a very brief summary of some of the major problems, philosophical and psychological, raised and discussed in the chapters in each of the five parts of the book.

Perhaps the heart of the chapter is to be found in section III, in which we indicate why we think progress in the nature of spatial representation must draw jointly on the resources of psychological and philosophical approaches to these issues. There we consider several examples gleaned from various chapters of different kinds and forms of interaction between particular psychological claims and concerns, on the one hand, and particular philosophical claims and concerns, including the concerns described in section I, on the other. Finally, in section IV, we highlight one of the issues raised, implicitly and explicitly, by a variety of chapters, which we believe provides particularly fertile ground for further joint psychological and philosophical work.

I

What is the connection between the capacity for spatial thought and grasp of the idea of a world out there, an external world we inhabit? The connection runs so deep the question may appear at first blush senseless. Surely to think spatially just is to think of the environment we inhabit! If this is so, the value of the question lies in its reminding us that this is what we mean, rather than the capacity to engage in formal or mathematical geometry. Engaging in geometrical calculations and proofs of geometrical theorems is not yet representing an external world. And this distinction, between mathematical geometry on the one hand, and thought about a spatial environment on the other, can be used to introduce the central issue that the chapters in this volume should be thought of as addressing. What makes what we call 'spatial thought' more than an exercise in mathematical computation but, rather, a way of representing the environment we inhabit, the external world 'out there', as we say? What does the capacity for this kind of spatial thought require, granted that it is more than and distinct from the capacity for formal geometry?

(1) By 'the world out there' we mean, at least, the physical world. So the question is, in part: What makes spatial thought thought about the physical world? One fundamental ingredient in the idea of a physical as opposed to merely abstract or mathematical space is that a physical space is, essentially, a space in which movement, change or, more generally, causal processes and events can occur. The phrase 'intuitive (or "naive") physics' has come to mean the physical principles we employ in our everyday perception, thought and action, which give substance and structure to this fundamental idea, by systematically linking geometrical properties such as length and volume with physical properties such as velocity, acceleration and mass. Predicting where a ball in flight will fall is one example of a primitive grasp of some such principle. Another is to be found in the way one regulates one's speed and acceleration when racing someone to a location one can see. So the question of how spatial thought concerns the physical environment is the question of what these spatio–physical principles are, and what is involved in grasping them.

(2) By 'the world out there', the external world, we also mean the world as it is independently of our interaction and engagement with it; it is what is 'there anyway'. The idea that we have such a conception of the world is sometimes expressed in the claim that we have a conception of the world as it is from no particular point of view, and that our conception of the world is, in this sense, objective. If spatial thought is thought about the world out there in this sense, then we should expect an account of spatial thought to explain how our spatial representation makes available such a disengaged, objective picture of the world. The claim that spatial thought does provide us with the framework for thinking in this way about the world is, in recent writings, often linked with the idea that we can and do use absolute or allocentric 'cognitive maps' of places and their spatial relations, where the contrast is with egocentric representations of our environment, that is, ways of representing places and their relations which are, in some way, subject-relative. So the question of what makes spatial thought thought about the world out there is, in part, a matter of the following questions. First, how exactly should we characterize the idea of an allocentric representation? Second, what precisely is the connection between allocentric representations of places and their relations on the one hand, and the idea of a detached or objective representation of the world, on the other? Finally, what, if any, is the connection between the answer we give to the latter question and the explanation of the way in which spatial thought is thought of a physical world – that is, does the characterization of what is involved in grasping an intuitive physics have any bearing on the way we characterize the relation between allocentricity and objectivity?

(3) But when we talk of 'the world out there' we mean not only the world as it is anyway, independently of our engagement with it. We mean in addition, though

possibly not in contrast, the world *immediately* out there, i.e. the world we inhabit, the world we perceive and in which we act, the world in which we are located. What this comes to can be illustrated as follows. When viewing a scene in a film, say a car chase, the way one represents places and their relations is completely independent of and has no bearing on one's own position in the represented world. Our representation of the world depicted in the film is in this sense detached. But by 'the world out there' we mean precisely a world which does contain us in it. If spatial thought is thought about the world in this sense, we should expect an account of spatial thought to explain the difference between the way we represent places when watching a film, for example, and the way of representing places which does, automatically, have bearings on where we are in that world. How exactly should we characterize the latter way of representing the environment? Here, it is natural to appeal to the notion of egocentric frames of reference for representing places and their relations. So the question of what makes spatial thought thought about the world out there is partly a matter of how we should characterize the notion of egocentric frames of reference. Is there a characterization that captures this engaged way of representing the world? How should we explain the interaction between egocentric and allocentric representations in giving a full account of the way we represent the world out there? What, if any, is the link between the egocentricity of a spatial representation and the various ways in which geometric and physical reasoning are interwoven?

The idea of 'the world out there', the external world, has, then, at least these three ingredients in it: it is the physical world; it is the world as it is anyway, the objective world; and it is the world immediately out there, the world we inhabit.

(4) Consider now the difference between a mouse who can find its way about a house, say, and a human subject watching herself in a video screen that shows her as she walks about a shop. Intuitively, the way the mouse and the film represent relations among places each lacks something the other has for making possible the self-conscious thought of oneself as located in the places represented.

Thus, while there is certainly a sense in which the mouse can locate itself relative to places it represents on its 'map' of the house, this sense does not involve the capacity to think of itself, from the outside as it were, as one object among others in the space represented, as one does when watching a film of oneself. It need not involve any kind of representation of itself as an object at all, and the capacity to do so is one ingredient in full self-consciousness. But the navigating mouse has something the film does not provide: for in watching the film of myself I may not realize that the person standing next to a shelf of baked beans is me. This way of representing what is in fact one's own location from the outside is not sufficient on its own to yield thought: 'I am in such and such a location'. Intuitively, one reason it is not sufficient is connected with the fact that the way spatial relations are represented is not hooked up with my navigational abilities, with my capacity to act in the space represented, where it is precisely such a hook-up that gives the sense in which the mouse can locate itself.

Thinking of oneself from the outside, as one object among others, is one ingredient in self-consciousness. The immediate link between one's perceptions and one's own actions is, arguably, another. The insight that there are deep and difficult connections between explanations of the nature of spatial thought and different ingredients in what is required for self-consciousness, goes back to Kant. Certainly the full import of the phrase 'the world out there' or 'the external world' has some connection with the idea that it is the world which one self-consciously thinks of oneself as inhabiting. The fourth very general question about the nature of spatial thought is, then: How do the ways in which spatial thought is thought about the physical world, the world as it is anyway, and the world immediately out there, link up with the capacity for self-consciousness? Which of these ingredients are needed for making possible such self-conscious self-locating thoughts, and how exactly do they do this?

Very generally and abstractly put, all the chapters in this interdisciplinary volume can be seen as contributions to unpacking the following questions. How are these four ingredients in the notion of a world out there related to each other? How do the capacities for spatial thought, perception (in the various sensory modalities) and action contribute to a grasp of these ingredients in our notion of a world out there? How should the fact that these spatial capacities are essentially interwoven with a grasp of one or more of these ingredients in the notion of a world out there, constrain the account we give of these capacities?

More specifically, each of the five parts of the book addresses a particular topic which bears directly on these very general questions in a way that begins to provide the materials for making them less abstract and more manageable. The individual contributions to each part, by psychologists and philosophers, serve simultaneously to illustrate how dense is the web of conceptual and empirical issues involved in explaining the nature of spatial representation, and, we believe, how much progress can be made by combining the resources of philosophical and psychological approaches to this issue, an issue that lies at the very heart of any explanation of the relation between mind and the world.

As we said at the outset, the particular point of this volume is that it brings together work in philosophy and psychology on central issues in spatial representation. We have so far been describing, in very generalized and abstract terms, one philosophical perspective on these issues, a perspective which, we believe, serves to illustrate why issues of spatial representation are so central to an account of the way the mind represents the world. We also believe that any progress with these questions can only be made by confronting them with independent psychological perspectives on issues of spatial representation. Conversely, we believe that progress with the latter requires explicit recognition and use of more specific formulations of the general philosophical questions we have been outlining. In section III we will give some illustrations of the kind of exchange of ideas we have in mind, drawing on the contributions to this volume by philosophers and psychologists. Before that, in section II, we provide a brief description of the topics covered in the book.

II

The book is divided into five parts, each part covering a topic central to an explanation of spatial representation. The divisions, which are the divisions that structured the workshop, roughly reflect traditional distinctions among topics in both psychology and philosophy. One of the virtues of the way issues are tackled by the psychologists and philosophers within each part is that their chapters serve to bring out the strong, multilayered connections among the five parts; connections that were brought out and made vivid in the discussion sessions during the workshop. There are many chapters that could have easily appeared under other headings from those that they appear under here.

What follows is a very summary description of issues tackled under each heading. Readers interested in more detailed introductions to specific issues addressed in the individual chapters may consult the introductory comments at the beginning of each part. These contain short summaries of central points made in each chapter, and comments aimed at highlighting some of what appear to us to be central issues and new questions that arise from a comparison of the particular psychological and philosophical problems raised by the individual contributors to these parts. Here we make do with a list of topics covered in each part.

Frames of reference

We are all familiar with the difference between being able to get to point *a* only via a certain route, and being able to get to it from any other location in its vicinity. The notion of 'frames of reference' is much used in psychological taxonomies of differences between spatial representations of places, of the kind that underpin, for example, the differences between the above two navigational abilities. However, there is considerable variation in the way the term 'frame of reference' is understood, which in turn leads to quite different accounts of how and whether the taxonomy of spatial frames of reference maps on to explanations of what is involved in grasping the idea of a world out there, in any of the senses we sketched.

The chapters in this part are all concerned with providing theoretically motivated distinctions between types of frames of reference used in representing places. Particular issues addressed include: How ought we to categorize the various different ways of representing places and spatial relations, and which of these categories generate connections with which, if any, of the ingredients sketched earlier in the idea of a world out there? And what exactly is the nature of this connection? To which of these categories do the spatial representations generated by various computational models proposed for explaining navigational abilities belong? How do these categorizations match with the developmental trajectory of children's spatial abilities? How do we discover the type of spatial representation being employed or required by particular organisms or particular tasks? What in general can count as evidence for the use of this or that kind of spatial representation?

Intuitive physics

Imagine rolling a ball into a hole in the grass, or predicting where a ball in flight will fall. The capacity to do so requires a grasp, on some level and in some way, of an intuitive or naive physics; that is, of the systematic interdependences that exist between semetric properties and magnitudes such as shape, size and distance; and physical properties and magnitudes such as velocity, acceleration and, at least in the first case, force. Indeed, the capacity to represent physical objects is one very fundamental way we have of putting physical flesh, so to speak, on our spatial concepts. It is one of the most basic manifestations of having a grip on the idea of a physical world out there. It is also arguable that thinking of a portion of the world as a physical object in contrast, say, to thinking of it as a place or a sensation or a hologram, turns at least partly on applying to it some of these physical principles.

In Part I 'Frames of reference' one issue to arise is how and whether grasping the kind of physical principles involved in object representation is linked to the frame of reference used in representing places. The chapters in this part focus on the very notion of a physical object. Some of the issues addressed include: What are the particular physical principles in fact used in visually based predictions and memories of trajectories? What are the principles infants use in carving up the world into physical objects, and how are they related to those used by adults? Are there particular physical principles subjects must grasp if they are to be credited with a grasp of the notion of a material object, and if so what are the grounds for making such claims? What is the difference between grasping linguistically formulated physical theories and the kind of grasp manifested in perceptual and action-based engagement with objects?

Spatial representation in the sensory modalities

Imitating a visually perceived gesture, without being able to see one's own body, requires matching visual input to the capacity to produce spatial movements. Feeling a shape blindfolded and then selecting the equivalent visually perceived shape requires matching haptic and visual input. Success in such tasks suggests but does not entail a common, amodal way of coding spatial and physical properties. Moreover, there are strong intuitions that for perceptions to makes us aware of a world out there, their content must, in some sense, be amodal. However, each sensory modality has, prima facie, its own very distinctive phenomenology; there are, that is, some modality-specific ingredients in the experiences typical of each modality.

The distinction between amodality and modality specificity is much used in psychology and is gaining currency in philosophy. These notions are, however, used in a rich assortment of ways which are not always distinguished. The chapters in this part provide the materials for distinguishing among various uses of the amodal/modality-specific distinction and for using this to address problems that arise when attempting to explain the nature of spatial perception and action. Particular issues addressed include: What precisely is the relation between the various senses of the amodal/modality specificity distinction and the account we give of what is involved in the representation of a world out there? In particular, are there ways in which perception can be said to be modality-specific which are consistent with the claim that perceptions make us aware of the world out there? Can young infants match across the modalities, and if so what does this tell us about the way they represent spatial and physical properties? What are the similarities and differences between visual and haptic shape perception and object recognition in adults? How should we account for the distinctive phenomenology of the way we are aware of our own bodies from the inside? What in general are the theoretical tools needed for describing the spatial phenomenology of perceptual experiences, in any modality?

Action

Our capacity to perceive objects in space, and to predict and explain their behaviour, is not the only way we have of manifesting our grip on the idea of physical space. Intuitively, our capacity to direct our actions towards places in the environment and to find our way around in it is at least as important. That is, the physical space we think about is at least as much the space in which we move and act as the space in which objects of perception move and interact. One issue to arise in Part I is whether there is a way of grasping the idea of a physical world which is constitutively linked with the capacity for spatial action.

The chapters in Part IV are concerned with the notion of spatial action itself. In particular, they focus on three questions. First, what is the relation between the way in which spatial information is coded and used in the control and co-ordination of spatial behaviour, on the one hand, and the form and content of the spatial representations involved in perception, on the other? Second, what exactly is the relation between explanations of intentional action in terms of the perceptions, beliefs, desires and so forth of the agent, and explanations of bodily movements in terms of patterns of motor cortex firing, muscle contractions, external forces on the limbs and so on. That is, what is the relation between *actions* and mechanical movements of the body? Finally, to what extent are answers to the first and second questions interdependent and, if they are, what form does such interdependence take?

What and where

In Part I an issue to arise is the question of whether the capacity to represent the connectivity of places in the environment in a disengaged way, depends in any way on the capacity to represent physical objects. The chapters in Part V are concerned with an analysis of the relationship between the *perceptual* representation of objects and that of their location.

Whilst at an intuitive level there seems to be a close interrelationship between the perception of objects within various frames of reference and the perception of their location, such a putative interdependence has been brought into question by neurophysiological and neuropsychological data. The hypothesis at the core of Part V is one put forward by Ungerleider and Mishkin in 1982 with reference to the visual system. They proposed that different brain systems were required for the perception of objects as compared with the perception of the location occupied by those objects. 'What' and 'Where' systems are segregated from the level of the retinal ganglion cells onwards.

A distinction between these two classes of information can be harnessed to constrain theories at the cognitive level, and as several of the chapters in Part V clearly show, the 'what v. where' dichotomy may be both acceptable and useful when used in this way. However, it is also clear that a number of central cognitive and constitutive issues remain and an uncritical transfer of a neurophysiological dichotomy into the realms of psychology and philosophy may well be both premature and inadvisable. The definition of what is 'what' and what is 'where' is by no means agreed by all writers, and questions addressed in Part V include: If it is granted that objects in the 'real world' are inherently spatial, how are the spatial properties of objects integrated with their identifying properties – particularly when such identifying properties may themselves depend on spatial organization and structure? If many of the properties constitutive of material objects are dependent on actual and potential spatial interactions (object–object and person–object) how can a perception, let alone a conception, of a material world ever develop on the basis of such segregated informational input? How in particular should we account for the immediacy and naturalness of our ability and very young children's abilities to utilize these classes of information in the course of the simplest daily activity?

III

Do infants perceive physical objects? Do they have the same concept of a physical object as adults? Do navigating animals use allocentric frames of reference for representing places? Do children? Do the blind code spatial properties in the same way as sighted subjects? All of these are empirical questions to which there should be empirically discoverable answers. They are among the questions addressed by the psychologists contributing to this volume. Answering them requires devising tests, the passing of which would count as evidence for possessing this or that ability; and intrinsic to the debate among psychologists about such tests are questions about whether this or that test does provide the requisite evidence. These debates, in turn, automatically raise what we may label constitutive, philosophical issues: What is it to represent a physical object, to employ this or that frame of reference? What is the difference between perceptual and conceptual representations of objects or places? And these questions in turn, questions in the philosophy of mind, inevitably bring with them metaphysical issues concerning the nature of space, physical objects and, ultimately, the relation between mind and the world.

Describing the relation between empirical and constitutive issues in this way might suggest the view that progress in empirical questions must await the resolution of these questions in the philosophy of mind and metaphysics, a resolution that can and should proceed independently of empirical questions of the kind mentioned in the preceding paragraph. Not only is such a view absurd given the extraordinary progress made in psychology with these and many additional questions, which we will soon return to, without such resolution; more importantly, it misses what seems to us to be the main point of bringing together philosophical and psychological approaches to issues in spatial representation, which turns on the essentially two-way traffic of ideas it provides. For while we believe that at least some philosophical concerns can and should lead to asking new empirical questions, we also believe that empirical and constitutive questions asked by psychologists raise new and important constitutive questions for philosophical approaches to spatial representation, questions that, in addition, challenge bundles of ideas cherished by centuries of transcendental arguments.

The following two examples will, we hope, give some indication of the kind of two-way traffic of ideas we have in mind. These are but two of very many to be found throughout the book, and the best illustration of the workings of this interplay is, of course, to be found in the chapters themselves (further elaboration of the examples discussed here, and of others, can also be found in the introduction to each part). Our first example focuses on the chapters written by two psychologists and one philosopher in Part I. These are all concerned with the constitutive question of how we explain the notion of an allocentric map. We will focus on bringing out, in very general terms, the mutually enriching effects of bringing together distinct psychological and philosophical perspectives on this constitutive issue. In our second example, we consider in some detail the relation between the constitutive claims made by one philosophical contribution to Part II, and empirical claims made by one of the psychological contributions to that part. We will focus here on bringing out the way in which examining the relation between constitutive and empirical claims made in these two particular chapters serves to raise new empirical and constitutive questions, questions we illustrate by reference to findings in the second psychological chapter in this part.

1 Place representation

A central theme of Part I is the question of how we should explain what makes a spatial representation of places in the environment detached, allocentric, map-like or objective. The terms vary, and at least part of the variation is due to different background interests into which this issue slots in psychology and philosophy respectively. A crude characterization of the difference between the psychological and philosophical points of departure for approaching the problem of the map-likeness, allocentricity and so forth of place representations might go as follows.

Ever since Tolman's initial distinction between place learning and response learning to distinguish between two kinds of navigational ability, and his coining of the term 'cognitive map' to describe the kind of representation underpinning the former ability, the attempt to characterize what makes a mental representation of places 'map-like' has been a going, and much debated, concern in psychology (Tolman, 1948). The common understanding is that map-like representations are in some sense detached from or independent of subjects' actual location and movements at any one time.

Two common themes in the debate about how to articulate what such detachment comes to can be singled out. First, the focus in this debate has been on different ways of coding spatial relations among places, and it is in this connection that the notion of 'frame of reference' is sometimes used, the idea being that this notion can be helpful in taxonomizing ways of representing places and their relations in such a way as to capture what we mean by a fully map-like representation. It is relative to this concern that cognitive maps are described as allocentric ways of representing relations among places.[1]

The second theme is this. A general trend in examining the notion of a map-like representation of places has been to begin with a minimal navigational ability, one that does not require the ability to represent places at all, and to work up from there, via increasingly sophisticated navigational abilities that require explanation by appeal to increasingly sophisticated forms of place representation, to those that require explanation by appeal to a way of representing places that might be termed fully allocentric, or map-like. Implicit or explicit in many such attempts is the idea that an account of what makes a representation map-like can be exhaustively explained in terms of the navigational abilities it underpins.[2]

The philosophical approach to spatial representation characteristically takes the notion of a fully map-like or detached spatial representation for granted, and goes on to ask how the capacity to employ such map-like representations of places is related to other abilities. For many philosophers, the impetus for this question is to be found in the Kantian idea mentioned earlier that there is some deep connection between the capacity to represent the world in an objective, unengaged way, on the one hand, and the capacity for full-blown self-consciousness, on the other. The source of the connection in Kant is to be found in one of the most difficult passages in his writing, the Transcendental Deduction (Kant, 1933, pp. 129–75). Very crudely, one of the questions asked there is what does unity of consciousness, explained as consisting in the possibility of thinking of my experiences as mine, require? More specifically, are there any conditions that experiences must meet if it is to make sense to describe the subject of those experiences as having the wherewithal to think of those experiences as her own? And the answer, Kant's answer, is that this requires that the experiences be connected in such a way as to represent a spatial world, where it is the spatial and temporal concepts employed that provide a subject with the idea of an objective, mind-independent world, where this, in turn, is said to be necessary for making sense of the idea of oneself as the subject of these experiences.

Kant's argument for this conclusion is notoriously difficult; attempts to understand it have led to various illuminating spellings out of the nature of the links between our grasp of spatial concepts, our grasp of the idea of an objective, mind-independent world, our grasp of the concepts of a physical object and causation and the capacity for reflective self-consciousness; that is, concepts that are intuitively connected in some way with our grasp of the idea of an external world.[3] For our immediate purposes, the point worth stressing is that a common theme in most of these accounts is the equation of spatial thought, the representations of places in the environment, with a largely unexamined notion of a map-like or detached representation of places and their relations. The focus of attention has been, rather, on attempts to spell out the connection between the notion of a map-like or allocentric representation, however explained, and these other concepts that go into our grip of the idea of an external world.

It is perhaps not surprising, therefore, that a first quick glance through many philosophical and psychological writings on the question of what is involved in allocentric/detached/ map-like/objective spatial representation might suggest that each discipline understands the question differently and has wholly tangential concerns here. Crudely, from each perspective, the other presupposes the existence of answers to precisely the questions that occupy it. Thus, from the psychologists' perspective, what exactly is involved in coding spatial relations in a map-like or detached way is not really engaged with by philosophers; they simply assume that there is some answer and move off to questions of the relation between that and various ingredients in the notion of an external world. For the philosophers, the psychologists may seem to be taking for granted the fact that spatial representations are important precisely because they are representations of the external world (though the concern with knowledge of the external world was what originally motivated Tolman's distinction between place and response learning), in that they do not ask what having a grip on the idea of an external world comes to, and what capacities other than the capacity for geometrical reasoning having such a grip requires.

It is clear that both kinds of question must be considered if we are to begin to have a better understanding of the nature of spatial representation. The chapters in Part I demonstrate the beginnings of such progress by showing, if nothing else, how philosophical and psychological approaches to the issue of spatial representation are mutually enriched by bringing the preoccupations of the other discipline to bear on their own. We begin with a brief sketch of the kind of benefit to what we may label the 'Kantian perspective' in being confronted with the psychological literature, and then consider two examples of the way philosophical preoccupations can lead to new psychological questions. We end by sketching the way in which one issue which philosophers treat as central to their interest in spatial representation is raised in its most stark form by one of the psychologists contributing to this part.

All three chapters in Part I describe kinds of behaviour that seem to require explanation by appeal to the capacity to represent places, but which, prima facie, do not require appeal to the capacity for entertaining disengaged or self-conscious thoughts. Thus, some animals and young infants have what it seems right to describe as the capacity to re-identify places as the same again as they move relative to them. Other, more sophisticated navigational performances, by older children and animals, would naturally be described as cases of re-identifying places from wherever they are relative to them. Surely at least some of these abilities would seem to require explanation by appeal to a capacity to represent places. And surely we have here some kind of grip on a world out there, in some sense of the phrase. Thus, the idea that places are the stable framework through which movement occurs is essential to our conception of a place in the physical world, indeed, partly constitutive of what makes the space physical. And surely, in at least some of the navigational abilities described in these chapters we have a basic, very primitive, manifestation of this way of representing physical space, namely, sensitivity to stability of places and their relations despite the subject's own movements relative to them.

Just as surely, such ways of representing places do not entail reflective self-consciousness, nor a grip on the idea of a world, explicitly conceived of as the world as it is independently of one's own experiences. For the philosopher, one immediate effect of confronting the Kantian perspective with the psychological attempts to characterize these ways of representing places is to make vivid the need for recognizing and taxonomizing weaker forms of spatial representation than those discussed from this perspective. To the extent that philosophers are wedded to the idea of a constitutive connection between spatial representation and grip of some kind on the idea of an external world, this requires matching these weaker forms of place representation to appropriate notions of a world out there. This is one strand in the philosopher John Campbell's chapter in Part I, in which he takes as his starting point developmental work on infants and children's navigational abilities, other examples of which are described in detail by developmental psychologist Herbert Pick, and the cognitive neuroscientist John O'Keefe's computational model for the working of the hippocampus.

However, perhaps the most important effect of such confrontation is to challenge the central Kantian theme to the effect that there is some connection between a kind of spatial representation (map-like or allocentric) on the one hand, and self-consciousness and objectivity, on the other. For what studying the psychological literature makes vivid to the philosopher is both how difficult in general it is to capture the notion of a map-like representation, and how much more difficult it is to capture it in such a way that it has any connection at all with either objectivity or self-consciousness. This, in effect, is how Campbell reads the challenge posed by O'Keefe's computational model, a model that was originally proposed as an explanation of the mechanism underpinning cognitive maps, the latter being explained by O'Keefe as capturing central Kantian ideas about spatial representation. For such mechanisms explain navigational abilities of rats and the like, creatures we certainly have no inclination to think of as either self-conscious or as explicitly conceiving of the world as independent of their experiences. More generally, there is nothing in the navigational abilities O'Keefe's model explains which would warrant the ascription of such capacities. That is, if O'Keefe's model does give an exhaustive account of what is involved in allocentric mapping of the environment, we would have everything we need for map-likeness or allocentricity without any hint of a connection with objectivity or self-consciousness.

Campbell thinks the challenge of saving the Kantian intuition to the effect that there is an important connection between a form of thinking about the spatial world, and objectivity and the capacity for self-consciousness, can be met. As he shows, careful examination of the precise sense in which O'Keefe's computational model does and does not meet intuitive criteria of what it is for a representation to be map-like leads to an original and fresh argument for a connection between a way of representing places in the environment and self-consciousness. The crucial ingredient in Campbell's argument involves pressing home the question of what is involved in grasping the idea of a physically structured, causally connected world, and what it is for a representation of places to be a representation of physically connected places. He suggests that the distinction between different frames of reference, as employed by psychologists for distinguishing types of spatial representation, should be constitutively linked to distinctions between different ways of grasping the physical connectedness of the places represented. He then argues that once we have this link in play we can establish a connection between the following four concepts: the representation of places as physically connected in a way that depends crucially on the capacity to represent objects as moving in that space; the notion of a map-like or allocentric representation of places; an objective or disengaged representation of the world; and the capacity for self-consciousness, that is, thought of oneself as one object among others.

This is a somewhat abstract and condensed description of one of the ways in which the philosopher Campbell draws on psychological literature to re-examine and challenge central philosophical preoccupations with the notion of objective spatial representation, simultaneously challenging the psychologist O'Keefe's constitutive account of what mapping involves. We shall say more about O'Keefe's response to this challenge, but let us now turn to the question of what the philosophical perspective can contribute to the psychological taxonomies of kinds of spatial representation.

As indicated above, we believe the potential benefit of bringing philosophical preoccupations to bear on psychological approaches to spatial representation lies mainly in the former's tradition of explicit consideration of the link between the grasp of spatial concepts and the grasp of other concepts. In particular, as noted above, the focus has been on the link between spatial representation, the grasp of the various ingredients in the notion of a world out there, and the capacity for self-consciousness, awareness of oneself as in that world. Such focus is not merely a reminder of why the issue of spatial representation lies at the heart of any account of the nature of mind. What it adds is a potential framework for providing a psychologically significant classification of kinds of spatial representation, a classification that will turn on the kind of grip on the idea of a world out there provided by a specified way of representing places and their relations. Or, rather, what it adds is explicit articulation of the materials for such a framework, and explicit recognition of their importance, materials which, we believe, anyway hover in the background of much psychological work in this area, and indeed underlie at least some of the disagreement in the psychological literature about how exactly the notion of a map-like representation should be characterized.

One example of such disagreement turns on the fundamental issue of whether cognitive maps can be defined in operational terms, as whatever provides for this or that kind of navigational ability. That they can and should thus be defined is a common presumption in much psychological writings, a presumption challenged powerfully by Pick, who shows that various behavioural tests in fact used for testing whether children have maps are neither necessary nor sufficient for securing a definitive answer. Campbell suggests that behavioural criteria could not capture the notion of a map-like representation, understood as detached or allocentric, because spatial representations whose significance is exhaustively cashed out in terms of their implications for an agent's actions (navigational abilities, in this case) provide for a kind of grip on the physical connectedness of places which, he argues, depends essentially on the subject's interaction with the world, and is therefore not detached, as required by a map. That is, precisely by focusing on the relation between spatial representation and grip of the idea of an external physical world, he provides one possible way of motivating a principled dissatisfaction with a purely behavioural criterion for allocentric spatial representation. Whether or not this argument is accepted, it at least serves to highlight one way in which the issue of the kind of grip on the idea of an external world provided by various forms of spatial representation touches on fundamental disagreements within psychology about how one should define the notion of a map-like representation. (An alternative reason for rejecting behavioural definitions of map-likeness is discussed by O'Keefe. See also p. 14 below and the Introduction to Part I, pp. 29–30).

One possible empirical benefit of bringing the connection between spatial concepts and grip of the idea of an external world to bear on psychological treatments of allocentricity is this. Pick, beginning with the question of whether evidence that has been taken to suggest that children do use allocentric maps in representing their environment does in fact do so, suggests a characterization of the kind of spatial information contained in such maps that is in many ways very close to that urged by Campbell when considering the question of what provides for a disengaged representation of the world. Now if Campbell is right, the ability to represent places in this way is constitutively linked, for us, with two other abilities, which Campbell himself argues are interrelated. First, there is the ability to operate with a primitive theory of perception whereby one explains one's perceptions as being the joint upshot of the way the world is and one's own position in it. This claim immediately leads to the developmental question of the connection between Pick's findings about the development of the ability to use allocentric maps, and the development of what has come to be labelled children's 'theory of mind', the ability to employ mental concepts.[4] The second ability Campbell suggests is constitutively linked to allocentric mapping is the ability to employ various forms of physical, causal reasoning in representing objects in the environment. The second developmental question turns on whether these particular forms of reasoning about objects which Campbell describes do indeed develop in tandem with the mapping ability described by Pick. These empirical questions can in turn, as Campbell would be the first to admit, lead to more detailed and perhaps new ways of articulating the basic Kantian insights Campbell seeks to defend.

Although we have been sketching what we take to be examples of the benefits of bringing together philosophical and psychological approaches to issues of spatial representation, it is not the case that the approaches should be or indeed are restricted to paid up members of either discipline. Thus Campbell engages directly with developmental data and with O'Keefe's and others' computational models. The latter engagement is, perhaps, unsurprising, as O'Keefe's computational model for the working of the hippocampus was originally inspired by the Kantian idea that spatial representation does provide an absolute representation of the world, and the model was intended as an explanation of the mechanism for achieving this way of representing the environment. It is this claim which was the original stimulus for Campbell's chapter.

O'Keefe agrees with Campbell that a map-like representation does involve the capacity to represent oneself from the outside as one object among others in the space represented by the map. He also agrees that his model for a cognitive map does not of itself explain this ability. He proposes an evolutionary model of how, armed with the ability to represent places in the way provided for by his model, a subject might develop the ability to represent itself and other objects as moving and interacting in the space represented by the original map. When we have this ability we have, he argues, full detachment or allocentricity.

However, he disagrees with Campbell on two fundamental and arguably connected issues. First, he does not agree that the capacity to represent objects, in particular the intuitive physics used in such representation, has any role at all in explaining what is involved in grasping the connectedness of space. This is related to his defence of one strand in Kant's ideas about space, namely that space is in some sense ontologically and indeed epistemologically prior to objects. O'Keefe, in defending this idea, rejects Campbell's suggestion that our grip on the idea of a physical world, via our grip on the notion of a physical object, has anything to do with explaining what it is for a map to represent places as connected. More generally, the notion of a physical world has nothing to do with the way in which our maps or spatial concepts provide us with a grip on the connectedness of space; whatever the latter comes to, it is independent of and in some sense prior to any conception of physical objects, causation and so forth.

However, it is his adoption of a second (arguably related) strand in Kant's thought about spatial representation that introduces one of the central philosophical issues to arise from this volume. We may get a sense of this debate by considering our reaction to someone claiming that science has shown there is no space, as we understand it. The strong resistance, not to say incredulity, with which one would naturally greet such a suggestion stems from the fundamental role played by the concept of space in structuring our thought about the world. Divesting our experiences and thoughts of the structure provided by spatial concepts is tantamount to making them unrecognizable. It is such thoughts that lead to the Kantian claim that space is an *a priori* condition of experience.

On what the philosopher Peter Strawson calls the 'austere' interpretation of what makes a feature of experience (or concept) *a priori*, it is such if it is an essential ingredient of any experience we make comprehensible to ourselves. On the 'transcendental idealist' interpretation, a feature is *a priori* if the presence of a feature in experience is attributable entirely to the nature of our cognitive constitution and not at all to the way things are in the world and to their effect on us (Strawson, 1966, p. 68). It is the latter interpretation that O'Keefe adopts with respect to spatial concepts and he holds that his arguments for it, based on appeals to the deliverances of non-Euclidean geometry and of quantum mechanics, strengthens his arguments for his nativist position, in which spatial concepts are innate rather than acquired. (If the world is not really spatial, they could not have been acquired by interaction with it.)

O'Keefe's original contribution to this debate is to argue that the various features we ascribe to the spatial world, including Euclidean geometry and the use of it to represent the spatial relations among places, are the products of properties of the vertebrate brain system (which he goes on to describe) and the way it organizes incoming sensory information, rather than features of the world out there which our brain faithfully represents. (The appeal to these features thus has a dual and connected role, for it is also O'Keefe's reason for rejecting behavioural criteria of map-likeness.) In arguing for this interpretation of the sense in which spatial representation is *a priori* he is knowingly taking a highly controversial stance, not merely relative to our everyday faith in the veridicality of our experiences, but also relative to established philosophical interest in examining the notion of spatial representation. For as Campbell makes clear, at least one reason for trying to spell out what is involved in the idea of a detached spatial representation of our environment, of the world out there, is that the possibility of doing so is closely related to the possibility of defending some form of realism, rather than idealism, with respect to our representations.

Certainly, the full force of the notion of the world out there is that it is a world that is independent of our constitution; and the idea that spatial concepts have a role in providing us with a grip on the world out there is normally understood as the idea that they do so by representing the world out there as it really is. It is this idea that O'Keefe challenges and Campbell defends. We shall not attempt to go into the various strands of their arguments here. For our purposes of showing the close intersection of interests among philosophical and psychological approaches to issues of spatial representation it suffices to note that the issue that lies at the heart of philosophical preoccupations with spatial representation is raised in its most vivid form by one of the psychological contributions to this volume.

2 *Object representation*

The question of what are the principles involved in object representation has a potential ambiguity about it. One question one might be interested in is the straightforwardly empirical question of what are the principles in fact used in segregation of the scene into units, in the prediction of trajectories of moving objects and in predictions of the results of physical interaction among more than one object. Here, the project is to see whether and how it is possible to infer the physical principles that inform and guide segregations and predictions from the segregations and predictions we in fact make. One very fruitful method of inquiry here, pursued with respect to memory and prediction by Cooper and Munger (ch. 4), is to see whether we make any systematic errors in remembering and predicting locations of moving objects, errors that tell us about the principles we in fact exploit.

This straightforwardly empirical concern is prima facie distinct from that which informs the following two constitutive issues, where it is the combination of both these constitutive issues which yields the second reading of the question: What are the principles involved in object representation? First, there is the issue of what are the essential properties of objects – what must be true of an entity if it is to count as a physical object? Second, there is the issue of how much of a grasp of the essential properties of objects, and of what kind, a subject must have if she is to be credited with the capacity to represent physical objects. Theories here vary along a variety of dimensions, some of these specific to the concept of an object, others reflections of claims about the nature of representation in general, to yield a somewhat complex network of positions one might take. When we combine this with the range of possible positions with respect to the issue of what are the essential properties of objects and how these should be determined, we have a map of possible positions one might take with respect to the constitutive reading of the question: What are the principles involved in object representation?

Evidently, whatever position one takes with respect to these constitutive questions it cannot be settled simply by appeal to findings about the principles we in fact use in making predictions, as the following analogy with the difference between explaining what is involved in grasping the concept 'cat', and explaining the ability to recognize cats, should make clear. Thus, it may be that in recognizing cats by touch, say, we use cues such as typical texture. But the ability to exploit such cues is not intrinsic to grasp of the concept 'cat'; we can imagine that someone has a perfectly good grip on the kind of animal cats are but who is incapable of recognizing cats by their texture. Conversely, it may well be that the cues we actually exploit in recognizing cats by-pass any appeal to the kind of categorical distinctions that are essential to a grasp of the concept 'cat' (for example, it may be that the cues we use would do as well for toy and real cats, and do not in any way exploit the difference between them). So, in explaining what is involved in grasping the concept 'cat' we cannot simply appeal to the cues we actually exploit in recognising them. And exactly the same applies to the concept of a physical object and the principles we actually exploit in segregating objects and predicting their movements.

But to say this is not to say that empirical findings and constitutive claims have no bearing on each other. In what follows we will discuss in some detail one particular question about the relation between the two, raised by the philosopher Christopher Peacocke's response to the possibility of conflict between his own claims about what it takes to represent a physical object, and the psychologists' Lynn Cooper's and Margaret Munger's claims about the intuitive physics in fact internalized by the perceptual system.

In chapter four Cooper and Munger distinguish between two kinds of principles used in classical physics to describe the behaviour of objects. Kinematic principles link position, velocity and acceleration to describe the pure motion of objects without regard to mass. Dynamical principles employ the concepts of force and mass to explain changes in the states of rest and movement of objects. Cooper and Munger suggest that the experiments they describe on the way we remember and predict the locations of apparently or really moving objects, show that in these situations we do not exploit dynamic principles, and that an appeal to kinematics alone can explain both what we get right, and the small but systematic errors we make.

They also suggest that their experiments, along with a host of others, lend support to the view argued by the psychologist Roger Shepard (in, for example, Shepard, 1984) that the principles internalized over the course of evolution and actually exploited by the perceptual system in making predictions about trajectories and remembering earlier positions of objects in motion may well be kinematic, rather than dynamic. They suggest that such a view has much to recommend it on evolutionary grounds, given the fact that the extra computational cost of taking dynamical variables into account, over and above kinematic variables, arguably outweighs the small increase in accuracy that would be gained from doing so.

In chapter six Peacocke argues that in order to represent physical objects a subject must employ precisely the kind of dynamical principles Cooper and Munger suggest may not be internalized or actually exploited by the perceptual system. More specifically, the thesis he argues for is the following. 'We experience objects specifically as material objects. Part of what is involved in having such an experience is that perceptual representations serve as input to an intuitive mechanics [dynamics] that employs the notion of force' (p. 169). This claim is based on two constitutive claims; one metaphysical claim and one linking principle between the philosophy of mind and metaphysics.

The metaphysical claim about the nature of matter says that 'for something to be a quantity of matter is for changes in its state of motion to be explicable by the mechanical forces acting on it, and for its changes of motion to exert such forces' (p. 170). The general connecting principle between the philosophy of mind and metaphysics, applied to the case of material objects, says that a thinker's mental representation of objects must be 'suitably sensitive' to all the substantial properties that objects necessarily have. More specifically, given the claim that the link with force is an essential property of space-occupying matter, and that physical objects are, at least, lumps of space-occupying matter, this means that a subject's mental representation of physical objects must be 'suitably sensitive' to the link with force.

On the strength of these two principles, Peacocke goes on to argue that if we found a subject who had no grip at all of any kind on the relevance of force to the changes in states of rest and motion of entities she perceived, and who employed only kinematic principles in reasoning about their movements, such a subject should not be credited with the concept of a material object. On the face of it, we might seem to have here an example of a straightforward clash between empirical findings about the physical principles we in fact employ in perception-based reasoning about objects, and constitutive claims about the kind of physical principles we must be capable of employing if we are to be credited with perceptual experiences as of physical objects. That is, it might appear that we are faced with the choice of assuming that the average and normal adults tested by Cooper and Munger do not have a grip on the notion of a physical object, on Peacocke's criterion of what the minimum having such a grip requires; or of querying Cooper's and Munger's empirical findings; or of rejecting Peacocke's requirement on what is essential to a grasp of the notion of a physical object.

In fact, the appearance of such a clash, interesting as it would be, is illusory, as Peacocke's own remarks about the relation between Cooper's and Munger's particular findings and his own particular claims make clear. Suitable sensitivity, as Peacocke presents it, does not require that the mechanical principles be correct or complete; and it does not require the ability to use mechanical principles in making systematic predictions about the behaviour of objects one observes. All that is required for such sensitivity is that the subject has 'some conception of the magnitude of force' and takes it that 'force is necessary to change the state of rest or of certain kinds of motion' of perceived objects (p. 173). So this requirement would even be consistent with a claim, which is in fact stronger than the one actually made by Cooper and Munger, that mechanical principles that employ the concept of force are never in fact exploited in making predictions about perceived objects.

Although there is no clash between their respective claims, the possibility of a clash between them, and the need to avoid it, turns on the delicate relation between constitutive and empirical claims about object representation. On the one hand, we expect a constitutive account of what is involved in representing physical objects to be of some use in deciding whether a particular subject, or group of subjects (such as a particular species of animal, or young infants) is capable of representing physical objects. A constitutive account should have some kind of normative force relative to empirical findings; not anything we find a subject doing in response to stimulation by what are in fact physical objects counts as grasping the notion of a physical object. This would suggest a kind of independence from empirical findings about the principles actually exploited by subjects.

On the other hand, we believe that normal mature adults do have the concept of a physical object. More strongly, it is arguable that their capacity for mature thought about the external world is partially constituted by the capacity for representing physical objects. That is, the concept of a physical object, like the concepts of space, time and causation, is arguably fundamental to the very possibility of a mature grasp of the idea of an external world. This suggests a certain kind of sensitivity to the principles actually exploited by mature normal subjects, and to the particular forms in which they are exploited. The scope of this sensitivity can be roughly indicated as follows. If a constitutive account requires the capacity to employ particular kinds of physical principles which normal mature subjects in fact have no grasp of; or if it requires that principles these subjects do have a grasp of be exploited in forms of reasoning in which they are never in fact exploited, then, rather than accept that these subjects do not have a grip on the notion of a physical object, we ought to adjust the normative, constitutive requirement in such a way as to avoid this consequence. Such sensitivity is not required of constitutive accounts of less fundamental concepts. Thus, if it turns out that a constitutive account of what it is to grasp the concept of an acid, or of polio, results in depriving most subjects today, and certainly at earlier times, of these concepts, it is far from obvious that we should adjust our constitutive account so as to avoid this consequence.

The exact form such sensitivity should take in the case of fundamental concepts is a difficult and open question. It can involve either adjustments to what counts as having the right kind of grip on a substantial principle claimed to be essential to object representation; or to the substantial principles claimed to be essential to the notion of an object. In the introduction to Part II we will raise the possibility of pursuing the second option. Here we will focus on the first, which in Peacocke's case involves spelling out what exactly suitable sensibility to the link with force involves. How this should be done is not a question pursued by Peacocke in his chapter. One option left open (neither endorsed nor rejected) is that doing so would involve the following kind of two-way exchange between empirical findings and constitutive concerns. The example we use to illustrate such exchange takes as its point of departure some findings of Elizabeth Spelke and Gretchen Van de Walle, described in chapter five.

Two of the principles they suggest dictate infants' reasoning about objects are the continuity and the solidity constraints. The solidity constraint says that no parts of distinct objects ever coincide in space and time. The continuity constraint says that objects move only in connected paths from one place and time to another. One of the experiments described in illustration of the workings of these constraints is the following. Infants are shown a brightly coloured surface, suspended above a floor. The surface and floor are both covered by a screen, and a ball is shown being dropped behind the screen. When the screen is removed, the ball is to be seen in one of the following two conditions. In the first, it is resting on the suspended, intervening surface. In the second, it is under this surface, on the floor. Infants are surprised by the second outcome. The authors write, in explanation of this surprise: 'Given that the upper surface neither ruptured nor moved, the ball could only have reached the lower surface by 'jumping over' or 'passing through' the upper surface, in violation of the continuity and solidity constraints' (p. 147).

Now, adults' explanations of the results of collision, such as displacement or rupture, are of course bound up with their understanding of primitive mechanical laws. Collisions are paradigmatic of the kind of events that require explanation by appeal to magnitudes such as force and mass. One question that could be asked about these experiments, which goes beyond the purposes to which they are put by the authors, is this. Should we take the phrase 'given that the surface neither ruptured nor moved' to imply infants have some kind of implicit understanding of such mechanical principles?

This depends, at least in part, on results of experiments of the kind the authors say have yet to be conducted. Thus, one thing we would want to know is whether, in this experimental set up, infants would be equally unsurprised to see the ball resting on the floor, so long as there was a sign of some kind that the intervening surface had been ruptured or displaced. Intuitively, we would need more than that to be convinced that we have any kind of sensitivity to force here. One thing we might want to know is whether they make any predictions about whether the falling object will be halted by the surface, displace it or rupture it. But even when and if we find such predictions, intuitively this would still not be enough. For the basis of such predictions may be past experience of collisions between those kinds of objects. Or, more abstractly, it might be based on features such as velocity, acceleration, size, shape, substance and so forth, rather than appreciation of force or mass.

One set of empirical questions to arise from asking about the development of infants' and children's grip on the concept of force is, then, this. When do infants begin to make any kind of reliable prediction about the outcomes of collisions, and what is the basis for such predictions? Does systematic sensitivity to force and mass ever enter into such predictions? A related question, of no small interest to the explanations of adults' predictive abilities as well, is how reliable are predictions about collisions based on features other than mass and force?

We have so far been focusing on visually-based predictions. It is hard to believe that haptic manipulations of objects do not involve some kind of developing sensitivity to force and mass. Another question is, then: What is the developmental trajectory of such haptic abilities? Are mechanical principles systematically deployed here at any stage, and if so what are they? At what point and how are these transferred to objects of sight, to the extent that visually-based predictions do not start out as sensitive to force and mass?

One central constitutive issue that arises from empirical questions such as these about the developmental trajectory of infants and children's predictions about the outcome of collisions can be brought out by means of the following speculation. It seems at least conceivable that a subject could get quite far in making predictions about such outcomes without appealing to force or mass. Suppose now that children, when they begin to offer explanations and justifications for their predictions, appeal to these other features, supplemented perhaps by appeals to volition, to take the place of force (as in the principle that objects continue to move until they want to stop or someone/thing else wants them to stop.) Alternatively, suppose that there is systematic sensitivity to force, say in haptic manipulation, but that appeals to force do not play any part in children's generalized explanations and theories of why objects behave as they do. The constitutive question is: When do we say that they are 'suitably sensitive' to force, though, perhaps, mistaken about its nature? And what is the rationale for drawing the line where one does?

The interest in testing the concept of 'suitable sensitivity' against the developmental trajectory of infants' and children's reasoning about objects is, perhaps, less in the actual answer we give to the above question but, more, in the fact that such testing simultaneously yields good empirical questions, and requires justifying intuitive distinctions about degrees and kinds of sensitivity to kinds of events whose true explanation does require appeal to mechanical laws. Thus, we charted a crude trajectory from mere absence of surprise at two possible outcomes of a collision, to systematic predictions about what such outcomes will be that are not based on the appreciation of force and mass, to systematic sensitivity to force and mass in one sensory modality only, through to explanations, mini-theories, that justify and unify such predictions with or without explicit reference to force. This crude trajectory requires refining, revising and motivating. Such a project depends simultaneously on empirical research and findings, and on finding the theoretical tools to motivate constitutive distinctions among ways of being sensitive to or representing important properties of physical objects. Even if the interest is to defend quite strong requirements on the kind of sensitivity that underpins a grasp of the concept of a physical object, considering weaker forms of sensitivity can at the very least provide new ways of arguing for and justifying such requirements.

IV

The chapters of this book, individually and collectively, raise at least as many questions as they answer, and suggest very many avenues for future work. We end these introductory remarks with an indication of one particular topic, explicit in some chapters, implicit in others, which, it seems to us, provides particularly fertile ground for future interdisciplinary work.

At several points throughout this Introduction we have mentioned the idea that explanations of spatial thought and explanations of self-consciousness must be intimately linked, an issue addressed directly, in different ways, by Campbell and O'Keefe. One of the reasons the nature of this link is so hard to articulate turns, of course, on how far we are from having anything like an adequate account of what it is to be self-conscious. Some of the questions raised in a variety of chapters, though not directly concerned with this issue, point to ways in which quite specific empirical and constitutive questions can help in breaking into the tangled web of issues surrounding the question of what exactly is the relation between self-consciousness and spatial thought. We end with some illustrations of the kinds of question we have in mind.

The experiments described by Cooper and Munger, designed to test the physical principles internalized by the perceptual system, use stimuli such as abstract, computer-generated shapes which suggest no particular category to which the objects belong. The stimuli used in Spelke's and Van de Walle's experiments, designed to test the principles used by infants in segregating the perceived scene into objects and predicting their movements, use only inanimate stimuli, brightly coloured cones, balls and the like. One question left open in both sets of experiments is to what extent are the principles they describe in fact applied to animate stimuli in general and to persons in particular?

This question is distinct from the question of what kinds of principles must a subject have a grip on if she is to be credited with perceiving an object *as* an animal in general or a person in particular. For it may turn out to be the case that for the purposes of predicting movements there are many situations in which both infants and adults ignore information about which category an object belongs to. This would be an interesting empirical discovery but it would not, of course, yield any conclusion about the principles involved in grasping the distinction between these categories, nor about whether the particular subjects examined have a grasp of these principles.

Spelke and Van de Walle are interested both in the principles infants in fact use in predicting movements, and in the principles that are constitutive of their concept of a physical object. One particularly interesting question that arises from their work is, then: Do infants have any primitive progenitor of the distinction between animate and inanimate objects and, within the latter category, between human and non-human ones? Or do they, rather, operate with a general notion of a physical object which encompasses all categories? If they do have some way of marking the distinction between these categories, what does this involve? And what are the tests that should be used to elicit the answers to this question? In particular, focusing on the progenitor of our capacity to perceive persons as persons, at what point and in what form does an intuitive psychology, analogous to an intuitive physics, get going? And how exactly is it related to the physics applicable to persons in virtue of their being physical objects? What are the basic concepts (or proto-concepts) that go into these principles?

These questions, about whether infants apply anything like an intuitive psychology in their perception of and interaction with persons is, in a sense, the complementary of questions raised by the philosopher Michael Martin's chapter, which focuses on what he calls the phenomenology of 'body sense', the way one is aware of one's own body from the inside. For exactly the same question of the relation between the grasp, on some level and in some form, of an intuitive psychology and the grasp of an intuitive physics comes into explaining what is involved in thinking of my body as mine. On the one hand, as Martin stresses, this involves awareness of oneself as a physical object. On the other hand, there is something inherently psychological in the way one is aware of one's own body, through acting, locating sensations and so forth. What exactly is involved in both these ingredients of bodily awareness, and how are they related to each other?

The question of the relation between the psychological and physical ingredients in self-consciousness, and the psychological and physical ingredients in awareness of others as subjects and objects, comes together in the developmental psychologist Andrew Meltzoff's intriguing suggestion that the developmental core of our capacity to think of ourselves as one physical object among others is to be found in the capacity to recognize that one is being imitated and to manipulate such interactions. Why exactly do we have something like the beginning of the capacity to think of oneself as an object? To what extent, if any, does this involve, or depend on, awareness of others as subjects? (For example, is infants' manipulation of others' responses a manifestation of some social and psychological principle of causation, distinct from causal principles operative with non-human objects and, if it is, is this crucial for the sense in which we might have here a manifestation of awareness of oneself as a physical object?) And last, but not least, what bearing does Meltzoff's theory have on the general idea that the grasp of the idea of an objective world requires the capacity for self-consciousness?

Finally, in the previous section we mentioned Peacocke's suggestion that the grasp of a primitive mechanics that employs the notion of force is essential for the grasp of the concept of a physical object. At least one way of bringing home the centrality of force to our thought of physical objects is to think of its role in our conception of ourselves as both physical objects and subjects. First, the capacity to exert force in the manipulation of objects, and to feel the effects of force, as in sensations of pressure, intuitively plays an extremely important role in giving us a sense of ourselves as one object among others. Second, exertion of force in action is one fundamental way in which psychology and physics, with respect to our own bodies, come together, in a way that is more primitive than any reflections about ourselves as objects and subjects. Can such intuitions about the dual role of force be spelled out and justified? What in fact is the developmental role of bodily awareness in the application of mechanics to other objects? What is the relation between empirical and constitutive claims here?

We list but a few of the very many questions suggested here about the relation between the intuitive physics and intuitive psychology that go into our grip on ourselves and others as subjects and objects, focusing particularly on questions about the developmental origins of this grip. Whatever we find here will, of course, only be the beginning or core of whatever it is that goes into being aware of oneself as one object among others in the space one inhabits. But one way of beginning to unpack what the full-blown capacity comes to is to focus on primitive progenitors and to ask: What is lacking here, and why, for both detached thought about the world and full blown self-consciousness?

NOTES

1 On frames of reference see especially Pick and Lockman (1981); and O'Keefe and Nadel (1978, ch. 1).
2 For a review of central concerns in psychological debates about the notion of a cognitive map and, more generally, place representation, which link these explicitly to Kantian themes, see O'Keefe and Nadel (1978, ch. 1); on the progressive build-up of increasingly sophisticated navigational abilities and the link between these and cognitive maps, see Gallistel (1990, chs. 3–6, 35–220.
3 Articulations of Kantian insights that suggest various ways of spelling out connections among these concepts in a way that is close to the concerns expressed here, include: Strawson (1968, Part II, pp. 47–152); Strawson (1959, Part I, pp. 15–134); Evans (1982, section 6.3, pp. 151–70; ch. 7, pp. 205–66); Evans (1985); Campbell (1984–5); Campbell (1986); Cassam (1989–90).
4 There is now a vast literature on this topic. A representative collection of influential approaches to children's theories of mind is to be found in Astington, Harris and Olson (1988).

REFERENCES

Astington, Janet W., Harris, Paul L. and Olson, David R. 1988: *Developing Theories of Mind*. Cambridge: Cambridge University Press.
Campbell, John 1984–5: Possession of concepts. *Proceedings of the Aristotelean Society*, vol. LXXV, 149–70.
Campbell, John 1986: Conceptual structure. In Charles Travis (ed.), *Meaning and Interpretation*, Oxford: Basil Blackwell, 159–74.
Cassam, Quassim 1989–90: Kant and reductionism. *Review of Metaphysics*, XLIII, 1, 72–106.
Evans, Gareth 1982: *The Varieties of Reference*. Oxford: Oxford University Press.
Evans, Gareth 1985: Things without the mind. In his *Collected Papers*, Oxford: Oxford University Press, 249–90.
Gallistel, Charles, R. 1990: *The Organisation of Learning*. Cambridge, Mass: MIT.
Kant, Immanuel 1933: *Critique of Pure Reason*, tr. Norman Kemp Smith, London: Macmillan.
O'Keefe, John and Nadel, Lynn 1978: *The Hippocampus as a Cognitive Map*. Oxford: Oxford University Press.
Pick, Herbert L., Jr and Lockman, Jeffrey J. 1981: From frames of reference to spatial representation. In L. S. Liben, A. H. Patterson and N. Newcombe (eds), *Spatial Representation and Behaviour Across the Life Span*. New York: Academic Press, 39–61.
Shepard, R. N. 1984: Ecological constraints on internal representations: Resonant Kinematics of perceiving, imagining, thinking, and dreaming. *Psychological Review*, 91, 417–47.
Strawson, Peter 1959: *Individuals*. London: Methuen.
Strawson, Peter 1966: *The Bounds of Sense*. London: Methuen.
Tolman, E. C. 1948: Cognitive maps in rats and men. *Psychological Review*, 55, 189–208.

Part I

Frames of reference

Introduction: Frames of reference

Bill Brewer and Julian Pears

Suppose that my glasses are now on my nose, just as they were an hour ago, but an hour ago I was in a different room in the house. How should we answer the question whether my glasses are in the same place that they were an hour ago? Given that my glasses are still on my nose, we might say that they are still in the same place. But given that I was in a different room in the house an hour ago, we might equally say that my glasses are not still in the same place. In relation to my nose, my glasses are still in the same place, but in relation to the house they are not. To evaluate whether or not my glasses are in the same place, we need to know how places are being individuated here. When places are individuated by their spatial relations to certain objects, a crucial part of what we need to know is what those objects are. As the term 'frame of reference' is commonly used, these objects would be said to provide the frame of reference. In the above example, either (parts of) my nose or (parts of) the house are to provide the frame of reference.

This common usage of the term 'frame of reference' is similar to the use of the term in physics. Consider, for example, how the rationale behind experiments such as the Michelson–Morley experiment might be described. On the aether theory, aether was taken to be the medium of propagation of light. Light was held to have a specific velocity in its medium of propagation, independent both of its direction of propagation and of the velocity of its source. The experiment was designed to establish the state of motion of a frame of reference at rest in the aether in relation to a frame of reference provided by the laboratory. Its basis was the prediction, based on pre-relativistic assumptions, that an observer in a laboratory which is moving relative to a frame of reference at rest in the aether, who measures the velocity of light by measuring the time light takes to travel from one place to another (places in the frame of reference provided by the laboratory), should measure light as having different velocities in different directions. (An analogy here would be to think of the light as a fish swimming at a constant speed through the sea and the experimenter as someone in a boat, measuring the fish's speed by measuring how long it takes for the fish to swim the length of the boat. If the boat is itself moving through the sea, then the time measured would be expected to depend on the direction in which the fish is swimming in relation to the direction in which the boat is moving.) The point of relevance is simply that what is called the frame of reference again has to do with how places are individuated. When the laboratory provides the frame of reference, places are individuated by their spatial relations to (parts of) the laboratory.

In the case where co-ordinates are used to represent spatial positions, this could be elaborated as follows. Two events are represented as being in the same spatial position if and only if they are assigned the same co-ordinates. Specifying a frame of reference would have to do with specifying how co-ordinates are to be assigned to events in the world, on the basis of their spatial relations to certain objects. These objects provide the frame of reference. When the way in which co-ordinates are to be assigned to events is thus fixed, we might talk of the co-ordinate system as being 'anchored' to the world.

A simple case would be one where places are individuated in such a way that sameness of place requires sameness of distance from the specific objects (or parts of objects) which provide the frame of reference. But more complex cases can be imagined. Suppose, for example, that places in a field are individuated by (a) their distance from the midpoint between two landmark objects; and (b) the angle between the line from the midpoint to the place in question and the line from the midpoint to one of the landmarks. In this case sameness of place would not require sameness of distance to either object. (Consider the midpoint itself: if the landmark objects move apart appropriately, something at the midpoint would be represented as remaining in the same place, despite being further from each of the objects.) Nevertheless, it is by the spatial relations to the two landmarks that places are individuated and sameness of place is fixed: sameness of place requires, *inter alia*, sameness of distance from the midpoint between the two objects.

Given this basic notion of a frame of reference for spatial representation, one interesting question to ask is whether the extent to which spatial thought or perception succeeds in representing a world 'out there', in each of the various senses outlined in the Introduction, has anything to do with the frame of reference employed. A central concern of this collection is to establish the connection between spatial thought and the grasp of the idea of an objective world out there. So it is natural to ask at this point what the relation is between the taxonomy of modes of spatial representation by frame of reference, on the one hand, and the hierarchy of senses in which spatial representation is representation of the independent physical environment which we perceive and act in, on the other. More generally, how useful is this basic notion of a frame of reference in characterizing a creature's ways of representing places: how it perceives and thinks of where things are in the world around it?

There is a widespread intuitive conviction that crucial features of the objectivity of spatial representation rests on its being *map-like* representation, from a detached point of view, from which the subject figures as one object among many, tracing a continuous spatio-temporal route through the persisting, independent world represented. The contrast here is with a more subjective or idiosyncratic way of locating the things around one, which is essentially dependent upon one's own perspective, position, orientation and engagement in the world. This is spatial representation somehow tied to one's own point of view, which therefore has difficulty in capturing the idea of a world which is there anyway, the way it is independently of the way it currently seems, a world in which one is located, and through which one moves. Following a recent trend in the literature, we label this the distinction between *allocentric* and *egocentric* spatial representation. For the moment these are simply theoretical place-holders awaiting substantial explication. A crucial task then, is to give a constitutive account of what lies behind this intuition.

Our initial discussion of frames of reference as fixing the essential background context which is required to make sense of the notion of sameness of place, provides us with one way of drawing a distinction of this general kind, at least to the extent that it can reflect the significance of the question of whether spatial representation is subject-relative in any sense. For we can distinguish between systems of spatial representation in which sameness of place is relative to the location of the subject and those in which it is not. If the system is such that things are to be represented as being in the same place only if they bear the same spatial relations to the subject's body, while things are to be represented as being in different places only if they bear different spatial relations to the subject's body, then the frame of reference involved is *body-centred*. Here places are individuated in such a way that sameness of place requires sameness of spatial relations to (parts of) the subject's own body. If, on the other hand, places are individuated in such a way that sameness of place requires sameness of spatial relations to (parts of) things in the environment other than the subject's own body (whether parts of the room the subject is in, familiar city landmarks, or whatever), then the frame of reference involved is *environment-centred*. Notice that it follows from this definition that representing one's whole body as moving from one place to another requires places to be represented in a non-body-centred frame of reference.

It is confusing, but must be noted, that precisely this distinction between body-centred and environment-centred frames of reference for spatial representation, is often given in the literature as the definition of a distinction marked by the terms 'egocentric' and 'allocentric'. This simply raises a terminological dispute, on which we have taken a stand which minimizes the confusion with respect to the following chapters. The substantial question, which remains, is what the connection is between the two distinctions we have identified.

Thus, we arrive at the question of whether our distinction between body-centred and environment-centred frames of reference yields an account of what makes spatial thought grasp of the idea of a world out there. Is the distinction between egocentric and allocentric spatial representation – the intuitive distinction between more-or-less successful engagement in the world based on idiosyncratic perspective-dependent spatial representation on the one hand, and detached reflective thought about the objective spatial world out there on the other – correctly characterized in terms of the distinction between body-centred and environment-centred frames of reference? If not, then what more, or what else, is required to capture the way in which spatial representation succeeds in representing an independent physical world through which the subject moves? In either case, we need a constitutive account of what is it for a system of spatial representation to fall on one side rather than the other of the preferred distinction: what is required for allocentric spatial representation; what is our evidence for its presence; and what, exactly, does it buy for us?

All three chapters are concerned with aspects of these fundamental questions concerning the nature of allocentric spatial representation. What are the necessary and sufficient conditions for representations to display the spatial organization of the world in this reflective or detached way, from no particular point of view? Are such representations possible/actual, and in what kinds of creatures? What does possession of the capacity for allocentric spatial representation confer with respect to thought of the world out there as objective, and about one's place in it? How are we to discover whether particular creatures' spatial representation is truly allocentric?

One thing which is clear from the outset, is that merely specifying whether the frame of reference involved is body-centred or environment-centred, construed as above, may leave much unspecified. Any way of thinking about places such that sameness of place is a matter of sameness of position in relation to oneself will qualify as involving a body-centred rather than an environment-centred frame of reference. But there might well be important differences between such ways of thinking about places: for example, one can be thinking about where something is in relation to something which is in fact oneself, though one does not know it, yet this might be quite different from the more normal way of thinking about where something is in relation to oneself. The following example helps to bring out this distinction. Think of watching a wasp hovering around a person one sees in a mirror, when one does not realize that the person seen in the mirror is in fact oneself. Seeing the wasp coming back close to the person's ankle again, even though the person has moved slightly, one might think 'It's coming back to that same place again'. But one would not reach down towards one's ankle to brush it away. The frame of reference is body-centred, but this particular way of thinking of places is not linked immediately with one's capacity for basic spatial action, in the way which is characteristic of the more normal way of thinking of where things are in relation to oneself. Nevertheless, specifying the frame of reference involved, in this basic sense of fixing the background framework only with reference to which the notion of sameness of place makes sense, is one step towards characterizing a way of representing places.

John Campbell develops this sort of point against the possibility of characterizing the egocentric/allocentric distinction in terms of the distinction between body-centred and environment-centred frames of reference for spatial representation. He then goes on to suggest an alternative approach to the egocentricity or allocentricity of spatial representation, which focuses attention specifically on this immediate link between egocentric ways of representing places, and basic action and perception. His account proceeds in terms of the kind of *physical interpretation* which is given their represented spatial relations. He argues that a geometry based on a particular origin, axes and primitive topological or metric notions, such as connected, straight line, closer than, between, distance, direction and so on, must be supplemented with an account of the *causal significance* assigned to such notions, before it constitutes a genuine system of representations of places in the external world. A mere geometrical description of this kind is insufficient to capture a particular way of representing spatial properties and relations, yet it is precisely ways of representing physical space that are at issue.

What is crucial here, is an account for each of the primitive geometrical, topological and metrical notions, no doubt also along with their relevant temporal derivatives of what it means to the subject to be representing the world in terms of that notion. What does it amount to *for the subject* to perceive one thing as *closer* than another, *behind* another, moving away at such and such a speed? What is the *significance for him* of one thing being *between* two others, or three things being on a *straight line*? And so on. Such pure geometrical terms are given a physical, causal significance by the subject, perhaps in terms of what's visible, what might come into view, whether things are within reach, too big to grasp, moving away slowly enough to catch, on a collision course, arranged in such and such a shape, on the shortest path from one to another, etc.

Campbell's suggestion is that the resultant system is *egocentric* just if this significance can only be given by reference to the subject's own capacities for perception and action, in what he calls *causally indexical* terms. Here the causal significance of things' standing in various spatial relations is essentially given in terms of its consequences for the subject's perception or action. It will be *allocentric* if and only if this significance can be given without appeal to the subject's perceptual and active abilities, *causally non-indexically*, in terms which give no single object or person a privileged position, which treat all the world's objects (of a given kind) as on a par with respect to their physical interactions.

Pick's concern is very much complementary to this: it is with the developmental trajectory of spatial representation in infants, in particular with respect to the egocentric/allocentric distinction. The standard picture here is of at least three developmental stages. First, the capacity for coding particular egocentric vectors to salient landmarks, which are suitably updated with the infant's relative movement. Second, the capacity to add and subtract such vectors to provide extended route-like knowledge. Third, the capacity to transcend personal involvement, and move from the egocentric to the allocentric, with full-blown configurational spatial knowledge. He admits that this is essentially correct, and aims to provide both the most illuminating characterization of what spatial knowledge consist in at these three stages, and a better understanding of what should count as evidence for children's possession of each such stage.

Perhaps as a result of all the unclarity about the distinction between the egocentric representation employed in the first two stages and the allocentric representation of the third, many experimenters are inclined to apply behavioural criteria for achievement of configural knowledge, which are not in fact sufficient. Although perhaps not in the most obvious, elegant or efficient way, these tasks can satisfactorily be performed on the basis only of successive manipulations of egocentric, route-based knowledge. Similarly there is the reverse tendency to adopt behavioural criteria which are not in fact necessary. The infants' inability may result from their failure to access the required configurational information in connection with the particular task at hand, perhaps due to lack of experience or understanding, rather than from ignorance. Sometimes the 'missing' allocentric representations can successfully be probed in different ways.

Pick and Campbell agree then, as does O'Keefe, that standard *behavioural* tests for allocentricity are almost bound to be inconclusive. Given perceptual or active tasks, there will surely always be a system of spatial representation with given causally indexical physical significance, which is nevertheless sufficient to underlie successful performance. When the criteria bring in the subject's perceptual and active capacities, what justification could there be for giving the spatial representation underlying these capacities a physical significance that transcends any consequences for the perception and action? No doubt there will be cases in which a fully allocentric spatial map appears to provide the most obvious, efficient and elegant solution, but this is not the point. The options seems always to be open to mimic performance by appeal only to egocentric representation. Perhaps language, in the form of sustained conversation with the subject about her conception of the nature of the world and her place in it, provides the only conclusive macroscopic evidence for genuine allocentricity.

O'Keefe believes that the crucial tests here are ultimately physiological, in the way our neural machinery codes spatial information. Over a number of years he has developed a detailed computational account of a crucial element of rats' spatial representation, based on single cell recording from the hippocampus, which is seen to constitute a strongly perspective-independent mapping system. This way of representing places uses an environment-centred frame of reference given by an anchored co-ordinate system which is such that sameness of place is fixed in relation to the centroid and slope of a set of landmarks. Here, as in one of the cases mentioned above, sameness of place does not require sameness of distance from any of the landmarks. (Consider the centroid itself: if the landmarks all move apart appropriately, an object at the centroid will remain in the same place, despite being further away from each of the landmarks.) Nevertheless, when it comes to filling in how places are individuated, it is the landmarks rather than the animal's own body which are crucial. Places are individuated by their spatial relations to the landmarks, rather than by their spatial relations to the animal's body. Sameness of place requires sameness of certain spatial relations to the landmarks rather than sameness of certain spatial relations to the animal's own body. Thus the frame of reference clearly qualifies as environment-centred.

Part of O'Keefe's motivation is the desire to solve once and for all the theoretical doubts about the possibility of genuinely allocentric spatial representation. It is a live question though, to what extent the present model succeeds in this, and what, if anything, must be added to it. O'Keefe himself admits a shortfall here, and goes on to sketch a way in which a truly allocentric system might evolve out of his own model, as its basis, in six plausible stages. In any case, his physiological discoveries are certainly a suggestive basis for an allocentric system of spatial representation. He also argues that his model should be interpreted as lending support to the Kantian thesis that our Euclidean spatial understanding of the external world is something we bring to the organization of our thought about things out there, in virtue of the operation of the relevant neural structures, rather than a record of the way things are in themselves.

To the extent to which possession of a genuinely allocentric map of one's environment is thought to transcend merely the involvement of an environment-centred frame of reference, one might ask what any such further requirements are supposed to buy, in the way of our disengaged idea of an independent, fully objective world, in which we are each simply one object, of a certain kind, among many. According to Campbell, this strongest form of allocentricity, in which the basic spatial notions are given causally non-indexical physical significance by the subject, over and above its simply being the case that the subject can re-identify stable, connected places over time, in the world through which she moves, is indeed required for a number of features of our objective conception of the spatial world. Firstly, he argues, it is our capacity for allocentric spatial representation, in this strong sense, which ultimately underlies our characteristically *realist* conception of the world as constituted by states of affairs whose obtaining may even in principle be beyond our capacities to discover or recognize. Secondly, it is again only allocentricity which enables the detached conception of ourselves as objects quite on a par with others, which is required for the fully self-conscious appreciation of our nature and place in the world, which has often rightly been connected with this realism. Thirdly, it is only allocentricity which necessitates our conception of some of the things represented as place-occupiers in the world *as persisting, mind-independent objects*, entities with an internal causal connectedness and unity which sustains determinate cross-temporal criteria of identity, and whose states at some time t' are the joint product of their states at an earlier time t, and the world's impact on them during the interval $[t, t']$. (See Part II for more on this notion of *object perception*.)

1
Organization of spatial knowledge in children

Herbert L. Pick Jr

How is our spatial knowledge organized? This question has been posed by many investigators of spatial cognition over the last 20 years. A classical answer by Siegel and White (1975) as well as others suggested a progression both in ontogenetic development and microgenetic. According to this formulation the initial organization includes mainly information about proximity to landmarks. Next there is a stage in which there is route knowledge, and finally achievement of an organization which is characterized by configurational knowledge. There is considerable research indicating such a trend over age with children and also reports of such a trend with adults as they acquire more experience with a space. Perhaps as a reaction to rigid behaviourism many researchers (including the present author) have been happy to speak of this final level in terms of mental metaphors, such as mental images, mental maps, cognitive maps, etc. In fact there are various kinds of specific operations which are accepted as evidence for organization at various levels. Thus, for the most complex or sophisticated form of spatial information, configurational knowledge, the ability to construct external representations such as models or maps is one kind of acceptable evidence. Another is the ability to make spatial inferences, i.e. to take short cuts or make detours. Still another is the ability to take a different perspective on a particular spatial layout as a whole. Generally all these types of evidence seem to include the common characteristic of having simultaneously available a large number of spatial relations about a layout.

The goal of this chapter is to examine the nature of this evidence for inferring configurational knowledge and suggest that it is not so compelling in two respects. First, some of the operations that seem to imply configurational knowledge may not, in fact, require it. Thus, errors may be made in attributing too much knowledge to people about a spatial layout. Secondly, and conversely, the failure to demonstrate a specific sophisticated form of knowledge may not necessarily mean its absence. It may be available if the situation facilitates its retrieval.

The plan of this chapter is to illustrate the typical ontogenetic trend with research examples. Then research will be discussed which shows the qualifications and limitations of the traditional interpretation. Finally, the implications of this interpretation of the evidence will be discussed.

1 Examples of evidence for configurational spatial knowledge in children

The first example (Hakke, Smith and Pick, 1984) is a study in which the behaviour of toddlers was examined when they were taken on a short walk and were asked to return to their starting point. These children were 16, 20 and 24 months of age. They were taken individually with one of their parents to the outside of a small square room which was constructed inside a larger activity room. The interior room had four doors, one in each wall by which it was possible to enter the room. The child left their parent at the first door and was guided by an experimenter on a short walk. The walk was to one of the four doors of the interior room and the child and experimenter entered the room and walked to the centre. When there, the child was asked to return to their parent. Some of the children entered the room at the door right beside their parent. The others walked a quarter, half or three quarters of the way around the room to one of the other doors and entered at that point. The paths the various groups of children took to the centre of the room are depicted in the top row of Figure 1.1, labelled as guided routes.

Of particular interest was the children's return path to their parent. These are illustrated for the various ages of children in the lower rows of Figure 1.1. Each return path of a single child is depicted by a single line from the centre of the room through one of the doors. Consider, for example, the return routes of the 16-month-old children who had been guided from the front of the room, three quarters of the way around before entering. These routes are depicted in the left-hand column of the second row. All but one of the ten children in this group left the room by the same door they entered and all but two of these took the long way around the room to get back to their parent. And in general the great majority of children of this age simply reversed their path to return, whatever their initial guided route, as indicated in the other columns of the second row of the figure. Next, consider the return paths of the 24-month-old children. These are depicted in the left-hand column of the bottom row. Seven of the ten children in this group returned to their parent by the shortest route, the door by which they started and left their parent. (As the depicted paths indicate a few of these children wavered a bit before choosing the most direct route.) And, in general, as the other columns of this row indicate these older children overwhelmingly chose the most direct route back to their parent no matter what their original guided path. At 20 months of age some of the children behaved like the younger children and some behaved like the older children. Thus, we see in this very simple space there appears to be a shift with age in how these young children return to their starting place. The young children reverse their routes and the older children take a short cut, which requires inferring the most direct route back to their starting point. This kind of spatial inference indicating knowledge of routes or directions which one has not directly experienced is one kind of evidence for configurational knowledge.

A second example (Hazen, Lockman and Pick, 1978) was a study of somewhat older children from three to seven years of age. These children were taught a path through a more complex set of rooms and then were given several tests to assess what knowledge they had acquired. The rooms were identical, square-shaped, with a curtained door in the centre of each wall. The rooms were arranged in a two by three configuration (two by two for the younger children) as depicted in Figure 1.2. The different rooms were distinguished by having a different toy animal on the floor of each room. The children were taught a path such as the zig-zag route illustrated in the figure. They were asked

Figure 1.1 Routes taken by toddlers to return to their parent left after a short walk around and into an interior room (from Hakke, Smith and Pick, 1984).

to lead an experimenter along the path telling in each room which door to go through to get to the next room and which animal 'lived' in that room. On the first trial the experimenter led the child, telling her which way to go and which animal would be in the next room. After that, the child led the experimenter and was corrected if a mistake was made. Trials were repeated until the child reached a learning criterion.

What knowledge did the child acquire about the space and what could it do with this knowledge? The child was asked several types of questions to assess this. First they

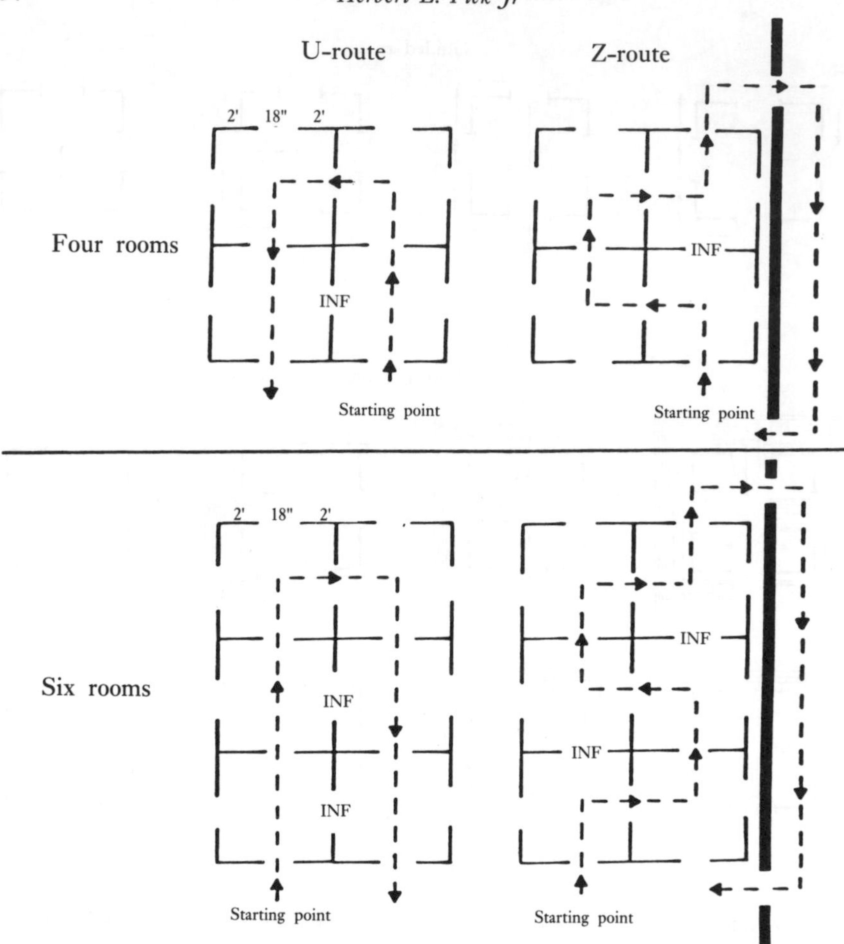

Figure 1.2 Routes learned by children through a complex of four or six rooms (from Hazen, Lockman and Pick, 1978).

were asked to reverse the route, to go through it backwards and while doing so, to name the animals they would find in the next rooms as they went through. They were also asked to make spatial inferences as to what room was behind doors they had not previously traversed while learning the route. These doors are marked 'inf' in Figure 1.2. The youngest children (the three-year-olds) could reverse the route but were unable to anticipate correctly the animals in reverse order, nor could they make the correct spatial inferences. Children in the middle of the age range (four to five years) could correctly anticipate the next animals but still couldn't make the spatial inferences. Only the oldest children were able to make the spatial inferences correctly. Again, the ability to make spatial inferences about the direct route to locations without having had specific experience with the route is taken as evidence for configurational knowledge of the layout of a space. Here we see the older children exhibiting ability to make such inferences and younger children not so. The younger children have route knowledge and can operate on that route knowledge but didn't demonstrate any configurational knowledge. The developmental trend found here is similar to that in the previous example; the age of transition is later, presumably because of the greater complexity of the space.

Let us briefly consider some of the other kinds of evidence for configurational knowledge of spatial layout. Ability to solve Piaget's three mountain perspective-taking task is often taken to imply configurational knowledge. In this task children are asked to identify a photograph of a model spatial layout they are looking at as it would be seen from a perspective different from their own. Piaget and many others found that not until late in childhood could children do this correctly. Similarly, ability to construct a model of a space or to draw a map is taken as evidence for configurational knowledge. Again, Piaget made seminal observations of this, asking children to draw maps in a sandbox of how to get from their home to school. He noted that young children could sometimes draw the layout correctly in the neighbourhoods of their home and of their school but often didn't have them oriented correctly with respect to each other.

2 Reservations with respect to inferring configurational spatial knowledge

Does the manifestation of these kinds of evidence necessarily imply configurational knowledge of a spatial layout? Inferring such knowledge must be done with caution. Consider as a case in point the construction of external models as evidence. Recall the study by Hazen et al. (1978), described above, of children learning a zig-zag path through a set of rooms. In addition to the tests previously described, the children were asked to construct a model of the 'animal house' out of small cardboard cartons representing the rooms. They were asked to put a miniature toy animal in each room as there were in the house. Paradoxically children in the middle of the age range who could not make correct spatial inferences were in some cases able to construct a correct model. However, the way they did this was instructive. They placed the cartons one by one on a table in the order and in the position as if they had been walking along the path. In their manner of doing this, the correct configuration emerged. They did not seem to have the configuration in mind from the beginning. The point is that the simple production of the external model may not necessarily imply configurational knowledge.

Similarly, the ability to make spatial inferences may be questioned as evidence for configurational knowledge with a thought experiment. A very simple kind of spatial inference involves the localization of something that is temporarily occluded as one moves. Suppose, for example, a person is moving down a corridor past an open door and notices an object in the room that then is occluded with further movement. Is the position of the now hidden object known to the observer? That is, does the observer know the relative direction or azimuth of the object from his or her own position even though it is out of sight? Suppose the observer had noticed at the same time as the original object had been seen through the open door, a second object symmetrically located with respect to the direction of movement, an object say in the corridor which remained in view. As movement continued in the same direction down the corridor the relative change of direction of the occluded object would be mirrored by the relative change of direction of the object which has remained in view. In short, there could be concurrent perceptual information for the location of out-of-sight objects, provided one had registered certain kinds of spatial relations when the objects were in sight.

Is there any evidence that people are sensitive to the changing relative direction of objects as they move about? In situations even more extreme than the hypothetical one just described, Rieser, Guth and Hill (1986) have shown that people walking without vision from one location to another seem to automatically update the relative direction of other positions. Interestingly they do not seem to do this as well or as efficiently when simply imagining moving from one location to another. Also, of interest, is the observation made by the same investigators that blind persons don't update very well or efficiently even when they actually locomote.

What might account for this difference in updating between real and imagined movement among sighted persons and for the difference in updating between sighted and blind persons with actual locomotion? Normally sighted people continuously experience optical flow patterns whenever they move their heads with their eyes open. These flow patterns specify the changing directions of all visible objects. This pervasive optical stimulation may serve to calibrate the proprioceptive and biomechanical stimulation of sighted persons as they locomote. This stimulation then provides information for keeping track of changing position when vision is occluded. That proprioceptive and biomechanical information, of course, is not available for imagined movement by sighted persons, nor is it calibrated as well for blind persons.

This interpretation is made somewhat plausible by evidence that biomechanical stimulation can be recalibrated by changing the normal relation between it and optical flow stimulation. A situation was created where persons walked biomechanically at one speed but were subjected to optical flow stimulation which specified moving at a different speed (Rieser, Ashmead and Pick, 1988). People were asked to walk on a treadmill turning at one rate while the treadmill was moved through the world at a different rate. A test of the effects of this treatment was given before and after a brief (15 minute) exposure to this situation. Participants were asked to look at a visual target, close their eyes and walk to it. Consider a condition in which the treadmill was turning at a relatively low rate while movement through the world was at a very high rate. Recalibration would imply that a small amount of biomechanical movement would achieve a great deal of actual movement through the world. In short, they should undershoot in walking towards the target with their eyes closed after the exposure relative to their accuracy before the exposure. Such results were obtained. (Conversely, if the exposure consisted of the treadmill turning at a very fast rate with movement through the world at a very slow rate they would be expected to overshoot in walking towards the target afterwards. This result was also found.)

In sum, at least for simple spatial situations, it seems possible to account for ability to make spatial inferences on the basis of updating the relative positions of non-visible locations by means of concurrent visual or biomechanical stimulation. It may not be necessary to resort to configurational knowledge or mental representations. The research supporting this possibility has been carried out with adults. However, there are some developmental implications of the difference between sighted and blind persons in their updating. If the interpretation about biomechanical stimulation being recalibrated is correct, this recalibration must be occurring during the growth of children when their body dimensions, particularly their leg-length, are changing. The suggestion of perceptual and perceptual-motor systems being recalibrated with one another is not new. Recent suggestions to this effect have been made by Bower (1982) and by Banks (1988).

3 Reservations with respect to denying configurational spatial knowledge

Let us turn now to the opposite side of the coin. There are failures of children in spatial tasks which might be interpreted to indicate that they are deficient with respect to configurational spatial knowledge. Such an interpretation may be too severe. Three examples will be presented.

The first example involves a difficulty of young children in making a spatial inference (Pick and Lockman, 1982). In this case children in their own multi-level dwellings were asked to point directly at locations occluded by the walls or floors of the dwelling. Thus, from the kitchen they might have been asked to point to a bathroom upstairs or to a sofa in the living-room, etc. Interest was focused on their directional accuracy. The children included three to four-year-olds, their older eight to ten-year-old siblings, and their parents served as a mature control group. All the children and their parents were approximately equally accurate at localizing non-visible objects on the same floor. When pointing to locations on different floors the pattern was quite different. The parents were equally accurate at pointing to locations on different floors and on the same floor. The older children were not quite as good as their parents, but the younger children were markedly deficient in localizing objects on different floors. The younger children often made gross errors in pointing, for example, to the front of the house when they should have pointed to the back. It was as if they didn't take into account the fact that in going upstairs or downstairs one also translated horizontally. Thus, these results might be taken to mean that the younger children did not have configurational knowledge of the layout of their house at least with respect to relations between the upstairs and downstairs. However, if the question were changed and they were asked, for example, whether the bedroom was in the front or back of the house, or what they would see when looking out of the bedroom window, they did know. They did have this knowledge. They simply didn't access it with the pointing task.

The second example involves an apparent deficit in children's ability to generate an external representation of a spatial layout in a meaningful way. In this case the deficient external representation was linguistic, a verbal set of instructions to a listener (Plumert, Marks, Pick and Wegesin, submitted). Six-year-old children helped an experimenter to hide a set of objects at different places throughout their three-storey homes. The experimenter guided the children along a spatially very inefficient route while hiding the objects. Thus, they might start by hiding the first object on the top floor, go to the basement to hide the second object, then back to the middle floor for the third object, to the basement for the fourth object, up to the top floor. . . and so on. The hiding route, in fact, involved 11 stair traversals to hide the set of nine objects. Half of the children were asked to tell a second experimenter how she might find the objects in an easy, quick manner. The route that they specified in their directions averaged approximately nine floor traversals for retrieval of the nine objects. Although this was better than the purposely inefficient hiding route, it was about what would be expected in retrieving the objects randomly.

Does this mean that the children did not have the knowledge to perform an efficient search? The other half of the children were asked to retrieve the objects themselves rather than direct another as to how to retrieve them. They retrieved the objects making on the average about four floor traversals, which was close to the minimum possible. They would gather all the objects on the same floor before going on to retrieve the objects on the next floor, and so on. At least, when searching for the objects themselves the children did have knowledge to search efficiently. It might be argued that this efficient search doesn't really imply configurational knowledge of the whole set of locations. Rather, when being on a particular floor and retrieving an object the visual environment provided cues for other objects on the same floor. These weren't available to the children giving verbal directions. However, when some extra children were run in the verbal condition and were asked specifically where the nearest hiding location was to a location they had spontaneously mentioned, they did give locations on the same floor. They just didn't do this under instructions to tell someone else an easy, fast way to find the objects. (It could be argued that the children didn't understand that someone would actually be running up and down stairs to get these objects and it didn't make any difference to anyone that they would have to traverse a lot of extra stairs. A condition to control for this possibility was run in which the children gave instructions through a walkie-talkie to an experimenter actually retrieving the objects. The results were the same kinds of inefficient routes.) In this situation adults direct others to retrieve objects with the same efficiency that they show when finding them themselves. Again, this overall pattern of results, particularly the fact that the children know which locations are close to any given location, suggests that children may at least have the basis of configurational knowledge even when they are not expressing it in a particular task.

The third example involves a perspective-taking problem like Piaget's three mountain task. Here a child is asked to select a picture which corresponds to the view of the layout which an observer at a different station point would have. Huttenlocher and Presson (1979; see also Huttenlocher and Newcombe, 1984) have shown, like Piaget, that this is very difficult for children up to nine or ten years of age. However, if the task is changed slightly to ask only which item in the array would be close or far or to the left or right of the other observer, the problem becomes doable by much younger children. (An opposite pattern of results is obtained if the task is changed to asking children to take a different perspective by mentally rotating the array. That is, if children are asked to imagine the layout as rotating around so that an object opposite them would be directly beside them, then choosing a picture depicting a view that they would see becomes relatively easy, but saying which item would be near, far, etc. becomes relatively difficult.) Thus, using perspective-taking tasks as evidence for configurational knowledge, the failure of children in these tasks should not be taken to indicate that they have no basis for such performance. If relatively minor changes are made in the task they can be shown to have some of the information thought to characterize configurational knowledge.

Conclusions

What is one then to make of this restatement of the traditional evidence for configurational spatial knowledge and its development? Let's pose this question in relation to the criterion for configurational knowledge suggested at the beginning, that is having available a large number of spatial relations simultaneously. It was argued early on that simply finding that children (or adults for that matter) could make spatial inferences or could produce external representations of complex spatial layouts did not necessarily mean that they had configurational knowledge. There might be other simpler ways they could achieve this behaviour. Thus, in one case it was suggested that an external configurational representation emerged when children tried to construct a model on the basis of route knowledge. In another case it was suggested that spatial inferences could be accomplished using concurrent perceptual information to specify the location of non-visible objects. In both these cases, the idea of inferring configurational knowledge is alright. We just have to be more cautious about when it is done. It is necessary to have an operational definition for when configurational knowledge is inferred. In that sense one could arbitrarily say that spatial inference or construction of external representations which have configurational properties is the criterion for configurational knowledge. However, it is necessary to be sure that the definition captures the properties of our intuitive idea, in this case the simultaneous availability of many spatial relations. To do this it is not enough that a configuration slowly emerge, but rather there should be evidence that all spatial relations are equally and readily available. Thus, one should examine the process of construction of the model and see that it could be done quickly and without many false steps in placing the individual rooms in relation to each other. In the case of the children in the experiment with the animal rooms this was not observed and it is reasonable to conclude that they did not have configurational knowledge in spite of the fact they produced a configurational model.

Deciding when spatial inference might or might not be evidence for configurational knowledge seems more difficult. It was claimed that, in principle, the location of a non-visible object could be specified by concurrent perceptual information. The examples of this were for particular out-of-sight objects whose relations to particular in-sight objects were known. This 'in principle' argument becomes more difficult if one considers a large number of objects or locations in the environment. If one is able to update the relative location of any arbitrary object in the environment as one moves in any arbitrary direction the proposed mechanism based on the idea of optical flow may break down. At the very least it would require noticing and storing a huge number of spatial relations among objects in view and remembering these for relatively long periods of time as some of them were occluded. And if the space was complex it would require transitivity between reference objects for occluded objects that themselves were subsequently occluded. The problem is where to draw the line or how to define when the process of localizing out-of-sight objects is accomplished with perceptual updating and when it is accomplished on the basis of configurational knowledge.

More convincing evidence for configurational knowledge would be accomplishment of a task which would be very difficult to accomplish without simultaneous availability of multiple spatial relations. Spatial integration of separately learned spatial layouts would be a good candidate for such a task. This is illustrated by the 'aha' experience one may have after learning one's way around one neighbourhood, say one's work place, and separately learning the layout of another neighbourhood, say one's home area, and then discovering the connection between them. In such a case there can be the experience of knowing the spatial relations among all the locations of both neighbourhoods. It has been possible to capture this situation experimentally. In one study, Sullivan et al. (in preparation) taught groups of children and adults the layout of a complex of rooms by having them walk around the rooms. The subjects were then blindfolded and taken circuitously around the building to another complex of rooms and learned that second layout. They were then shown that the two complexes were, in fact, adjacent to each other and joined by a common door. The subjects' ability to integrate their spatial knowledge of the two layouts was assessed by asking them to point from locations in one complex in the straight-line or Euclidean direction of various locations in the other complex. Even the children could do this far better than chance. (They were, however, not as accurate in these between-space judgements as they were in pointing directly at locations within the same complex.) Similar results have been obtained by Montello and Pick (in press) with spatial layouts on different levels of a building.

The other side of the coin is the argument made above that children and others may often have more configurational knowledge than we would want to acknowledge. The examples supporting this assertion were children's ability to answer specific questions about the spatial relations of particular target locations after they had failed to demonstrate more global knowledge of a spatial layout. While this ability is intriguing and not completely understood, it really can not be considered evidence for the simultaneous availability of multiple spatial relations. If they did have configurational knowledge, they would be expected to be able to answer such questions, but the converse is not necessarily true. They could be generating or computing the answers to such questions at the time they were posed. One would need to perform a more analytic examination of how the questions were answered than has so far been done.

A final issue to raise in conclusion is how the kinds of experience we have affects the organization of our spatial knowledge. In particular, it may be that some sorts of spatial experience would facilitate acquisition of configurational information or the organization of spatial information to capture configurational properties. Other sorts of experience may facilitate route-like or other organization. In spite of the practical and theoretical importance of this issue there has not been a great deal of research and there are few answers.

The issue of experience has two aspects. One concerns experience with space *in general*; the other aspect concerns the nature of experience with a specific space. The first aspect of space in general is reminiscent of nineteenth century discussions about whether people raised on egg-shaped surfaces would have the same geometric concepts as we do. In the current discussion this aspect arose in relation to the role of the presence or absence of early visual experience in the extent to which a person updates relative spatial relations as they move around (Rieser, Guth and Hill, 1986). Attention here will be focused on the second aspect of specific spatial experience where there has been somewhat more research, with the aim of posing questions and suggesting promising areas for investigation.

One can acquire spatial knowledge directly through experience in a space and indirectly through some form of encoded information. In all the research cited above, spatial information was acquired by actual experience in spatial layouts. In most cases the nature of experience in the space was not manipulated. There is a limited amount of research on children's spatial cognition in which experience was manipulated. For example, Herman (1980) found that children reconstructed spatial layouts more accurately after they had walked around the periphery than after they walked about within the space. However, they were even more accurate when they had walked within the space and their attention was called to the spatial relations (see also Cohen, Weatherford and Byrd, 1980). In becoming acquainted with a space the complexity of the path would undoubtedly be an important variable but it may be difficult to define this in a principled way. The complexity of the path would also undoubtedly interact with the complexity of the space itself. A related issue is distinguishing between whether a space is large-scale or small-scale. The distinction has been made in terms of whether the space can be apprehended as a whole from one viewpoint or whether it requires multiple viewpoints. The assumption is that acquisition of configural knowledge is more difficult in multiple viewpoint (large-scale) spaces.

The indirect acquisition of spatial knowledge is through symbolic representation of spatial information. How to present such spatial information so as to facilitate particular kinds of organization of knowledge has not been well studied. Models and pictures of actual spaces can be used to depict spatial information about other spaces. Indeed, recent work by DeLoache (1989) has compared these two modes and surprisingly found that pictures were more effective than models in depicting spatial information for young children. However, her studies were primarily concerned with communication about the specific location of an object as indicated by a landmark and not with configural knowledge or spatial relations. An earlier study by Bullinger and Pailhous (1980) provided spatial information about the layout of a model village to children by means of closed circuit TV. They compared two conditions, one in which the camera lens was wide angle and the other in which the lens had a narrow field of view and the camera had to be panned to get information about the spatial relations among the locations. Reconstruction of the village was better in the former condition.

Spatial knowledge can, of course, be acquired from maps, a cultural invention, whose purpose is to convey such information. An investigation by Thorndyke and Hayes-Roth (1982) with adults examined specifically whether actual experience in a space or experience studying a map resulted in more configurational knowledge. Orientation to non-visible objects, the spatial inference kind of task described earlier, was better after moderate experience actually moving around the space than moderate experience studying a map. However, marking the location of target objects on a piece of paper relative to a pair of reference locations was more accurate for those with map experience. (With more extended experience the differences disappear). This pattern of results suggests that we need to be more analytic about the nature of configurational knowledge, and that the way we acquire spatial information can affect different aspects of configural knowledge differently. Although considerable map reading research has been carried out with children, by psychologists and especially geographers, it has not been directed towards understanding the acquisition of spatial, particularly configural, knowledge as such. See Liben and Downs (1989) for an excellent general review and an example of one kind of project, especially concerned with children's concept of a map. They did show in one case that children were sensitive to the intrinsic spatial relations among locations on a map but not to the locations represented in relation to the frame of reference on the map.

A final symbolic means of presentation of spatial information is linguistically. There has been very little research concerned with children's acquisition of spatial knowledge on the basis of linguistic material. There is some literature on development of understanding of spatial terms and ability to follow simple spatial directions but not of acquisition of information about spatial layout. Language can be considered a highly encoded form of representation of spatial information. And, in general, one might consider spatial representations varying along a dimension of encodedness. At the end of less encoding would be models and moving pictures, extending to still pictures, to maps and to language, the most encoded form. What are the strengths and weaknesses of each of these for conveying spatial information and how do they compare with actual experience within the spatial layout?

REFERENCES

Banks, M. S. 1988: Visual recalibration and the development of contrast and optical flow perception. In A. Yonas (ed.), *Perceptual Development in Infancy: The Minnesota Symposium on Child Psychology*. Hillsdale, NJ: Erlbaum, vol. 20, 145–96.

Bower, T. G. R. 1982: *Development in Infancy*, (2nd ed.). San Francisco: Freeman.

Bullinger, A. and Pailhous, J. 1980: The influence of two sensorimotor modalities on the construction of spatial relations. *Communication and Cognition*, 13, 25–36.

Cohen, R., Weatherford, D. L. and Byrd, D. 1980: Distance estimates of children as a function of acquisition and response activities. *Journal of Experimental Child Psychology*, 30, 464–72.

DeLoache, J. S. 1989: The development of representation in young children. In H. W. Reese (ed.), *Advances in Child Development*. New York: Academic Press, vol. 22, 1–39.

Hakke, R. J., Smith, R. and Pick, H. L. Jr 1984: Spatial orientation and wayfinding in an unfamiliar environment. Paper presented at the International Conference on Infant Studies, New York.

Hazen, N. L., Lockman, J. J. and Pick, H. L. Jr 1978: The development of children's representation of large-scale environments. *Child Development*, 49, 623–36.

Herman, J. 1980: Children's cognitive maps of large-scale spaces: Effects of exploration, direction, and repeated experience. *Journal of Experimental Child Psychology*, 29, 126–43.

Huttenlocher, J. and Newcombe, N. 1984: The child's representation of information about location. In C. Sophian (ed.), *Origins of Cognitive Skills*. Hillsdale, NJ: Erlbaum, 81–111.

Huttenlocher, J. and Presson, C. 1979: The coding and transformation of spatial information. *Cognitive Psychology*, 11, 375–94.

Liben, L. S. and Downs, R. M. 1989: Understanding maps as symbols: The development of map concepts in children. In H. W. Reese (ed.), *Advances in Child Development and Behavior*. New York: Academic Press, vol. 22, 145–201.

Montello, D. R. and Pick Jr, H. L. (In press): Integrating knowledge of large-scale spaces above and below ground, (*Environment and Behavior*).

Pick Jr, H. L. and Lockman, J. J.: 1982: Development of spatial cognition in children. In J. C. Baird and A. D. Lutkas (eds), *Mind, Child, Architecture*. Hanover, NH: University Press of New England, 48–63.

Plumert, J., Marks, R., Pick Jr, H. L. and Wegesin, D.: The ability of children and adults to organize efficient searches and route directions, (submitted).

Rieser, J. J., Ashmead, D. and Pick, H. L., Jr 1988: Perception of walking without vision: Uncoupling proprioceptive and visual flow. Paper presented at the meetings of Psychonomic Society, Chicago.

Rieser, J. J., Guth, D. A. and Hill, E. W. 1986: Sensitivity to perceptive structure while walking without vision. *Perception*, 15, 173–88.

Siegel, A. W. and White, S. H. 1975: The development of spatial representations of large-scale environments. In H. W. Reese (ed.), *Advances in Child Development and Behavior*. New York: Academic Press, vol. 20, 9–55.

Sullivan, C. N., Montello, D. R., Pick Jr, H. L. and Somerville, S. Integration of spatial knowledge, (in preparation).

Thorndyke, P. W. and Hayes-Roth, B. 1982: Differences in spatial knowledge acquired from maps and navigation. *Cognitive Psychology*, 14, 560–89.

2
Kant and the sea-horse: An essay in the neurophilosophy of space

John O'Keefe

Introduction

In a previous publication, Nadel and I (O'Keefe and Nadel, 1978) argued that there were several ways in which the brain represented space. Some of these representations were egocentric, locating entities within spatial frameworks fixed to body parts: the eye, the head, the body; at least one was allocentric, incorporating a framework centred on the environment itself. The former neural representations of space were to be found in several different parts of the brain, most notably in neocortical areas such as parietal lobes, while the latter was centred on a more primitive cortical structure, the hippocampus.

The postulation of one or more neural spatial representation systems leads naturally to the question of their relationship to the physical world. Here I am using the term 'neural' or 'psychological' space to refer to spaces represented by neurons or networks of neurons; 'physical world' means the world which exists independently of cognitive representing beings or which would exist in the absence of such beings; the study of this domain is the province of physics. One might ask whether physical space exists and whether the neural systems reflect this physical space with a greater or lesser degree of faithfulness. At one extreme the representations could be exact replicas of the physical world, capturing all relations and interactions of physical entities with absolute fidelity. At the other extreme one could imagine that physical space did not exist, that physical entities occupied an unextended point-like realm, and that neural spaces are entirely a mental or biological fabrication rather than a true 'representation' as such. The modern version of this latter 'idealist' position might suppose that the physical world acted like a virtual reality machine in which all entities occupied the same 'place' although perhaps not at the same time, and that there were rules according to which the entities could be called up, e.g. in fixed orders or with definite time lags between them. On this view, *neural* spaces would represent not a space in the physical world, but the orderliness of the activation process, and the dimensions of the neural space would map onto the degrees of freedom in the ordering. The degree of match between neural and physical space has implications for our understanding of whether neural representations of space are learned or innate. A good match leaves both options open, since it is plausible that each organism learns anew to represent the spatial relations in the physical environment, as empiricists such as Berkeley would have it; equally plausibly, organisms might have evolved genetic and epigenetic mechanisms which build neural structures designed to represent spaces closely corresponding to physical space, as the nativists claimed.

One line of argument in favour of the nativist position would flow from the demonstration of the universality of the Euclidean metric in psychological spaces. As we shall see, this was used by Kant as an argument for the synthetic *a priori* nature of space when it was believed that only one geometry was possible. With the discovery of non-Euclidean geometries, the argument is turned on its head. If these geometries are all equally plausible on mathematical grounds, whence the universality of three-dimensional Euclidean space in neural representations? I will deal with this argument in greater detail in section three.

A poor or non-existent match between physical and neural spaces would also tend to shift the balance towards the nativist position. While it is plausible to suppose that evolutionary pressures exerted over eons could incrementally select for neural structures which correlated poorly with constant environmental variables, it seems less likely that such structures would be created anew by each organism on the basis of its individual experiences. The existence of a serious mismatch between the Euclidean spatial representations of the brain and the physical world is strongly pointed to by recent experiments in quantum physics, and these will be described briefly in section four.

Finally, perhaps the strongest argument in favour of the nativist position stems from the properties of the neural spatial systems themselves. If it transpires that the structure and function of neural spatial systems is such as to preclude their having been learned, then the nativist case is strengthened. As I shall point out in section five, current ideas about the way that spatial information is stored in one of these systems, the allocentric mapping system, do not seem to be compatible with the structure itself having been built by interactions with the physical world.

I shall argue, then, for the neo-Kantian position that at least one neural space, the hippocampal allocentric cognitive map, is built primarily on the basis of genetic and epigenetic rules with perhaps a small number of parameters left open to environmental calibration. This raises the interesting possibility that the form of the neural spatial representation is due to the inherent limitations on the representational capabilities of neurons and neural networks on the one hand, and to the ability of evolutionary processes to build representational structures, on the other. This latter refers to the well-recognized constraint that complex biological structures can only be assembled incrementally one step at a time, and that each step must make a plausible contribution to the survival capacity of the organism in which it occurs. Specifically, I shall suggest that the three-dimensionality and the Euclidean metric of neural space are due to the neural basis of the system rather than to any aspect of the environment.

In order to make this case it will be necessary to set out a plausible neural model for a neural spatial representation system in section five. Here I shall rely on our latest model for the hippocampal cognitive map. Then in section six I shall speculate on a plausible scenario for the step by step process by which this system might develop from the elemental mapping system found in rodents and other small mammals to the fully fledged allocentric spatial system which represents the environment from any location and includes within itself a representation of the subject-as-object. I have previously suggested that *consciousness* and *self-consciousness* are related to the operation of such a system (O'Keefe, 1985) and I will briefly refer to some of these ideas in passing. Finally, I will try to see whether it is possible to begin to realize some of the dividends implicit in the Kantian programme. One of these is the suggestion that our spatial representation system gives us synthetic *a priori* knowledge about the physical world because that knowledge comes from the structure of the representing system itself. (For example, our intuitive grasp of Euclidean geometry is the paradigmatic example of such knowledge.) In this final section (section seven) I will examine whether it is possible to ground some of our intuitions about space and some of the definitions and postulates of Euclid in the properties of the mapping system, i.e. whether one can conclude that the intuitive naturalness of these postulates is merely a demonstration of the fact that they are representations of the principles by which the mapping system works. If this argument is valid, it follows that we may be able to learn something about the representational properties of the nervous system by exploring the limits of our spatial abilities. For example, I will assume that it is a fact that we cannot represent psychological spaces of higher than three dimensions and that this is not due to any intrinsic aspect of physical space, but to the inability of the nervous system to represent vectors of higher than three dimensions.

1 Nativist versus empiricist accounts of space

In our book (1978) Nadel and I surveyed the ideas of philosophers on the properties and ontogeny of neural space and its relationship to physical space. Of particular interest to us here is the difference between Kantians and empiricists such as Berkeley on the ontological priority of space and objects and the related ontogenetic development of the two. Kantians hold that space is experienced as an entity in its own right which is the presupposition for the existence of objects and which acts as a container for the apprehension of these objects. In contrast, empiricists view space as derived from objects, either as a shorthand summary of the relations between objects or as an abstraction from those relationships. Related to this ontological question is the developmental one. Nativists hold that each member of a species enters the world with a spatial framework which has evolved for that species and which need not be learned by each individual anew. In contrast, empiricists usually view each individual as having to construct its own spatial framework on the basis of its experience with the world. I will offer arguments for the nativist view, in particular that which is naturally aligned to a Kantian position.

2 Kant's theory of space

Kant divided the mind into two major categories, the sensibility and the understanding (Kant, 1963; all quotes are taken from this edition) . In his *Transcendental Philosophy*, he sought to elucidate the necessary properties of these two faculties. Of particular concern to us here is his discussion of the sensibility, which he claimed was the faculty which brought us into contact with the external world and the structure of which he claimed was a source of knowledge in itself. This idea that the structure of an innate faculty could be a source of knowledge about the external world is a difficult concept which it has only become possible to understand scientifically in the light of the subsequent development of Darwinian ideas.

In his own terms, Kant argued that propositions can be classified into three different types: analytic *a priori*, synthetic *a posteriori* and (of particular importance for the present discussion) synthetic *a priori*. Analytic *a priori* statements were propositions in which the predicate was contained in the meaning of the subject and which therefore could be validated by the process of unpacking the meaning of the subject and comparing it with the contents of the predicate. Dictionary definitions were the archetypal form of *a priori* analytic. In contrast, synthetic *a posteriori* propositions were assertions in which the subject did not imply the predicate and which therefore could not be verified by the semantic unpacking process. Instead, their validation relied on empirical methods, on information derived from experience with the external world.

In addition to these two uncontroversial types of proposition, Kant proposed a third, the synthetic *a priori* which, while conveying information which went beyond the semantically-based *a priori* analytic, could not be verified by recourse to empirical investigation. These synthetic *a priori* notions were the properties which must be attributable to the mind in order to understand its role as a knowledge-processing mechanism. Since they were the very basis for the comprehension of empirical objects themselves, they could not be verified by reference to these objects. These were 'concepts to which all objects of experience necessarily conform, and with which they must agree' (Kant, 1963, p. 23) which constituted 'knowledge that is. . . independent of experience and even of all impressions of the senses' (Kant, 1963, p. 42).

Foremost amongst these non-empirical intuitions which formed the basis for apprehending the objects of the physical world was that of space. 'If we remove from our empirical concept of a body, one by one, every feature in it which is (merely) empirical, the colour, the hardness or softness, the weight, even the impenetrability, there still remains the space which the body (now entirely vanished) occupied and this cannot be removed' (Kant, 1963, p. 45). Space was a mode of apprehension of the external or physical world. It was a property of the faculty which put us in contact with the external world, the sensibility.

> Space is nothing but the form of all appearances of outer sense. It is the subjective condition of sensibility, under which alone outer intuition is possible for us. Since, then, the receptivity of the subject, its capacity to be affected by objects, must necessarily precede all intuitions of these objects, it can readily be understood how the form of all appearances can be given prior to all actual perceptions, and so exist in the mind *a priori*, and how, in a pure intuition, in which all objects must be determined, it can contain, prior to all experience, principles which determine the relations of these objects. (Kant, 1963, p. 71)

The idea is that this notion of space is a construct of the mind which organizes the incoming sensory data and that space itself does not necessarily exist in the physical world. Clearly the idea of space and spatial relationships must mirror some aspect of the macroscopic organization of the physical world reasonably faithfully or it would not be useful to the organism. Kant, however, insisted that the *Ding an sich*, the physical world external to the knowing organism, was essentially unknowable and could at best be represented.

It is the contention of the present chapter that modern research into the functions of an area of the forebrain called the hippocampus provides support for Kant's position. Following our earlier proposal (O'Keefe and Nadel, 1978), I shall suggest that this part of the brain provides vertebrates with a Kantian spatial mapping system. Specifically, I shall propose that

1 space is not a feature of the physical world,
2 but instead a property of vertebrate brain systems which organizes incoming sensory information;
3 that these structures have evolved to reflect important aspects of the physical environment and provide animals with powerful survival advantages, and
4 that the organization of these brain systems depends on several simple genetic and epigenetic principles which instantiate deep Platonic/mathematical insights.

3 Is space a property of the physical universe, or is it a construct of the mind? Evidence from non-Euclidean geometry

The apparent power of Euclidean geometry to represent aspects of the physical world combined with its apparent innateness provided Kant with a strong argument for his view. The subsequent discovery of non-Euclidean geometry by Lobachevsky, Gauss and Bolyai (see Rosenfeld, 1988 and Greenberg, 1974) was taken as evidence against the Kantian position, since it was clear that the Euclidean metric, *a priori* or not, was not an infallible guide to the structure of physical space, which might equally well or better be represented by spherical or hyperbolic geometry. While this argument clearly punctures the half of the Kantian position that asserts that infallible knowledge of the physical world derives from our spatial representations, it leaves even more puzzling the *origins* of that spatial representation. If the physical world is not Euclidean, whence does this strong intuition arise? One line of argument would be to admit that the physical world may be non-Euclidean or even non-spatial, but that we extract a Euclidean representation from interacting with this world since the Euclidean is in some sense simpler or more natural. This is certainly not true mathematically, so we need to examine the exact force of the terms *simple* and *natural*. What seems to be intended is that the way in which the mind operates is such as to make it easier to entertain a Euclidean space as opposed to the alternatives. The analogy is perhaps the ease with which a particular computer can run different programs. This view, however, assumes a certain architecture to the mind pre-existing the learning experiences. Detailed examination of this architecture will undoubtedly reveal it to contain many of the features attributed by the nativists to the innate spatial sense, such as the assumption of continuity, homogeneity, the Euclidean definition of the straight line and the validity of the parallel postulate. I suggest that evidence *against* a congruence between psychological and physical space is evidence *for* an innate view of psychological space.

4 Is space a property of the physical universe, or is it a construct of the mind? Evidence from quantum mechanics

On the present view, much (if not all) of the activity of modern science is bound up with the properties of psychological spaces and their projection onto the physical world. A typical causal explanation of a phenomenon involves the construction of a causal story within which the event to be explained is identified as one of a cast of characters set within a spatio-temporal context and the causal interactions amongst these characters delineated. One of the important properties of the spatial framework is to provide the distance metric over which causal forces operate. In general, the farther away two entities or events are, the lower their potential for causal interaction. It is one of the main theses of this chapter that the entire panoply of discrete entities set within a spatio-temporal framework and interacting through causal forces or fields are properties of the psychological spatial system which may at best mirror the broad spectrum of physical reality in a weak and tenuous manner.

Recent developments in the field of particle physics demonstrate how inadequate the standard spatio-temporal model of physical reality is. Although the ideas and experiments deal primarily with the microscopic world of elementary particles, I shall argue that they clearly have implications for our understanding of the macroscopic world of everyday tables, chairs and conscious beings. The first problem occurs when elementary particles such as photons or electrons are described in terms of the classic properties of physics, such as position and momentum. The wave properties of these particles are such that properties such as exact location within a spatio-temporal framework do not exist until the particle is forced to 'reveal' them when subjected to interrogation during an experiment. Furthermore, properties which at the macroscopic level can be treated as independent of each other, e.g. position and velocity or momentum, no longer admit of independence. Designing an experiment to maximize information about the momentum of a particle reduces the amount of information one can extract about its location (Heisenberg's uncertainty relationship). The wave properties attributed to the particle are such that accuracy of assessment of momentum increases with the number of wave cycles measured, but as these spread out over space there is a commensurate loss in locational accuracy. It would appear that the imposition of the notions of discrete particles moving in space and time is strained by the data. While the mathematics are entirely adequate to describe events in this realm, no consistent classical spatio-temporal representation can be given to these events.

This situation has been exacerbated by recent experimental results designed to test a prediction made by Einstein, Podolsky and Rosen (1935),[1] henceforth EPR. EPR wanted to show that certain predictions of quantum mechanics were incompatible with it being a complete description of reality. The reality of an entity was defined for them at least in part by its separability from other entities. If it was not possible to interact with one entity without influencing another, then in some sense they were not independent and therefore did not exist as independent real entities. EPR noted that the Schrödinger equation which governed the behaviour of interacting particles as well as individual particles would seem to predict that once two particles had interacted they were never in future entirely independent, and hence could not be said to exist as independent aspects of reality. In a thought experiment, they imagined a situation in which two particles, e.g. an electron and a positron, or two photons, interact at a particular time and place and then fly off in opposite directions. The interaction, according to quantum mechanics, results in a superposition of the wave functions of the particles which are inextricably linked thereafter, so that neither particle alone has a well-defined wave function, but the composite system does. Under these circumstances (called *entanglement*), many properties of the particles, e.g. their spins, are opposite. For example, the spin of any individual particle is not determined but can be calculated to have a certain probability of being either up or down relative to a particular direction. Once it has been empirically measured, however, the spin of the other particle must be in the opposite direction. Thus, if one always oriented the measuring devices in the same direction with respect to each other, one would always register a +1 (for up) on one and a –1 (for down) on the other but which detector registered which direction would vary in a probabilistic way. This result could be explained if it turned out that there were a set of subtle influences which acted on both particles to keep them in an antithetical state, but that these influences were not detected by available experimental techniques. These *hidden variables* would act on all of the particles crossing each detector, but would not constitute an influence of one measurement on the other. Quantum mechanics, on the other hand, predicts that the actual measurement process on one particle has an effect on both particles, and furthermore that this effect is a function of the relative orientations of the two measuring devices. At certain angles the correlations between the measurements will be greater than would be predicted if the effects of the measurements were independent. The logical basis for experimental tests of the hidden variables theory was proposed by Bell, who showed that the correlations between pairs of particles measured by detectors set at different angles to each other would show one pattern if the particles were being influenced by a hidden variable acting locally on each particle of the pair and a different pattern if the quantum mechanical entanglement conditions held. According to Shimony (1989), experiments designed to test these predictions using Bell's inequalities have supported quantum mechanics in six out of eight experiments. Now it might be argued that there is still scope for the transmission of influences via hidden variables controlled by the settings of the instruments themselves. For example, the settings of the detectors might influence the properties of the particles themselves at the point of separation. EPR noted that the distance between detectors could be allowed to be great enough such that any effect of the measurement on one particle would need to travel faster than the speed of light to influence the other.

Aspect and his colleagues (Aspect, Dalibard and Roger, 1982; Aspect, Grangier and Roger, 1982) have recently performed an experiment in which the polarization of photons, a quantum system with the requisite entanglement conditions, was measured in a search for these hidden variables. Polarization can be detected by setting a detector to a specific angle and measuring whether the particle passes through it or is blocked. The probability of a particle passing is a function of the angle between the detector and the 'polarization' of the wave associated with the particle. The measurement forces the particle to declare itself and influences the polarization of the sister particle. In the Aspect et al. study, the detectors were far enough apart and their orientation settings varied independently at such a high speed (50MHz) that any signal (a hidden variable) transmitted from the first measurement to influence the second would have to travel faster than the speed of light. And yet the predictions of quantum mechanics held. Measurements on one particle affected the measurement on the second.

Faced with this result, it would appear that only the postulation of faster than light causal influences between particles can save the classical notion of a physical spatial framework. Regardless of how far the particles separate spatially, there is an instantaneous 'knowledge' of the world shared between them. Since one of the primary properties of physical space is the enforcement of a temporal and causal separation between spatially separated entities, the existence of physical space is called into question by these results at least at the microphysical level of fundamental particles.

One reasonable response to this argument might take the form that this is all very well, but what has it to do with the macroscopic level of tables, chairs and everyday life? This argument has been considered in detail and countered by d'Espargnat (1983, at various places, but most trenchantly pp. 122–4). He notes that there is a fundamental undecidability about the definition of a macroscopic object. Since all macroscopic objects are extended in space, they must be conceived of as partitionable into systems of microscopic particles. At what point, he asks, does a system of microscopic particles which obey the quantum laws become a macroscopic body which does not?

> Most certainly it cannot be a physical system composed of two or three particles (or atoms). In fact, experimental data show with certainty that such small systems still obey the laws of quantum theory (for example, they can be diffracted, just as their components can). They also show that measurements can be performed on such systems, whose results, according to quantum theory, cannot be reconciled with the assumption that such systems possess, on any occasion and at any time, well-defined – or at least macroscopically well-defined – positions. However, for a possible reconciliation of physical realism with quantum theory to be achieved, it would be necessary that such an assumption be tenable at least with regard to the various parts (pointer, dial, etc.) of the instruments. But, then, what, in terms of the number of microcomponents, is the minimal complexity an object should have in order that the assumption in question be valid? Is that number equal to 37, 75, or to 40 billion? Clearly it is impossible for any such answer to be correct, for such a choice is arbitrary. Nowadays, in fact, the theorists who think the reconciliation is to be achieved along these lines almost all agree (on the basis, again, of elaborate calculations) that the number in question must necessarily be infinite. (d'Espargnat, 1983).

An alternative radical response is to alter the equations which govern the behaviour of fundamental particles, the Schrödinger wave equations, to incorporate a factor which would not be detectable at the microscopic level of the EPR experiment but which would summate over large numbers of particles to emerge at the macroscopic level. One such attempt is that of Ghirardi and colleagues (Ghirardi, Rimini and Weber, 1988) who have inserted a term into the equation which causes the wave equation to collapse spontaneously to a localized position for the particle at an infinitesimal rate for individual particles (of the order of once every 10^8 years), but which sums over the number of particles in macroscopic objects so that for an object of 10^{23} particles the localization would take place 10^7 times per second. Further, the probability of the particle localizing in distant locations would diminish at a commensurate rate, resulting in the apparent spatial stability of macroscopic physical objects.

A version of Bohr's complementarity principle which fits with the present view

Bohr's complementarity principle (1949) explained wave-particle dualism as the expression of an entity which undisturbed was neither, but which when subjected to a measurement was *forced* to express itself as a particle or as a wave. Which form it took depended on the details of the experimental set-up. Thus the entire situation, including the measurement apparatus, needed to be taken into account. He further emphasized that the measurement aspect of the experiment was necessarily described in terms of the concepts of classical physics. In terms of the present schema, measurement involves the creation and use of cognitive instruments which instantiate a measuring scale and which would appear to be set (of necessity) within a spatio-temporal framework. Printers, rules, balances and the rest are the imposition of the cognitive psychological framework onto the physical. It follows that the problems of quantum physics may be considered to result from the attempt to impose a three-dimensional spatial framework on a physical world where it does not exist. If there were no physical spaces, then the EPR paradox would evaporate, since there would be no need to treat the two sister particles as spatially separate, and thus no problem about the influence of one on the other.

5 The hippocampal cognitive map as a plausible Kantian *a priori* spatial system

One of the arguments against the nativist account of psychological space has been the difficulty in conceiving how a representational framework such as space could be instantiated in networks of neurons, and how such a network could be built on the basis of genetic and epigenetic rules. A strong counter to this argument would be a description of one such possible neuronally-based framework. It is a contention of this chapter that the cognitive map theory fulfills this function. In the next sections I will set out the features of this theory in skeletal form as a demonstration of how such a system might work.

The hippocampus as a spatial mapping system

It is becoming clear that there are several parts of the mammalian central nervous system which represent the location of objects within spatial frameworks. Many of these spaces are egocentric ones in which objects are located in a framework which is referenced to a sensory receptor or body surface, e.g. retinal axes of the visual cortex, head-centred axes of the parietal cortex. One of these neural spatial frameworks, however, appears to be referenced to the environment and not to the animal. It is this allocentric spatial mapping system which Nadel and I (1978) have sought to identify with the Kantian *a priori* space (O'Keefe, 1990; 1991).[2] We argued that a region of the limbic system, the hippocampus, and surrounding areas has all the components which would enable it to function as a spatial mapping system. In the next section I will briefly describe the main features of the system.

A short summary of the main anatomical and physiological features of the hippocampus

The hippocampus is the paradigmatic example of archicortex, a primitive three-layered version of the six-layered neocortex. It consists of three sheets of identical cells, the granule cells of the dentate gyrus and the pyramidal cells of the CA3 and CA1 fields of the hippocampus proper. Within each sheet there are interneurons which appear to modulate the activity of the main cell types (see below). The primary inputs to the system come from two sources, the entorhinal cortex and the septum. The entorhinal cortex in turn receives inputs directly or indirectly from many neocortical areas and is believed to be the major conduit of sensory information into the hippocampus. In contrast, the septum gets its inputs primarily from the hypothalamus and brainstem, and is thought to convey information about actual or intended movements and about the animal's bodily states and needs. Part of the septum transforms the movement information into a sinusoidal signal which synchronizes the activity across large areas of the hippocampal sheets, and which provides the clock signal against which temporal patterns in the hippocampal cells are measured. This clock runs at a rate of 6–12Hz in the rat and at slightly lower rates in larger animals such as rabbits and cats. The most obvious manifestation of this clock is the quasi-sinusoidal oscillations of the hippocampal EEG called the *theta pattern*. The amplitude and the frequency of this sinusoidal wave vary as the animal moves around the environment. In unpublished studies Michael Recce and I have shown that in a familiar environment the frequency of this wave correlates highly with the speed with which the animal moves. There is a lower correlation between the amplitude and speed. The main cell type, the pyramidal cell, fires when the animal enters a small region of the environment. This area has been called the place field, and it has been suggested that these cells are signalling some aspect of the animal's location and hence they have been called *place cells* (O'Keefe and Dostrovsky, 1971; O'Keefe, 1976; 1979). Other neurons in nearby brain regions signal the direction in which it faces in an environment irrespective of its location in that environment (Taube, Muller and Ranck, 1990) and as we have noted above there are neurons and slow waves which signal the speed with which the animal is moving through an environment. In summary, the system contains the three types of information required to form a spatial mapping system: places, directions and speeds or distances (see O'Keefe and Nadel, 1978, for a more detailed description).

The suggestion that this part of the brain contains a spatial mapping system is further reinforced by the effects of damage on the abilities of animals. The most obvious effect of damage is a profound and selective deficit in the animal's ability to solve place learning tasks, tasks where it is required to go to a particular place as opposed to a particular cue in an environment (O'Keefe et al., 1975; Morris et al., 1982; Rasmussen et al., 1989).

I have recently suggested that the neuronal elements are the constituents of a vector space in which places are represented as vectors in a polar space whose centre is the centroid or geometrical centre of the cue distribution and whose axis is a direction determined by the overall slope estimated by the average of the vectors joining those cues (O'Keefe and Nadel, 1978; O'Keefe, 1990; 1991).

At any point in an environment, the animal's location and direction are given by a vector to the centroid whose length is the distance to the centroid and whose angle is the deviation from the slope (360-γ) in Figure 2.1. Other places (A and B) are similarly represented.

Movements in an environment are coded as vectors representing the distance and direction moved within this space (T). As the animal moves around a known environment the mapping system continually adds the translation vector to the current location matrix; the resultant vector represents the location expected at the end of the movement. Comparison of this internal navigation representation with the representation derived from the sensory inputs when the animal arrives at the next location provides a powerful mismatch detection system which gives constant confirmation of the adequacy of the stored internal representation and a means for signalling deviations from that representation (as, for example, when a cue in a familiar environment has been moved, Figure 2.2).

This mismatch system provides the internal signal for exploration, a behaviour designed to originally construct or subsequently modify the representation of the

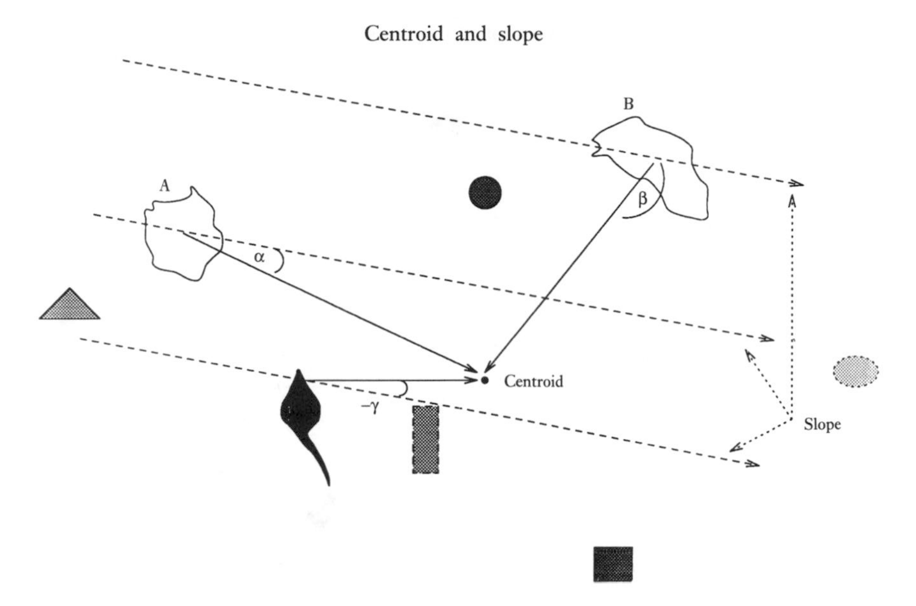

Figure 2.1 Use of the centroid and the slope as the origin and the 0° direction of a polar co-ordinate system fixed to the environment. A rat is in an environment with five cues (shaded). Two other places (A and B) are also shown. Each location in the environment is designated by a vector which points to the centroid and has a length and angle (α, β, γ) relative to the slope. The method by which the centroid and slope can be calculated from any location in the environment has been discussed in O'Keefe (1990; 1991).

current environment. This method of learning does not depend on biological drives or bodily needs but is a purely cognitive impulse. It assumes that there has been an evolutionary gamble that the acquisition of a particular type of knowledge for its own sake will prove to be an efficient survival mechanism for the individual, since information acquired at a time when biological needs such as hunger or thirst are absent might be useful at some subsequent time when that need arises. This type of learning principle cannot in principle be learned itself by any individual since rewards, drive reductions and other reinforcers could not reinforce the formation of maps or the learning of potentially important places unless the system was already in existence.

A separate system is used to generate the required translation to get from the present location to a goal. It is supposed that when an animal encounters a goal object such as food or water within a familiar environment (e.g. at location A) it stores this location in an incentive location store regardless of whether or not it is, e.g. hungry at the time. These incentive addresses have inputs from the motivational systems and can be activated by hunger as well as by food in a particular location. When the animal subsequently finds itself in the same environment and motivated by hunger, the system can calculate the translation and rotation required to reach the goal from its current location. The current vector, when subtracted from the incentive goal location vector, gives the required vector (T in Figure 2.2).

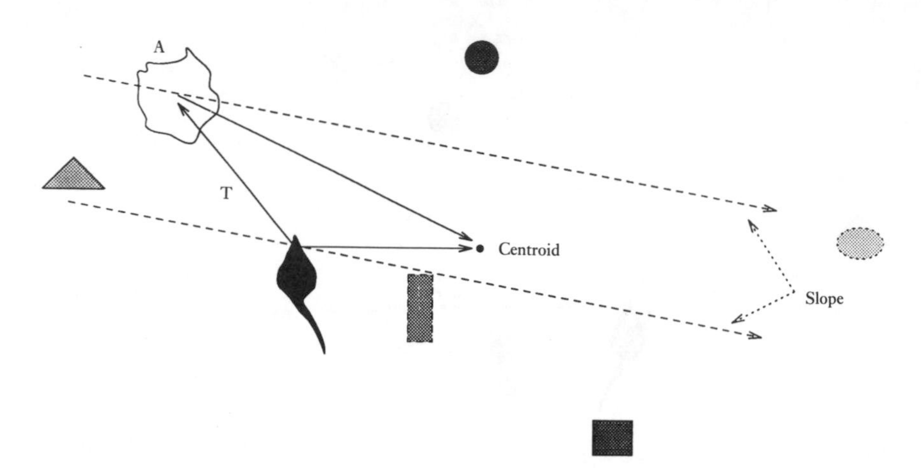

Figure 2.2 Use of the movement translation vector (T). As the animal moves around a known environment, the mapping system calculates a translation vector (T) based on information derived from the motor, proprioceptive and vestibular systems. The addition of this translation vector to the current location vector yields the vector expected from the next location (place A). Mismatches between the actual vector and this expected vector trigger exploration, a behaviour designed to investigate the source of the discrepancy and correct it by the incorporation of the changed or new information into the map. Another part of the mapping system calculates the translation vector needed to move the animal from its current location to a desired location (e.g. one containing food when the animal is hungry). In the figure, this might be location A and would involve the subtraction of the A vector from the current location vector to obtain the required T vector which is sent to the motor programming systems to control movement.

The neural representation of vectors

What we need, then, is a way of representing two-dimensional (or three-dimensional) vectors within the nervous system. Vector representation requires two quantities, either the X and Y in the Cartesian framework or the magnitude and phase angle in a polar framework. It is important that these quantities be bound together in some natural way so that they cannot be easily separated or misplaced, and this would seem to require that they be represented in the activity of the natural unit of the nervous system, the single neuron. The most obvious variables to use are the rate and pattern of firing of neurons to represent each vector. Furthermore, if the system is to be used to do computations on the vectors, there needs to be some co-ordination across the neurons. These two considerations lend support to the idea that there is a clock within the system for co-ordinating the cross-neuronal activity and against which the intra-neuronal patterns can be timed. Given such a clock, the time of firing after the clock pulse would represent one variable and the number of spikes would represent the other. For example, if the clock rate were 10Hz, then the interpulse interval could be divided into ten segments of 10ms each, with the segment representing one variable and the number of spikes within that segment, the second variable.

Phasor theory for a cognitive map

One way in which the hippocampus could code for vectors, then, is to use the number of spikes in a burst to code for the amplitude of the vector, and the timing of the onset of the bursts relative to the theta clock cycle to code for the phase of the vector in either egocentric or allocentric polar co-ordinate system. I have described this theory in some detail elsewhere (O'Keefe, 1991) and will give only a brief outline here (see Figure 2.3). The basic idea is that each pyramidal cell acts as a harmonic oscillator where the amplitude of each oscillation represents the current distance to an object or to a location, and the phase shift of the wave relative to the septal clock theta wave represents the angle of the object or location in polar co-ordinate space. Addition of two or more vectors is performed by addition of their sinusoidal representations, subtraction by the addition of one to the 180° phase reversed representation of the other. Vector averaging is accomplished by the division of the resultant by the number of input vectors by the interneuronal system.

The map is built from a master clock, a large number of identical oscillators and their interconnections

We are now in a position to explore, at least in outline form, the mechanism by which the sequential patterning of gene switching could build a system which acted as a spatial framework. The components required are a large number of identical harmonic oscillators to represent the individual vectors and a clock oscillator which broadcasts its signal in such a way as to reach all of those oscillators simultaneously despite their locations over a wide area of brain tissue. This allows the system to act as a unified whole despite its spatial extent and makes it a good candidate for the neural basis of the unified experience of consciousness (see O'Keefe, 1985, for a discussion of the properties of consciousness and how they might relate to hippocampal function).

Harmonic oscillators are such an intrinsic aspect of living organisms that they might be part of the definition of life itself. All or almost all living things have an oscillator which is roughly of the duration of the earthly day (usually 25 hours) and some mechanism for synchronizing this clock to the rotation of the earth, usually via a zeitgeber such as the sunrise, the period of the light/dark cycle, or other (see, for example, Moore-Ede et al., 1982). Many other oscillators, such as the four-day estrus cycle of rats, have been shown to depend on the circadian rhythm. In addition, longer cycles such as the annual migration of birds appear to be timed by the relative phases of two independent circadian clocks which are phased to different zeitgebers (such as light and temperature) and are therefore free to change phases relative to each other. For example, Arctic terns fly south in the autumn, when a light-related corticosteroid oscillator is 180° out of phase with a temperature-related prolactin one. Time to fly home in the autumn is signalled by a synchronism of the two oscillators (Meier and Fivizzani, 1980). Shorter period oscillators are the one-hour period hypothalamic system which controls the burst release of hormones from the hypothalamus, the motor oscillator controlling sleeping, grooming and other rhythmical activity, the respiratory

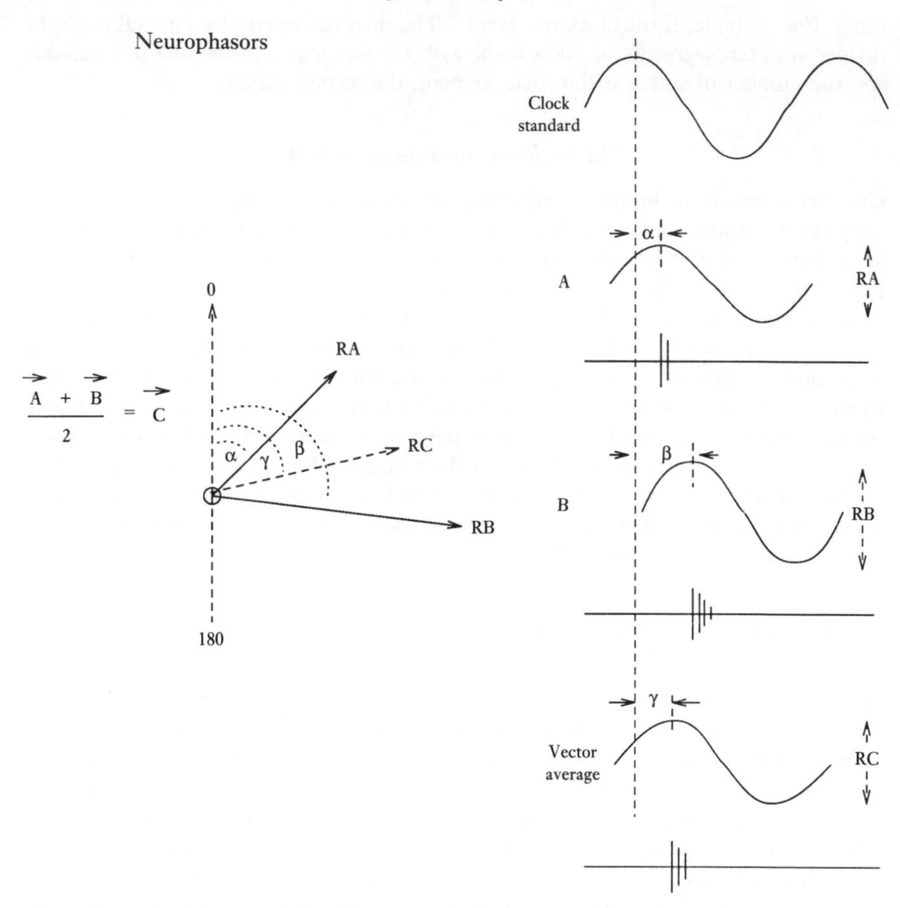

Figure 2.3 Neurophasor method for vector arithmetic. The left side of the figure shows each vector in standard notation with length (RA, RB) and angle (α, β). Vector averaging of A and B = C. One way in which this non-linear operation could be accomplished in neurons is shown on the right. Each vector is represented by a sinusoidal oscillation in a neuron. The length (e.g. RA in A) is represented by the amplitude of the sinusoid and the angle (α) by the phase angle of the oscillation relative to a clock standard, shown at the top. Addition of two vectors is accomplished by the linear addition of the two sinusoids. The average is taken by a division operation carried out by interneurons. Under each sinusoid is shown the assumed firing pattern in a hippocampal complex-spike neuron, which results from the above oscillation and which conveys the result of the computation to the next neuron.

oscillator, the heart-rate pacemaker and the 40Hz oscillations of the olfactory system which have recently been shown to occur in the visual cortex as well (Jacklet, 1989; Freeman, 1991; Eckhorn et al., 1989; Gray et al., 1989).[3] What may be distinctive about the hippocampal oscillators is the suggestion that they are used to represent spatial vectors, although it might be useful to think of these other systems as representing vectors in (non-physical) spaces as well.

The genetic system is very good at generating multiple copies of cells, and the only additional requirement for the hippocampal system is that these cells line up in a particular orientation along the sheet so that the afferents can stream at right angles to their dendrites in an orderly fashion. There is evidence that the size of the hippocampus or some subdivision of the hippocampus correlates with an animal's spatial ability, and that these differences in size can be genetically determined (Krebs et al., 1989; Schwengler et al., 1988).

Perhaps the most difficult part of the story to pin down is the correlation of the theta frequency with speed of movement which provides the metric for the map. Could this be laid down genetically? Let us remember that this metric must continuously change during development as the animal grows, since a step by a small animal will not equate with the same step by a larger animal. Similarly, variations in wind resistance, load carried and other factors must be compensated for. That such compensation occurs was shown in experiments (Morris and Hagan, 1983) where the frequency of theta was a function of the distance jumped but not the amount of additional weight carried. One possibility is that the system has an internal calibration mode which tests some aspects of the system itself as a substitute for the physical world. For example, the size of the body could be estimated from the size of some part of the hippocampus. It is of interest here that the dentate gyrus continues to generate new cells throughout life, and might be influenced by body size. An alternative is a calibration mode for the whole system which might, for example, estimate the time for a signal to propagate from one end of the hippocampus to the other end, using the length of the hippocampus as a measure of distance. When the rat is not engaged in behaviours which change its location in an environment there are no sinusoidal theta rhythms and large numbers of pyramidal neurons fire in synchronous bursts. The size of hippocampus could be estimated by the variance amongst the total population burst.

6 Evolution of the cognitive map

It is clear that the model of the cognitive map presented above does not qualify as a truly allocentric spatial system (see also Campbell, ch. 3). It will be the contention of this section that we can start with the model of the cognitive map appropriate for an animal such as the rat and show how, by a sequence of plausible steps, it could develop into an allocentric system which would be completely independent of the animal's location in an environment, represent the animal itself as an object in the environment, represent the animal as a member of the subset of objects which contain internal principles of action (minds) and therefore as a subject, and which could be used to predict not only the location, movements and interactions of objects, but of subjects as well.

The principles I am exploiting here are that certain abilities of a system can act as a platform on which others can be predicated. Having a map of the type available to the rat opens up a set of possible developments which would not happen in its absence.

One of the driving forces here is the notion of an evolutionary-pregnant situation. This is one where a particular development results in a gap or tension which can be seen to be filled or satisfied by a subsequent development. Although I will not try to assign exact neural correlates to each development, it is my contention that the explosive development of the temporal and prefrontal neocortices is 'driven' in part by the existence of the hippocampal map. For example, the notion of agents and the development of a self-referenced body image are probably related to the development of the frontal cortex. The development of language needs to be fitted into this context but I will not do so in this chapter.

1 The first step forward from the simple rat model would be the incorporation of objects into the mapping system. The present model does not rely on the ability of an organism to construct the notion of an *object* but will work with *features* as the sensory input. A primitive definition of an object might be a spatio-temporal bundle of sensory stimuli which maintains its integrity as the animal moves or as it itself moves through a succession of places within the map. Objects, in turn, may make it easier to construct maps since they can be identified as the same entity from all (or at least several different) directions.

2 A second step forward is the development of the ability to move the point of view in the map without *actual* physical movement in the environment. This step could be accomplished by the uncoupling of the motor system from the system which sends movement information to the hippocampus and controls the EEG theta activity. This would allow the animal to 'imagine' what it would see if it were to go to another place in the environment and can be considered to be the genesis of the contrary-to-fact conditional.

3 This is an evolutionary-pregnant situation, since there is a mismatch between the actual situation of the animal and the imagined situation, due to the fact that there is nothing in the map corresponding to an entity in the location of the place where the animal actually is. There is no representation of the animal in its own map. This vacancy should be filled by an object marker. In the first instance this would probably be a representation of the animal's body as a particular shape and species-specific set of characteristics. The location of this objective bodily representation would be assigned the location in the map corresponding to the subjective point of view. Thus there would be two points of view: the current perceived one which would be assigned to the objective self representation, and a subjective one which would often correspond to this location but would be free to move to other imaginary locations. We have then precursors of the 'objective' me, the subjective 'I' and the current sensory inputs. With this development, the animal can assess each situation on the basis of a spatial representation which contains itself as an element.

4 As part of this development, or earlier or independently, the distinction between agents and non-agents needs to be made. Agents are objects which can change location in the map without external influence. Non-agents need to have action applied to them (pushed, etc.) to change location, and normally do not do so. The distinction is important, since there is a strong preference for use of non-agents as landmarks in the mapping system. If the animal is going to use an object token as a place marker for itself in the map of an environment, it will use the agent-marked form, rather than the non-agent form, when this becomes available.

5 The mapping system can now be used as the basis for the development of a primitive mechanics/kinematics. For example, movement of non-agentive objects on a

trajectory which will intersect with the location of another non-agentive object will lead to a collision and possible change in the location/behaviour of both. If either of the objects is an agent, this does not follow. So there is another evolutionary-pregnant situation for the development of a 'folk' psychology – the development of a mental model of the agent as an attempt to predict the behaviour of this class of object. For example, if the kinematic interaction is between an agent and a non-agent, it is likely that the agent will try to act so as to avoid the collision. This has a limited predictive power since it will happen only if the agent has 'observed' the impending collision and therefore greater predictive power can be built into the system by incorporating the notion of an observing response. This analysis does not apply to interactions between two agents since there are many instances in which one or both might be acting to promote the contact (predator/prey; mating; aggression). Again, there exists an evolutionary-pregnant situation in which there is pressure to discriminate and classify the agent class into these categories.

6 The next stage is the elaboration of the model of an agent, in particular the development of mental models of agents. Here I will follow Strawson in his view that this takes place as a single venture in which the organism imputes the same general mental model to itself as agent and to other agents (Strawson, 1959). The development of the self then is intimately linked to the development of a mental model which is assigned to the physical object that stands for the animal in its own map. There are three sources of information which can contribute to this model: the observation of the behaviour of other agents; the observation of the animal's own behaviour in so far as it is accessible; and the availability of signals about the internal state of the animal, such as, for example, autonomic sensations, etc. Note that these three sets of inputs need not fit together in any simple way, and that the relationship between them may be learned in individual or social contexts.

I would like to argue that at this stage we have a fully fledged representation which fulfills all of the criteria for an objective allocentric spatial system. It enables the organism to view the environment from any vantage point; to find its way around that environment; to identify the goals in an environment and calculate the trajectories required to reach them; to predict the interactions between non-agentive objects; and to a lesser degree to predict the behaviour of agents, including itself. The process upon which this latter capability depends involves the development of a theory of mind to explain behaviour and the attribution of such an entity as mind to the organism itself.

The present chapter follows Kant in giving to the spatial system, and its constituent *places*, ontological (and phylogenetic) primacy above objects. John Campbell (ch. 3) has argued that *some* of the properties of absolute spatial systems, such as the connectedness of places, could be derived from the causal connectedness of objects. In the next section I will try to show how this and other properties of space can be attributed directly to the properties of the cognitive mapping system.

7 Transcendental properties as emergent brain properties

One of the chief consequences of the argument advanced in this chapter is that the nature of psychological space is due to the properties of the brain systems which construct the representation of space, and not to anything about the physical world in itself. In this section I will discuss several of these properties and suggest how they relate to the computational aspects of the hippocampal system. Specifically, I will try to show how the continuity, homogeneity, three-dimensionality and Euclidean metric are derived from the properties of this system.

A corollary of this conclusion is that the self-evident axioms of Euclidean geometry are the expression of the fundamental properties of this system. This approach, following Kant's transcendental philosophical method, sees the examination of the axioms of geometry as a search for the ontological presuppositions embedded in a spatial framework or, according to the present argument, the neural system underpinning that spatial framework. I will allude where appropriate to the relationship of some of these postulates to the present theory. In particular, I will examine the fifth or parallel postulate in the light of the recent discovery of a directional system in the postsubiculum by Taube, Ranck and their colleagues (1990).

Continuity or connectedness of space

This property assumes that objects which move from one place to another can only do so by moving through a set of contiguous places which abut on the starting place at one end and the finishing place at the other. A related concept is that of the geodesic, which is the path between two places which covers the minimum distance. Furthermore, this distance is the Euclidean straight line. This contrasts with the alternative view that space is quantized or discontinuous and therefore there are gaps between the places. In the present cognitive map model, the continuity of the space is guaranteed by the fact that the allocation of places within an environment to a particular set of place cells is done in competition with its neighbours in such a way that the fields of each small group of cells spread across the surface of an environment to fill the entire environment. We conclude this on the basis of our observation that any small group of neurons has place fields which tend to cover the entire surface of an environment. This is presumably done by the inhibiting interactions amongst them which result in the dominance of each over its neighbours within a restricted domain. This mechanism implies that the constriction or even loss of a particular neuron will result in the expansion of the fields of neighbouring neurons so that the entire surface is invariably completely covered. One ineluctable consequence is that there will always be a continuous set of place cell representations within an environment, and therefore between any two place cell fields within that environment.

The idea that there is a direct or shortest distance between two places comes from the mechanism for getting from one place to another, which generates the distance/direction vector from the current location to other desirable or undesirable locations in that environment (see the section above). By combining these vectors by vector arithmetic, the shortest path to any given goal can be calculated and the appropriate behaviour planned.

Multimodality of spatial contents

Space is populated with multimodal entities, whether they be collections of features or objects. This aspect of space is explained by the multi-sensory inputs to the hippocampus from all of the neocortical sensorium. Much of the sensory information analysed in the neocortex maintains its modality specificity as it is processed. The output of these analyses is then combined in several areas into multimodal representations. In addition to the hippocampus, other multimodal areas are found in the prefrontal cortex and the amygdala.

Three-dimensionality of space

One of the greatest problems in the philosophy of space is the limitation of dimensionality to three. Why not two or four? The author of *Flatland* (Abbott, 1952) found it perfectly plausible that two-dimensional beings could exist and thrive in their limited world. The present position attributes the dimensionality of psychological space to the limitations on the representational properties of single neurons. In the present model, each neuron makes its independent contribution to the calculation of the centroid and the co-operation between neurons is limited to allocation of places to be represented and computation of trajectories. On this account, the number of dimensions which can be represented must be limited to those which can be represented by a *single* pyramidal cell and these are three. In the rat we have been able to identify only two dimensions, represented by each neuron in the polar co-ordinate notation of the phasor system – the distance by the amplitude of the wave or number of spikes in a cycle and the angle relative to the reference frame direction by the phase shift of the cell burst relative to the clock waves.

It is not obvious how the second angle of elevation is represented but I will postulate that this exhausts the three degrees of freedom available to the individual neuron and explains the limit of spatial representation to three independent dimensions.

Euclidean metric of space

The first postulate of Euclid states that for every point P and for every point Q ≠ P there exists a unique straight line that passes through P and Q. This is seen as an intrinsic part of the operation of the cognitive mapping system. For example, the part of the system which calculates the trajectory vector between the current location and a desired location returns a vector which points directly from the current location to the desired location. It does not return a family of possible directions, nor does it return a route which takes into account obstacles and barriers. Within the present theory these latter are represented as a set of vectors with their origins along the barrier. The overall route is planned by calculating the vector sum of the goal-oriented vector and the barrier avoidance vector to get a composite vector. The importance for the present purpose is that the primitive vector returned by the mapping system is always a single geodesic from current location to goal, and it is claimed that this property of returning only one vector between any two places underlies the intuitiveness of the first postulate.

The third postulate states that it is possible, given two points not equal to each other, to draw a circle with the centre at one of the points and the radius R = length between the two points. This postulate is one of the fundamentals of vector theory, since it underpins the assumption that the length of the vector does not change with rotation around the origin, i.e. that the length and angle of a vector are independent quantities. It may have the further function in the phasor version of the model of providing the basic translation between vector representation and phasor representation.

Recall that the amplitude of the vector is represented by the radius of the circle which in turn is represented by the amplitude of the phasor.

Finally, the fifth postulate of Euclid, the parallel postulate, states that parallel lines never meet. Following a definition of parallel lines, 'two lines are parallel, if they do not intersect', i.e. 'no point lies on both of them', there follows the postulate, 'that, if a straight line falling on two straight lines makes the interior angles on the same side less than two right angles, the two straight lines, if produced indefinitely, meet on that side on which the angles are less than two right angles' (Ryan, 1989). This postulate was examined with great interest by numerous mathematicians, since it was felt that it might be derivable from one of the other four. The conclusion that it was not derivable, and that it was not essential, led to the development of non-Euclidean geometries in the nineteenth century. The importance of the postulate, and the feature which sets it apart from the others, is its appeal to infinity (*indefinitely*, above) which places its verification beyond the usual limits of human perception. In the context of the present discussion, it is difficult to see how such a far-reaching principle could be learned from experience with a finite world. Why did this idea seem so natural to generations of mathematicians? The answer offered by the present view is that it is an underlying operating principle of the cognitive mapping system, in particular the directional system. Recall that this system enables the animal to compute a direction on the basis of the cue distribution, and that this direction is independent of location within the environment. Looked at from the point of view of the environment, this system superimposes upon each environment a set of parallel lines which never meet. Nowhere in the computation is there a boundary condition limiting the extent of the range over which the system operates. In principle, the lines can be conceived as extending into infinity. The parallel postulate is the computational basis for the notion of direction, one of the elements of the mapping system.

ACKNOWLEDGEMENTS

I should like to thank the members of the King's College, Cambridge Research Group, and in particular John Campbell and Naomi Eilan, for discussion and comments on an earlier version of this chapter. I should like to thank Jeremy Butterfield of Jesus College, Cambridge for his comments on the sections on quantum mechanics. The research from my own laboratory referred to in this chapter was supported by the British Medical Research Council.

NOTES

1 Good discussions of the implications of these ideas and experiments can be found in Bohm (1981, pp. 65–110); d'Espargnat (1983) and Bell (1987, chs. 16 and 20). See also the Berkeley physics course (vol. 4), Wichmann (1971) and Shimony (1989) for good technical introductions to quantum physics and EPR.
2 The cognitive map theory has been set out in detail in O'Keefe and Nadel (1978) and more recent extensions relevant to the present discussion can be found in O'Keefe (1990, 1991).
3 A recent book which covers some of these topics and serves as a good introduction to the neurobiology of rhythms in different systems in different species is Jacklet (1989). The 40Hz oscillations in the olfactory system are described by Freeman (1991); those in the visual cortex by Eckhorn et al. (1989) and Gray et al. (1989).

REFERENCES

Abbott, E. A. 1952: *Flatland*. New York: Dover Books.

Aspect, A., Dalibard, J. and Roger, G. 1982: Experimental test of Bell's inequalities using time-varying analyzers. *Physical Review Letters*, 49, 1804–7.

Aspect, A., Grangier, P. and Roger, G. 1982: Experimental realization of Einstein-Podolsky-Rosen-Bohm *gedankenexperiment*: A new violation of Bell's inequalities. *Physical Review Letters*, 49, 91–4.

Bell, J. S. 1987: *Speakable and Unspeakable in Quantum Mechanics*. Cambridge: Cambridge University Press.

Bohm, D. 1981: *Wholeness and the Implicate Order*. London: Routledge and Kegan Paul.

Bohr, N. 1949: Discussion with Einstein. In P. Schilpp (ed.), *A. Einstein*, vol. I. New York: Harper and Row.

d'Espargnat, B. 1983: *In Search of Reality*. New York: Springer Verlag.

Eckhorn, R., Reitboeck, H. J., Arndt, M. and Dicke, P. 1989: A neural network for feature linking via synchronous activity. In R. M. J. Cotterell (ed.), *Models of Brain Function*. Cambridge: Cambridge University Press, 255–72.

Einstein, A., Podolsky, B. and Rosen, N. 1935: Can quantum-mechanical description of physical reality be considered complete? *Physical Review*, 47, 777–80.

Freeman, W. J. 1991: Nonlinear dynamics in olfactory information processing. In J. L. Davis and H. Eichenbaum (eds), *Olfaction*. Cambridge, Mass: MIT, 225–49.

Ghirardi, G. C., Rimini, A. and Weber, T. 1988: The puzzling entanglement of Schrödinger's wave function. *Foundations of Physics*, 18, 1–27.

Gray, A. M., König, P., Engel, A. K. and Singer, W. 1989: Oscillatory responses in cat visual cortex exhibit inter-columnar synchronization which reflects global stimulus properties. *Nature*, 338, 334–7

Greenberg, M. J. 1974: *Euclidean and Non-Euclidean Geometries* (2nd ed.). New York: W. H. Freeman & Co.

Jacklet, J. W. (ed.), 1989: *Neuronal and Cellular Oscillators*. New York: Marcel Dekker Inc.

Kant, I. 1963: *Critique of Pure Reason* (2nd ed.), tr. by N. K. Smith. London: Macmillan & Co.

Krebs, J. R., Sherry, D. F., Healy, S. D., Perry, V. H. and Vaccarino, A. L. 1989: Hippocampal specialization of food-storing birds. *Proceedings of the National Academy of Sciences USA*, 86, 1388–92.

Meier, A. H. and Fivizzani, A. J. 1980: Physiology of migration. In S. A. Gautheaux (ed.), *Animal Migration, Orientation and Navigation*. New York: Academic Press, 225–82.

Moore-Ede, M. C., Sulzman, F. M. and Fuller, C. A. 1982: *The clocks that time us: Structure and function of circadian timing system*. Cambridge, Mass: Harvard University Press.

Morris, R. G. M., Garrud, P., Rawlins, J. N. P. and O'Keefe, J. 1982: Place navigation impaired in rats with hippocampal lesions. *Nature*, 297, 681–3.

Morris, R. G. M. and Hagan, J. J. 1983: Hippocampal electrical activity and ballistic movement. In W. Seifert (ed.), *Neurobiology of the Hippocampus*. London: Academic Press, 321–31.

O'Keefe, J. 1976: Place units in the hippocampus of the freely moving rat. *Experimental Neurology*, 51, 78–109.

O'Keefe, J. 1979: A review of the hippocampal place cells. *Progress in Neurobiology*, 13, 419–39.

O'Keefe, J. 1985: Is consciousness the gateway to the hippocampal cognitive map? A speculative essay on the neural basis of mind. In D. A. Oakley (ed.), *Brain and Mind*. London: Methuen.

O'Keefe, J. 1990: A computational theory of the hippocampal cognitive map. In 'Understanding the brain through the hippocampus', *Progress in Brain Research*, 83, 287–300.

O'Keefe, J. 1991: In J. Paillard (ed.), *Brain and Space*. Oxford: Oxford University Press, 273–95.

O'Keefe, J. and Dostrovsky, J. 1971: The hippocampus as a spatial map. Preliminary evidence from unit activity in freely moving rats. *Brain Research*, 34, 171–5.

O'Keefe, J. and Nadel, L. 1978: *The Hippocampus as a Cognitive Map*. Oxford: Clarendon Press.

O'Keefe, J., Nadel, L., Keightley, S. and Kill, D. 1975: Fornix lesions selectively abolish place learning in the rat. *Experimental Neurology*, 48, 152–66.

Rasmussen, M., Barnes, C. A. and McNaughton, B. L. 1989: A systematic test of cognitive mapping, working memory and temporal discontiguity theories of hippocampal formation. *Psychobiology*, 17, 335–48.

Rosenfeld, B. A. 1988: *A History of Non-Euclidean Geometry*. New York: Springer Verlag.

Ryan, P. J. 1989: *Euclidean and Non-Euclidean Geometry, An Analytic Approach*. Cambridge: Cambridge University Press.

Schwengler, H., Crusio, W. E., Lipp, H.-P. and Heimrich, B. 1988: Water-maze learning in the mouse correlates with variation in hippocampal morphology. *Behavioural Genetics*, 18, 153–65.

Shimony, A. 1989: Conceptual foundations of quantum mechanics. In P. Davies (ed.), *The New Physics*. Cambridge: Cambridge University Press, 373–95.

Strawson, P. K. 1959: *Individuals*. London: Methuen.

Taube, J. S., Muller, R. U. and Ranck, J. B. 1990: Head-direction cells recorded from the postsubiculum in freely moving rats. *Journal of Neuroscience*, 10 (2), 420–47.

Wichmann, E.H. 1971: *Quantum Physics*. Newton, Mass: Education Development Center, Inc.

3
The role of physical objects in spatial thinking

John Campbell

1 Physical objects and the connectedness of a space

In *Individuals*, Strawson held that the re-identification of places depends upon the re-identification of things. We re-identify places by their relations to things; we see that we are once again in front of the blackboard, for example. There is also a dependence of the re-identification of things upon the re-identification of places. But this mutual dependence is only a reflection of the fundamental character of reference to things and places in our thinking (Strawson, 1959, pp. 36–8). In consequence of the dependence of the re-identification of places upon the re-identification of things, Strawson said, 'the fact that material bodies are the basic particulars in our scheme can be deduced from the fact that our scheme is of a certain kind, viz. the scheme of a unified spatio-temporal system of one temporal and three spatial dimensions' (Strawson, 1959, p. 62; cf. Wiggins, 1963 and Woods, 1963).

It can hardly be denied that our reference to places is densely interwoven with reference to things, and that reference to things greatly enriches the capacity we have for reference to places. But that does not mean that reference to things is essential for reference to places. Indeed, the very notion of a 'thing' invites scepticism. Suppose, for example, that a creature re-identifies a place by using the shell of a room, or the structure of a maze. Is it then re-identifying the place by reference to things? It is often said that there is a level of language-use that is more primitive than the ability to refer to physical things.[1] This 'feature-placing' level of discourse can be given a preliminary characterization in terms of the distinction between mass terms and count nouns. We can distinguish between mass terms such as 'pandemonium', which do not admit the question 'How many?', and count nouns, such as 'tiger', which do. But there may be a use of 'tiger' as a mass term which is prior to its use as a count noun. This use of 'Tiger!' would be merely a response to the presence of tigerhood, by someone who might be quite incapable of making the distinction between one tiger and two being present, or having the idea of its being the same tiger again as was here previously.[2] We can distinguish between 'Mud!' at one place and 'Mud!' at another, without making any appeal to physical things. Since features do not typically spread all over all places in an environment, we can sometimes uniquely identify a place in the environment as 'the place which is F', for example. So we can use located features to re-identify places, given a certain stability in the layout of features in the environment. Yet it must certainly be true that there is some crucial role for physical objects in our ordinary spatial thinking. I want first to elaborate upon the point that place-re-identification does not depend upon physical objects, and then try to find what role they do play. We shall see that their distinctive role is to be found not in the re-identification of places, but in establishing the connectedness of a space: the fact that every place in a particular environment is spatially related to every other place in the environment.

We can begin with an extremely elegant and simple experimental paradigm used with infants by the developmental psychologist Linda Acredolo, to find whether and in what way they are identifying places (Acredolo, 1990). Her experimental space is an enclosure ten feet square, with two identical windows across from each other. There is a round table in the centre of the room, with a buzzer under it, and a long, movable rod attached. At the end of the rod is a seat on wheels, which can be rotated around the table. On top of the seat is an infant. In the training phase, the buzzer sounds in the centre. About five seconds later, an adult appears at one of the windows, calling the child's name and generally entertaining it for five seconds or so. Of course the child turns to look; and the pairings of buzzer and event, always at the same window, continue until the child has developed an expectation that the event will follow the buzzer; that is, on hearing the buzzer, the child turns towards the window before the adult appears. The chair was rotated around the table to the other side of the room after the training phase was over. The buzzer was sounded, and the experimenters watched to see towards which window the child looked in expectation of the event. Obviously, if the child has merely learnt a spatial response, such as 'look to the left' it will look towards the wrong window. There certainly are spatial behaviours which are more primitive than the ability to identify places – for instance, the ability to reach to the left or the right, or to jump out of the way of an oncoming object. Even if the infant has only a particular response, such as looking to the left, its behaviour may still be properly described as spatial. It may vary the type of muscular movement in many different ways, depending on the starting orientation of its head and body when the buzzer is sounded, so as always to achieve the result of looking to the left. So it may be impossible to describe the response as a non-spatial muscular movement. It might indeed be said that in the case in which the child is only giving spatial responses, it is using a notion of 'place': one in which no sense can be assigned to the idea that it is itself in motion, or capable of movement – it has an array of places, such as 'just within reach and to the right', which it carries with it through the world. Using this frame, something would be said to be in 'the same place' at one moment as at another moment, if at both times it was just within the subject's reach, and to the right, whether or not the subject had, as an observer using a more standard frame of reference, such as the walls of the room, might say, 'moved' in the meantime. We can certainly imagine a subject for whom this way of thinking is a possibility. For example, an oriental despot might so arrange matters that however and whenever he moves, there is always Turkish delight just within reach and to the right. At this very primitive level, there would evidently be no dependence of place-re-identification upon thought of physical objects.

We have also to consider the case in which the child manages to use the information available to it through the rotation to keep track of the right window, and look towards it, even though this means giving a different spatial response, such as looking to the right rather than looking to the left. Acredolo found a gradual transition from giving dominant weight to spatial responses, to giving dominant weight to this kind of place learning as the infants grew older; or more precisely, as the time in which the child had been capable of self-locomotion increased. When we reflect on just how the child which operates not as a despot, but using a more standard criterion of identity, manages to re-identify places, it is hard to believe that it must be doing so by recognizing physical things. Perhaps it simply keeps its eye on the place as it is moved around. And there are many more sophisticated ways in which, without using physical objects, the child who succeeds in the Acredolo paradigm might re-identify the place, otherwise than by simply keeping an eye on it. For example, it might do it by keeping track of its own movements as it is rotated. If this is what is happening, then it does not depend upon any ability to think in terms of physical objects. It is also evident that the whole class of animals which manage to find their way back home simply by keeping track of their own movements – the directions and distances of their travel from moment to moment – then using path integration to find the direct route home – are re-identifying places (see for example Müller and Wehner, 1988). In response to that point, it might be acknowledged that this kind of re-identification of places does not depend on reference to physical things, for here thought about places does not depend on the use of landmarks. But, it might be said, there plainly are instances in which landmarks are used to reidentify places over time. And in such cases, the landmarks used must be enduring things. For a landmark to function as a landmark, it must stay in the same place, or be at the same place, over a period of time. It must be a constant feature of the place. And, this line of thought continues, an object just is a spatio-temporally continuous feature, or group of features. So the enduring character of the places requires the enduring character of the things.

At this point we have to scrutinize the notion of a material body, and the contrast between material bodies and features. The contrast I want to pursue is not simply one between possession and lack of criteria of identity. It has to do rather with the type of criterion of identity that physical objects have. I do not want here to try to give a full analysis of the notion. I want, rather, to highlight one difference between thinking in terms of features and thinking in terms of objects. In contrast to features, the condition of a thing at any one time is thought of as being causally dependent upon its condition at earlier times. One of the determinants of its properties at a given time is which properties it had earlier, and this is so no matter how much it has moved around. It might, indeed, be held that internal causal connectedness is more basic than spatio-temporal continuity, in our notion of objecthood (cf. Shoemaker, 1984; Slote, 1979). Once we acknowledge this causal dimension of the ordinary notion of an object, we can find room for the possibility of landmarks being used which are not objects. We can certainly make sense of someone finding their way around using stably located features, or groups of features, which they do not think of as having that internal causal connectedness over time. Someone might think in a way that was more primitive than any that involved thought of internal causal connectedness, and *a fortiori*, more primitive than any that involved thought of physical objects. They could still be re-identifying places.

Having put the point in this way, though, it might be wondered whether we ever think about physical objects at all; for surely this notion of 'internal causal connectedness' is more abstract than anything we ordinarily use. But the abstractness is simply a product of the generality of the description of an ordinary mode of thought. The notion of 'internal causal connectedness' is presupposed in our grasp of the way in which objects interact. For if we are to have any appreciation at all of the effect that one object can have upon another, in a collision, for example, we have to understand that one central determinant of the way the thing is after the collision will be the way that very thing was before the collision. We have to understand the dependence of objects on their earlier selves, to grasp that their earlier selves are only partial determinants of the way they are now, and that external factors may have played a role. We need a distinction, then, in describing our ordinary thought, between the causality that is, as it were, internal to an object, and has to do with its inherent tendency to keep its current properties, or for them to change in regular ways, and the causality that has to do with the relations between objects, and the ways in which they act upon each other (Shoemaker, 1984, p. 241). We can, then, say a little more about what it means to be using stable features as landmarks, rather than enduring things. Someone who is not operating with the conception of physical things is someone who does not have the conception of 'intrinsic causality'; the way in which the later stages of a thing are causally dependent upon its earlier stages. This person does not have the conception of the earlier stages of the thing as a partial determinant of the upshot of any causal interaction between it and something else. So this person cannot have the conception of 'causal interaction' at all – there really is no saying what would happen if we were somehow to engineer a collision between two of these groups of features. To try to give an answer, one would have to appeal to the 'causal inertia' of the group of features, and that just is an appeal to their intrinsic causality, which is importing the notion of a physical object which has those features.

One might try to provide criteria of identity for located features, but this will simply serve to emphasize the distinction between features and physical things. For instance, one might say that at any given time, two located features are identical just if they occupy the same place. And we might try to allow for the possibility of movement by a feature. For we surely do want to be able to acknowledge that there may be a given qualitative feature at each of a continuous series of neighbouring places over a given time; and it seems hard to resist the description of that as a case in which a single particular feature is continuously changing its position. If we talk in this way, then we have to acknowledge that the total spatial career of a particular feature, over a period of time, is essential to it; if we assume that it had some quite different spatial career, then we simply lose our grip on it being that very feature that we are thinking about. In contrast, we can make sense of a particular object's having had a quite different spatial career, partly because of the extra structure in our conception of the object afforded by our grasp of its internal causality. That is, we can, for example, hold constant the origin of the thing, even though we vary the route that it takes.[3] This is not possible in the case of something we are thinking of simply as a feature. Just because we are not appealing to the idea of its internal causal relatedness, we have no way of making sense of the idea of the origin of a particular feature. So there is nothing we can hold constant, as what identifies it as that particular feature, while varying its route. The spatial career of a located feature is essential to it; in contrast, the location of an object at a time is the paradigm of a contingent fact.

Once we have explained the distinction between features and material objects in this way, it is hard to see why the reidentification of places could not use distinctive features rather than physical objects. Certainly this seems to be the right way to describe the reasoning of many animals. Homing pigeons, for example, are certainly capable of reidentifying places. But supposing them to be capable of reidentifying places by using some features which are distinctive of them, does not depend upon supposing the pigeons having any conception of the landmarks as objects with the internal causal connectedness characteristic of objects. The pigeons may have no expectations at all as to what would happen in the case of a collision between two landmarks.[4]

Success in the Acredolo paradigm shows how an infant might be able to re-identify one particular place over time. As we have seen, this does not seem to depend on reference to physical things. But success in the paradigm does nothing to display any capacity one might have to think of places as all spatially related to each other. It does not display the connectedness of the space. The paradigm simply tests for the re-identification of a single place over time. It does nothing to establish that the child has any conception of the place as one of a network of spatially related places; it does nothing to establish that the child is thinking in terms of a unified, or connected, space. The conception of a place as one among a network of spatially related places is at least as fundamental to our thought about places as the ability to re-identify places over time. Any appreciation of the connectedness of the space will have to involve the background method which the child uses to keep track of places over time. As we shall see, it is when we attend to the distinctions between various ways in which we can register the connectedness or unity of the space, that we find the role in our spatial thinking of the notion of a physical object. The thesis is still in play that if the child is to have a certain type of appreciation of the connectedness of the space it is in, it must be capable of referring to physical things. It would have to be made out that there is a conception of connectedness which depends upon the conception of physical things.

What does it take to grasp the spatial connectedness of a space? To answer this, we have to consider the relation between grasp of spatial and grasp of causal concepts. On one view, we need spatial concepts to elucidate the notion of cause (Salmon, 1984, ch. 5). But this view should not be pressed so hard as to imply that we can ascribe spatial representations to animals in a way that outruns their capacity to give causal significance to them. What makes spatial reasoning reasoning about one's environment, about the space one is in, rather than a purely formal exercise in mathematical computation, is the ability to grasp the physical significance of the spatial representations one forms. So we must look for an account of the laws or regularities, however probabilistic or open to exception, which connect spatial properties with other physical properties. To do so is not to demand definitions of spatial concepts in terms of other physical concepts. There is no presumption that the spatial concepts are less primitive than any others; there is no presumption that we can separate out, from the flux of physical thinking, some regularities which alone deserve to be elevated into definitions. There may be no definitions to be had (Friedman, 1983, pp. 264–339). Still, insofar as spatial reasoning is to be understood as reasoning about one's actual environment, rather than as pure geometry, it is theoretical. It is only its figuring in an 'intuitive physics' of one's environment, through regularities connecting the spatial properties with other physical properties, that makes it reasoning that is not purely mathematical, but rather, reasoning about the space in which one lives.

Notice also that we have to explain why an animal's capacity to engage in spatial reasoning might have been selected for under evolutionary pressure. And if the reasoning has no physical significance, then it has no selection value; it can be of no assistance to the animal in coping with its surroundings. We ought to be reluctant to ascribe spatial reasoning whose use by the organism defies explanation in terms of selection pressures. If we subscribe to an evolutionary–teleological view of content-ascription, we will not simply be reluctant to do this but regard it as incoherent to do so.[5] Of course, there are views on which spatial reasoning is simply a form of causal reasoning: spatial notions can in some sense be reduced to or explained in terms of causal notions (van Fraassen, 1985; Sklar, 1983). But we can accept the need to relate the ability of a creature to represent space to its capacity to give physical significance to those spatial representations, without insisting on any such reductionist thesis.

If we ask how one could exercise grasp of the fact that every place in one's environment is spatially related to every other place in it, the answer is that we have to look at the physical significance one assigns to this relatedness; at the way one grasps the causal connectedness of the space. Because of their internal causal connectedness, physical things have a special role in registering the connectedness of space. Specifically, reference to physical objects is used to grasp the connectedness of the space through spatial thinking at a certain 'absolute' or 'objective' level; for their internal causal connectedness precisely means that the possibility of their travel through space can give physical significance to its spatial connectedness. Features could not do this work, because of their lack of internal causal connectedness, and the impossibility of their having routes other than the routes that they actually take; the possible 'trajectories' of actual features could not capture the full modal force of the connectedness of the space. To delineate the exact role that physical objects play, though, we have first to remark that there are many ways in which the conception of the space as connected could be grasped which do not depend on physical objects. At first, this may seem quite impossible. As I have described it, thought at the level of feature-placing is, as it were, causally inert. So how could a creature reasoning at this level possibly register the physical significance of anything, let alone the connectedness of a space?

Within psychology, the question of the connectedness of a creature's conception of the space it is in has become familiar in the context of discussion of whether one or another animal has and uses cognitive maps. For example, consider an instructive recent dispute over whether honey bees use maps – 'instructive' because it puts pressure on the notion of a 'map'. In effect, the protagonists agree in defining 'map' in terms of the kinds of connectedness represented among the places represented. Thus Wehner and Menzel write:

> Even though the term *cognitive map* is often defined rather vaguely and applied to various kinds of animal orientation, we hope to be in line with most workers in the field – and especially with Tolman (1948), who coined the term – when we use it in the way a human navigator does. Seen in this light, a cognitive map is the mental analogue of a topographic map, i.e. an internal representation of the geometric relations among noticeable points in the animal's environment. In operational terms, this means that an animal using such a map must be able to compute the shortest distance between two charted points without ever having travelled along that route. More generally, it must be able to determine its position, say, relative to home, or to any other charted point, even when it has been displaced unexpectedly to an arbitrary place within its environment (Wehner and Menzel, 1990, pp. 403–4).[6]

The reference to metric properties ('shortest distance') here seems to make this definition insufficiently general: we need a definition which relates to the topology of the representation only, rather than any metric properties it might have; we should certainly want to allow that a map inscribed on rubber, or a city Metro map, or their psychological counterparts, are maps, even though they do not represent metric properties at all. But we could keep the focus on connectedness in the representation without appealing to metric properties, while still acknowledging that even an unmetricized map will allow some notion of 'optimum route' – Metro maps are not entirely useless. It is through its possession and use of such a map, metric or not, that an animal gives physical significance to whatever grasp it has of the connectedness of the space it is in: the fact that any place is spatially related to every other place. Once we put things in this way, though, it seems apparent that there will be many sophisticated navigational systems that need make no essential use of physical objects at all, rather than of features as landmarks. In section 3 I shall consider the distinctions that can be drawn among different navigational systems. Then, in section 4, I shall introduce a notion, 'causal indexicality', that characterizes all the ways of thinking about space described up to that point. In section 5 we shall see that physical objects may have a foundational role to play in the operation of a way of thinking about space that is not causally indexical.

2 Egocentric space

Before proceeding to the topic of maps and connectedness, there is a fundamental objection which has to be addressed here. The child in either phase of the Acredolo paradigm could be said to be identifying places egocentrically: it identifies them as 'to the right', 'in front and up a little' and so on. What is different between the child in the early stage and the child in the late stage is the criterion of identity that it uses for places over time. But in either phase, the place-identification that it uses in controlling which way the child looks is egocentric. And this suggests that the child is after all identifying the places around it by reference to a physical object, namely itself. We could put the point by asking how we are to distinguish the class of egocentric frames of reference – that is, egocentric methods of identifying the places in one's environment. On the face of it, an egocentric frame of reference is a body-centred frame of reference; or one which is centred on a part of the body. The developmental psychologists Herbert Pick and Jeffrey Lockman put the idea as follows. They define a 'frame of reference' to be 'a locus or set of loci with respect to which spatial position is defined'. 'Egocentric' frames of reference then are those which 'define spatial positions in relation to loci on the body'. They are contrasted with 'allocentric' frames of reference, 'which simply means that the positions defining loci are external to the person in question' (Pick and Lockman, 1981, p. 40). This definition seems indeed to give a reductive account of the notion of an egocentric frame, defining it in terms of notions which genuinely seem to be more fundamental than it. If the definition is correct, then even the very simple place-identifications we have been considering, by the child in either phase of the Acredolo paradigm, depend upon reference to a thing, namely the body.

It is worth reflecting on the general form of the definition. In trying to say what is characteristic of an egocentric frame of reference we are not dealing with a problem in pure mathematics. It is not, for example, on a level with the question whether a frame of reference uses polar or Cartesian co-ordinates. Thinking in purely formal terms, the best we could do would be to say that it must be possible, using an egocentric frame, to specify spatial relations to a single privileged point; but that would not separate an egocentric frame from one centred on the sun, for example. We have to say something about the physical significance of the origin of the frame: we want to say, for example, that it must be centred on the subject. This notion of 'the subject' is not a purely formal notion of pure mathematics. But saying where the frame is centred is only one way of giving physical significance to the formal notions. An alternative would be to consider the physical significance of the axes of the frame of reference, and to take them as fundamental: an egocentric frame would then be one whose axes had a particular kind of physical significance. It would then be a substantive thesis, rather than a definition, that egocentric frames are invariably centred on the body, or a part of the body. And it would be quite wrong, on that approach, to suppose that in using an egocentric frame one must be identifying places by their relations to a body already identified.

The definition of an egocentric frame as a body-centred frame takes for granted the general notion of an object-centred frame of reference, and says that the egocentric frames are a particular class of object-centred frame, namely, those which are centred on the body or a part of the body. The general notion of an object-centred frame is certainly legitimate. Consider an object such as a table, or a bus. We can think of the internal spatial relations between its parts. We can use this system of internal spatial relations to identify points within the object. There may be natural axes that the object has. For example, given a pillar-box, we could define a set of axes by reference to its long axis, its 'line of sight' through the slot and its 'coronal plane', through which the door moves when it is opened. So far, what we have is a way of identifying points internal to the object. But the system of spatial relations that we have set up between the parts of the thing could be further used to identify points external to it. We could, for instance, identify a coconut on a palm tree as lying on a line through the bottom of one leg of the table, and the top of another, and a hundred yards distant in the direction going from bottom to top. This way of identifying places need not be used only with inanimate objects as its basis. One could equally well take the internal spatial relations between the parts of a horse, or its natural axes, and use them to give fully allocentric identifications of the places around it. One could do the same with a human body. One could do the same with what is in fact one's own body. And then, by Pick and Lockman's definition, what we have is an egocentric frame. But evidently, there is a finer distinction that we want to make here. For it is not as if we can assume extensionality: not just any way of thinking of the subject will do. The notion of egocentric space is a psychological notion; the reason why we want it is to explain why the infant, for example, turns one way rather than another. In particular, perceptual knowledge of the body will not do. Merely seeing one's own body in a mirror, for example, and using it to set up a system of axes, will not provide one with an egocentric frame.

The obvious proposal is that subjects have to be using the direct, non-observational knowledge of their own bodies constituted by their possession of a 'body-image'. In one use of the phrase, 'the body image', it has to be thought of as referring to a relatively long-term picture of one's own physical dimensions. So someone's body-image might be changed as a result of their having a skin graft, or the loss of a limb, or simply by growing up. In this use of the term, the 'body image' provides one with a general sense of what kinds of movement are possible for one. It assigns a particular structure to the creature, which underlies its possibilities of movement (O'Shaughnessy,

1980, p. 242). We cannot directly use that to set up a system of axes – it assigns no particular shape to the body. What we need, rather, is what O'Shaughnessy calls the 'here-and-now' body-image, which 'is given by the description or drawing or model one would assemble in order to say how the body seems to one *at a certain instant*. For example: torso straight, right cylindrical arm stretched out from body, crooked at right angles, etc.' (O'Shaughnessy, 1980, p. 241). Given its possession of such a body-image, it can plot the spatial relations between the various parts of its body, and use them to construct a body-centred set of axes which will indeed be the egocentric axes. This proposal relies on a direct relation between the subject's body-image, and its ability to act. We have to think of the body-image as giving the subject a practical grasp of the ways in which it is possible to act; the possibilities of movement open to it. Of course, there must be some relation between these two conceptions of the body-image (O'Shaughnessy, 1980, p. 247). The immediate problem is, though, to understand why this shift, from 'outer perception' of the body, such as seeing it in a mirror, to 'inner perception' as provided by the short-term body-image, should be thought to achieve anything. After all, as we saw, simply managing to use the spatial relations between the parts of the body to set up a system of axes does not in general secure one an egocentric frame. Why should we think that an egocentric frame is guaranteed if one relies on the spatial relations between the parts of the body given in 'inner perception'? The point here is that there is in general no direct connection between the mere use of an arbitrarily chosen body to set up co-ordinate axes, and the subject's capacity for directed spatial action. What the present proposal relies on is a direct relation between the subject's short-term body-image, and its ability to act. We have to think of the short-term body-image as giving the subject a practical grasp of the ways in which it is possible to act; the possibilities of movement open to it. The reason why this seems promising is that the short-term body-image has direct connections with action, of the type possessed by the egocentric axes. The proposal is that we can view the direct connection between action and egocentric space as a product of the direct connection between action and the short-term body-image. But now we have to ask how it can be that the body-image has this direct connection with action. And we immediately face a dilemma. For how are the spatial relations between the parts of the body given in the body-image? One possibility is that they are given in egocentric terms – that one foot is represented as to the right of another, below the rest of the body and so on. But then it can hardly be held that the subject uses the natural axes of its body to set up the egocentric axes; rather, it has already to use the egocentric frame to grasp the spatial relations between the parts of its body. Alternatively, suppose that the spatial relations between the parts of one's body are given in non-egocentric terms. Then there is no prospect of using the axes of one's body to set up an egocentric frame; one is in no better position to do this with respect to the body of which one has 'inner perception' than one would be with respect to a body of which one has 'outer perception'. In both cases the problem is the same. One's grasp of egocentric spatial axes, with their immediate connections to moving and acting, cannot be generated from a grasp of spatial relations which are given non-egocentrically. Grasp of egocentric spatial axes must be taken as primitive.

This means that a certain kind of reductive ambition for the definition of an egocentric frame as a body-centred frame has to be abandoned. We cannot view this definition as explaining the notion of egocentricity in more fundamental terms. We cannot see it as defining egocentricity in terms of (a) the generic notion of an object-centred frame of reference, plus (b) the notion of a body-centred frame. For when we inquire into the needed notion of body-centredness, it turns out that it already appeals to the notion of the body as given in the body-image, with its spatial relations given egocentrically. In particular, then, we cannot take the body-image to be more fundamental than the egocentric axes: we cannot derive them from it. The egocentric axes have to be taken as primitive, relative to the body-image.

It might be asked whether the body-image is not at any rate co-ordinate with the egocentric axes, so that they have to be taken to be equally fundamental in the direction of spatial action. But while some egocentric reference frame is evidently essential if we are to have spatial action – otherwise the action could not be regarded as spatial at all – it does not seem that a subject needs to have a body-image in order to be capable of egocentric spatial action – action we would want to explain by appealing to its possession of an egocentric frame of reference. The co-ordination and direction of spatial action may be achieved by purely distal specifications of the locations which are the endpoints of the actions, without the subject having a single central body-image at all (Scott Kelso, 1982). If a body-image is superimposed on the subject egocentric axes, that is additional to the requirements for it to be thinking about places egocentrically. So when the subject is identifying places egocentrically, it cannot be thought of as doing so by first identifying a physical thing – itself – through a body image, and then identifying places by their relation to its body. Rather, its capacity to use the egocentric axes is more fundamental than its capacity to think in terms of a body-image. The egocentric identification of places does not depend upon a prior identification of a body. The notion of the egocentric frame is more fundamental than the relevant notion of body-centredness. It is only when we have elucidated the notion of an egocentric frame that we are in a position to say what this notion of body-centredness is.

How then are we to characterize egocentric frames of reference? One alternative approach would be to say that an egocentric frame is one defined by the axes 'up', 'down', 'left', 'right', 'in front' and 'behind', with the origin identified as 'here'. Places cannot be identified by directions from a single origin alone. We should have to add something about the way distances are measured, using this frame of reference; or at least we need some kind of order relation. Even so, this approach would not give us enough to say in general what an egocentric frame is. We want to be able to allow as intelligible the hypothesis that humans may use many different egocentric frames. Consider, for example, the axes defining the movements of the hand in writing. There is no reason to suppose that this will be the very same set of axes as is used to define the movements of the whole body. Nevertheless, it is still an egocentric frame. So an approach which tries to define what it is for a frame to be egocentric by simply listing a particular set of axes will not work. Again, there is no reason to suppose that all species will use the same egocentric axes. For example, creatures which are differently jointed to us, or which live deep underwater, may use different axes. Finally, even if we could, by listing a suitable set of axes, give an extensionally correct identification of the egocentric frames, we would still have the explanatory work to do. We should still have to explain what it was about the terms 'left' and 'right', for example, that made them particularly connected to moving and acting, for instance. Even so, it may still be the case that the right way to give a general definition of the notion of an egocentric frame of reference is by defining a class of axes, rather than by making a general demand about where the frames must be centred. And, of course, we would expect that an extensional approach here would not succeed: we have to grasp how the subject is apprehending those axes.

The axes that are distinctive of an egocentric frame are those which are immediately used by the subject in the direction of action. They may include, but need not be confined to, the natural axes of the body. In the case of 'in front' and 'behind', we have a distinction defined in terms of the body and its modes of movement and perception. Its application to us depends on exploiting ways in which we are not symmetrical. If we were symmetrical, being double-jointed and able to look either way, then our current notions of 'in front' and 'behind' simply could not be applied to ourselves, could not guide our actions in the way that they do. But we are not symmetrical in this way, and the distinction does guide our actions. In the case of 'up' and 'down', it does not seem that we have here a distinction which is defined in terms of asymmetries of the body. It has to do rather with orientation in the gravitational field. The extensive apparatus we have to tell us how we are oriented in the gravitational field is precisely an apparatus to tell us which way is up. Of course, the reason this matters to us is the pervasive influence of gravity on every aspect of our ordinary actions. So here we have an egocentric axis which is not defined as a natural axis of the body. Of course, there is such a thing as the long axis of the body, but that is not the same thing as 'up' and 'down', which continue to be defined in terms of the gravitational field even if one is leaning at an angle. The case of 'right' and 'left' does not follow either of these models. The fundamental distinction here does not have anything especially to do with the bodily axes at all. It is not, as in the case of 'in front' and 'behind', that there is any bodily asymmetry that the distinction labels, since animals are generally right/left symmetric. Nor does it, like 'up' and 'down', label some external physical magnitude which is of general importance for action. None the less, it is evidently an axis which is used in the direction of action.

I said that egocentric axes are those which are 'immediately' used in the direction of action. It may be that no very precise definition can be given of that notion of 'immediate' use, and that the notion of an egocentric reference frame must to that extent remain a rough and intuitive one. But we can get some sense of the required conception by contrasting egocentric frames, such as used by the infant in the Acredolo paradigm, with more complex dead reckoning systems: that is, systems which enable one to keep track of where one is by keeping track of how fast one has been moving, in what direction and for how long. The point about such systems that matters here is their use of a compass which is external to anything used in the immediate direction of action. For example, an animal might use the position of the sun, together with its knowledge of the time of day, as a compass. It can use this to keep track of each of its various swoops and sallies, and so to plot the direct route home. But before it can actually translate this into action, it has to know the direct route home not merely in terms of direction specified in terms of the external compass: it has to know which way to point itself to travel in that direction. It is in this sense that the egocentric axes are 'immediately' used in the direction of action, whereas the external compass is not. Of course, a dead reckoning system could also use the egocentric axes themselves, though in practice this would mean a considerable loss in accuracy. Notice, incidentally, that these dead reckoning systems are body-centred; what makes them not egocentric is the axes that they use, and the indirectness of their role in guiding action.

It thus appears that the re-identifications of places achieved by children in the Acredolo paradigm can be managed without any reference being made to physical things. But all that we have so far is that ability to home back in on a single place, to re-identify it. We do not yet have any ability to grasp the physical significance of the connectedness of the space: the fact that every place in the space is connected to every other. As I said, it is here that we find the role of physical objects in spatial thinking. As we saw, though, it must also be acknowledged that one way in which the connectedness of a space can be grasped is through possession and use of a cognitive map; and there is, on the face of it, no reason why that should require the conception of an object. But how can the causal significance of the connectedness of a space be grasped by a creature which is thinking only in terms of causally inert features?

3 What is a map?

In the literature on cognitive maps, a distinction is sometimes drawn between 'absolute' and 'relative' modes of spatial thinking. The notion of 'absolute' space that is being used here has to do with the distinction between ways of thinking that involve an explicit or implicit dependence upon an observer or agent; and those which have no such dependence. These latter are the 'absolute' modes of spatial thought. It has often been maintained that this notion of an 'absolute' spatial conception is at best a kind of limit case, derived from increasing attenuation in the dependence on the observer or agent. The idea that one might actually attain this limit is regarded as subject to empiricist or pragmatist critique. Thus Poincaré wrote: 'Absolute space is nonsense, and it is necessary for us to begin by referring space to a system of axes invariably bound to our body' (Poincaré, 1946, p. 257, cf. 244–7). More recently, though, in work in the tradition begun by Tolman, this idea has been challenged, most powerfully by John O'Keefe and Lynn Nadel:

> Most authors attempt to derive all psychological notions of space from an organism's interaction with objects and their relations. The notion of an absolute spatial framework, if it exists at all, is held by these authors to derive from prior concepts of relative space, built up in the course of an organism's interaction with objects or with sensations correlated with objects.
>
> In contrast to this view, we think that the concept of absolute space is primary and that its elaboration does not depend upon prior notions of relative space... [there] are spaces centred on the eye, the head, and the body, all of which can be subsumed under the heading of *egocentric space*. In addition, there exists at least one neural system which provides the basis for an integrated model of the environment. This system underlies the notion of absolute, unitary space, which is a non-centred stationary framework through which the organism and its egocentric spaces move (O'Keefe and Nadel, 1978, pp. 1–2).

The point of the empiricist or pragmatist critique is to insist that this 'absolute' conception cannot legitimately be ascribed to an organism; we cannot make any sense of any such conception. Now we cannot resolve this issue simply by pointing out that animals have navigational skills. For there are many types of navigational system which do not have this 'absolute' character. We shall shortly consider two such systems. What we would like is to see a detailed account of what an 'absolute' mode of spatial thinking would look like, together with some account of how it could be that a creature could be said to grasp such a mode of thought. In some recent papers, John O'Keefe has set out a model which is held to be of precisely this 'absolute' or 'allocentric' character (O'Keefe, 1990; 1991). As we shall see, the crucial feature of this model is the way in which it captures the connectedness, or unity, of the space mapped; that is, the spatial relations that any one place bears to any other place.

On this model, the animal has to find what O'Keefe calls the 'slope' (or 'eccentricity') and the 'centroid' of the environment. Finding the slope is a way of using the distribution of cues in the animal's environment to provide it with compass directions. That is, it defines a direction for the environment, which does not depend on which way the animal itself is pointing. Or, to put it another way, as the animal is rotated, its angle with the slope of the environment changes, so that it can use its angle with the slope as a way of defining which way it is pointing. On the other hand, if the animal is moved from one place to another, without any change in which way it is pointing, its angle with the slope remains constant. The centroid of the environment is 'the geometric centre or centre of mass of the cues in the environment' (O'Keefe, 1990, p. 306; 1991, p. 283). It can be thought of as a point which has the following characteristic. If each cue in the environment is assigned a mass of one unit, and rigid rods used to connect the centroid to each cue, then the resulting construction will balance evenly around the centroid.

The details of how the slope and centroid are computed are not crucial to the present discussion, but I include them for the enthusiastic reader. We begin with the animal's ability to keep track of its own displacements, and its perceptual knowledge of the vectors from it to each of a number of landmarks in its environment. These vectors specify directions in terms of the axes of the animal's own body. The problem then is to use these egocentric vectors to generate a non-egocentric representation of the places around it. On this model, the animal has now to calculate the slope and the centroid of the environment. The slope is defined as 'the deviation from symmetry or isotropism of the cue configuration in different directions.' Suppose we have a pair of landmarks, with their egocentric positions given – so their direction and distance from the subject is given. The line connecting them has a particular gradient in egocentric space. We can take the gradients of all the lines connecting each pair of cues in the environment. We can then take the average of all these individual gradients; that gives us the slope of the environment. The way in which this slope is identified in the animal's egocentric space will of course vary depending on what direction it is facing in; so rotations of the animal will change the measure of the slope. But the slope is invariant with translations of the animal which change its position without changing the direction in which it is pointing. This means that the slope can be used as a non-egocentric measure of direction. The centroid is defined as 'the geometric centre or centre of mass of the cues in the environment.' One way to find the position of this place is by taking the average of all the egocentric vectors to the landmarks around the subject.

Now, once the slope and the centroid for a particular environment have been established, we can use them to define positions in it. We can define a position by giving the vector to it from the centroid: that is, its distance from the centroid, and its direction from the centroid, given as the angle of the straight line connecting it to the centroid, with the slope. So the animal can use this to store and remember the location of a particular target: the vector to it from the centroid. It can also find out where it is itself, by finding the current vector from it to the centroid. One type of activity that the model can be used for is to enable the animal to predict its next location on the basis of its current location and movement. The animal's displacement can be represented by a vector giving distance and angle with the slope; this, together with its knowledge of its original position – that is, the vector from its original position to the centroid – means that it can keep track of where it is even if landmarks are removed (O'Keefe,

1990, p. 307; 1991, pp. 286–90). Also, the system can be used to find the movement needed for the animal to get from where it is to a target location. Given two vectors, one representing its current location – the vector from it to the centroid – and the other the vector from the centroid to its target, the system can subtract the two and calculate the movement vector required to get directly to the target (O'Keefe, 1990, p. 308; 1991, p. 290). The computational power of the system enables the animal to give physical significance to the connectedness of the space, in quite a strong sense: no matter where the animal is, it can represent the vector from itself to any other location. But although it can represent the connectedness of the space in this strong sense, there is no evident dependence of the system on physical objects being used as cues. For all that we have so far, the cues might well be located features. The crucial issue, as we shall see, has to do with the sense in which the connectedness of the space is represented. There is a sense in which the animal using this system does not have a fully 'disengaged' or 'objective' grasp of the connectedness of the system; and it is that strong notion of connectedness that defines the intuitive notion of a map-like, or absolute representation of space, and ordinarily depends upon the use of physical objects.

Philosophers sceptical about the idea of the animal representing vectors to and from the centroid should reflect on O'Keefe's startlingly concrete proposal about how this is being achieved. The suggestion is that these vectors are being represented by sinusoidal waves found across the hippocampus. The distance to the cue corresponds to the amplitude of the wave; the angle of the vector corresponds to the phase of the sinusoid. Addition and subtraction of vectors is accomplished by addition or subtraction of the sinusoids (O'Keefe, 1991, pp. 284–5; 1990, pp. 309–11). I am not concerned to question whether an animal could use the kind of scheme described by O'Keefe; it is evident that the thing is possible. What concerns me is whether this is a system which constitutes an 'absolute' or 'objective' way of thinking about space. The crucial point about this model, given our concerns, is that there is evidently no reason why the cues to which it appeals should be taken to be physical objects, rather than stable located features. If this is an 'absolute' or 'objective' mode of spatial thinking, then it is one that has been achieved without any ineliminable appeal being made to the notion of a physical thing.

Do we have here a system which constitutes an 'absolute' or 'objective' way of thinking? Consider, by way of contrast, the Wilkie–Palfrey 'triangulation' model of the behaviour of rats in the Morris swimming task. Here, rats are placed in a swimming pool filled with an opaque liquid. There is a submerged platform to which they learn to make their way. The platform, being submerged in an opaque liquid, cannot be seen by the rats. But they can reliably make their way to it, from any starting point in the pool, so long as it keeps its relation to the distinctive landmarks they can see around the pool. The 'triangulation' model supposes that once on the platform, the animal stores the distances to each of the cues it can see. Then when it next tries to get to the platform, it notes the distances from where it is to each of the landmarks around it. If the distance to a particular landmark is currently greater than it was from the goal platform, the animal swims towards it. If the distance is less than it was from the goal, the animal swims away from it. Its movement is the resultant of all these calculations (Wilkie and Palfrey, 1987). This model is one that O'Keefe himself cites as not truly 'objective', in contrast to the model of slope and centroid. What, then, is the difference between these models? If we look at the matter formally, the striking thing is the extent of the similarity between the models, rather than differences between them. On the slope/centroid model, the vectors to various cues from the centroid are recorded. On the triangulation model, the distances to various cues from the goal platform are recorded. In fact, to improve the parallel, we could consider a version of the triangulation model in which the animal also uses some external compass, such as a gradient in the lighting across the room, to find and record the directions of various cues from the goal platform, as well as their distances. O'Keefe would certainly regard such a model as still falling short of the genuinely objective; but it is strikingly parallel to the slope/centroid model.

It might be said that the parallel is a fake, because in the triangulation model, the vectors to cues from the goal platform are given egocentrically, whereas in the slope/centroid model, the vectors to cues from the centroid are given non-egocentrically. But this way of drawing the contrast cannot be sustained. In the triangulation model, the animal has (a) memory of vectors to cues from the goal platform, and (b) a capacity to put that memory, together with its current perceptions, to use in directing its actions. The memory of vectors to cues from the goal platform is not itself an egocentric presentation of those locations. Rather, it gives information about the positions of things only when put together with the animal's current perceptions and its method of integrating the two. Of course, the memory may be precisely a memory of an earlier sighting of those cues from the goal platform. And that sighting at the time carried egocentric spatial information. But the spatial information that the original sighting carried is quite different to the spatial information that the memory carries. As the memory is used now, it tells the animal where the target is only when it is coupled with information about the current distances to the landmarks. The original perception, however, directly carried information about the location of the target with respect to the cues, without there being any need for such manipulation. We could approach the parallel from the other direction, by considering how the animal using the slope/centroid model might operate. Suppose, for example, that it proceeds as follows. It calculates the slope and centroid of its environment, in the way indicated. It then makes its way to the centroid, and looks about it to find the vectors from it to its various potential targets. In finding these vectors it in effect logs in the content of its current perceptions, proceeding in the same way as the rat on the platform in the water maze on the triangulation model. It might be protested that the animal using the slope/centroid model uses an external compass, namely the slope of the environment, in recording these vectors, rather than its egocentric axes. But we already agreed that the animal using the triangulation model might also use an external compass to log in the vectors to cues from the goal platform, without this transforming its system into an 'objective' representation of space.

It might be acknowledged that there is this parallel between the two models, but a contrast noted too. In the case of the slope/centroid model, the animal uses the vector from it to the centroid, and the vector from the centroid to the cue, to compute the direct vector from it to the cue. In the case of the triangulation model, the animal uses the vector from it to the cue, and the vector from the goal platform to the cue, to compute the direct vector from it to the goal platform. This means that there is a contrast between the role of the centroid and the role of the goal platform. The centroid functions as a means of getting the animal to any target it likes. On the triangulation model, though, the other cues function only as means whereby to get the animal to the goal platform. So unlike the slope/centroid model, the triangulation model is organized around a single-goal destination. This is a relevant contrast between the two models, and one that O'Keefe stresses. But it is hard to believe that it will bear the weight of explaining the notion of an 'absolute' or 'objective' representation. We cannot plausibly say that an 'objective' representation is just a method of navigation which is not dedicated to any single goal. For there are models which are not dedicated to a single goal, yet which cannot plausibly be classified as 'objective'. For example, a system which takes momentary egocentric presentations of the places of things, and uses knowledge of the subject's movements to continually recalculate their current egocentric positions, is of this type (O'Keefe, 1988).

We might press this line of thought in a somewhat different way, though. Perhaps the point is not that the one system is goal-centred while the other is not, but rather that certain types of computation are possible on the slope/centroid model which are not possible on the triangulation model. The slope/centroid model allows one to consider a wider range of spatial relations than does the triangulation model. For example, the triangulation model will not allow one to represent the direct vector from oneself to a currently unperceived cue, even if one has recorded the vector from the goal platform to the cue. But the slope/centroid model will enable one to represent the vector from oneself to a currently unperceived cue. So perhaps the relevant contrast is that the triangulation model is limited in which spatial relations it can represent between the places it represents. Following up this line of thought, we might propose a formal criterion for a way of representing space to constitute a 'map', an 'absolute' or 'objective' mode of thought:

the system must be capable of representing, perhaps after operations defined within the system, the direction and distance between any two arbitrary represented places.

Or, more generally, since we want to allow for the possibility of 'objective' representations which do not use a metric, but perhaps a more primitive relation of spatial order:

the system must be capable of representing, perhaps after operations defined within the system, the spatial relation between any two arbitrary represented places.

We can pursue this line of thought by considering the constraint we remarked already on the representation of places. For a creature to be representing places, it must have some grasp of the criterion of identity for places over time. But we can also ask whether it appreciates that the places it represents are all spatially related to each other. And then we can raise the question at what level this conception of the places as all spatially related to each other receives its physical interpretation: How does the animal assign causal significance to the spatial connectedness of its environment? The conception of places as all spatially related to each other has implications for the causal interconnectedness of a region; that there can in principle be causal relations, perhaps of very complex types, between the items represented as being at various places. And now the question is: At what level does the animal register the causal interconnectedness of the space? It can, of course, register the causal interconnectedness of the space through its perception and action. From any point in the space, it can act upon any other. This is the way in which causal interconnectedness is registered in the slope/centroid model. We simply do not have such a full registering of the causal interconnectedness of the space, at any level, in the triangulation model. What we have is (a) from any point in the space, the animal can act with respect to the goal location, and (b) the navigational system being used is one which can, in principle, log in any arbitrary location as the goal location. This is not at all the same thing as a registering of the simultaneous causal interconnectedness of all places in the space. And I think that this is the theoretically interesting distinction between the slope/centroid model and the triangulation model.

Though the triangulation model does not meet the full force of this 'connectedness' constraint, neither does the slope/centroid. What the animal using this system can do is represent the vector from it to any arbitrary target recorded in the system. But it cannot simply represent the vector between any pair of arbitrary places, regardless of whether or not it is thinking of itself as at either of those places. We could, indeed, increase the power of the system, by allowing it to solve 'travelling salesman' problems, in which the animal has to compute the optimum route by which to reach a number of destinations. But this evidently will still not enable it to meet the full force of this 'connectedness' constraint. The animal still has to be thinking of itself as at the start of the itinerary, and as heading for the destination. Further, even meeting the constraint as stated above does not guarantee that the model will capture the full force of the connectedness of the space. The model allows the animal to represent only the geodesics between places. It cannot represent spatial relations of arbitrary complexity between any two places. Finally, the animal is incapable of representing configurational properties of numbers of places: it cannot register the fact, for example, that a particular group of places is configured as the vertices of a regular polygon. It is easy to miss these points because, given a statement of the spatial relations grasped by the animal using the slope/centroid model, a geometer could infer deductively all the other spatial relations that hold in the environment. The animal using the slope/centroid model need not be able to extract any of these further relations from the information it has. The reason why it does not is that its grasp of the physical significance of the spatial information it represents has to do entirely with its concerns in practical navigation, where these further spatial relations have no role to play. So although the slope/centroid model registers a much wider range of spatial relations than does the 'triangulation' model, it does not register the full spatial connectedness of the area represented, because the only way in which causal interconnectedness can be registered by this system is through the creature's use of it in navigation. On the face of it, though, it ought to be possible to represent the causal interconnectedness of a space at another level than through one's own engagements in the space. It ought to be possible to represent the causal interconnectedness of the space by having a disengaged picture of what is going on there. This cannot be done using the slope/centroid model, but it would give us an 'absolute' representation which would be capable of registering the full spatial interconnectedness of a region.

O'Keefe says of the slope/centroid system that it defines 'an allocentric position which is independent of the animal's movements but which can be used to locate the animal's current position' (O'Keefe, 1990, p. 306). Does this mark a contrast between that system and the triangulation model? Here we have to attend carefully to the sense in which the current location of the animal is marked. Even on the triangulation model, the animal using it can be described as finding its own position with respect to the target; realizing that it is to the left of the target, and so on. These are all just ways of describing the the animal's grasp of the vector from it to the target. There is no explicit representation of the animal itself in relation to its surroundings. In particular, there is no representation of the causally significant properties of the animal, such as its capacities for perception and action, and no attempt to explain its current perceptions or what is going on around it in terms of its position and movements. But the picture is exactly the same in the slope/centroid model. There is a sense in which the animal can be said to know where it is with respect to any of its targets; or, indeed, with respect to the centroid. But there is, again, no explicit representation of the animal itself in the system, no explicit representation of its causally significant properties. For there to be such a representation, the animal would have to be thinking of itself as a physical thing. But the animal using O'Keefe's model is not registering the connectedness of the space it is in by thinking in terms of the possible routes of physical objects, such as itself, through that space. The animal is registering the connectedness of the space through the fact of its own engagement in the space, not through reflective thought about its engagement in the space.

4 Causal indexicality

To understand the idea of an 'absolute' representation, a disengaged conception of the connectedness of the space, we need to introduce a new notion. I want to spend the whole of this section explaining this notion. I shall then take up its bearing on the representation of places. We are familiar with the idea that terms may be spatially or temporally indexical, in that their content depends upon where or when they are used. But there is a deeper phenomenon, of causal indexicality, which is what is often being discussed in accounts of indexicality. For many terms which are spatially or temporally indexical are also causally indexical; only this last tends not to be discussed as a separate phenomenon.

A first shot at isolating the phenomenon would be to say that causally indexical terms are those whose causal significance depends on who is using them. But that does not isolate just the class of terms with which I shall be concerned.

Many concepts are causally significant. That is, judgements made using them have some significance for the ways in which the world will behave, and for how it would behave in various possible circumstances. Often, one's grasp of the significance of such a judgement will be, in the first instance, a matter of how one reflectively expects the world to behave, and what counterfactuals one explicitly takes to be true. In such cases, one's grasp of the judgement has to do with the detached picture one builds up of how things stand around one. But there are cases in which one's grasp of the causal significance of a notion has to do not with any detached picture, but rather consists, in part, in one's practical grasp of its implications for one's own actions. These are the causally indexical terms.

I shall give some examples first, before more general description of the class. Although I am ultimately concerned with spatial notions, I shall begin by looking at some non-spatial classifications, to put the spatial cases into context. The easiest way to construct causally indexical terms is by the use of the first person. For example, consider the notion, 'is a weight I can easily lift'. Judgements made using this notion, about whether one object or another is a weight I can easily lift, have immediate implications for my own actions, for whether I will bother to attempt to lift this or that thing. Or again, a judgement such as 'this is too hot for me to handle' has immediate implications for my own actions with respect to the thing; the notion 'is too hot for me to handle', is again causally indexical. These examples might suggest that the distinction here is between notions whose causal significance varies from subject to subject. For example, whether the predicate 'is too hot for me to handle' applies to a thing depends upon who the subject is. But the application of the predicate, 'is magnetic' is not relative to a subject in this way. This, though, does not give us quite the contrast we want. The predicate, 'has the same mass as I do' varies in application depending upon who is using it. But it has no immediate implications for action. The significance for one's own actions of something's having the same mass as oneself depends entirely upon further beliefs that one has, such as whether things having the same mass as oneself will be easy or difficult to lift. We want to separate off those predicates whose causal significance has to do with the immediate implications for one's own actions and reactions to the world.

Let us consider our examples a little further. The predicates 'is a weight I can easily lift', and 'is too hot for me to handle' both use the first person. But intuitively, the use of causally indexical predicates does not depend upon self-consciousness. Even a creature which did not grasp the first person could use causally indexical representations. So there ought to be other examples of causally indexical predicates. Further, these predicates, 'is a weight I can easily lift' and 'is too hot for me to handle', use notions of 'weight' and 'temperature' which we ought to pause over. They need not themselves be causally indexical; they may rather be on a par with a notion such as 'magnetic'. Whether two things are the same weight or temperature, and what their particular weights or temperatures are, may be definable entirely in non-indexical terms. The complex predicates we have constructed may be using non-indexical physical notions to define causally indexical terms. But on the face of it, one might grasp causally indexical terms without having any grasp of these non-indexical notions.

If this is correct, then there ought to be more primitive examples of causally indexical terms. These would not be defined in terms of non-indexical physical notions, and they would not use the first person. It seems immediately obvious that there are such more primitive terms for weight and heat. Unstructured uses of 'is heavy' and 'is hot' may relate to the causal impact of the thing upon the subject, rather than being uses of some observer-independent system of classification. Moving closer to the spatial cases which concern us, a notion such as 'within reach' seems to have immediate implications for the subject's actions. The most immediate effect of judgements made using this notion is that the subject will try to contact things which are within reach, but will not try to contact things which are judged to be out of reach. This predicate is not first-personal. A creature could use representations of things as within reach or out of reach without having the ability to think using the first person.

In ordinary English, there is a certain 'social pull' in our use of these terms. If we say simply that something is 'heavy', that may be heard as having to do not with its relation to one's own powers, but rather to do with its relations to the powers of a normal human; or indeed to its comparison with some reference class of objects of the same general type. I want to set aside this phenomenon. We often do need to consider causal indexicals which have to do simply with the relation of the thing to the subject's own powers. In English, the way in which we make this explicit is by use of the first person. But we may want to use a term so restricted when reporting the reasoning of an animal which is not self-conscious. For example, an animal may reliably perform a task even though we vary a number of parameters, but then give up at some point. The ball has been thrown too far for the dog to recover, the stick is too heavy for it to lift. In such cases we can test for just why the animal is not attempting the task, and we have to be able to say, for example, that it is because the ball seems to have been thrown too far, or the stick seems too heavy. And these notions, 'too far' or 'too heavy', have to do only with a relation to the subject's own powers, even though the animal may not be self-conscious.

In these cases, it is not just that a grasp of the term requires the ability to register when it applies. It is rather that one's grasp of the causal significance of the term is exercised in the way one reacts to recognition that it applies. So, for instance, grasp of the notion 'within reach' is exercised not simply by differentially responding to cases where something is within reach, which might be done by simply looking confident, for example, just when something was within reach, but by the way one moves and acts. Similarly, grasp of the notions 'heavy' and 'hot' is exercised not simply by responding differentially to heaviness and heat, but by differences in the ways in which one prepares to lift something heavy, or to touch something that is hot. We might put the point by saying that one's behaviour makes it evident that one knows heavy things take more effort to lift, or that putting two heavy things together will make the resultant package impossible to lift, rather than less heavy. But this gives an excessively reflective account of one's grasp of the causal significance of heaviness. For there may be a certain lack of generality in a creature's grasp of causal significance here. It may be that it grasps the significance of weight for its own actions, but that there is no possibility of applying the notion in connection with the actions of other creatures. This tying of causal significance to the creature using the notion is characteristic of causally indexical terms.

In the *Physics*, Aristotle gives some early examples. We can contrast a theoretical understanding of the causal properties of particular types of wood, for example, or different metals, such as iron or silver, with the understanding possessed by the carpenter or metalworker. The artisan's grasp of causal properties is not a matter of having a detached picture of them. Rather, it has to do with the structure of his practical skills; the particular way in which he deals with various types of wood, or how he uses different metals. The detached theorist need not have these skills. It is in characterizing the propositional knowledge of the carpenter or metalworker that we have to use causally indexical notions. The subject's grasp of such a notion has to do with his practical grasp of its implications for his own actions. Do we really need to ascribe contentful states to the artisan at all? Surely, it might be said, all that we have here is a set of complex behavioural skills. We should just describe the behaviour and let the content go. This would be the correct procedure if there were just one, or a small set, of concisely describable routines through which the artisan can go. But this is not the case, for someone reasonably skilled. Exactly what the person does depends upon exactly what their goals are. Of course, it is not that the carpenter can do anything he likes with a piece of wood, or that the metalworker can bend his metal to any purpose whatever; but there is no bound on the number of goals that they can in practice achieve, operating in different ways to achieve them. This purpose-relativity of the practical skills means that we cannot give a simple behavioural reduction of them. We do here have genuine content.

The most striking family of causally indexical terms is those which have to do with egocentric spatial classifications – notions such as 'within reach', 'to the right' and so on. And we can contrast them with spatial relations such as 'between', 'adjacent to' or specifications of location using latitude and longitude. A parallel point seems to hold for temporal classifications, for much of what we have said about spatial concepts also applies to temporal concepts. Temporal concepts are also 'theoretical'; they must be given some physical significance, there must be some regularities connecting them with other physical concepts, if they are to be recognizable as concepts relating to the world in which we live. And they can be given this physical significance through a practical grasp of their implications for one's interactions with one's surroundings, though this is not the only way in which it can be done.

There are, however, many other categories across which we can apply this distinction. For instance, we can contrast different ways of thinking of weight. On the one hand, I can classify weights in terms of how well I can lift them, and in what ways – perhaps the weight I can lift depends upon exactly how I try to lift it, or what the shape and size is of the thing which has that weight. If I am trying to carry a bookcase, for example, I might think that I could manage the weight were it not so large and awkward. That is, even within the realm of the causally indexical, there may be some analysis of my abilities to lift things, in terms of causally indexical specifications of weight, cumbersomeness and so on. In contrast, there is the kind of specification of weight used in giving recipes, where there is no reference to the causal powers of the subject. The general point to note here is that causally indexical terms may be theoretically related to each other: they may be connected to action as a body, rather than being related to action on any simple one-by-one-basis.

This does not eliminate the distinction between causally indexical and causally non-indexical terms, though it does mean that we can distinguish between those which have a greater or lesser degree of theoretical interconnectedness. Suppose, for example, that we have a creature which will reach for things it wants which are within its grasp, but which will not reach for these things if they are far beyond its grasp, even if it can see them perfectly well. Suppose someone held that this ability, to reach only when it is worth doing so, must be underpinned by: (a) a non-egocentric representation of the distance and angle between the thing and what is in fact oneself. This representation of distance and angle will not be causally indexical; its causal significance will not be especially for one's own actions with respect to the thing; (b) a representation of the volumetric properties of one's own body, and its reaching abilities, so that one can tell whether that body could reach a thing at the distance and angle from it displayed in (a); and (c) the ability to initiate movements towards a desired object, depending upon whether it was within reach of one's own body, as established under (a) and (b). The problem with this line is that it ascribes a certain generality to the creature's representations, and there may be no warrant for doing this. In the case I just described, the creature is supposed to be representing its own direction and distance from the thing, and using knowledge of its own characteristics to find whether the thing is within reach. There is a generality implicit in this, because the creature is being ascribed knowledge from which it could find whether other creatures could reach for this or that object, given knowledge of their volumetric properties. But there may be no prospect of the creature being able to use its knowledge base in this way. It may be that the notion of something being 'within reach' that we should want to use in characterizing its knowledge is dedicated to its own capacities for movement, and has no potential for application to the movements of other creatures. Its representation of something as 'within reach' may be quite directly tied to its own initiation of movement. The reason for saying that the representation is precisely a representation of something as 'within reach' is entirely its direct relation to the creature's actions.

An egocentric framework for place-identification, such as used in guiding simple actions and defined by the axes left/right, up/down, in front/behind, will evidently be causally indexical: grasp of the causal significance of particular distances and directions will have to do with their practical implications for action. In the case of the 'triangulation' model we considered earlier, the frame is plainly not egocentric – it would most naturally be thought of as centred on the target platform – but grasp of its causal significance is still plainly indexical, and has to do with the animal's understanding of the implications for its own movements.

Suppose we have an animal which is capable of thinking in terms of egocentric frames centred on places other than its own current location, such as the places occupied by other animals, or places which it occupied in the past. Is this animal thinking in causally indexical terms? It depends on how it gives physical significance to the spatial relations it represents. If it did this in a disengaged way, thinking only in terms of the causal relations between the places represented and the creature occupying the place on which the frame is centred, then we would have here a non-indexical mode of thought. If, however, its grasp of these causal relations is ultimately a matter only of the pragmatic implications for its own actions of the way things are with the creature at the other location, or the lessons to be learnt from its own past self, then the mode of thought that we have here is still causally indexical.

Finally, consider the use of a way of identifying places that has to do only with what the subject expects to perceive, not with its action upon the world at all. For example, the child in the Acredolo paradigm considered earlier might be thinking in this 'visually indexical' way. If there is such a 'purely perceptual' mode of spatial thought, it too will be causally indexical: it uses the fact of the subject's interaction with the world rather than the subject's thought about it, even though the 'action upon' component of that interaction has been eliminated. But it may anyhow be impossible to separate grasp of perceptual content from grasp of its implications for action.

I began by posing the question how a creature which did not think in terms of physical objects, but only in terms of causally inert features, could possibly grasp the causal significance of the connectedness of its environment, the fact that every place in it is spatially related to every other place in it. We then remarked that a creature using the slope/centroid model in practical navigation approximates towards achieving precisely that. But this still leaves us without an explicit answer to the initial question. We can now supply that. In our present terms, the point is that for the creature using the slope/centroid model, its grasp of the causal significance of the spatial relations it represents is causally indexical. It is a matter of its practical grasp of the implications for its own actions and perceptions. There is the characteristic lack of generality in its use of the model; it may be quite incapable of ascribing use of the model to any other animal. And the animal need not be capable of representing its own causally relevant properties in a 'disengaged' way, though it may be operating with a 'body-image' directly tied to its own capacity for harmonious movement. So although the features are causally inert, the animal's own interactions with its surroundings can constitute the needed grasp of theoretical significance. So far, incidentally, all we have explicitly considered as giving physical significance to the connectedness of a space is that the way things are at one place should be causally dependent upon the way things were at another place. And in the models we have been considering, that is achieved by the animal navigating itself from one place to another. On the face of it, though, these are extremely rudimentary considerations. Of itself – without looking at the specific manner of causation – this bare causal connectedness would not enable us even to introduce an order relation upon the space, much less a metric. All that we have is a potentially endless number of connections between pairs of places. But in fact, a great deal more structure than this is implicit in those models of animal navigation. This structure is imposed by the fact that we are looking at the career of a single animal, in these causally indexical models. The constraints and structure in the way the animal gets itself around automatically constrain and structure the physical significance of the connectedness of the space. For example, an order relation is imposed by asking, for instance, whether the animal has to pass through *b* when taking the designated route from *a* to *c*. In fact, though, it may be a mistake to look for the most basic kind of spatial reasoning in the use of purely topological concepts. Perhaps the ability to measure spatial intervals is really the primitive ability, and the capacity to reason in terms of purely topological concepts is much more sophisticated. In any case, we can see in outline how an animal could in its use of a navigational system give physical significance to metric properties by using them simply in computing the exact angle to take to get to a particular target, or to find in planning, for example, how much time and effort a journey will consume.

As we saw, the slope/centroid model secures the connectedness of the environment in a stronger sense than does the 'triangulation' model. It can represent the direct route from any place the animal is at to any other place in the environment, not just the spatial relation between the animal and the designated target. But, because of its causally indexical character, there is still a limitation on its capacity to represent the spatial relations among the places in its environment. It is capable of recognizing only the geodesics between places; it cannot represent the spatial relations between any two arbitrary places in its environment, irrespective of whether the organism itself is being supposed to be at one of them; and it cannot represent the configurations of arbitrarily large numbers of places. These limitations are evidently tied to the causally indexical character of the model. It is because of the direct tie of the model to action that the animal is confined to thinking of the spatial relations of places at which it itself might be, and to thinking in terms of geodesics, for example. So if the creature is to transcend all such limitations, it will have to think about the connectedness of its environment in a way that is not causally indexical.

This notion of causal indexicality cuts across the classical distinction between concepts of primary qualities and concepts of secondary qualities, which we could draw somewhat as follows. Terms such as 'electron' are theoretical, in that one could not understand them unless one grasped something of the theory in which they are employed. So, too, grasp of a primary-quality term, such as 'cylindrical', is theoretical; it is just that the theory in question is a much more primitive one, involving such points as 'a cylinder will roll along its main axis but not along any other'; 'cylinders cannot be stacked together without leaving some gaps at the sides'; 'the amount one can get inside a cylinder depends upon its length and breadth'. And so on. The difference between a term such as 'electron' and a term such as 'cylindrical' is that in the case of 'electron' the theory is sophisticated, whereas in the case of 'cylindrical' it is primitive; and one cannot in general spot electrons by unaided observation, whereas one can usually tell immediately when one is confronted by a cylinder. In contrast, in the classical story, in the case of secondary-quality terms, such as 'yellow', there is no associated theory.

Grasp of the term just is having the ability to spot yellow things when one sees them. Yet causally indexical terms are obviously not primary-quality concepts, in this sense, because grasp of their physical significance does not consist just in a reflective grasp of a theory in which they are embedded; even though there are theoretical interconnections among causal indexicals. It consists in a practical grasp of their implications for unreflective action and perception. But the causally indexical terms cannot either be classified as secondary-quality concepts, because grasp of them is not simply a matter of being able to register the presence of a perceptible property. One must have a whole complex of appropriate reactions to the presence of the property. The immediate conclusion is that causally indexical terms are neither primary nor secondary. The question to which I now turn is how to characterize frames of reference which have the genuinely disengaged character of primary-quality concepts as traditionally conceived; which approximate towards constituting an 'absolute' conception, free from any relativity to an agent or observer.

5 Physical objects and intuitive physics

What is the most primitive level at which we could expect to find causally non-indexical notions? One suggestion would be that we have to consider operations with a developed scientific theory, one which somehow resists an instrumentalist interpretation. But this, it seems to me, comes much too late. We can look for causally non-indexical notions at an earlier point. In particular, we can consider the way in which a creature can have and use a simple theory of perception and action; a simple theory of the way in which the world acts upon it and the way in which it acts upon the world. A creature which has such a theory has some reflective understanding of what is going on as it moves around. It has some reflective understanding of the causal relations between it and the happenings it perceives, rather than simply a practical ability to interact with those events. In describing that reflective understanding, we have to bear in mind that the causally significant properties of sentient creatures are rather different to those of inanimate objects. We have to take account of the perceptual systems and capacities for deliberate action the creature has. The most striking causally significant aspects of location are their implications for whether and how the place can be perceived by the subject, and for whether and how it is possible for the creature to act with respect to the place, to avoid it or to reach it, for instance. There are also, of course, the opportunities for mechanical interaction with whatever is at the place, which the creature shares with any other physical thing. This provides a reflective counterpart to the animal's causally indexical thinking; a 'theory' of the animal's own interactions with its environment. But although thinking in terms of a simple theory of perception and action may provide one with causally non-indexical modes of thought about one's relation to one's surroundings, it evidently does not of itself provide one with a richer conception of the spatial connectedness of the places in one's environment than one had already. For after all, these ways of thinking are only reflective counterparts of the causally indexical modes of thought, so they can be expected to register exactly the same range of spatial relations. If we are to capture a richer range of spatial relations than those that are available at the causally indexical level, then we shall have to consider thought not just about the causal relations between the subject and what it interacts with, but about the causal characteristics of what is in its environment, and their relations to each other. In particular, we shall have to look at its thought about the physical objects around it.

Even the most abstract theorist must acknowledge that there are many different types of physical thing. One reaction to the complex diversity of the everyday world is to suppose that our 'intuitive physics' of our environment must be a patchwork of a million different pieces, in which there is nothing identifiable as 'the' role of physical objects: everything depends upon what sort of physical thing we are talking about. Alternatively, it might be held that there is, as it were, a core to our intuitive physics, a central conception of how physical things behave, which becomes overlain with endless more specialized pieces of knowledge (for discussion of this issue see Kaiser et al., 1986, and chapter five). But this issue need not concern us here. For all that we are looking at is the way in which we give theoretical significance to the connectedness of the space we are in, within a reflective intuitive physics. And here it does seem that we can point to structural features of the very notion of a physical thing. Suppose we begin with the rudimentary idea that one way to register the physical significance of the connectedness of a space is through the idea that the way things are at one place may be causally dependent upon the way they were at another place. The internal causal connectedness of physical objects, which is what differentiates them from features, means that they can be used to give physical significance to the connectedness of a space. In particular, the possibility of movement by an object from one place to another means that we can see how the way things are at one place could causally depend upon the way they were at another place. Here we see the role of the ability to re-identify physical things; for the way things are at one place is causally dependent upon the way things were at another place, through object movement, if it is one and the same object that is at the destination as was at the starting-point. To underline the point, we might once again contrast them with located features. As we saw earlier, we can make some limited sense of the notion of the movement of a located feature through a region. We can talk about there being a feature F at place p at time t, and a continuous series of transitions in which there is a feature F at place p' at time t', through a continuum of neighbouring places and times. And we might ask whether we can use this kind of idea to fill out the conception of the way things are, at the place at which the feature ends up being causally dependent upon the way things were at the place from which the feature started. But this would miss a defining characteristic of feature-placing talk: that features are precisely not thought of as having that kind of internal causal connectedness. So this kind of transition does not give us a way of registering the physical significance of the causal connectedness of a space. Just because features lack the internal causal connectedness characteristic of physical things, this series of transitions does nothing to establish a causal connection between one place and another. As we saw, a grasp of the causal significance of the connectedness of a space can be achieved by a creature thinking only in terms of located features, but this is achieved only through the use of causally indexical thinking. At the moment, what we are trying to understand is how the physical significance of connectedness could be grasped by a creature using a causally non-indexical intuitive physics – an 'absolute' representation of its surroundings. And it is here that located features cannot help.

Another aspect of the role of physical objects here is to say that they are, as it were, the 'units' of causal connection and interaction in our intuitive physics. This is brought out by the fact that we suppose physical objects move, in general, independently of one another; movement together is, on the face of it, evidence that we are dealing with but one thing (see chapter five on 'the principle of contact' and the 'principle of cohesion'). When we consider the movements of objects through a space as causally connecting one place with another, we can, by further considering the details of that movement, see how to give physical significance to the metric for the space, within our intuitive physics. The crucial notion here will be the time taken for the thing to reach a particular destination from a particular starting-point, given what sort of thing it is and what causes are affecting its motion. There are, indeed, further constraints which are imposed merely by the fact that the object movement is continuous. If it could happen that objects moved discontinuously from starting-point to destination, it could still be that differences in the times taken for them to do so enabled us to give some theoretical significance to the use of a metric. The continuity of object movement means that an order is imposed on the places between the starting-point and the destination, depending on what trajectory we assign to the object; and this in turn is responsible to our conception of the causes of its movement. These remarks only begin on the structure given to the space of our intuitive physics by the role of physical things. There are, of course, many further phenomena which in diverse ways transmit the effects of things being thus and so at one place, to their being thus and so at another place. For example, there are the everyday phenomena of magnetism, heat and cold, the flow of liquids, and the winds. One fundamental range of alternatives to physical objects emerges if we consider mariners navigating in a vast circuit of tides, whirlpools, eddies and currents. It is, in principle at any rate, open to them to register the physical significance of the spatial connectedness of the region they are in without exploiting the fact of their own navigation through the space, or introducing the notion of a physical object. The waves themselves, propagated throughout the space, and interacting with one another in endlessly complex ways, demand for their understanding a rich grasp of the connectedness of the space. And to some extent we can, in our common-sense understanding of the world, use this kind of causal thinking on land, if we watch the effects of an earthquake, or the impact of a hammer-blow on the wall of a house. But these phenomena are not sufficiently pervasive in ordinary experience to provide the full strength of our grasp of the theoretical significance of the connectedness of the space we occupy.[7]

The kind of physical-object reasoning I have described shows how we can give theoretical significance to the connectedness of the space we are in at an 'absolute' or causally non-indexical level. As I said above, the obvious way to achieve a causally non-indexical representation is through reflection on one's own interactions with one's surroundings. The question that now arises, though, is whether there could not be a creature which had a causally non-indexical representation of its surroundings, through physical-object reasoning of the kind just indicated, but which did not reflect on its own interactions with the environment. The question is whether there could not be a subject who thought in a causally non-indexical way about physical things, as outlined above, but who had no conception of its own causally significant properties. So this subject would have no simple theory of perception and action. This subject would be in a rather peculiar position. We want causally indexical spatial thinking, of the kind which one might use in simple navigation, to be capable of serving as evidence in the construction of a causally non-indexical description of the layout of one's surroundings. And the causally non-indexical representation itself must ultimately be capable of affecting one's actions. But the subject we are considering would be using the causally indexical evidence to construct a non-indexical representation, without using the information to build up any picture of its own relation to its surroundings, though it could still be interacting with them. We can form some sense of the way the world would seem to this subject by using the analogy of someone watching a documentary film. Such a film is not ordinarily shot to display the autobiography of a single person. Indeed, in watching the film one might not even raise the question of how it was shot: how the camera angles were achieved or, indeed, whether a camera was used at all. One might simply watch what is happening. And one might build up some picture of the space in which the action is taking place, and the interactions of the objects in it. The subject we have to consider is someone who views his or her own perceptions as being in some ways like such a documentary. He builds up a picture of the space, and the kinds of objects in it. And he can use his perceptions in guiding his actions, in visually guided reaching for example. But at the reflective level, he simply does not raise the question how he is acquiring this information. He simply watches what is going on, and builds up an 'absolute' or causally non-indexical representation of it all.

This issue is connected to the question whether we can in the end make sense of the conception of a causally non-indexical or 'absolute' level of thought at all. To anticipate, the connection is that the physical-object reasoning only has any claim to be reckoned as causally non-indexical if it is embedded in a simple theory of perception and action; but the point needs some elaboration. Let us begin with the question whether we can make anything of the idea of a causally non-indexical or 'absolute' level of thought. In one way this is a familiar issue. One formulation of realism is precisely as the thesis that there is such a level of thought. And one way to formulate the anti-realist challenge is as the claim that there is no such 'absolute' level of spatial thinking. It is often said that we can characterize realism as the thesis that a statement may be true even though it is in principle impossible for us to tell that it is true (McDowell, 1976; Wright, 1987; Peacocke, 1986, p. 86). This has the characteristic deficiency of modal formulations of philosophical theses: that what matters for the truth of the thesis is not the modal claim itself, but the ground of the possibility. What matters for the truth of realism is how it might happen that a statement could be true even though we could not tell that it is. On the face of it, an anti-realist might hold some such modal thesis. One might hold that reality is a human construction, while maintaining that it is not, point by point, a human construction; rather, as an artefact of the general construction process used, it may be that some statements have their truth-values determined, even though we have in principle no way of finding out which truth-values we have determined them to have (Kreisel, 1969, esp. p. 148). Or again, consider the position of a realist who holds a radically externalist view of the mind, on which it is sufficiently receptive to its environment that it can find out about any environment in which it is embedded. This view can hold that the plasticity of the mind means that any truth can be known; but it is not thereby an anti-realist position. The notion of an 'absolute' or 'objective' way of thinking about space, used in the context of a theory of perception, gives us a distinctively realist way of describing the ground of the possibility of recognition-transcendence, because after all, a statement may be true even though no one was appropriately positioned to observe its truth. Our 'film' subject, who thinks at a causally non-indexical level about what is going on, without having the wherewithal to formulate its independence from his perception of it, may nevertheless still seem to have grasped something of the categorical basis for the possibility of recognition-transcendence, in a distinctively realist way (Dummett, 1991).[8]

The really deep issue here is whether one's grasp of causal relations has to be understood ultimately as deriving its content from its connection with causally indexical terms; whether the notion of causality has to be understood exhaustively in terms of its connections with one's own actions and reactions. The thesis that it does is the same as the thesis of anti-realism about the spatial world. For the realist, the general philosophical interest of the notion of 'absolute' space, and the correlative notion of causal non-indexicality, is that it seems to be here, in their use in the context of a simple theory of perception and action, that we have our most primitive form of distinctively realist thinking. This way of formulating the issue of realism gives a new role to the anti-realist argument which Michael Dummett has developed. This argument presses hard such questions as: 'In what does our grasp of meaning consist?' (see Dummett, 1976). Suppose that we are considering whether we have a grasp of causation that does not consist in a practical grasp of how one ought to behave. Here, it is exactly the meaning-theoretic argument that seems most pertinent. What would it be to grasp this 'objective' causality? Would there be any behavioural difference between a creature which thought only in causally indexical terms, and a creature which had a grasp of causally non-indexical notions? If not, can we say that there is any difference between these creatures at all? I do not think that this line of argument can in the end be correct; but it is obviously urgent. The shift away to thinking in terms of causation has the further advantage, from the point of view of a proponent of the meaning-theoretic argument, that the argument need not keep the restriction to considering only assent/dissent behaviour; it is quite implausible that all our grasp of meaning can be characterized in terms of assent/dissent behaviour. Shifting to the formulation in terms of causation, the proponent of the meaning-theoretic argument can consider the full range of ways in which we interact with our environment, while maintaining urgent pressure on the realist.

It does not seem essential to thought of physical objects as such that one should be thinking in causally non-indexical or 'absolute' terms. Consider a cat chasing a mouse, for example, or watching a bird flap its wings and fly off. It is unquestionable that the internal causal connectedness of the thing shows up in the cat's representation of it – it is thinking of the thing as an object – but the internal causal connectedness shows up in the way in which the cat itself interacts with the thing, in perception of it or in acting upon it. There is no need for the cat to have any disengaged picture of the causal characteristics of the thing. Even though we can grasp the internal causal connectedness of other objects while thinking in causally indexical terms, however, the strategy seems much harder to apply to one's own case. When one is, in causally indexical terms, putting to work the internal causal connectedness of the things around one, the way in which one does it is by exploiting one's own interactions with those things; although one need not think of these interactions in fully self-conscious terms. On the face of it, this strategy will not work for one's own case. The kind of interaction described above cannot show that the cat is self-conscious; that it conceives of itself as one physical object among many. The point about the 'film' subject is that this subject cannot in this way make out its claim to be engaging in thought which is causally non-indexical. The 'film' subject has no way of resisting the charge that his thought about his environment is not after all causally non-indexical, for he lacks the conception of himself as a physical thing in interaction with his surroundings. And the very hardest case for the anti-realist reduction of the causally non-indexical to the causally indexical is the case of one's reflective grasp of the interaction between oneself and one's surroundings.

On one conception of it, self-consciousness depends upon the capacity for thought at a causally non-indexical level. On this conception, to be self-conscious is to be capable of thinking of oneself as having the internal causal connectedness characteristic of physical objects (Neisser, 1988, esp. pp. 46–50). And that conception of oneself cannot be exercised through unreflective interaction with one's surroundings. This is, so even though the conception of other things as physical objects can be exercised through one's unreflective interactions with them, as we saw in the case of the cat and its prey. It is evident that no amount of such unreflective interaction would enable the creature to exercise the conception of itself as internally causally connected. To exercise that conception one must be capable of thinking reflectively about one's own interactions with one's surroundings. One must have some grasp of one's own causally significant properties, and how they interact with the things around one. Centrally, one must have some conception of how where one is affects what one can perceive. And one must have some conception of how it is that one affects one's surroundings – what kinds of actions one is capable of. One must, in short, have at least a simple theory of perception and action in one's grasp. Kant's insight, to which Wittgenstein was faithful throughout his career, was to recognize that realism and self-consciousness are inseparable problems. One aim of this chapter has been to display something of why that is so.

ACKNOWLEDGEMENTS

Thanks to Bill Brewer, Jeremy Butterfield, David Charles, Naomi Eilan, Roz McCarthy, Ian McLaren, Hugh Mellor, John O'Keefe, Christopher Peacocke, Nick Rawlins, Michael Redhead and Tim Williamson. Earlier versions of this chapter were presented to discussion groups and seminars in Cambridge and Oxford, and I am indebted to participants for their comments.

NOTES

1 For discussion of this more primitive 'feature-placing' use of language, see Ayer, (1973, pp. 89–93); Dummett (1973, pp. 232–4 and ch. 6 'Identity', esp. pp. 562–83); Dummett (1981, pp. 216–19); Evans (1985a and 1985b); Quine, (1974, §15, 'Individuation of Bodies'); Strawson (1959, ch. 7) and Strawson (1971).
2 For a discussion of the mass/count distinction see Quine (1960, §19).
3 On origin and identity cf. Kripke (1980, pp. 110–15) and Forbes (1985, ch. 6) for subsequent discussion.
4 For a review discussion of pigeon homing see for example Gallistel (1990, pp. 144–8).
5 For this view of content-ascription see Millikan (1984).
6 See also Wehner, et al. (1990, pp. 479–82); Menzel et al. (1990); Gallistel (1990, pp. 123–40), for a review of Gould's work on the map hypothesis for bees. The original paper is Tolman (1948).
7 For further discussion of spatial aspects of our intuitive physics, see Hayes (1985, esp. pp. 18–30); Hayes (1990, esp. pp. 187–95); McCloskey (1983a); McCloskey et al. (1980) and McCloskey (1983b).
8 The deep and difficult question which remains is to relate this conception of realism to the formulation given by Michael Dummett (1991).

REFERENCES

Acredolo, Linda 1990: Behavioural approaches to spatial orientation in infancy. In Adele Diamond (ed.), *Annals of the New York Academy of Sciences*, 608, 596–612.

Ayer, A. J. 1973: *The Central Questions of Philosophy*. London: Weidenfeld and Nicolson.
Dummett, Michael 1973: *Frege: Philosophy of Language*. London: Duckworth.
Dummett, Michael 1976: What is a theory of meaning? (II). In G. Evans and J. McDowell (eds), *Truth and Meaning*. Oxford: Oxford University Press, 67–137.
Dummett, Michael 1981: *The Interpretation of Frege's Philosophy*. London: Duckworth.
Dummett, Michael 1991: *The Logical Basis of Metaphysics*. London: Duckworth.
Evans, Gareth 1985a: Identity and predication. In his *Collected Papers*, Oxford: Oxford University Press, 25–48.
Evans, Gareth 1985b: Things without the mind. In his *Collected Papers*, Oxford: Oxford University Press, 249–90.
Forbes, Graham 1985: *The Metaphysics of Modality*. Oxford: Oxford University Press.
Friedman, Michael 1983: *Foundations of Space–Time Theories*. Princeton, NJ: Princeton University Press.
Gallistel, Charles R. 1990: *The Organisation of Learning*. Cambridge, Mass: MIT.
Hayes, Patrick J. 1985: The second naive physics manifesto. In Jerry R. Hobbs and Robert C. Moore (eds), *Formal Theories of the Commonsense World*. Norwood, NJ: Ablex, 1–36.
Hayes, Patrick J. 1990: The naive physics manifesto. In Margaret A. Boden (ed.), *The Philosophy of Artificial Intelligence*. Oxford: Oxford University Press, 171–205.
Kaiser, M. K., Jonides, J. and Alexander, J. 1986: Intuitive reasoning about abstract and familiar physics problems. *Memory and Cognition*, 14, 308–12.
Kreisel, Georg 1969: Mathematical logic. In Jaakko Hintikka (ed.), *The Philosophy of Mathematics*. Oxford: Oxford University Press, 147–52.
Kripke, Saul 1980: *Naming and Necessity*. Oxford: Oxford University Press.
McCloskey, M. 1983a: Intuitive physics. *Scientific American*, 114–22.
McCloskey, M. 1983b: Naive Theories of Motion. In Dedre Gentner and Albert L. Stevens (eds), *Mental Models*. Hillsdale, NJ: Erlbaum, 299–32.
McCloskey, M., Caramazza, A. and Green B. 1980: Curvilinear motion in the absence of external forces: Naive beliefs about the motion of objects. *Science*, 210, 1139–41.
McDowell, John 1976: Truth-conditions, bivalence and verificationism. In Gareth Evans and John McDowell (eds), *Truth and Meaning*. Oxford: Oxford University Press, 42–66.
Menzel, R., Chittka, L., Eichmüller, S., Geiger, K., Peitsch, D. and Knoll P. 1990: Dominance of celestial cues over landmarks disproves map-like orientation in honey bees. *Zeitschrift für Naturforschung*, 45c, 723–6.
Millikan, Ruth 1984: *Language, Thought, and Other Biological Categories*. Cambridge, Mass: MIT.
Müller, Martin and Wehner, Rüdiger 1988: Path integration in desert ants, *cataglyphis fortis*. *Proceedings of the National Academy of Sciences, USA*, 85, 5287–90.
Neisser, Ulric 1988: Five kinds of self-knowledge. *Philosophical Psychology*, 1, 35–59.
O'Keefe, John 1988: Computations the hippocampus might perform. In L. Nadel, L. A. Cooper, P. Culicover and R. M. Harnish (eds), *Neural Connections and Mental Computations*. Cambridge, Mass: MIT, 225–84.
O'Keefe, John 1988: A computational theory of the hippocampal cognitive map. *Progress in Brain Research*, 83, 301–12.
O'Keefe, John 1991: The hippocampal cognitive map and navigational strategies. In Jacques Paillard (ed.), *Brain and Space*. Oxford: Oxford University Press, 273–95.
O'Keefe, John and Nadel, Lynn 1978: *The Hippocampus as a Cognitive Map*. Oxford: Oxford University Press.
O'Shaughnessy, Brian 1980: *The Will*, vol. 1 Cambridge: Cambridge University Press.
Peacocke, Christopher 1986: *Thoughts: An Essay on Content*. Oxford: Basil Blackwell.
Pick, Herbert L., Jr and Lockman, Jeffrey J. 1981: From frames of reference to spatial representation. In L. S. Liben, A. H. Patterson and N. Newcombe (eds), *Spatial Representation and Behaviour across the Life Span*. New York: Academic Press, 39–61.
Poincaré, Henri 1946: *The Foundations of Science*, tr. George B. Halsted. Lancaster, Pa: Science Press.

Quine, W. V. O. 1960: Divided reference. In his *Word and Object*. Cambridge, Mass: MIT, 90–5.

Quine, W. V. O. 1974: Individuation of bodies. In his *The Roots of Reference*. La Salle, Illinois: Open Court, 55–9.

Salmon, Wesley, C. 1984: Causal connections. In his *Scientific Explanation and the Causal Structure of the World*. Princeton, NJ: Princeton University Press, 135–57.

Scott Kelso, J. A. 1982: *Human Motor Behaviour*. Hillsdale, NJ: Erlbaum.

Shoemaker, Sydney 1984: Identity, properties and causality. In his *Identity, Cause and Mind*. Cambridge: Cambridge University Press, 234–60.

Sklar, Lawrence 1983: Prospects for a causal theory of space-time. In Richard Swinburne (ed.), *Space, Time and Causality*. Dordrecht: Reidel, 45–62.

Slote, Michael, A. 1979: Causality and the concept of a 'thing'. In Peter A. French, Theodore E. Uehling, Jr. and Howard K. Wettstein (eds), *Midwest Studies in Philosophy vol. IV: Studies in Metaphysics*. Minneapolis: University of Minnesota Press, 387–400.

Strawson, P. F. 1959: *Individuals*. London: Methuen.

Strawson, P. F. 1971: Particular and general. In his *Logico–Linguistic Papers*. London: Methuen, 28–52.

Tolman, E. C. 1948: Cognitive maps in rats and men. *Psychological Review*, 55, 189–208.

van Fraassen, Bas 1985: The causal theory of time and space-time. In his *An Introduction to the Philosophy of Time and Space*. New York: Columbia University Press.

Wehner, R., Bleuler, S., Nievergelt, C. and Shah, D. 1990: Bees navigate by using vectors and routes rather than maps. *Naturwissenschaften*, 77, 479–82.

Wehner, Rüdiger and Menzel, Randolf 1990: Do insects have cognitive maps? *Annual Review of Neuroscience*, 13, 403–14.

Wiggins, David 1963: The individuation of things and places (I). *Proceedings of the Aristotelian Society*, suppl. vol. XXXVII, 177–202.

Wilkie, D. M. and Palfrey, R. 1987: A computer simulation model of rats' place navigation in the Morris water maze. *Behavioural Research Methods, Instruments and Computers*, 19, 400–3.

Woods, Michael 1963: The individuation of things and places (II). *Proceedings of the Aristotelian Society*, Suppl. vol. XXXVII, 203–16.

Wright, Crispin 1987: Anti-realism and revisionism. In his *Realism, Meaning and Truth*. Oxford: Basil Blackwell, 317–41.

Part II

Intuitive physics

Introduction: Intuitive physics

Naomi Eilan

All the chapters in Part II and portions of Campbell's chapter in in Part I are concerned with the connection between the representation of physical objects, in perception and thought, and grasp of an intuitive [naive] physics. They all share the view that the connection is deep; that both perceiving and thinking of an entity as a physical object are inseparable from grasp of an intuitive physics – principles that systematically link geometrical properties such as distance and size with physical properties such as velocity, acceleration and mass.

However, the phrase 'intuitive [naive] physics', as it is used in the psychological literature (and, increasingly, in the philosophical literature) covers a wide range of different kinds of principles, and different ways of grasping them. These differences are reflected in the substantially different claims made in each of the four chapters about the content of the physical principles that are linked to object representation, and about the nature of the link. Our aim in these introductory remarks will be to bring out a few of the distinctions and issues that may be of use in assessing the relations among these claims; where they potentially conflict, where they are complementary and so forth. We introduce some of these issues in somewhat abstract form in section I. In section II we will give a very brief summary of some of the major claims made in each chapter. In section III we then return to the distinctions discussed in order to illustrate some of the problems raised by a comparison of claims made in the chapters, in particular those that turn on questions about the relation between empirical findings and constitutive claims in this area.

I

(1) The first issue, or set of questions, that bears on all four chapters may be introduced by means of the following, possibly apocryphal, story about students in a London college who were asked what happens to the level of water in a glass when the glass is tilted. According to the story, a substantial number in the study maintained that the water would remain parallel to the bottom of the glass. We may assume these students

had progressed beyond the stage at which drinking from a cup without a lid is hazardous. This would suggest that on some level, and in some form, they knew very well what happens to water in a tilted glass, their considered theory notwithstanding.

Intuitively, then, we have here an example of two different ways of grasping or exploiting physical principles and, potentially, a difference in kind among the principles exploited in each case. The first set of questions raised by all and addressed by some of the chapters we are considering turns on the problem of how we should develop the intuitive distinctions that seem to be called for. Two distinctions introduced by Cooper and Munger in their brief survey of types of experiments designed to tap our intuitive physics are potentially relevant here. First, there is the distinction between internalized or innate principles and acquired ones. Second, there is the distinction between explicitly articulated and implicit principles. As Cooper and Munger note (ch. 4), these distinctions are, in experimental work, often combined such that at one extreme there are experiments designed to elicit explicit, possibly acquired naive theories by asking subjects to make explicit judgements about, say, the motion of objects released from circular motion or from moving trains. At the other extreme are experiments designed to elicit deeply internalized, not explicitly articulated uses of principles, in perception and in action. Examples here include experiments which severely degrade visual information and examine how the perceptual system completes it, as in the case of the experience of apparent motion. (See chapter four for an account of the range of experiments here. See also Hayes (1985; 1989); McCloskey (1983a; 1983b)).

An additional distinction and set of questions is also potentially relevant to the story. This is the distinction between different types of content. Is there a difference in type of content between principles embedded, in some sense, in perception and action and principles that figure in articulated physical mini-theories? If there is a difference in type of content, how exactly should this be explained? And how, if at all, does this distinction relate to the distinctions between implicit and explicit, innate and acquired principles?

The four chapters, as we noted, focus on substantially different physical principles when spelling out the link between object perception and conception and grasp of an intuitive physics. One question is whether any of these differences can be resolved by appeal to the distinctions among innate and acquired, implicit and explicit principles, and principles with different types of content. This, in turn, raises the problem of how exactly these various distinctions are themselves to be explained, a problem which is explicitly addressed by some of the chapters.

(2) A second distinction, which cuts across the distinctions mentioned in (1), is certainly relevant to assessing how the claims made in the four chapters are related. This is the distinction between claims about principles in fact exploited in reasoning about objects, and claims about principles constitutive of grasp of the notion of a physical object. Although the distinction was introduced in some detail in the General Introduction it is perhaps worth repeating its bare outline here, as it is the fact that the claims made in the four chapters fall into both camps that makes comparison among them simultaneously difficult and particularly interesting.

As noted in the General Introduction, the question of what are the principles involved in object representation has a potential ambiguity about it. One question one might be interested in is the straightforwardly empirical question of what are the principles in fact exploited in segregation of the scene into units, in the prediction of trajectories of moving objects and in predictions of the results of physical interaction among more than one object (on any of the intuitive levels distinguished in (1))? Here, the project is to see whether and how it is possible to infer the physical principles that inform and guide our reasoning from predictions and so forth we in fact make. One very fruitful method of inquiry here, pursued with respect to memory and prediction in chapter four, is to see whether we make any systematic errors in remembering and predicting locations of moving objects, errors that tell us about the principles we in fact exploit.

This straightforwardly empirical concern is prima facie distinct from that which directs the following two constitutive issues. First, there is the issue of what are the essential properties of objects – what must be true of an entity if it is to count as a physical object? Second, there is the issue of how much of a grasp of the essential properties of objects, and what kind, must a subject have if she is to be credited with the capacity to represent physical objects? Theories vary along a variety of dimensions, some of these specific to the concept of an object, others reflections of claims about the nature of representation in general, to yield a somewhat complex network of positions one might take. When we combine this with the range of positions with respect to the issue of what are the essential properties of objects, and how these should be determined, we have a map of possible positions with respect to the constitutive reading of the question: What are the principles involved in object representation? (For more on the difference between constitutive claims about what is essential to grasp of the notion of a physical object, and empirical claims about principles in fact exploited in reasoning about objects, see the General Introduction, pp. 15–20).

The principles singled out by Peacocke and Campbell are said to be constitutive of the notion of a physical object. The principles singled out by Cooper and Munger are said to be in fact exploited in reasoning about objects. The principles singled out by Spelke and Van de Walle are said to be both constitutive and in fact exploited in reasoning. Some of the differences in these principles can certainly, at least initially, be explained by appeal to the distinction between constitutive and empirical claims. And, very generally, care should be taken when comparing the principles, especially where they seem to clash, not to make unwarranted inferences from claims about constitutive principles to claims about empirical ones, and vice versa. That said, however, perhaps the most interesting question is what positive account we should give of the relations among the particular empirical and constitutive claims made in these chapters, and as we shall see, there are several distinct strategies for approaching this issue.

(3) Finally, a third issue, which cuts across the distinctions introduced in (1) and (2), turns on the generality of the principles said to be linked to object representation. There are many examples of local knowledge about particular objects, or object kinds, that are not based on grasp of any generally applicable physical principles or laws. Thus, to quote an example of Peacocke's, one's knowledge of how much pressure to exert on the gear lever in order to move from second to third gear may be based on experience with that particular lever, rather than on the exploitation of any general mechanical principles.

As Spelke and Van de Walle note, on a view that is gaining some currency in psychology, our intuitive physics just comprises a patchwork of different example- or kind-specific local pieces of knowledge. We do not, in representing objects or making predictions about them, exploit any general physical principles. All the chapters we are considering reject this view, and maintain that there are some wholly general principles involved in object representation. The claims each chapter makes about the principles involved in object representation are claims about such general principles.

It is because they all focus on general principles that there is a prima facie possibility of conflict among the claims they make. However, all the distinctions mentioned in (1) and (2) are relevant here: that is, general principles can be held to apply on all the intuitive levels distinguished in (1) and can be either constitutive or empirical. Moreover, not every claim about general principles involved in object representation need be an exclusive claim, that is, a claim to the effect that *only* these principles are actually exploited or constitutive. All of this means that the relations among the various claims made in these chapters are not nearly as straightforward as might appear at first glance; however, keeping these distinctions in mind can, we believe, lead to fruitful questions about the bearings of the various claims on each other.

II

(a) Cooper and Munger suggest that a large number of experiments on perception lend support to the view that our perceptual system has, over the course of evolution, internalized general physical principles that reflect particular invariances in the environment. Their general concern is to examine the precise nature of the invariances reflected in these principles. More specifically, they are concerned with the question of whether these principles are dynamic or kinematic. Kinematic principles link position, velocity and acceleration to describe the pure motion of objects without regard to mass. Dynamical principles employ the concepts of force and mass to explain changes in the states of rest and movement of objects.

In order to get a sense of the level of representation they are tapping when they address this issue, it will help to have before us examples of the kinds of test they use for addressing this question. In one, a memory test, subjects are shown three successive discrete views of an object undergoing a transformation, such as a rotation, the nature of which must be inferred from the pattern of presentation. They are subsequently presented with a fourth view and the question they must answer, drawing on their memory, is whether the position of the object in the fourth view is the same as that presented in the third view. In another, a prediction test, subjects are presented with a continuous animated display of an object transforming in a well defined manner. At some point, the object disappears, and then reappears, continuing its previous transformation. Subjects are asked whether its point of reappearance is at the correct location on the transformational trajectory, supposing the transformation continued unseen during the interval of disappearance.

One remarkable finding of these experiments is that subjects are generally accurate in both kinds of test. These and a host of other experiments on the way we infer and predict motion in similar situations, present compelling evidence for the internalization and semi-automatic use of quite specific physical principles that yield generally very accurate representations of the trajectories of objects. Cooper and Munger go on to make a powerful case for appealing to purely kinematic variables, velocity and change in velocity, rather than to dynamic variables such as mass and friction to explain both what we get right, and the small but systematic errors we make in these experiments.

They also suggest that these experiments lend support to the idea, argued for by Shepard, that, because the computational cost of taking dynamical factors into account outweighs any small increase in accuracy, kinematic principles offer an elegant and efficient set of principles for the cognitive system to internalize for completing partial information about the movement of objects.

(b) One of Spelke and Van de Walle's central general claims is that the way infants (as young as three months old) segregate the perceived scene into units is bound up with systematic and wholly general expectations about the movement of those units. The general principles that guide these expectations are the intuitive physics in virtue of which infants can be credited with perceptions that represent physical objects. They also suggest that the principles infants use in segregating the perceived scene into units, and in making predictions about their behaviour over gaps in perception, are retained in adults and form the core of our concept of a physical object. Finally, they argue that there is no clear cut off point between reasoning and perception, in that the principles involved in both should be thought of as generated by a single system of knowledge. In particular, they argue against the view that perception is independent of the exploitation of physical principles.

The principles they single out as those exploited by infants and which provide the core of adults' concept of a physical object, are, in their most abstract formulation, the following: the *principle of cohesion*, which says that surfaces lie on a single object if and only if they are connected; the *principle of contact* which says that surfaces move together if and only if they are in contact; the *principle of continuity* which says that an object traces exactly one connected path over space and time. (Each of these principles is the product of more specific constraints. Thus, for example, the principle of continuity is the product of two constraints, the continuity and solidity constraints. The continuity constraint says that objects move only on connected paths from one place and time to another. The solidity constraint says objects move only on non-intersecting paths, such that no parts of distinct objects ever coincide in space and time.)

These are very general and abstract formulations of more specific versions of the principles which, the authors suggest, guide infants' expectations about the behaviour of surfaces they perceive. For example, the principle of cohesion is an abstraction from infants' expectations about the behaviour of surfaces perceived to be spatially connected. An experiment used to extract one ingredient in this principle is this. Three-month-old infants, are first habituated to the sight of a hand periodically appearing and tapping a conically shaped toy they can see. They are then presented with one of the following two conditions. In one, a hand appears and lifts the whole toy. In the other, the hand appears and lifts only the top of the toy. Infants are surprised by the second event (they look at it longer). This experiment suggests that infants treat stationary connected surfaces as single units and that they expect them to retain their connectedness as they move. Other experiments suggest that when connectivity is lost the surfaces are no longer regarded as lying on single units. Additional experiments further elaborate the conditions under which infants treat surfaces as lying on single units, all of which are captured in the abstract formulation of the cohesion principle. The other principles singled out by Spelke and Van de Walle are, similarly, abstractions from expectations about the behaviour of surfaces in perception and over gaps in perception, revealed in series of experiments designed to uncover the most general characterization of the principles governing those expectations.

(c) In the first half of his chapter Peacocke seeks to answer the general question of how we should explain what it is for general laws or principles of the kind that yield specific equations for computing the relation among spatial and physical magnitudes to be exploited in perceptual reasoning. In particular, he is concerned with providing an explanation that makes such exploitation psychologically real, but distinct from explicit articulation of such principles in thought. He argues that an account of the way physical principles are intrinsic to perception must meet the following constraint. The distinctive mark of the content of perceptual as opposed to conceptual representations of magnitudes, such as weight or length, is that they are represented in unit-free ways (thus, we do not, for example, see distances in conventional units such as feet, or feel weights in units such as grams). Any account of the entrenchment of physical principles in perception must, he suggests, take this way of representing magnitudes as a datum, which creates a particular problem for specificity. Peacocke goes on to propose a detailed account of the content of perceptual representations of physical magnitudes which meets this constraint.

In the second half of his chapter Peacocke goes on to argue for a constitutive link between the notion of a physical object, and principles of intuitive physics. More specifically, a grip on one particular kind of principle is essential for, and hence partially constitutive of, grasp of the concept of a physical object. It is in this connection that he argues for the following thesis. 'We experience objects specifically as material objects. Part of what is involved in having such an experience is that perceptual representations serve as input to an intuitive mechanics that employs the notion of force' (p. 169). Peacocke grounds his claim on two principles. First, a metaphysical principle about the nature of matter – that 'for something to be a quantity of matter is for changes in its state of motion to be explicable by the mechanical forces acting on it, and for its changes of motion to exert such forces' (p. 170). Second, a connecting principle between the philosophy of mind and metaphysics which, in its most abstract and general formulation, says the following: 'If an account of what is involved in something's having a certain property makes reference to some substantial condition which must be met by things which have it, a thinker's mental representations of that property must be suitably sensitive to the existence of this substantial condition' (p. 000).

When that property is the property of being a physical object, then the metaphysical claim combined with the linking principle jointly imply that to represent physical objects a thinker must be suitably sensitive to the fact that certain motions of material objects exert, and are caused by, mechanical force. The general rationale for the linking principle is that we need some account of what makes it the case that a subject has one kind of entity in mind, say a physical object, rather than another, say a hologram or a shadow.

(d) Campbell's general approach to the issue of object representation is informed by the question: what is the role of object representation in providing us with a grasp of the physical connectedness of the space they move in, which, at its most abstract, involves grasping the idea that how things are at one place may depend causally on how they are at another? That is, how do objects contribute to grasp of the idea of a physical world? The features he isolates as providing the general core of our notion of a physical object all have a role to play in this respect.

The ingredients he mentions as providing this general core are abstractions from more specific principles actually exploited in reasoning. He suggests, first, that in representing an entity as a physical object a subject must have some grip on its 'internal causal connectedness', that is, some grip on the idea that its states at any one time are dependent on its earlier states. (This kind of grip is an abstraction from expectations and reasoning of the following kind: identifying an object by a scratch one saw on it earlier; assuming that the velocity at which an object is moving will effect its velocity several moments hence; making specific predictions about the outcome of a collision, which involves understanding that the state of an object after the collision is a function not only of the properties of the object it collided with, but also of its own properties prior to the collision.)

Other core features of our concept of a physical object are the ideas that objects are units of causal interaction and of movement, and Campbell gives examples of the kinds of more specific principles these are abstracted from. It is the particular physical principles that constitute our substantial intuitive physics, such as kinematic and dynamic principles, from which these features are abstractions, which provide the details of the systematic physical structuring of the space in which we represent physical objects as moving.

A general theme running through Campbell's whole chapter is his insistence on the importance of distinguishing between indexical and non-indexical representations of spatial properties (see the Introduction to Part I, pp. 28–9 for more on this distinction). In particular, he distinguishes between ways of representing the spatial connectedness of space that rely essentially on the subject's own capacity to move in that space (these are causally indexical egocentric representations) and ways of representing such connectedness that do not, and are therefore causally non-indexical (disengaged or objective). The abstract features Campbell isolates as providing the core of our concept of a physical object are, he suggests, essential, but not sufficient for a subject's representation of objects to be exploited in a disengaged representation of the causal connectedness of space.

III

We will not, in these concluding remarks, attempt to relate all the distinctions introduced in section I to the claims made in the four chapters; nor will we attempt to list, let alone suggest, ways of engaging with the very many questions that might be raised by a comparison of the chapters. Rather, we will give just three illustrations of how the distinctions in section I can be used to highlight key questions about the relation between some of the empirical and constitutive claims made in the four chapters. It goes without saying that there are very many other issues that could just as fruitfully be pursued.

Before turning to these illustrations, a note about issues we will not be raising here. In the General Introduction we discussed in some detail the relation between, on the one hand, Cooper's and Munger's suggestion that the notions of force and mass do not enter into the calculations on which vision-based predictions and memories of trajectories are made, and, on the other hand, Peacocke's constitutive claim to the effect that object representations must be input to an intuitive mechanics that does employ the concept of force. In particular, we discussed the possibility that their claims might clash in a way that would force us to question either Cooper's and Munger's findings or Peacocke's constitutive requirement, or whether the subjects tested by Cooper and Munger have a grip on the notion of a physical object. We also examined the possible bearing of empirical findings on how Peacocke's notion of 'suitable sensitivity' might be developed. Readers interested in these issues may wish to consult section II in the General Introduction, as they will not be further pursued here.

(1) The highly abstract features Campbell isolates as essential to our concept of a physical object, in contrast to the substantial principles referred to by Peacocke and by Spelke and Van de Walle, are all ones that are said to be essential to the concept in virtue of the role of object representation in providing us with a grip on the idea of a physically structured world, a world in which how things are at one place depends causally on how they are at another. Campbell does not purport to give an exhaustive account of the concept of a physical object. One of the options left open by his stated position (an option he neither endorses nor rejects) can be used to introduce one of the central issues that a comparison of the four chapters brings to the fore.

This option would involve making the following claims. First, the only essential ingredients in our concept of a physical object are abstract ones, of the kind isolated by Campbell. These are the only ones that are implicated in all cases of object representation. Secondly, and correlatively, there is no one set of substantial physical principles that is essential to grasp of the notion of a physical object; a token representation of an entity counts as a representation of a physical object so long as it is linked with grasp of some substantial physical principles, a grasp that manifests a grip on the kind of abstract ingredients isolated by Campbell. But there is no one set of such substantial principles that is essential to grasp of the concept. Finally, on this view, the question of whether there are substantive physical principles uniformly employed in all cases of object representation is not constitutive but, rather, wholly empirical.

What this option comes to, and how it relates to the other claims we are considering, can be illustrated as follows. Beginning with the idea that the only essential ingredients in our concept of a physical object are the kind of abstract ingredients isolated by Campbell, one might ask the following question about Cooper's and Munger's suggestion that kinematic principles alone are exploited in a wide variety of perception-based predictions. What kind of grasp of the idea of a physical world could kinematic principles alone, without any appeal to force, deliver? Intuitively, a subject operating with kinematic principles alone would be exploiting some kind of grip on the internal causal connectedness of the entities it perceives (in assuming that velocities at one time effect velocities at another, for example), and certainly would have materials for assigning a metric to the space in which they move. As noted in the General Introduction, if we supplement kinematic principles with various generalizations about how size, shape and so forth are likely to determine the outcome of collisions, we have some kind of grip on objects as units of causal interaction, some material for giving substance to the idea of the causal connectedness of space without appealing to force.

So far, then, the claim might be a subject operating on kinematic principles alone would have a grip of some kind on the features Campbell isolates as essential for having a grip on the notion of a physical object. Intuitively, however, such a subject would have a very diminished understanding of what, in general, causes objects to move, which in turn would affect its grip on the causal connectedness of space. Mechanical principles have, intuitively, an important role to play in providing for just this kind of understanding, or at least are the right kind of principles to provide for such understanding. In addition, it is hard to see what a subject's grip on itself as physical object might come to if this did not involve some understanding of the relevance of force to its own movements.

But even if we accept that mechanical principles are important in these and many other respects, we cannot immediately infer that mechanical principles which exploit the concept of force are essential to grasp of the concept of a physical object, as Peacocke, for example, suggests they are. Making the latter claim requires that for a token representation of an entity to count as a representation of a physical object, a subject must have some grip of the relevance of force to the behaviour of the entity represented. From the perspective of ensuring maximal grip on the idea of a causally connected world, however, an equally valid alternative is to say that we have everything we need for the contribution of mechanical principles to grasp of the causal connectedness of space if there are some entities to which mechanical principles are, in fact, applied. As to object representation in general, the claim would be that nothing in the general importance of mechanical principles undermines the idea that, so long as one applies to the entity one represents some substantive principles, such as kinematic principles, the application of which manifest grip on the essential features isolated by Campbell, one counts as representing the entity as a physical object. In other words, the general importance of mechanical principles does not undermine the claim that in representing an entity as a physical object one must either be employing local, example-driven knowledge, or one of several different kinds of general substantive principles (mentioned in the other three papers), but need not be employing any one of these on each occasion of representing a physical object. On this kind of account there would be no one set of substantive principles essential to grasp of the notion of a physical object. Rather, whether there is one kind of substantive principle involved in all cases of representing an entity as a physical object is a wholly empirical question.

The claim that no particular set of substantive physical principles is essential to our notion of a physical object would, on the face of it, conflict with both Peacocke's and Spelke and Van de Walle's claims. They both argue that particular substantive principles are, in some sense, intrinsic to the notion of a physical object. The point of drawing out the possibility of this claim is not to endorse it, nor to suggest that Campbell would endorse it. Rather, the kind of argument just sketched serves to highlight how strong a claim is the constitutive claim with respect to substantive principles. It is not merely the claim that there is one or more substantive physical principles which are in fact involved in all cases in which objects are represented; it is the claim that this link is essential to a representation counting as a representation of a physical object. Such a claim requires justification of a principled kind, one which would, *inter alia*, aim to show why the line of reasoning just sketched is to be rejected; and why particular substantive principles are more than just ways of manifesting grip on abstract features claimed to be essential. The principles evoked by Peacocke, which link claims about substantive physical principles with claims about metaphysically necessary properties of physical objects and with principles in the philosophy of mind, could and would, presumably, be used to this effect. One possibility left open with respect to Spelke and Van de Walle is that the claims they make about the substantive principles that provide the core of our concept of a physical object should, contrary to first appearances, be thought of as claims about principles in fact involved in object representation, rather than about principles essential to grasp of the notion of a physical object. On this interpretation, their general approach fits better with the kind of argument just sketched than with arguments that would be marshalled against it. This is the issue we pursue in (2).

(2) The principles Spelke and Van de Walle suggest provide the core of our concept of a physical object are identical to the principles they suggest are in fact exploited by infants in segregating the perceived scene into units and in predicting the movements of those units. In what sense exactly do these principles provide the core of our concept? This is not a question pursued in much detail by the authors, but it is of particular interest given the issue raised in (1).

We may approach this question by considering one of the principles they suggest both guides infants and provides the core of adults' concept of a physical object, the principle of cohesion. At its most abstract, it is formulated as the principle that surfaces lie on a single object if and only if they are connected. As stated, a natural reading of the principle is that spatial connectedness of surfaces, of any kind, retained over movement, is not only necessary but also *sufficient* for those surfaces to count as surfaces of a single object. But, on this reading, the principle is arguably not constitutive of objecthood, on adults' concept of an object. A ring a person is wearing is spatially connected to the person yet distinct from the person; a baby on her father's shoulders is spatially connected to her father yet a distinct object. This is true even if adults share infants' expectations that stationary connected surfaces will retain connectedness over movement. The existence of these expectations can live side by side with an understanding that spatial connectivity of surfaces is not sufficient for objecthood.

For adults, connectivity of surfaces at rest and over movement is one important source of evidence for the surfaces lying on a single object. This evidence, though important, is defeasible by a host of other considerations, including sortal considerations, considerations of origin, behaviour in counterfactual circumstances and so forth. These further considerations can be thought of as part of the primitive theory of objects that informs the considered judgements we make on the basis of our perceptions. Once this theory is in play, we can give the principle a constitutive reading, as one which determines, relative to these other considerations, what kind of connectivity of surfaces is necessary and sufficient for them to be on a single object. But independently of the theory, the principle as stated would appear to be a principle of evidence for adults, rather than a constitutive principle.

Similar arguments might be mounted against taking the other principles, as stated, as referring to essential properties of objects. For example, it has been argued that solidity is not an essential property of matter (see Peacocke, ch. 6, p. 171); similarly, it might be argued that it is not at all obvious that it is essential to the very idea of an object that it move in continuous paths (see Campbell, ch. 3, p. 90). Even if such arguments were to be accepted, however, and they are certainly debatable, the principles singled out by Spelke and Van de Walle are, nevertheless, as they suggest, central to our grasp of the notion of a physical object. How might their centrality be explained if we do not think of them as referring to essential properties of objects?

One way of bringing out the centrality of these principles is implicit in Campbell's appeal to some of them as exemplifications of grasp of more abstract features in our concept of a physical object, features he argues are essential to it. Thus, for example, he suggests that the principles of cohesion and contact, taken together, yield the assumption that physical objects move, by and large, independently of each other;

movement together is prima facie evidence that we are dealing with one object. And this assumption, Campbell suggests, is one manifestation of our treatment of objects as units of causal interaction and connection. A similarly important role is assigned to the continuity constraint, the assumption that objects trace continuous paths. This principle is one ingredient in the intuitive physics we in fact use to describe the movements of objects, a physics which, in general, provides for grasp of the connection between distance and time, thereby underwriting grasp of the physical significance of the metric we use to describe the spatial relations among places.

The details matter less from the perspective of the question we are asking than the general strategy implicit in them for bringing out one sense in which Spelke and Van de Walle's principles do home in on the core of our concept of a physical object, which is this. Although, as stated, the principles should not, on this strategy, be regarded as referring to essential, constitutive properties of physical objects, they can be seen as concrete ways of grasping more abstract principles, principles which do, the claim would be, refer to essential properties of objects. Thus, they can be seen as particular ways of grasping the idea of objects as units of movement and causal interaction. On this strategy, the way to bring out their centrality is to ask in what way do they provide for grasp of the abstract features in our concept of a physical object, features that are argued to be constitutive of the concept on the grounds that they play an essential role in providing for the idea of a causally connected space? Conversely, if this way of bringing out their centrality were to be adopted, it would suggest that the justification for claiming these particular substantive principles lie at the heart of our concept of a physical object, rather than others that might give substance to the abstract features mentioned by Campbell, will lie in empirical investigation of principles in fact exploited, in the manner suggested in (1).

It goes without saying that this is only one way of reading the principles selected by Spelke and Van de Walle. An alternative is to see them as claiming that the substantive properties their principles refer to should be thought of as essential properties of objects; and that the substantive principles they refer to are constitutive of grasp of the notion of a physical object, for that reason. This would be to see them as making the same kind of claim as is made by Peacocke. This, in turn, would raise a whole set of new questions about, among other things, the relations between the properties they and Peacocke single out, and about differences between explanatory theories on the one hand, and collections of individual principles on the other.

The point of drawing out the possibility of the interpretation just sketched, and of the claims made in (1), is to suggest that one of the issues highlighted by comparing the claims made in these chapters is that they either explicitly assert, or implicitly suggest, different ways of interpreting the status of substantive principles involved in object representation, and, correlatively, different kinds of justification needed for giving a particular set of physical principles a primary role in object representation. Ultimately, the question of where one should look for such justification turns on the kind of concept one thinks the concept of a physical object is. One issue here is the relative weights one gives, in fixing the essential ingredients of the concept, to the role that object representation plays in giving us a grip on the idea of a physical world, and to scientific theories and categorizations of entities.

(3) Finally, we turn very briefly to the views expressed about the nature of the link between physical principles and the perception of objects. One assumption implicitly

shared is that perception yields immediate access to physical objects, rather than merely to sensations, and that its doing so is related to the employment of physical principles in the perceiving of objects, rather than merely in the formation of post-perceptual beliefs about objects.

The two psychological chapters, in different ways, make vivid the difficulty and importance of explaining what such entrenchment of physical principles in object perception involves. Thus, they both strongly suggest that the principles they argue are exploited in perception are, in some sense, implicit, rather than explicitly articulated. What does such a claim amount to? What is it for principles to be implicit, and psychologically real? What does it mean to say that principles are woven into, or intrinsic to, perception? In particular, are the contents of the perceptual representations into which they are woven of the same kind as the contents of judgements and beliefs based on perception? And if not, how does this affect the account we give of the sense in which physical principles are exploited in perception, in distinction to the way they are employed in our naive beliefs about the behaviour of objects?

Peacocke's account of what it is for general principles to be exploited in perception aims to address all these questions. Although the examples he uses are drawn mostly from dynamical physics, the kind of reasoning that guides the predictions and memories of positions explored in Cooper's and Munger's experiments is very much the kind of reasoning his account is intended to explain. In particular, there are two features about the way in which physical principles seem to operate in the Cooper and Munger cases which Peacocke's theory may be thought of as especially suited to explaining. First, there is the precision of the predictions, a precision that seems to be independent of our capacity to represent magnitudes by means of conventional units. Second, there is the fact that subjects would probably have no way of articulating the kinematic principles which Cooper and Munger suggest guide these predictions, which in turn suggests that the principles guide perceptual reasoning in a way that is independent from explicit articulation of them in thought. Peacocke's account of what it is for unit-free representations of magnitude to be input to implicit but psychologically real physical principles draws together and articulates both the features that intuitively characterize the reasoning examined by Cooper and Munger cases.

Finally: Campbell's distinction between causally indexical and non-indexical ways of representing spatial and physical properties also enables us to distinguish two ways in which physical principles may be exploited in reasoning. This is in many ways orthogonal to the kinds of concern that motivate Peacocke's distinction between perceptual and conceptual content. What, if any, is the relation between Peacocke's appeal to the unit-free nature of perceptual representations of magnitudes and the way they serve as input to implicit physical principles, and Campbell's notion of indexical contents, in which distances, say, are represented as within or out of reach, shapes as graspable or not, weights as liftable or not. This is worth addressing, especially in relation to the various kinds of psychological concerns with which they may be thought of as suited to engaging. More generally, the way we account for contents other than the contents of judgements, and the general question of how we should categorize different types of contents, is as yet largely unexplored territory, or more accurately territory that is only beginning to be explored. This is one example in which combining the resources of empirical and theoretical psychological work with the constitutive concerns of philosophers, is not merely a good idea but, arguably, essential for making substantial progress.

REFERENCES

Hayes, P. J. 1985: The second naive physics manifesto. In Jerry R. Hobbs and Robert C. Moore (eds), *Formal Theories of the Commonsense World.* Norwood, NJ: Ablex, 1–36.

Hayes, P. J. 1990: The naive physics manifesto. In Margaret A. Boden (ed.), *The Philosophy of Artificial Intelligence.* Oxford: Oxford University Press, 171–205.

Kaiser, M. K., Jonides, J. and Alexander, J. 1986: Intuitive reasoning about abstract and familiar physics problems. *Memory and Cognition*, 14, 308–12.

McCloskey, M. 1983a: Intuitive physics. *Scientific American*, 114–22.

McCloskey, M. 1983b: Naive theories of motion. In Dedre Gentner and Albert L Stevens (eds), *Mental Models*, Hillsdale, NJ: Erlbaum, 299–324.

4
Extrapolating and remembering positions along cognitive trajectories: Uses and limitations of analogies to physical motion

Lynn A. Cooper and Margaret P. Munger

Introduction

Both our everyday experience in the world and certain aspects of skilled performance suggest that our perceptual systems are remarkably accurate in extracting and using information about the motion of objects in space. Behavioural expressions of this competence include our abilities to locomote without colliding with stationary or moving objects, and to anticipate the trajectories of transforming objects in order to intercept, to follow or to avoid them. These ordinary perceptual abilities become more impressive when we realize that predictions about the current and continuing motion of objects are generally based on quite incomplete information about the structure of objects and events. Movement of our bodies and our eyes, as well as relationships of occlusion among objects, limit the quality of information sampled even over extended periods of time. From observations like these, a number of questions naturally arise: Just how reliable are our intuitions concerning the competence of our perceptual systems in anticipating the continuing structure of visual events? When are predictions about transformations on objects accurate, and under what, if any, conditions do they err? What sources of information about objects and events are used in generating perceptual anticipations? How do our perceptual systems select, among the many possible interpretations of the partial information available at any moment in time, which is consistent with the global structure of an ongoing event?

With respect to this latter question, Shepard and his associates (e.g. Carlton and Shepard, 1990; McBeath and Shepard, 1989; Shepard, 1984) have proposed that our perceptual systems have evolved to take advantage of certain invariant sources of information in the environment. In particular, invariants in the form of descriptions of the motion of objects in the world may have been internalized and used to constrain selection among the multitude of possible interpretations of the partial information available. These investigators have, further, suggested that such internalized external invariants can be effectively explored by degrading the information provided to the perceptual system. That is, when faced with severely limited input, the system will rely heavily on internalized constraints to interpret what information *is* presented. Investigation of how perceivers interpret such degraded input thus becomes a technique for determining just which external invariants have been internalized.

The suggestion, then, is that there may be a 'mental physics', internalized by our perceptual systems over the course of evolution, that reflects certain external physical regularities. The challenging empirical task is to determine which aspects of the regular behaviour of moving objects might constitute external invariants that would be most useful to internalize. In classical physics, two general types of information are used to describe the behaviour of moving objects. Kinematic information describes the pure motion of bodies without regard to mass – the position, velocity and acceleration of an object. Kinetics (or, more generally, dynamics) describes the forces causing movement, or acting on objects with mass. One way to formulate the question of what physical principles the perceptual system may have internalized is to ask whether observers are sensitive to and able to use these kinematic and dynamic sources of information in making perceptual judgements about ongoing visual events.[1]

Experimental work that has addressed this question is difficult to summarize briefly, in that a wide variety of tasks and levels of sophistication of perceivers have been examined. In general, though, the tasks can roughly be categorized according to both the cognitive level of the judgement required, and the extent of perceptual information made available to the subject. At one extreme – in terms of the reduced quality of perceptual information provided – are the elegant studies of apparent motion done by Shepard and his associates (McBeath and Shepard, 1989; Robins and Shepard, 1977; Shepard, 1984; Shepard and Zare, 1983). The pattern of results from these experiments is consistent with the idea that constraints of kinematic geometry are internalized and revealed in the paths of apparent motion experienced between two successive views of an object differing in position and orientation.

At another extreme are tasks requiring considerable conscious assessment on the part of subjects given various amounts and types of external information. These include the so-called 'naive' or 'intuitive' physics studies of McCloskey and his associates (e.g. McCloskey and Kohl, 1983; McCloskey, Washburn and Felch, 1983), as well as studies in which direct judgements of properties of object collisions are required (e.g. Gilden and Proffitt, 1989; Kaiser and Proffitt, 1984, 1987; Todd and Warren, 1982). The McCloskey group's procedure involves the selection of a trajectory for the continuing motion of an object released from, say, an initial circular motion (McCloskey and Kohl, 1983). The level of perceptual support for the judgement is varied by using static diagrams of the motion, a real-time computer display of object movement, or sometimes requiring the subject to interact directly with a moving object. Regardless of the extent of perceptual information provided, the majority of subjects are able to make trajectory judgements accurately; however, a subset consistently select an inappropriate path – in the case of the circular motion problem, a curved path instead of the correct straight path.

In the collision judgement experiments of Proffitt and his collaborators, subjects are asked to determine whether animated collisions are natural or anomalous, and to judge the relative masses of the colliding objects. Interestingly, not only can observers accurately determine whether or not collisions are natural, but they can also apparently use information about the relative velocities of objects after impact to assess their relative masses. It would appear, then, that kinetic or dynamic information can be computed from full perceptual exposure to visual events. Proffitt et al. (Gilden and Proffitt, 1989; Kaiser and Proffitt, 1987) conjecture that the route to such judgements about dynamics might well be through the use of heuristics based on kinematic information (e.g. relative velocity and/or angle of trajectory after collision could provide information about relative weight).

The situations that we will consider in detail for the remainder of this chapter fall somewhere between these two extremes, in terms of both degree of external perceptual support provided and cognitive level of the required response. One case provides successive, discrete views of an object undergoing a transformation that must be inferred from the pattern of presentations. Observers must judge whether the queried position is the same as the final presented view. The second case provides the subject with a continuous, animated display of an object transforming in some well-defined fashion. At some point in the transformation, the object disappears momentarily and then reappears, continuing to transform in the manner depicted before the disappearance. Observers must judge whether the point of reappearance is at the correct location on the transformational trajectory, had the transformation continued 'unseen' during the disappearance interval. For both types of stimulus situations, we pose a number of questions: (1) How accurate are subjects in predicting or in remembering the locations of moving objects? (2) How are judgements affected by manipulation of kinematic and/or kinetic properties of the depicted events? That is, is there evidence for the extraction and use of these sources of information? (3) How might we account for overall patterns of performance on these tasks? In particular, under what (if any) sets of conditions will analogy to the physical motion of objects provide a useful conceptual framework?

1 Remembering an object's final position in a transformation implied by a sequence of static views

We turn, now, to the first case described above, in which the trajectory of object motion must be inferred from successively-presented static views. This, of course, is the stimulus situation for producing the phenomenon termed 'representational momentum' by Jennifer Freyd and her collaborators (e.g. Freyd, 1983, 1987; Freyd and Finke, 1984, 1985; Freyd and Johnson, 1987; Kelly and Freyd, 1987). In Freyd's canonical experiment, subjects are presented with kinematic information via a sequence of static views. Specifically, three pictures of a simple, two-dimensional object (generally, a rectangle) are shown in a pattern that implies a rotation in the picture plane at some constant rate (generally, 34 deg/sec). A fourth view is then presented, and observers judge whether or not it depicts the object in the same position as that shown in the immediately-preceding (third) display. The position of the fourth, test view is systematically varied about the location of the third view in both the forward and backward direction along the implied path of motion.

The central results of this basic experiment are, first, that performance is generally quite accurate. That is, subjects correctly accept test stimuli presented in the actual position of the third view about 90 per cent of the time (cf. Freyd and Johnson, 1987, figure 3). However, there is a marked asymmetry in the tendency to false alarm to probes presented in positions close to the 'true same' location. Probes in positions that are slightly further along the path of the rotation implied by the first three views are accepted as 'same' substantially more often than are probe positions backward along the implied rotation. Freyd and her collaborators interpret this asymmetry as evidence that the perceptual system embodies a principle analogous to physical momentum, or the continuing steady movement of an object until acted on by a force. Thus, the phenomenon of distortion in memory for the position of an object in the direction of its implied motion has been termed 'representational momentum'. Similar distortions in the direction of implied translatory motion have been reported by Finke, Freyd, and Shyi (1986; see also Finke and Freyd, 1985; Finke and Shyi, 1988), and for animated translatory motion by Hubbard and Bharucha (1988; see also Hubbard, 1990).

This analogy between physical and mentally-represented momentum has a number of testable consequences. First, the magnitude of the perceptual/cognitive effect should be influenced by variables analogous to those that affect physical momentum, such as velocity and mass. Second, some internal mechanism analogous to a force must be postulated to explain why the motion incorporated in the mental representation stops at all (i.e. why memory distortions are seen primarily in probe locations extremely close to the true 'same' position of the test object). Freyd and Johnson (1987) have suggested that a braking procedure – called 'cognitive resistance' – must be applied to stop the motion included in the mental representation, and that such a process is like the operation of frictional forces in the physical world. Under this analysis, the mental representation of a moving object parallels the behaviour of a physical object that starts with a constant velocity (implied by the sequence of three static views) and suddenly encounters a large frictional force (the cognitive resistance applied following the third displayed view). The obtained memory distortion thus results from the inability of the braking procedure to stop instantly the represented movement of the object. The size of the memory shift, then, should be influenced by both the object's represented momentum and the effectiveness of the cognitive analogue of frictional force.

The influence of velocity on the magnitude of the memory distortion – predicted by the momentum analogy – has been obtained by various investigators under a number of conditions. Freyd and Finke (1985) report that the size of the estimated memory shift increases linearly with the velocity implied by the static displays. This result is exhibited when velocity is varied as the amount of time between static views, and also when velocity is manipulated by changing the relative positions of the views in a sequence implying translatory motion (Finke, Freyd, and Shyi, 1986). In this latter situation, implied acceleration, or rate of change of velocity, also affects the magnitude of the memory error, with final instantaneous velocity – as opposed to average velocity – being the effective predictor of performance.

The first of our own experiments to be described here investigates the contribution of another source of dynamic information – depicted mass of an object – to the magnitude of distortion effects in the implied motion paradigm. If such distortions result from a cognitive analogue to physical momentum, then variations in apparent mass should influence the size of the experimental effect in a manner similar to variations in implied velocity. We manipulated perceived mass by using stimuli displayed as three-dimensional objects. The objects were all modifications of pyramids, with the same square base, but constructed so as to create the appearance of different volumes. In the main experiment, the objects were presented as if viewed from the top in order to equate them for two-dimensional area. To provide an appreciation of the depth-cued, three-dimensional structure of each object, it was shown in side view at the beginning of an experimental trial and then rotated smoothly 90 degrees, as if 'bowing' into the top view. In a preliminary experiment, an independent group of subjects was shown a large number of different objects in these bowing displays and asked to rate the perceived mass of each object. The four objects used in the main experiment had been rated as significantly different in this preliminary study. Figure 4.1 shows the top and side views of each object, along with its relative mass rating.

Given the implied rotation of the objects, the simple mass of the object is not the proper dynamic variable to consider. In fact, it is the object's moment of inertia about the particular axis of rotation that would affect the momentum of a rotating object. Moment of inertia depends on the mass and the squared distance of all points of the object from the axis of rotation. Specifically, it is the sum of each point of mass in the object multiplied by the squared distance that point of mass is from the axis of rotation. The four objects used in this experiment have dramatically different moments of inertia. Further, the ordering of the objects is identical whether simple mass, moment of inertia or mass rating is used. For example, the object with the smallest mass and smallest moment of inertia was also judged to be the smallest in the preliminary, rating experiment.

The structure of the main experiment paralleled closely the 'representational momentum' studies of Freyd and her associates. Each trial began with a brief presentation of the rotation of the object about the horizontal axis into a top view. Then this top view of the object was presented sequentially in three orientations differing by 17 degrees, and implying a rotation in the picture plane about the object's centre. A fourth, test view was then presented in either the same position as that of the immediately-preceding third view or in a position varying about the third by +/– 2, 4 or 6 degrees, with equal probability. The subject's task was to determine whether the position of the final display was the same as or different from that of the previous third view.

The central results of this experiment, shown in Figures 4.2 and 4.3, can be summarized as follows: We successfully replicated the finding of a systematic distortion in memory for the final position in the sequence in the direction of the implied rotation. This effect is evident on inspection of the average data across the four objects, displayed in Figure 4.2. A weighted average was calculated from the percentage of 'same' responses for each test position using the test positions as weights. This number is positive if test positions displaced in the direction of the implied rotation are accepted as 'same' more often than are test positions displaced in the opposite direction. The weighted averages for the four masses ranged from 0.42 to 0.53 (peak shifts, calculated from quadratic regressions of the percentage 'same' responses against the test positions, ranged from 1.55 to 1.83 degrees). The uniformly positive signs of the overall weighted averages indicate memory errors in the direction expected from the analogy to physical momentum. Note, in addition, that these memory distortion effects, though statistically reliable, are quite small (under 2 degrees), as they are in Freyd's original experiments (cf. Freyd and Johnson, 1987).

The important, novel finding of the present experiment is that objects differing in perceived mass do *not* result in memory distortions of differing magnitudes. The significant linear relationship between extent of memory distortion and the masses (or moments of inertia) of test objects that would be predicted by the analogy to physical momentum is clearly absent in Figure 4.3, which shows weighted averages plotted as a function of relative mass ratings (which are monotonically related to the moments of inertia). While the least-squares linear regression does produce a positive slope, it does not differ significantly from zero. The results of this experiment, then, fail to support one particular consequence of the notion that the representation of a moving object has internalized momentum.

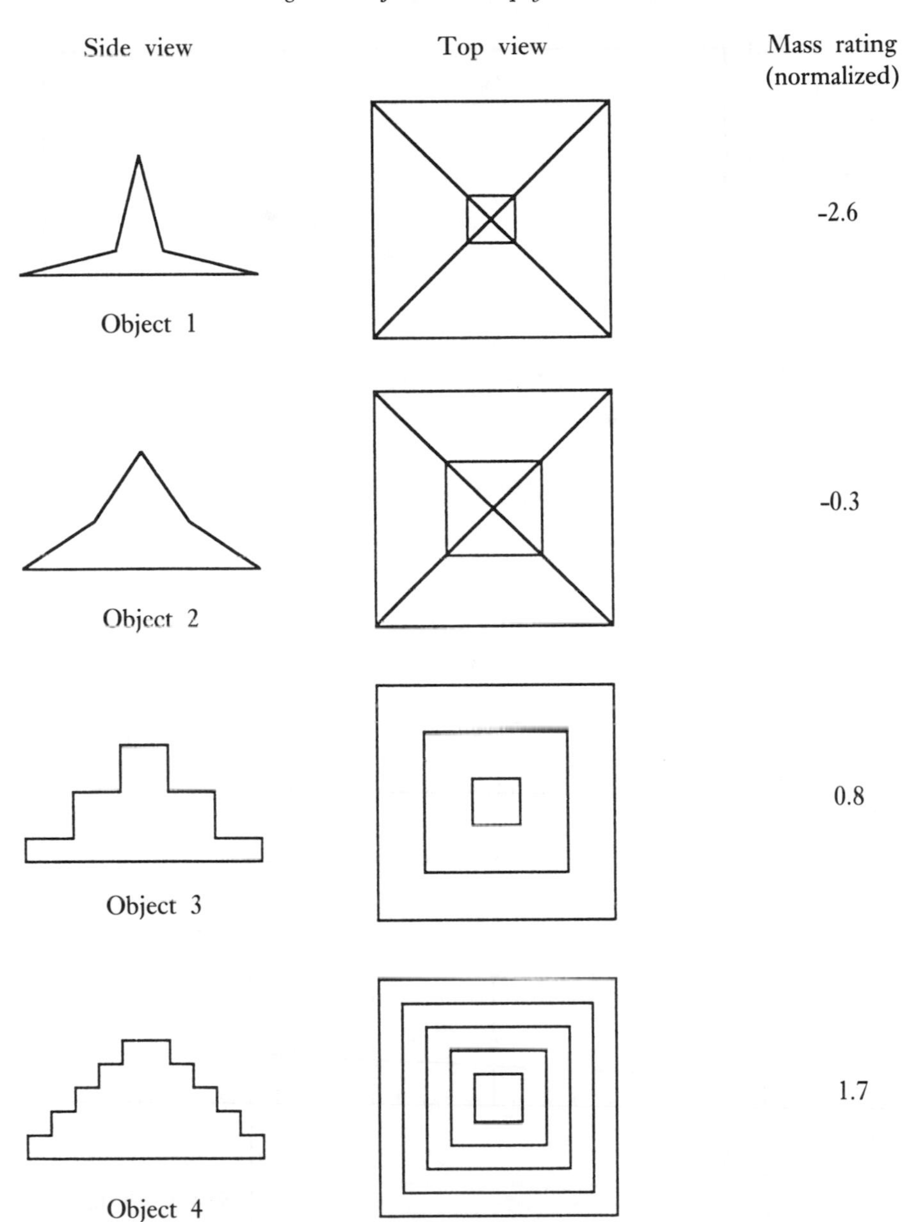

Figure 4.1 Schematic illustration of each of the four objects used in Experiment 1, in both side and top views, along with their (normalized) rated masses. See text for further explanation.

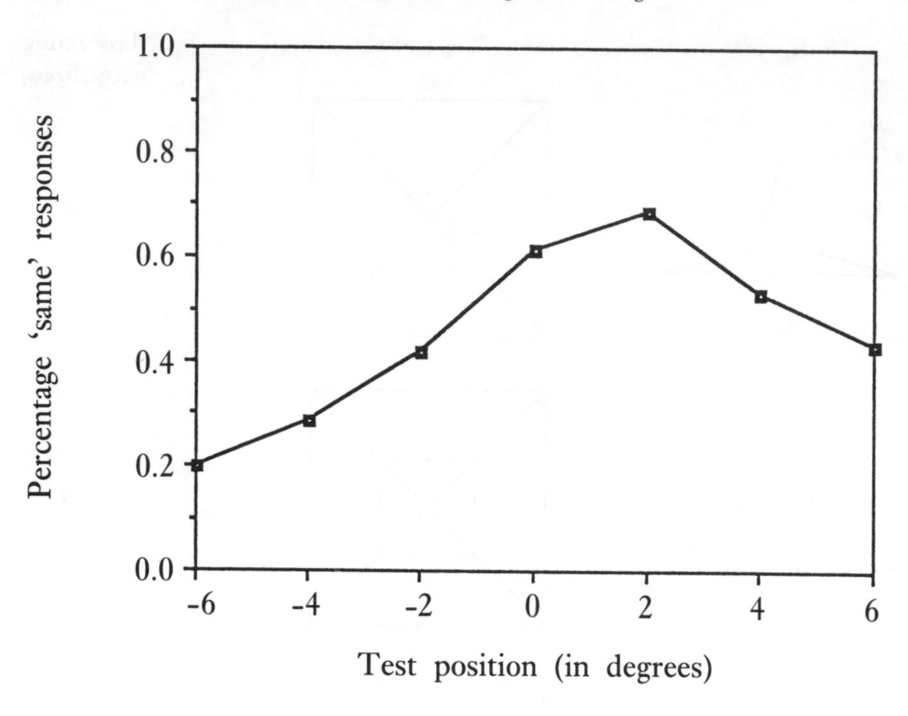

Figure 4.2 Percentage of 'same' position responses to test stimuli in Experiment 1, plotted as a function of departure of the test stimulus from the position of the third displayed view.

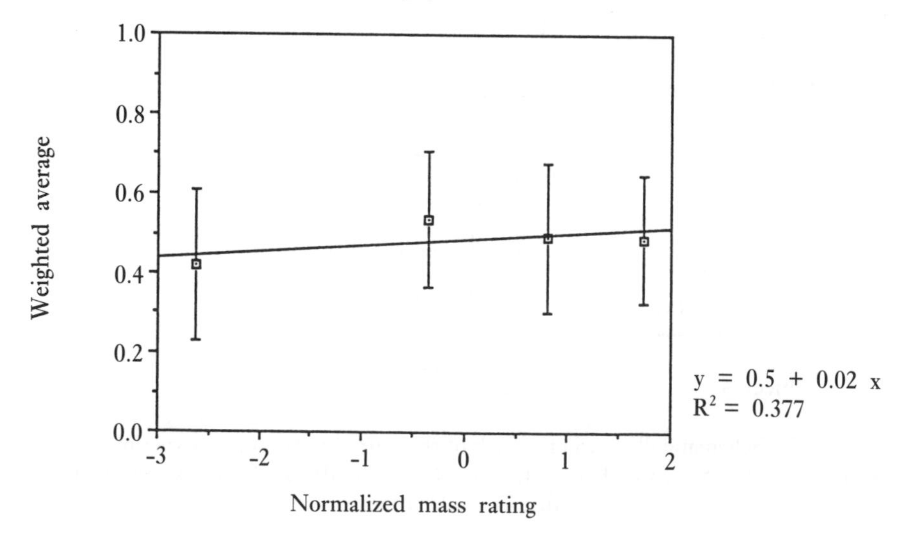

Figure 4.3 The derived weighted average measure, plotted as a function of the (normalized) rated masses of the objects in Experiment 1. Confidence intervals (.95) and equation for the least-squares fit are shown.

Another testable prediction of the analogy to dynamics, or physical forces acting on objects, concerns the influence of cognitive resistance on the size of the memory distortion. Recall that cognitive resistance, by analogy to frictional forces in the physical world, is the force applied to the represented movement in order to stop it. Such a braking procedure would be necessary to allow a comparison between the final depicted position in the implied motion of the object and the position of the test probe. The frictional forces that first come to mind are those involving the interaction between a surface and an object moving over that surface. However, the interaction between a medium and an object moving through it can also be characterized as a frictional force. We chose to focus on this second type of frictional force for two reasons. The first is that the canonical experiment of Freyd and her collaborators does not depict a surface. A surface would have to be depicted if surface friction were going to be varied experimentally. In addition, technical difficulties arose when we tried to depict a surface. If cognitive resistance indeed parallels frictional force, then the shape of the object should affect the magnitude of the directional error. That is, for objects with the same area, more aerodynamic or streamlined shapes should encounter less drag or resistive force when moving through the same medium.

We tested this prediction by showing objects with different shapes but the same area presented in static, successive views to imply translatory motion. Figure 4.4 displays the four objects, along with the calculated drag for each, assuming a medium made up of stationary, uniformly distributed, perfectly elastic particles.[2] The motion of each object across the screen was depicted in three successive, horizontally-displaced views differing by 14 degrees of visual angle. Following presentation of the sequence, a fourth test shape was displayed in either the same position as the third, or differing by +/– 2, 4 or 6 mm, with equal probability. The subject's task was to determine whether the position of the test shape was the same as or different from that of the final object in the simulated translation sequence. Each of the four objects in Figure 4.4 was shown moving in both left-to-right and right-to-left directions. Note that in the case of the asymmetric objects, this permits some critical comparisons to be made. In the analogy between cognitive resistance and frictional force, a triangle moving in the direction of its point should be more difficult to stop than the same triangle moving in the direction of its base. Thus, sequences depicting motion in the direction of the pointed end should produce larger memory distortions than sequences showing motion in the opposite direction.

The results of this experiment, illustrated in Figures 4.5 and 4.6, fail to confirm predictions based on an analogy between cognitive resistance and frictional forces in the world. Figure 4.5 displays percentage of 'same' responses as a function of position of the test shape, separately for the left-to-right and right-to-left directions of implied motion. As in our initial experiment, we replicate the directional memory distortion error generally associated with 'representational momentum'. However, as Figure 4.6 indicates clearly, the magnitudes of these errors are not systematically related to the different positions suggested by the model of a medium. The theoretical position takes into account both the force of resistance offered by the medium, and the velocity of the initial display. In general, the sizes of the distortion errors – estimated by both weighted averages (ranging from 0.03 to 0.37) and by the peaks of quadratic functions fit to the data (ranging from 0.07 to 0.78 mm) – are small, but (except in two cases) reliably different from zero.

In summary, the results of this line of work lead us to question the usefulness of models or analogies based on dynamics for explaining the highly reliable distortions in

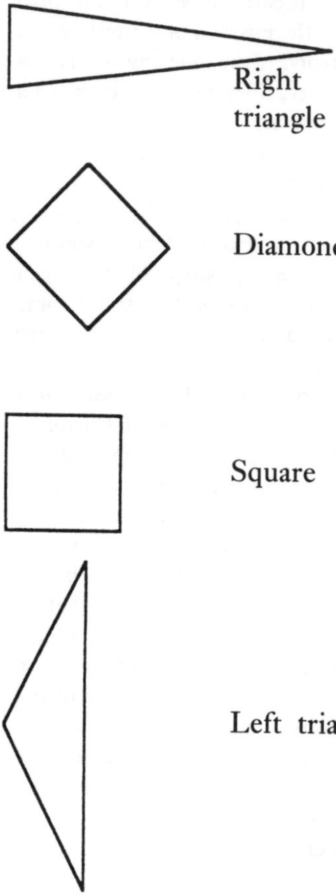

	Drag	
	Left-to-right	Right-to-left
Right triangle	0.06	0.50
Diamond	0.71	0.71
Square	0.71	0.71
Left triangle	2.00	1.78

The calculated drag takes into account
the force of resistance offered by the medium
and the length of the leading edge of the object.
$D = L * F_r = L * (-kV (\cos 2\theta - 1))$
For simplicity, kV = 1 in these calculations.

Figure 4.4 The four objects used in Experiment 2, shown with their calculated drag in both left-to-right and right-to-left directions. See text for further explanation.

memory for the position of an object presented so as to imply directional motion. According to the analysis offered in classical physics, an object with greater mass should exhibit more momentum than an object with smaller mass, because more force is required to change the current pattern of motion. In addition, more streamlined objects should take more time to stop when moving through the same medium. In our experiments, neither rated object masses nor object shapes (with resulting differences in calculated drag) affected the magnitude of the 'momentum effect', or directional memory distortion. Nevertheless, we have replicated the basic error in memory for position. This effect is robust, in that it emerges under manipulation of a number of experimental variables. At present, the only factors that appear to change the *size* of this effect in a principled way are kinematic ones, i.e. variations in the inferred velocity or change in velocity of the sequence of static views. This suggests that the system producing the error is sensitive to parameters of object motion, or kinematic information, but not to factors associated with the causes of that motion, or dynamic information.

A number of additional findings suggest that the momentum analogy may be misleading as a description of this basic phenomenon. Verfaillie and d'Ydewalle (1991) have recently reported that errors in memory for an object's position are determined by the global structure of the event implied in a sequence of static views, rather than by local characteristics of the implied motion in views immediately preceding the test display. Kelly and Freyd (1987) and Freyd, Kelly and DeKay (1990) have found directional errors in memory for continuous changes along non-visual dimensions, e.g. the pitch of a tone. Analogy to momentum of physical objects is clearly inappropriate in this latter situation. It is intriguing to speculate that the distortion error may be a manifestation of some basic cognitive tendency to misalign the position of successive, ordered, discrete events in a direction that anticipates their continuation or conclusion. Further specification of the perceptual/cognitive continua along which such errors occur, as well as of factors that influence their magnitude, should indicate whether this speculation has merit.

2 *Extrapolating an object's position of reappearance in an interrupted continuous transformation*

We consider, next, a series of studies in which observers are asked to predict, rather than to remember, the position of a transforming object. What can this research tell us about the perceptual/cognitive system's sensitivity to kinematic or to dynamic information? In the canonical experimental situation (cf. Cooper, 1989; Cooper, Gibson, Mowafy and Tataryn, 1987; Gibson and Cooper, 1988; Gibson, Bernstein and Cooper, 1989), observers view a computer-generated drawing of a three-dimensional object undergoing a continuous rigid rotation at a constant rate. At some unpredictable moment during the ongoing transformation, the rotation is interrupted and the object disappears. After an unpredictable amount of time, the object reappears and continues to rotate at the same rate as displayed before the blackout.

The subject's task is to judge whether or not the point of reappearance of the object is at the correct position in the transformational trajectory, had the rotation continued smoothly during the blank period. Thus, the procedure requires observers to extrapolate or to generate predictions concerning the continuing appearance of a transforming object that is momentarily obscured from view. Figure 4.7 schematically illustrates the experimental situation, as well as the object used in the initial experiments – a parallel

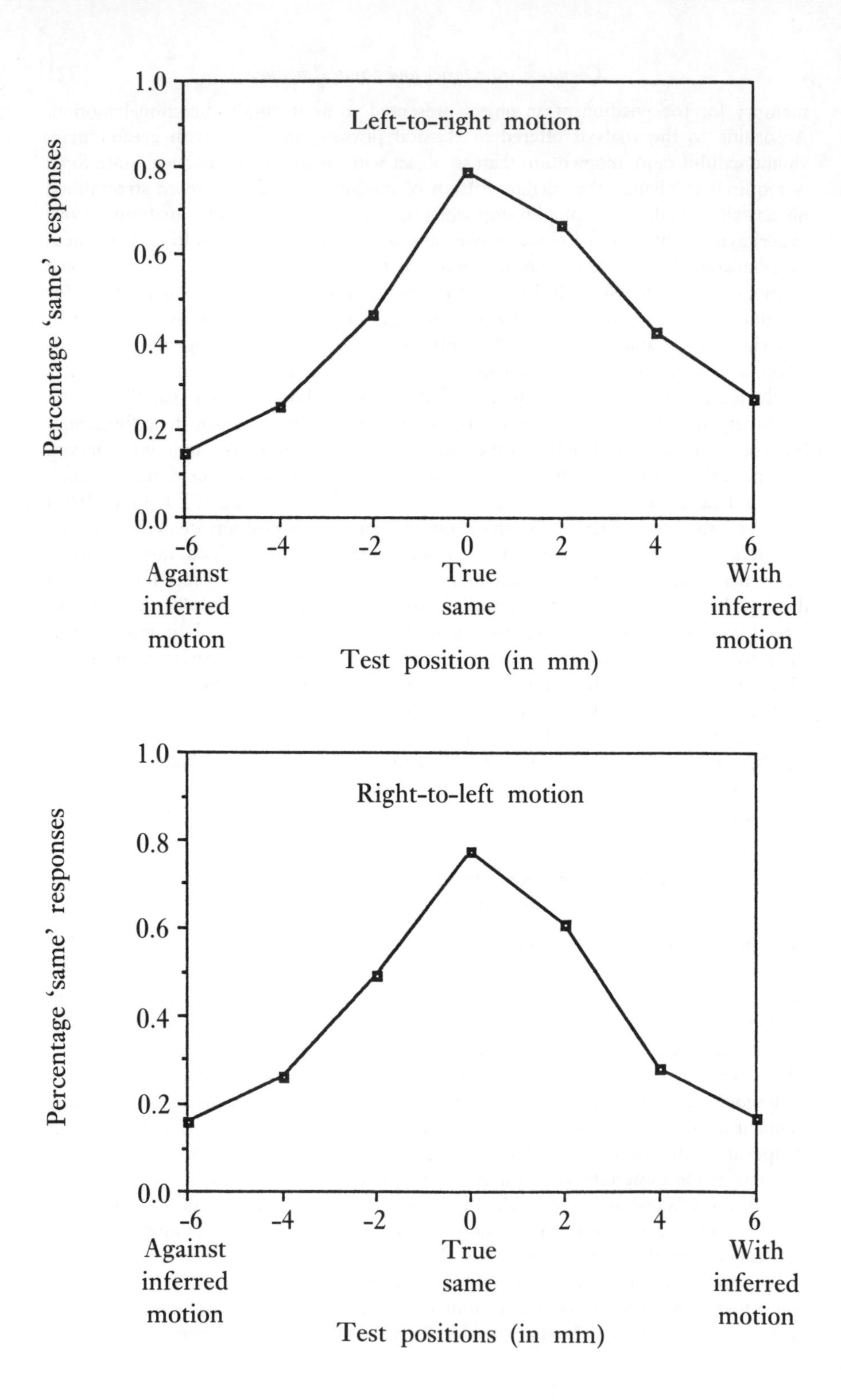
Left-to-right motion
Percentage 'same' responses
1.0
0.8
0.6
0.4
0.2
0.0
-6
-4
-2
0
2
4
6
Against inferred motion
True same
With inferred motion
Test position (in mm)
Right-to-left motion
Percentage 'same' responses
1.0
0.8
0.6
0.4
0.2
0.0
-6
-4
-2
0
2
4
6
Against inferred motion
True same
With inferred motion
Test positions (in mm)

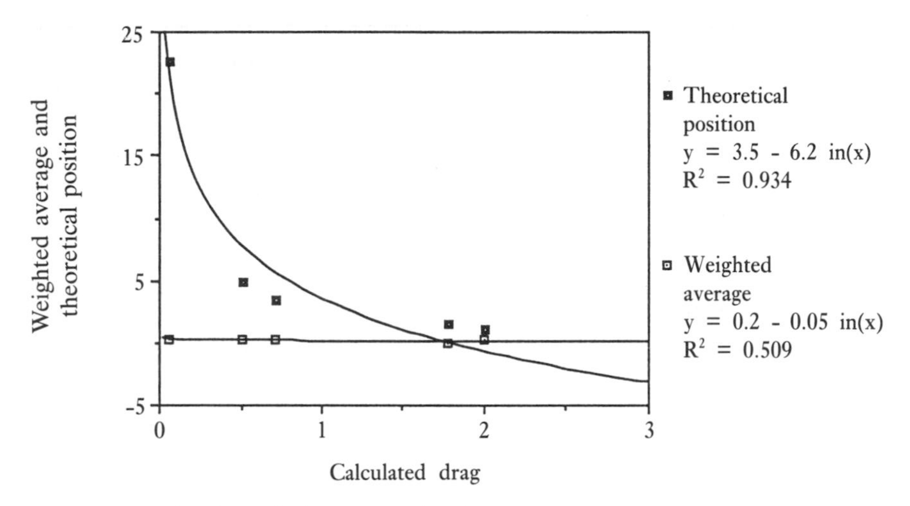

Note: The theoretical position takes into account the resistive force offered by the medium and the velocity of the inducing display.
$X(t) = X_0 + (V_0 / \alpha)(1 - c^{-\alpha t})$
$\alpha = ((L * k) / m)(1 - \cos 2\theta)$
For simplicity, k = m = 1 in these calculations.

Figure 4.6 Derived weighted average measure and theoretical position, plotted as a function of the calculated drag of objects in Experiment 2. See text for further explanation.

projection of a cube, with hidden lines removed, displayed as tilted 15 degrees on the x-axis and 5 degrees on the z-axis. As in the experiments on motion implied from successive static views, the point of reappearance of the object was varied to be either correct (0 degrees), or +/– 6, 16, 26 or 36 degrees forward or backward of the correct reappearance position. In addition, the duration of the blank interval was manipulated.

The basic experimental results – expressed as percentage of judgements that the object reappeared in the correct location, as a function of the angular displacement or 'shift' of the reappearance position from the objectively correct location – are shown in Figure 4.8. They can be summarized as follows: First, overall performance when the point of reappearance is shifted from the correct location becomes increasingly accurate as the angular size of the displacement increases, in either the forward or backward direction. Second, there is a marked asymmetry in the function relating percentage of 'correct reappearance' responses to test object position. Specifically, *undershoots* – i.e. cases in which the object reappears *before* the correct position – are uniformly less

---

Figure 4.5 Percentage of 'same' position responses to test stimuli in Experiment 2, plotted as a function of departure of the test stimulus from the position of the third displayed view. Separate functions are shown for left-to-right and right-to-left motion.

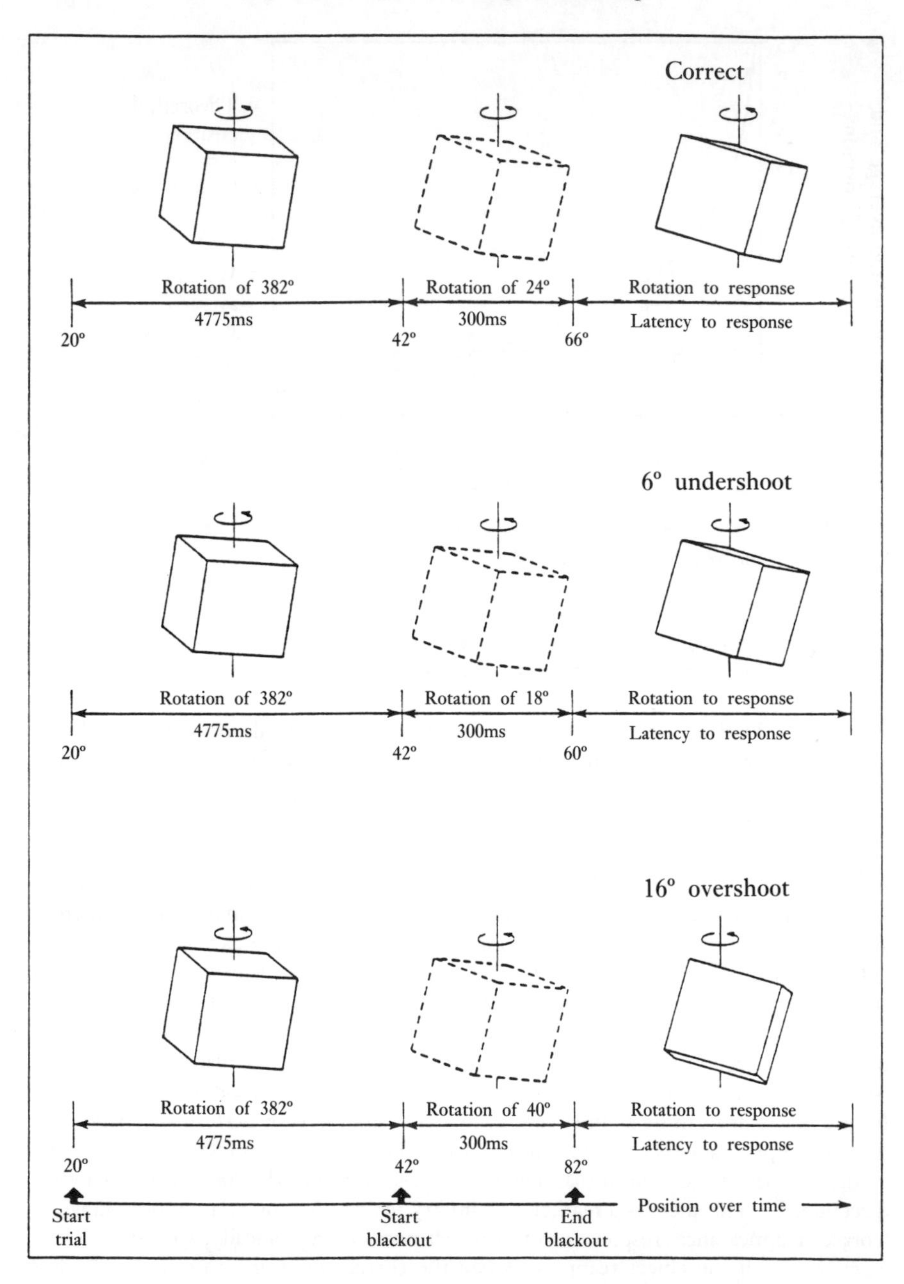

Figure 4.7 Schematic illustration of typical experimental trials in the 'interrupted transformation' experiments. From Cooper (1989).

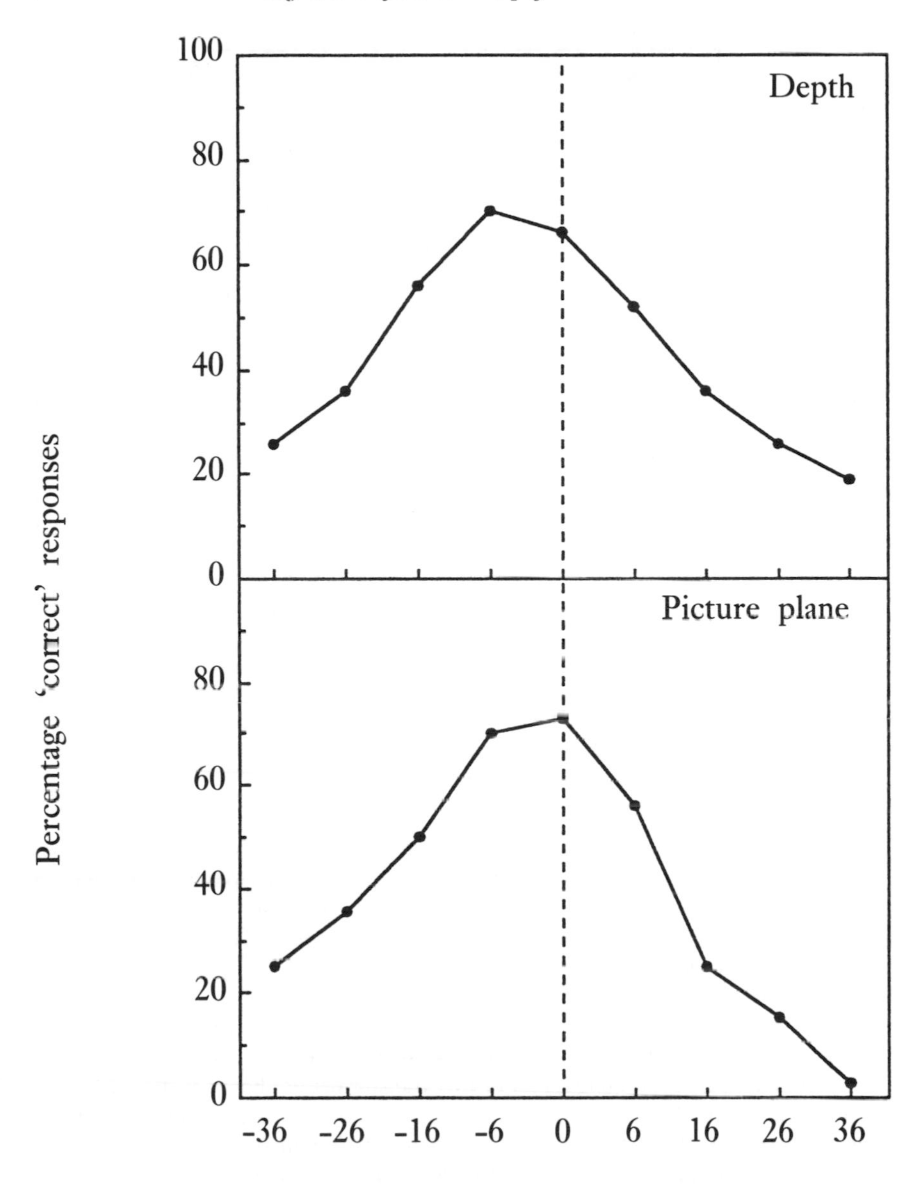

Figure 4.8 Percentage of 'correct' (reappearance position) responses, plotted as a function of angular displacement from the correct reappearance position. Separate functions are shown for objects rotating in depth and in the picture plane. Adapted from Cooper (1989).

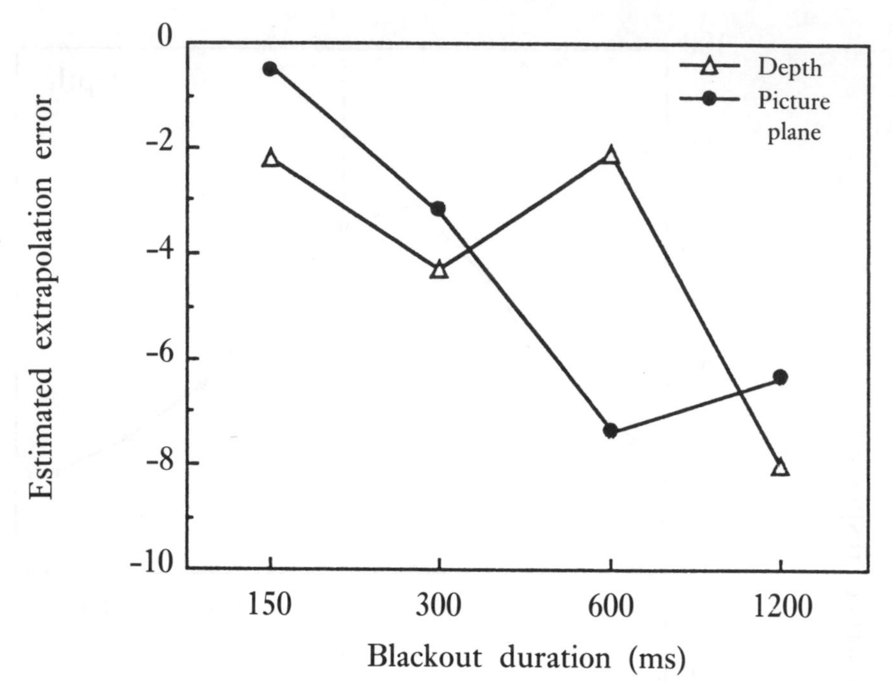

Figure 4.9 Derived extrapolation error measure, plotted as a function of duration of the blackout or blank period. Separate functions are shown for objects rotating in depth and in the picture plane.

detectable than *overshoots*, i.e. cases in which the object reappears in a position *beyond* the correct location. Third, the average percentage of 'correct reappearance' judgements is greater to small (6 degrees) undershoots than to objects presented in the objectively correct (0 degrees) position. That is, the peak of the response curve is shifted in the direction of undershoot or 'backward' errors.

As indicated in Figure 4.8, the same basic pattern of results is evident both for rotations about the y-axis in depth and for rotations in the picture plane. This finding is surprising, since the projected structure of an object rotating in the plane remains the same before and after an interruption of that transformation; however, the projection of an object before, during and after the interruption of a transformation in depth can be radically altered. In addition, for both types of rotation, increasing the length of the extrapolation period, or blackout interval, accentuates the tendency to accept slight negative shifts as 'correct reappearance' positions and generally leads to more variable performance. After Freyd and Johnson (1987), we estimated the magnitude of the 'extrapolation error' by fitting quadratics to the response functions and computing the derivative of the predicted curve. Figure 4.9 shows this derived measure, plotted separately for depth and picture-plane rotations, as a function of the duration of the blackout interval. The magnitude of the estimated undershoot becomes larger as interruption time increases, from a minimum of about 1 degree to a maximum of over 8 degrees in the negative or backward direction.

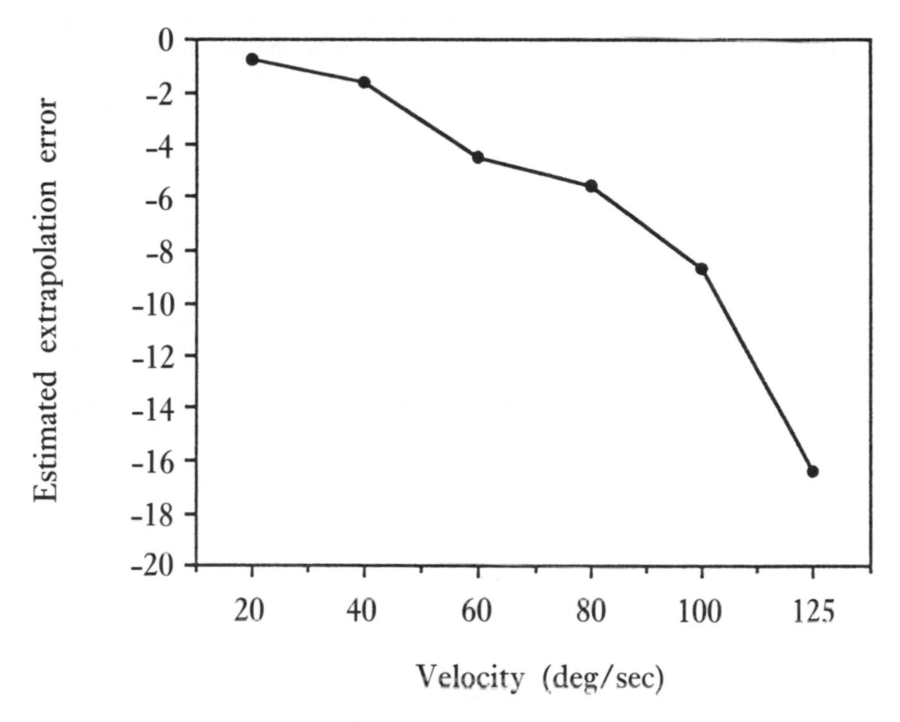

Figure 4.10 Derived extrapolation error measure, plotted as a function of velocity of the object's depicted rotation.

We and our collaborators (in particular, Lori Bernstein and Bradley Gibson) have explored several classes of variables that seem to be good candidates for influencing the occurrence and magnitude of this undershoot error. The general picture that is emerging is similar to the findings for memory errors for the final position of a sequence of static views. That is, kinematic factors seem to affect the magnitude of the extrapolation error, whereas dynamic variables have little or no influence on the extent of undershoot. As shown in Figure 4.10, velocity of the interrupted rotation (ranging from 20 to 125 deg/sec) has a substantial effect on the size of the extrapolation error, with faster velocities producing larger errors. However, a series of experiments done by Lori Bernstein places qualifications on the conclusion that this kinematic variable is controlling level of performance. Bernstein examined the independent contributions of rotational velocity, interruption time and distance travelled during the extrapolation period to the magnitude of the undershoot effect. The results indicate that the distance over which extrapolation was required is the most effective predictor of both overall accuracy and extent of directional error, with velocity and blackout duration sometimes making independent contributions to performance.

Stimulus variations corresponding to dynamic factors that have been assessed at present include object *size* – with projections of wire-frame objects – and object *mass* – with solid, coloured depth-cued objects. Object size, varied in one experiment (Cooper, Gibson, Mowafy and Tataryn, 1987) as 3.6 or 7.2 degrees of visual angle, had no effect on the extent of undershoot, nor did size interact with other factors that influenced this error. Apparent mass of shaded triangular and rectangular prisms also fails to change the pattern of results, despite additional slight modifications of the experimental situation. In summary, variation in object characteristics that parallel changes in dynamic factors have no influence on the magnitude of extrapolation errors, but there is some evidence that variations in kinematic information do affect the accuracy of the position prediction judgement.

The experiments summarized above, requiring the extrapolation of interrupted real-time transformations – like the studies described earlier, involving memory for the final position in an implied transformation – yield strong, replicable and systematic errors. The two situations, however, produce errors in different directions. In the case of transformations suggested by successive static views, memory distortions are in the direction of the continuing implied motion. In the case of extrapolating perceptually-driven transformations, errors in predicting reappearance position are in the backward direction – as if the extrapolation process were 'slowing down' during the interrupted portion of the event. We are far from understanding either the sources of these perceptual/cognitive errors or the relationship, if any, between them. In order to advance toward the desired understanding, it is important to highlight the differences between the situations in which each sort of error is generally obtained.

One central dimension of difference concerns the nature of the depicted transformations. In the studies on memory for final position by Freyd and her collaborators, the transformations are implied in a sequence of discrete, static views, and they depict translational or rotational motion only in the picture plane. In the extrapolation experiments of Cooper and her collaborators, the transformations are portrayed as continuous rotations in depth or in the plane of the picture. The objects presented in the two situations also generally differ in their dimensional structure, i.e. two-dimensional shapes in the implied motion experiments and perspective projections of solid three-dimensional structures in the continuous transformation studies. Finally, the perceptual decisions required in the two paradigms are substantially different, although information about final or continuing object position is probed in both tasks. One of the objectives of the ongoing programme of research in our laboratory is to create experimental situations combining the dimensions of difference between procedures. By decomposing the two canonical tasks and then reorganizing the components in novel ways, we hope to gain some understanding of the conditions that both *produce* and *influence the magnitude of* errors in predicting and remembering the position of an object undergoing a spatial transformation.

3 *Concluding remarks*

We began this chapter by observing that our ability to anticipate the consequences of transformations on objects in space is generally good, allowing us to locomote in an unimpeded fashion and to make predictions about the structure of ongoing events. We went on to ask whether such competence could be explained by our perceptual systems having incorporated descriptions of object movement like those offered in classical physics. If rules concerning the behaviour of objects in space have been internalized, which ones are they, and how are they applied in different situations?

Somewhat paradoxically, the research that we have presented shows that people reliably make systematic, directional errors when asked to remember the final position of an object inferred to be moving or to extrapolate the trajectory of an object's movement when a continuous transformation is momentarily interrupted. The occurrence of these errors is important, and they certainly provide a window into the architecture of our perceptual and cognitive systems. Nonetheless, their magnitudes are small, and the conditions under which they are exhibited may be highly circumscribed. Indeed, some researchers using experimental tasks quite different from those described here (e.g. Jagacinski, Johnson and Miller, 1983; Rosenbaum, 1975) have demonstrated reasonable accuracy in judgements about when a moving object will arrive at a target position. Furthermore, studies of purely imagined rotations on representations of objects (e.g. Cooper, 1976; Shepard and Cooper, 1982) suggest a remarkable degree of correspondence between the time courses and trajectories of the mental transformation, and the external rotation of an object that it simulates.

All of this is as it should be, and the experimental results are consistent with the requirements facing our perceptual and cognitive systems. Our perceptual systems must incorporate information that we need to know about the world, i.e. about certain changing relations between the perceiver and the environment. Our cognitive systems must internalize at least simple aspects of these relations in order to serve our perceptual needs. In particular, we must have internal procedures for keeping track of moving objects, for directing our gaze and for monitoring alignment and occlusion relations *based on currently incomplete and not-yet-sampled information*. Thus, any formula that suggests only that our systems are sensitive to specified physical invariances (e.g. Gibson, 1979) ignores the problem of determining just which aspects of the continually changing relations between the viewer and the world are most useful in serving our perceptual needs.

Such a formula also ignores the reliable, if small, discrepancies – highlighted in this chapter – between the movement of objects in the world and our cognitive abilities to simulate that movement. These discrepancies and the factors that control them can provide insight into the nature of the underlying mechanisms. Our experimental results suggest that analogies to dynamics, or the forces causing the motion of objects, may not provide a useful framework for characterizing the basic, unelaborated principles internalized by our cognitive systems. This finding makes good sense in that mental simulations of object movement based on dynamics could prove computationally cumbersome. For example, different 'overshoots' or 'undershoots' of remembered or anticipated trajectories would have to be computed over wide variations in object masses and coefficients of friction.

Kinematics, however, as Shepard (1984) has suggested, offers an elegant and efficient set of principles for the cognitive system to internalize in completing the partial information about external objects and events provided perceptually. And it is kinematic information, whether presented in a sequence of static views or provided by a continuously transforming object whose trajectory must be extrapolated, that influences how accurate we are in remembering and predicting the position of a moving object.

ACKNOWLEDGEMENTS

Experiments 1 and 2 reported in this chapter are based on a Master's thesis by Margaret P. Munger. We thank Julian Hochberg and Lori J. Bernstein for comments on this chapter and extremely helpful discussion about the entire line of research. Roger N. Shepard provided extensive comments and suggestions regarding the models of physical processes. James L. Park and Charles E. Wright helped develop the physical models, and were especially helpful with various calculations.

NOTES

1. We are confining our analysis to rapid, perceptual judgements where only preliminary analysis is performed by the subject rather than more cognitive, interpreted judgements of the motion of familiar objects. Clearly, in the latter situation additional knowledge about the particular object and its characteristic ways of moving (e.g. birds fly, trains have engines, frogs hop) would be used in making judgements about the object's continuing trajectory.
2. The drag of the objects was calculated using a simple model of a medium made up of stationary, uniformly distributed, perfectly elastic particles. The object must be travelling in line with one of its axes of symmetry. The force of resistance, F_r, is the amount of resistance the medium exerts on the object. In other words, F_r is the amount of force necessary to move an encountered particle out of the way of the object. F_r is proportional to the cosine of the angle of incidence for the particle, which is equal to half the angle of the object's protruding point. For example, the diamond has a protruding angle of 90 degrees which would make the angle of incidence of an encountered particle, and θ, 45 degrees. Specifically, $F_r = -kv(\cos 2\theta - 1)$. In this experiment, kv, where k includes the mass and v is the initial velocity, is equal for all four shapes (and taken to be 1 for the calculations in Figure 4.4). In order to compare different objects, the F_r was multiplied by the length of the leading edge of the object. For example, the square and two triangles all have one side that is vertical. The Fr for all three would be equal, but the number of particles each would encounter would depend on the length of the leading surface.

REFERENCES

Carlton, E. H. and Shepard, R. N. 1990: Psychologically simple motions as geodesic paths: 1. Asymmetric objects. *Journal of Mathematical Psychology*, 34, 127–88.

Cooper, L. A. 1976: Demonstration of a mental analog of an external rotation. *Perception and Psychophysics*, 19, 296–302.

Cooper, L. A. 1989: Mental models of the structure of visual objects. In B. E. Shepp and S. Ballesteros (eds), *Object Perception: Structure and process*, Hillsdale, NJ: Erlbaum, 91–119.

Cooper, L. A., Gibson, B. S., Mowafy, L. and Tataryn, D. J. 1987: Mental extrapolation of perceptually-driven spatial transformations. Paper presented at the 28th annual meeting of the Psychonomic Society, Seattle, Washington, 6–8 November.

Finke, R. A. and Shyi, G. C.-W. 1988: Mental extrapolation and representational momentum for complex implied motions. *Journal of Experimental Psychology: Learning, Memory, and Cognition*, 14, 112–20.

Finke, R. A., Freyd, J. J. and Shyi, G. C.-W. 1986: Implied velocity and acceleration induce transformations of visual memory. *Journal of Experimental Psychology, General*, 115, 175–88.

Freyd, J. J. 1983: The mental representation of movement when static stimuli are viewed. *Perception and Psychophysics*, 33, 575–81.

Freyd, J. J. 1987: Dynamic mental representations. *Psychological Review*, 94, 427–38.

Freyd, J. J. and Finke, R. A. 1984: Representational momentum. *Journal of Experimental Psychology: Learning, Memory, and Cognition*, 10, 126–32.

Freyd, J. J. and Finke, R. A. 1985: A velocity effect for representational momentum. *Bulletin of the Psychonomic Society*, 23, 443–46.

Freyd, J. J. and Johnson, J. Q. 1987: Probing the time course of representational momentum. *Journal of Experimental Psychology: Learning, Memory, and Cognition*, 13, 259–68.

Freyd, J. J., Kelly, M. H. and DeKay, M. L. 1990: Representational momentum in memory for pitch. *Journal of Experimental Psychology: Learning, Memory, and Cognition*, 16, 1107–17.

Gibson, B. S. and Cooper, L. A. 1988: Perceiving and extrapolating continuous spatial transformations. Paper presented at the 29th annual meeting of the Psychonomic Society, Chicago, Illinois, 10–12 November.

Gibson, B. S., Bernstein, L. J. and Cooper, L. A. 1989: Explorations of the mental mapping of three-dimensional object motion. Poster presented at the 30th annual meeting of the Psychonomic Society, Atlanta, Georgia.

Gibson, J. J. 1979: *The Ecological Approach to Visual Perception*. Boston, Mass: Houghton Mifflin.

Gilden, D. L. and Proffitt, D. R. 1989: Understanding collision dynamics. *Journal of Experimental Psychology: Human Perception and Performance*, 15, 372–83.

Hubbard, T. L. 1990: Cognitive representation of linear motion: Possible direction and gravity effects in judged displacement. *Memory and Cognition*, 18, 299–309.

Hubbard, T. L. and Bharucha, J. J. 1988: Judged displacement in apparent vertical and horizontal motion. *Perception and Psychophysics*, 44, 211–21.

Jagacinski, R. J., Johnson, W. W. and Miller, R. A. 1983: Quantifying the cognitive trajectories of extrapolated movements. *Journal of Experimental Psychology: Human Perception and Performance*, 9, 43–57.

Kaiser, M. K. and Proffitt, D. R. 1984: The development of sensitivity to causally relevant dynamic information. *Child Development*, 55, 1614–24.

Kaiser, M. K. and Proffitt, D. R. 1987: Observers' sensitivity to dynamic anomalies in collisions. *Perception and Psychophysics*, 42, 275–80.

Kelly, M. H. and Freyd, J. J. 1987: Explorations of representational momentum. *Cognitive Psychology*, 19, 369–401.

McBeath, M. K. and Shepard, R. N. 1989: Apparent motion between shapes differing in location and orientation: A window technique for estimating path curvature. *Perception and Psychophysics*, 46, 333–7.

McCloskey, M. and Kohl, D. 1983: Naive physics: The curvilinear impetus principle and its role in interactions with moving objects. *Journal of Experimental Psychology: Learning, Memory, and Cognition*, 9, 146–56.

McCloskey, M., Washburn, A. and Felch, L. 1983: Intuitive physics: The straight-down belief and its origin. *Journal of Experimental Psychology: Learning, Memory, and Cognition*, 9, 636–49.

Robins, C. and Shepard, R. N. 1977: Spatio-temporal probing of apparent rotational movement. *Perception and Psychophysics*, 22, 12–18.

Rosenbaum, D. A. 1975: Perception and extrapolation of velocity and acceleration. *Journal of Experimental Psychology: Human Perception and Performance*, 1, 395–403.

Shepard, R. N. 1984: Ecological constraints on internal representation: Resonant kinematics of perceiving, imagining, thinking, and dreaming. *Psychological Review*, 91, 417–47.

Shepard, R. N. and Cooper, L. A. 1982: *Mental images and their transformations*. Cambridge, Mass: MIT, Bradford.

Shepard, R. N. and Zare, S. L. 1983: Path-guided apparent motion. *Science*, 220, 632–4.

Todd, J. T. and Warren, Jr., W. H. 1982: Visual perception of relative mass in dynamic events. *Perception*, 11, 325–35.

Verfaillie, K. and d'Ydewalle, G. 1991: Representational momentum and event course anticipation in the perception of implied periodical motions. *Journal of Experimental Psychology: Learning, Memory, and Cognition*, 17, 302–13.

5

Perceiving and reasoning about objects: Insights from infants

Elizabeth S. Spelke and Gretchen A. Van de Walle

Introduction

The human environment is populated by a rich variety of material objects, from rocks, to spoons, to pocket calculators, to animals and people. To act on our surroundings effectively, we must apprehend these objects and anticipate their behaviour. Some of this task appears to be accomplished by perceptual mechanisms: a cup stands before us and we see its shape, colour and texture. Even in this case, however, perception goes beyond the immediately visible, for we appear to 'see' the complete cup, not just the surfaces reflecting light to our eyes. To act on the cup, moreover, we must apprehend object properties that are not obviously visible at all, such as the cup's weight and centre of mass. As our encounters with objects become extended over time, object perception appears to become increasingly inferential in character. We look at the cup at different places and times and apprehend a body that has persisted between those encounters. We drop the cup and anticipate its behaviour. The cup falls from view, and we infer how it will continue to move and where, approximately, it will come to rest. These last activities seem to reflect not an ability to perceive visible objects but an ability to represent hidden objects and to reason about their behaviour.

How do humans perceive surrounding layouts of spatially extended, bounded objects? How do we reason about objects so as to anticipate their future behaviour or infer their unseen states and motions? Are object perception and physical reasoning related activities? How does each depend on, and illuminate, human conceptions of the physical world?

Studies in computational vision and philosophy suggest discouraging answers to these questions. There may be no basic process of object perception, dependent on general constraints on objects' behaviour. Rather, perception of objects may depend on processes of object recognition. Based on a computational analysis of human vision, for example, Marr (1982, pp. 270–1) suggests that perceptual mechanisms, attuned to general constraints on the arrangement and behaviour of the visible environment, cannot segment the surface layout into objects.[1] Based on an analysis of human intuitions about object persistence over change, Wiggins (1980) arrives at a similar conclusion.

Research in cognitive and educational psychology further suggests there is no basic process of physical reasoning, guided by knowledge of general constraints on objects' behaviour. High school and college students show striking inconsistencies in their common-sense reasoning about objects (e.g. Halloun and Hestenes, 1985; McCloskey, 1983). For example, one person may judge that a ball set in horizontal motion by striking will move both forward and downward after losing its support, whereas a ball set in motion by a carrier will move only downward (McCloskey, 1983). Similarly, one person may judge that water leaving a curved tube will follow a straight path, whereas a ball leaving a curved tube will follow a curved path (Kaiser, Jonides and Alexander, 1986). These and other findings suggest that humans reason about objects by exploiting a collection of expectations about the behaviour of particular kinds of objects in particular circumstances.

In philosophy and in psychology, little consensus has emerged concerning the relation between perception and reasoning, or concerning the existence and nature of general conceptions of the physical world. In the light of the above characterization of object perception and physical reasoning, the apparent intractability of these questions is not surprising. If perceiving and reasoning each depend on a wealth of acquired information about particular kinds of objects, then one would expect these processes to resemble one another in some respects (because the objects we must perceive and reason about are the same) and to diverge in other respects (because the appearance of an object, so useful for purposes of recognition, is not always the most useful guide to reasoning about its behaviour). One also would expect the quest for general conceptions of the physical world to remain unfulfilled, either because basic conceptions are buried under a wealth of more specialized conceptions, or because no basic conceptions exist.

Against these conclusions, we will sketch a different picture. There is a basic process of object perception, according with general and pervasive constraints on the behaviour of the material world. This process begins to operate before children have developed knowledge of particular kinds of objects; it enables children and adults to single out the things about which they develop knowledge. There is also a basic process of reasoning about objects, according with the same physical constraints. This process enables children and adults to trace objects through time and to anticipate their future states and positions. It provides a framework within which humans can gain further knowledge about the behaviour of objects of particular kinds.

If these suggestions are correct, then studies of early development may shed light on processes of object perception and physical reasoning. Studies of infants and young children could serve to reveal the operation of these processes at the time when the processes are most needed, and before they are overlaid by a wealth of specific knowledge.

In what follows, we review some research on infants' perception and reasoning about certain simple, inanimate objects.[2] For purposes of exposition, we consider studies of object perception and studies of physical reasoning separately, but we will suggest that this division is artificial. Neither perceiving objects nor reasoning about objects fits comfortably within the common-sense distinction between observing the world and thinking about it. They fit poorly within these categories, we believe, because object perception and physical reasoning are aspects of a single human competence, centring on a single system of knowledge.

1 Object perception in infancy

Psychologists' understanding of perception in infancy has grown greatly over the last three decades, owing in large part to the development of a set of useful experimental methods. In particular, preferential looking methods, focusing on systematic differences in infants' looking time to different visual patterns (Fantz, 1961), have served to assess a broad spectrum of perceptual abilities, including perception of hue (e.g. Teller and Bornstein, 1987), orientation (Atkinson, ch. 14, Braddick, ch. 15 of this volume), form (e.g. Schwartz and Day, 1979), depth (e.g. Kellman, Van de Walle, Hofsten and Condry, 1990; Slater, Mattock and Brown, 1990) and intermodal correspondences (e.g. Meltzoff, ch. 9 of this volume; Streri, 1990). A number of investigators have used these methods to investigate infants' ability to divide the perceived layout into unitary, bounded objects. Because these studies have been reviewed elsewhere, we summarize only a sampling of studies here (see Spelke, 1990, for a more detailed summary).

1.1 Infant perception of the unity and boundaries of visible objects Let us begin with a simple situation: A three-dimensional object is presented in front of a uniform background. Can infants perceive the unity of this object and its separateness from the background? Experiments using four different methods provide evidence that infants perceive object unity as early as three months of age (Spelke and Born, 1983; Spelke, 1985a; Hofsten and Spelke, 1985; Kestenbaum, Termine and Spelke, 1987; Spelke, Hofsten and Kestenbaum, 1989; Spelke, Breinlinger, Jacobson and Phillips, 1993a). The most recent of these studies (Spelke et al., 1993a) serves as an example.

In this experiment, three-month-old infants were presented with one of four conically shaped objects in an otherwise empty display (Figure 5.1, a and b). The experiment used a habituation of looking time method, in order to investigate whether infants perceived each of the objects as a unit that should move as a whole. On a series of trials, infants in an experimental condition saw a hand enter the display and tap the object, which remained at rest. Looking time was recorded until the infant looked away from the display, ending the trial. Trials continued until the infant's spontaneous looking time declined to half its initial level: the criterion of habituation.

A sequence of alternating test trials followed. On each trial, the hand entered the display and then grasped and lifted the top of the object. On the three whole-object test trials, the object rose as a single body; on the three half-object test trials, the object broke apart and only its upper half rose into the air (Figure 5.1, c and d). Looking time to the event outcomes was recorded, beginning when the display ceased to move and continuing until the infant looked away.

Looking times to the two event outcomes were compared to the looking times of infants in a separate baseline condition, who viewed the raised half- and whole-object displays with no prior exposure to the objects. Since the infants in the baseline condition viewed exactly the same displays as those in the experimental condition throughout the time that looking was recorded, the baseline condition controlled for differences in the intrinsic attractiveness of the two displays.

In habituation studies, infants tend to look longer at novel or surprising events (see Baillargeon, 1986; Bornstein, 1985; Spelke, 1985b; Atkinson, ch. 14, Braddick, ch. 15 of this volume). If the infants in the experimental condition perceived each of the four objects as a unit, then the event outcome in which the object broke apart should have

Habituation

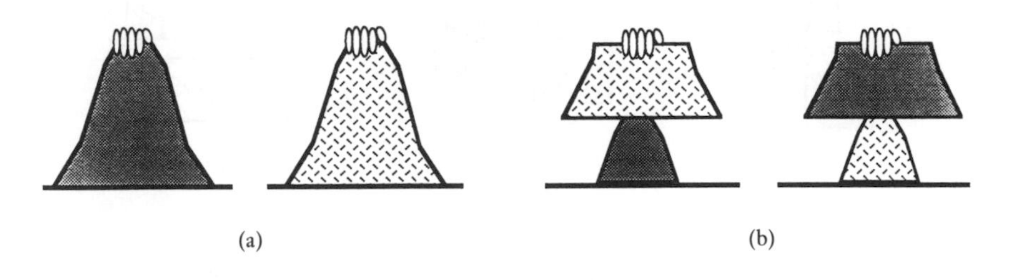

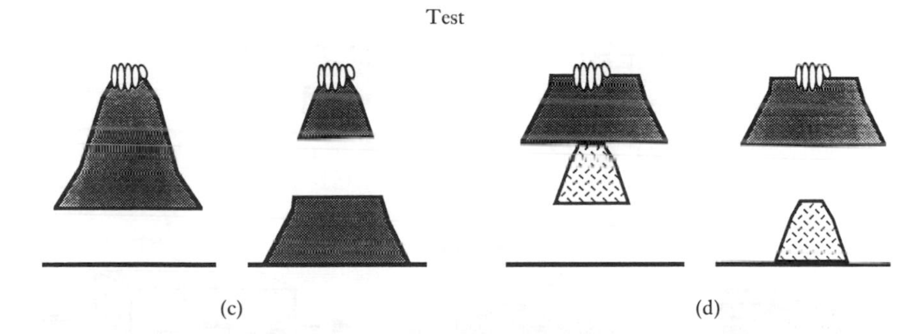

Figure 5.1 Schematic depiction of displays from an experiment on infants' perception of object boundaries. (After Spelke et al., 1993a.)

appeared more novel or surprising to them. The results supported this prediction: the infants in the experimental condition showed a significantly greater looking preference for the half-object display than those in the baseline condition, providing evidence that infants perceived the objects as unitary wholes.

Let us now complicate the situation and ask whether infants perceive object boundaries of scenes containing several objects. Adults typically are able to perceive the distinctness of each object in a complex scene. Studies of young infants provide evidence that they also can perceive the distinctness of objects, under two conditions. First, young infants perceive two objects as distinct if the objects are spatially separated by a gap (Figure 5.2, a–c). Two objects are perceived as distinct not only when the objects are separated vertically so that the gap is visible (Spelke et al., 1989; Kestenbaum et al., 1987), but also when the objects are separated in depth, so that the gap cannot be seen directly (Hofsten and Spelke, 1985; Kestenbaum et al., 1987). Second, young

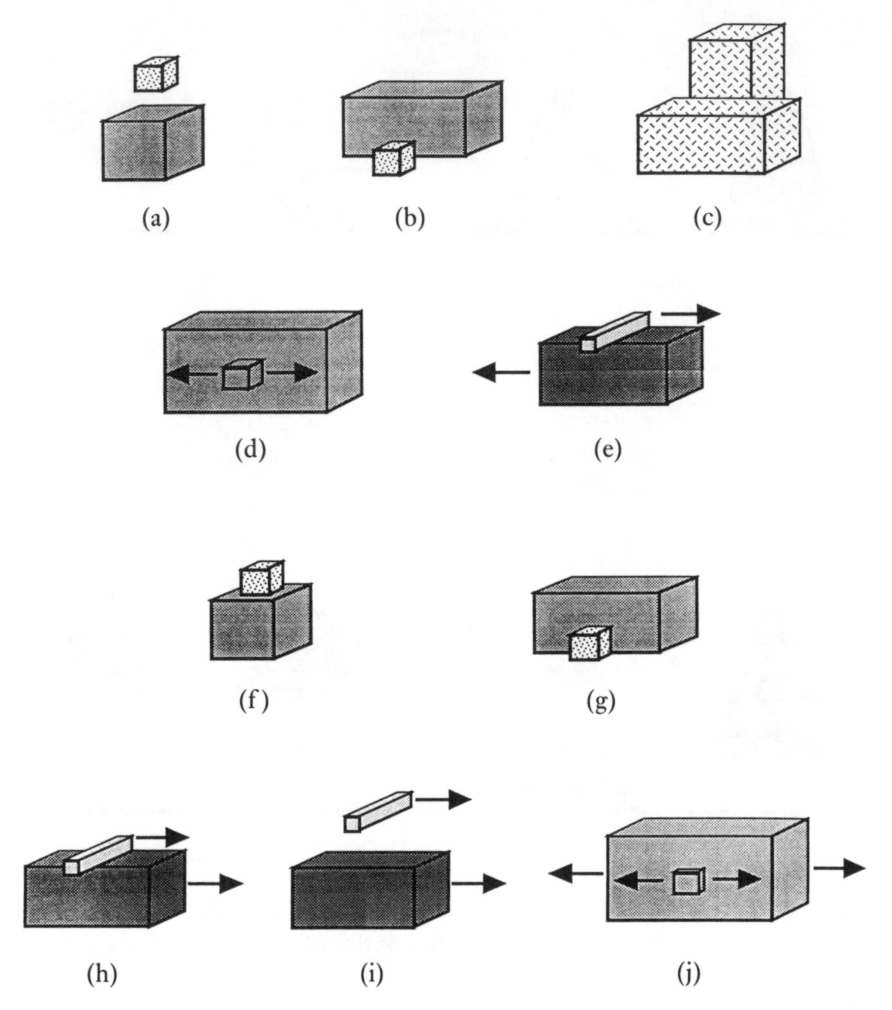

Figure 5.2 Schematic depiction of displays from experiments on object perception in infancy. Arrows indicate the direction and extent of object motion. The objects in (b), (c) and (j) were separated in depth; those in (d) and (g) were adjacent in depth. (After Hofsten and Spelke, 1985 (d, j); Kestenbaum, et al., 1987 (a, b, f, g); Prather and Spelke, 1982 (c); Spelke, et al., 1989 (e, h, i)).

infants perceive two objects as distinct if the objects undergo separate motions, even if the objects remain in contact throughout the time that they move (Figure 5.2, d and e). Two adjacent objects are perceived as distinct if one object moves while the other is stationary (Hofsten and Spelke, 1985) and also if each object moves rigidly in a different direction (Spelke et al., 1989).[3] In contrast, infants do not appear to perceive the boundary between two objects that are stationary and adjacent, even if the objects differ in colour, texture and form (Figure 5.2, f and g). In the experiment by Spelke et al. (1991a), for example, three-month-old infants' perception of object unity appeared to be equally strong, regardless of whether infants viewed an object with a uniform colour and simple shape (Figure 5.1a) or an object with a top and bottom that differed in colour and were irregular in shape (Figure 5.1b). The Gestalt relationships that specify the boundaries of stationary objects for adults – colour similarity, smoothness of edges and figural goodness – do not appear to be effective for infants.

The above findings suggest that young infants are sensitive to two symmetrical constraints on object motion: *cohesion* and *boundedness*. First, a single object is a spatially connected body that retains its connectedness as it moves. When two surfaces can be seen *not* to be connected (because they are separated by a detectable gap or because their connectedness is broken as they move), the cohesion constraint dictates that the surfaces lie on different objects. Second, distinct objects are not connected, and they do not become connected when they move. When no spatial gap or relative motion can be seen to separate two surfaces, the boundedness constraint dictates that the surfaces lie on a single object. Infants' sensitivity to these two constraints can be encompassed by a single *principle of cohesion*: surfaces lie on a single object if and only if they are connected. This principle accounts for all the findings described above.

Let us turn to the case in which two objects undergo a common rigid motion. Infants' perception of commonly moving objects has been studied with configurations similar to those described above (Figure 5.2, h–j). Perception of object boundaries has been found to depend on how the objects are arranged in space.

When two objects are adjacent and move together, infants appear to perceive one connected body (Hofsten and Spelke, 1985; Spelke et al., 1989). This finding is not surprising, since the objects also are seen as one body when they are stationary. When two objects are separated by a visible gap and move together, infants appear to perceive two distinct bodies, despite the common pattern of motion (Spelke et al., 1989). This finding follows from the cohesion principle: two parts of a single object cannot be wholly unconnected. When two objects are separated in depth, however, a different finding is obtained. Although the objects are perceived as distinct when they are stationary, they are perceived as a single body when they move together (Hofsten and Spelke, 1985). This perception does not follow from the cohesion principle alone. What accounts for it?

One possible account is suggested by the intuitions of adults. If two partly hidden surfaces move rigidly together, we infer that the surfaces are in contact somewhere out of view. This inference follows from the physical constraint of *no action at a distance*: distinct objects do not move together if they are separated by a gap. Conversely, if two partly hidden surfaces move independently, adults infer that they are separated by a hidden gap. This inference follows from the physical constraint of *action on contact*: objects do not move independently when they are in contact. Both constraints can be captured by a single *principle of contact*: surfaces move together if and only if they are in contact. Once two partly hidden surfaces are inferred to be continuously in contact, the display is perceived as a single object, like the display in Figure 5.2h, in accord with the cohesion principle. Therefore, these cohesion and contact principles together specify that commonly moving, partly hidden surfaces lie on a single object.

Do infants perceive partly hidden objects in this way? This possibility may appear remote. In order for infants to perceive two commonly moving, partly hidden surfaces as adults do, they must perceive occlusion: infants must represent surfaces as hiding parts of other surfaces. In addition, infants must make inferences about the surfaces

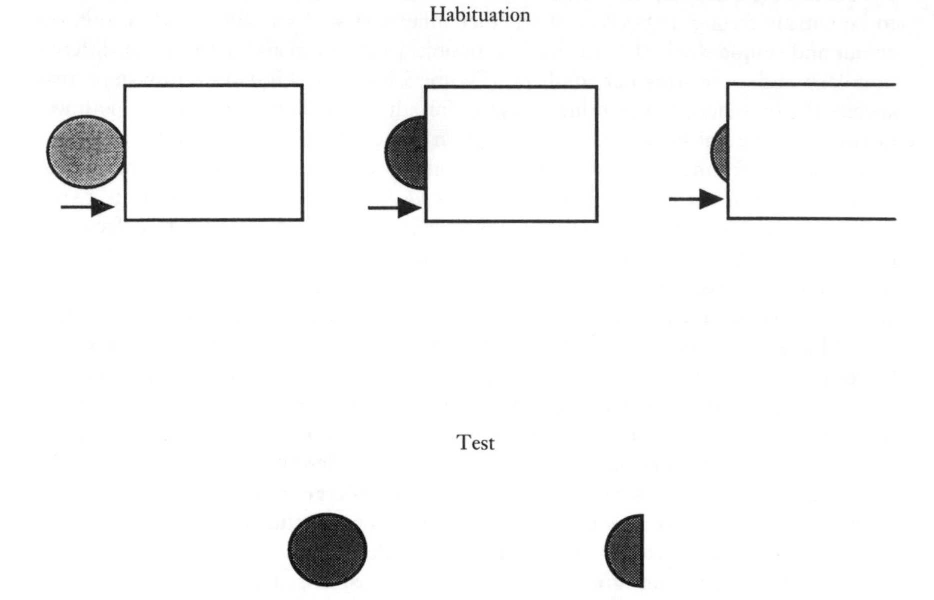

Figure 5.3 Schematic depiction of displays from an experiment on infants' perception of a progressively occluded object. Arrows indicate the direction of motion of the disk, which continued moving until it was fully hidden. (After Craton and Yonas, 1990.)

that lie in occluded parts of the layout. Finally, infants' inferences must follow from an analysis of the motions of visible surfaces, in accord with both the contact and the cohesion principles. We now turn to research that provides evidence for all these abilities.

1.2 Infant perception of partly occluded objects When a visible object moves partly out of view, young infants appear to perceive a persisting, unchanging body. Evidence for this ability comes from a variety of experiments (e.g. Baillargeon, 1987a; Bower, 1967; Craton and Yonas, 1990; Kellman and Spelke, 1983; Leslie, 1991; Spelke, Breinlinger, Macomber and Jacobson, 1992). An experiment by Craton and Yonas (1990) serves as an example. Four-month-old infants were habituated to a disk that moved in and out of view behind a screen (Figure 5.3). The disk underwent a fairly complex motion such that it was rarely fully visible or fully hidden: for most of the habituation period, the disk was partly occluded by the screen. Following habituation, the infants were shown test displays consisting of either a half disk or a whole disk. Four-month-old infants looked significantly longer at the half disk. This finding suggests that the infants did not perceive the occluding edge of the screen as a boundary of the disk, and that they did not perceive the disk to change shape as it moved from view. Rather, the infants appeared to perceive an object of constant form that was progressively hidden and revealed.

What do infants perceive when an object moves rigidly behind a surface that occludes its centre: can they infer that the visible surfaces are in contact behind the occluder throughout the event, and therefore perceive the object as one connected body? A number of experiments provide evidence for this ability. In one study (Kellman and Spelke, 1983), four-month-old infants in an experimental condition were habituated to a rod moving laterally behind a central occluding block. Following habituation, the infants were shown two test displays: a connected rod and a broken rod consisting of two aligned parts of the rod with a gap where the occluder had been (Figure 5.4). Their looking times were compared to those of infants in a baseline condition, who viewed the same test displays. The infants in the experimental condition looked significantly longer at the broken rod, relative to baseline. This experiment thus provided evidence that infants perceived the partly hidden rod as a connected object.

Subsequent experiments further investigated the conditions under which infants perceive the unity of a centre-occluded object. First, the object must move: infants do not infer contact between the visible surfaces of an object that is stationary (Kellman and Spelke, 1983). Second, any rigid motion in three-dimensional space leads infants to perceive object unity: motion in depth and vertical motion are as effective as lateral motion (Kellman, Spelke and Short, 1986). Third, retinal displacements produced by an object's motion are neither necessary nor sufficient for perception of object unity. If infants are in motion and view a centre-occluded object that moves conjointly with them, effectively eliminating any retinal displacement caused by the motion of the object, they infer contact between the object's partly occluded surfaces. Infants fail to infer contact between the surfaces of a stationary centre-occluded object during self motion, despite the fact that their own displacement produces substantial retinal displacement of the display (Kellman, Gleitman and Spelke, 1987). It appears that real, three-dimensional surface motion is necessary for perception of partly hidden objects. Fourth, in all the above studies, the Gestalt relations of colour similarity, smoothness of edges and figural goodness have no detectable influence on infants' perception of centre-occluded objects. In stationary displays, Gestalt relations fail to specify, for infants, that surfaces lie on one connected, partly hidden object (Kellman and Spelke, 1983; Schmidt and Spelke, 1984; Schmidt, 1985; Schwartz, 1982). In moving displays, these relations fail to influence either the strength of infants' perception of object unity (Kellman and Spelke, 1983) or the form of the objects that infants perceive (Craton and Baillargeon, personal communication). Infants' perception of centre-occluded objects accords only with the principles of contact and cohesion.

The above studies were conducted with infants at least three months old. It is interesting to ask whether younger infants would perceive partly occluded objects in the same way. Recent experiments by Slater, Morison, Somers, Mattock, Brown and Taylor (1990) suggest that they do not. Unlike four-month-olds, newborn infants may perceive the surfaces of a rigidly moving, centre-occluded object as separated by a gap.

In these studies, newborn infants were habituated to a rigidly translating, centre-occluded object (either an outline square or a rod). Following habituation, the infants were presented with paired test stimuli consisting of a broken and a connected figure. With both the outline square and the rod, newborn infants showed a significant preference for the connected rather than the broken test display, suggesting that they perceived the centre-occluded object as two spatially separated bodies bounded by the occluder. Additional studies provided evidence that these looking patterns were not produced either by a preference for a familiar object or by a failure of attention to the occluder

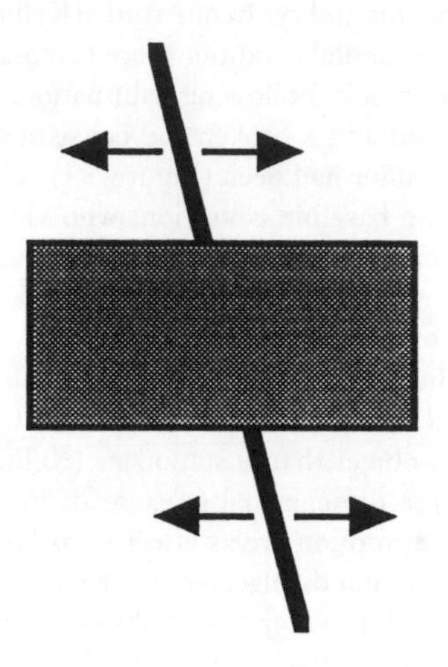

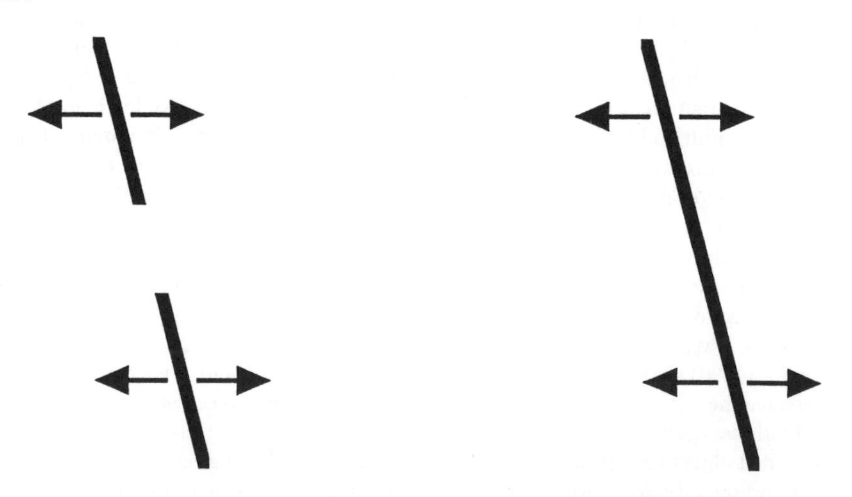

Figure 5.4 Schematic depiction of displays from an experiment on infants' perception of partly occluded objects. Arrows indicate the direction and relative extent of the rod's motion. (After Kellman and Spelke, 1983.)

during habituation. A final study directly compared newborn and four-month-old infants' performance using displays differing only in size. This study replicated Kellman and Spelke's (1983) original findings with four-month-old infants; newborns again exhibited the opposite visual preferences.

The above findings could be interpreted in at least four ways. First, newborn infants may be insensitive to the motion relations that are critical to the perception of partly occluded objects in infancy. This possibility is consistent with research on the neural mechanisms of motion processing and their development in infancy (Johnson, 1990). Second, developmental changes in depth perception may allow four-month-old infants, but not newborns, to perceive the correct depth relations between the occluder and the partly hidden object. Very young infants may fail to see a centre-occluded object as occluded because they fail to perceive its surfaces as standing behind the occluder. Consistent with this possibility, a wealth of research provides evidence that sensitivity to depth increases dramatically over the first four months (e.g. Held, Birch and Gwiazda, 1980; Kellman et al., 1990). Third, the neural pathways allowing cognitive control over visual attention may mature over the first four months. Newborn infants therefore may fail to exhibit, in their looking preferences, cognitive capacities that are present and otherwise functional (Johnson, 1990). Fourth, young infants may perceive occlusion, depth and motion relations correctly but may fail to perceive object unity in accord with the contact and cohesion principles. Current research is attempting to test these possibilities.

Whatever processes account for developmental changes in the first months of life, it appears that three- and four-month-old infants perceive visible objects in accord with the principles of cohesion and contact. For such infants, perception of visible objects evidently depends on processes that operate quite late in perceptual analysis. These processes take as input a representation of arrangements and motions of surfaces in the three-dimensional visible layout, not lower level representations of arrangements and displacements of images in the two-dimensional visual field. We now ask whether these processes are more central still: does object perception depend on distinct visual mechanisms, haptic mechanisms and the like, or does it depend on a single mechanism operating on representations of the layout obtained from any perceptual mode? If perception of objects depends on an amodal mechanism, then infants might perceive objects when they encounter surfaces through other perceptual modes, and their perception should accord with the same principles as perception of visible objects.

1.3 Infant perception of haptically presented objects Research on haptic perception provides evidence that infants perceive object unity and boundaries from active touch (Streri and Spelke, 1988). Infants aged four and a half months were given two rings to hold, one in each hand, under a bib that blocked their view of the rings and of their bodies. In one condition, the rings were constrained to move rigidly together; in another condition, the rings could be moved independently.[4] Infants explored the rings at will, without touching the area between the rings or bringing the rings into view. The experiments investigated whether, in each condition, infants perceived one connected object or two separate objects.

Perception of the unity or distinctness of the objects was tested through a haptic habituation and visual transfer test (Figure 5.5). In past research, infants were found to look more at a visible object that differed from the object they had felt during habituation (Streri and Pêcheux, 1986; Streri, 1987). After exploring the rings haptically to a

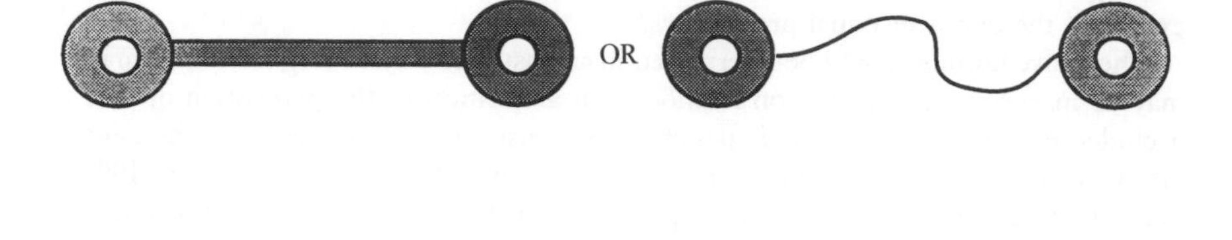

Figure 5.5 Schematic depiction of displays from an experiment on infants' haptic perception of object boundaries. (After Streri and Spelke, 1988.)

criterion of habituation, infants were presented with two visual displays consisting of two rings that were either connected or separated by a gap. Their looking preferences between the displays were compared to the preferences of infants in a separate baseline condition, who were presented with the same visual displays after no habituation sequence. The infants who had explored the rigidly movable rings looked significantly longer at the separated rings, providing evidence that they perceived the haptically presented rings as connected. Conversely, the infants who explored the independently movable rings looked longer at the connected rings, providing evidence that they perceived the rings as distinct. These perceptions accord with the contact principle and with the findings of studies of object perception in the visual mode.

Further experiments varied properties of the rigidly movable assembly such as the similarity of the two rings in form, texture and weight, and the simplicity of the overall shape of the assembly (Streri and Spelke, 1989). In contrast to adults, infants' perception of the rings was not affected by these properties. As in the case of vision, Gestalt relations evidently fail to influence infants' perception of haptically presented objects.

Most recently, experiments have investigated infants' perception of haptically presented objects that remain in contact but that undergo distinct rigid motions (Streri, Spelke and Rameix, 1993). In different conditions, the two rings either underwent separate horizontal motions (they could by pushed together and pulled apart) or separate vertical motions (they could be slid up and down with respect to one another). Infants either explored the rings actively, producing the relative motions, or they held the rings passively while an experimenter produced the motions. Like adults (Gibson, 1962), infants perceived the objects effectively only when they manipulated the assembly actively. In the active motion conditions, infants' looking preferences provided evidence that each haptic assembly was perceived as two distinct objects. These findings accord with the cohesion principle and with the findings of studies of visual object perception (Hofsten and Spelke, 1985; Spelke et al., 1989).

Experiments on haptic object perception therefore provide evidence that infants perceive haptically presented objects in accord with the cohesion and contact principles and not in accord with the Gestalt principles of similarity and figural goodness. Object perception appears to accord with the same principles in the haptic and the visual modes. It may depend on a single, amodal mechanism.

1.4 Principles of object perception In summary, infants appear to organize the perceived layout into bodies that are cohesive, that are bounded, that move independently of bodies from which they are spatially separated, and that move together with bodies with which they are in contact. Infants' perception of objects can be encompassed by two principles: the principle of cohesion (surfaces lie on one object if and only if they are connected) and the principle of contact (surfaces move together if and only if they are in contact).[5]

The above studies suggest that object perception accords with physical constraints on objects' behaviour. The research described in the next section investigated whether sensitivity to such constraints also allows infants to reason about objects. We ask whether infants can use knowledge of physical constraints in objects' behaviour to make sense of the motions of objects that are visible, and to make inferences about the motions of objects that are hidden.

2 Physical reasoning in infancy

2.1 Infants' representations of hidden objects Perceptual encounters with objects are sporadic: objects enter and leave the field of view whenever the perceiver, or other objects, move. Adults nevertheless represent each object as existing and moving continuously over space and time. These representations partly reflect our sensitivity to a basic physical constraint on object motion. According to the *continuity constraint*, objects move only on connected paths from one place and time to another. We now ask whether infants represent hidden objects in accord with this constraint.

First, consider whether infants represent the continued existence of an object that leaves their view, in accord with one aspect of the continuity constraint: objects exist continuously. Studies using three different methods, and quite different situations, provide evidence for this ability. In one series of studies, Clifton, Rochat, Litovsky and Perris (1991) presented six-month-old infants with two visible objects that differed in size and made distinctive sounds. Infants were allowed to reach for the objects. As is often observed (e.g. Bruner and Koslowski, 1972), infants reached in different ways for the large and small objects. After this familiarization, the room lights were extinguished, such that no object could be seen, and the sounds that had accompanied the large and small objects were presented in alternation. Presented with each sound, subjects tended to engage in the reaching movements elicited by an object of the appropriate size. Analyses of the reaching patterns of individual infants suggested that these reactions were not attributable to response learning during the period when the object was visible. Rather, infants appeared to represent each object's position and size, and these representations guided infants' reaching.

Further evidence for object representations comes from research using preferential looking methods and objects that move fully from view behind an occluder. Although we have already described one such study (Craton and Yonas, 1990), the most extensive studies come from the laboratory of Baillargeon (1986; 1987a; 1987b; Baillargeon,

Spelke and Wasserman, 1985). In Baillargeon's first experiments, infants were familiarized with a screen that rotated 180° on a table, in the manner of a drawbridge. After habituation to this event, a stationary object was placed behind the screen. The position and dimensions of the object were such that the object was fully visible when the screen lay flat on the table, was progressively occluded as the screen rose and was fully hidden by the time the screen had rotated 60°. The screen's rotation then continued. On alternating trials, the screen stopped when it reached the place occupied by the object (a novel but possible motion) or it rotated 180° through the place occupied by the object, revealing nothing in its path (a familiar but impossible motion). Infants as young as four and a half months looked longer at the latter motion (Baillargeon, 1987a). This finding provides evidence that infants represented the existence and location of the hidden object.

Given only the above findings, one could question the richness, robustness and accessibility of young infants' representations (Fischer and Biddell, 1991). Further studies by Baillargeon provide evidence that five- to seven-months-old infants' representations of objects are quite rich, incorporating information not only about the existence of a hidden object, but also about its location, size and rigidity (Baillargeon, 1987b; Baillargeon and Graber, 1987). In one series of studies, infants were presented with objects of different heights behind a rotating screen. Preferential looking to screen rotations of different extents depended reliably on the height of the hidden object: the taller the object, the less extensive the screen rotation that elicited a novelty reaction (Baillargeon, 1987b). This finding, like the findings of Clifton et al. (1991), of Craton and Yonas (1990) and of a further experiment by Baillargeon (Baillargeon and Graber, 1987), provides evidence that infants represented the size and shape of the hidden object and inferred that these properties would not change while it was hidden. In other studies, infants were presented with objects differing in rigidity. An extensive screen rotation evoked novelty reactions only for the rigid object (Baillargeon, 1987b). This finding suggests that infants can represent the permissible transformations of a hidden object: in this case, change in shape (see also Baillargeon, Graber, DeVos and Black, 1990).

If an object moves fully out of view in one location and returns to view in another location, it must have traced a connected path from the first location to the second. An experiment by Spelke and Kestenbaum (1986) suggests that four-month-old infants, like adults, represent the identity or distinctness of successively hidden objects in accord with this aspect of the continuity constraint.

Infants were presented with a display in which two objects passed successively behind two spatially separated screens (Figure 5.6): an object that stood to the left of the left screen moved behind it, and after a pause, an object emerged to the right of the right screen. At the right end of the display, the second object changed direction and the same motions were presented in reverse. This event occurred repeatedly, as long as the infant looked at it. Although only one object is visible at a time in this event, adults describe the event as containing two objects moving in succession, one on each side of the gap between the screens. This perception follows from the continuity constraint: objects move only on connected paths.

To investigate infants' perception of this event, a group of infants was habituated to it. Other infants were habituated to an event in which a single object moved continuously behind the two screens, to an event in which one or two objects moved behind a single wide screen, or to no event (baseline condition). Then all the infants were presented, on alternating trials, with displays of either one or two fully visible objects. Infants

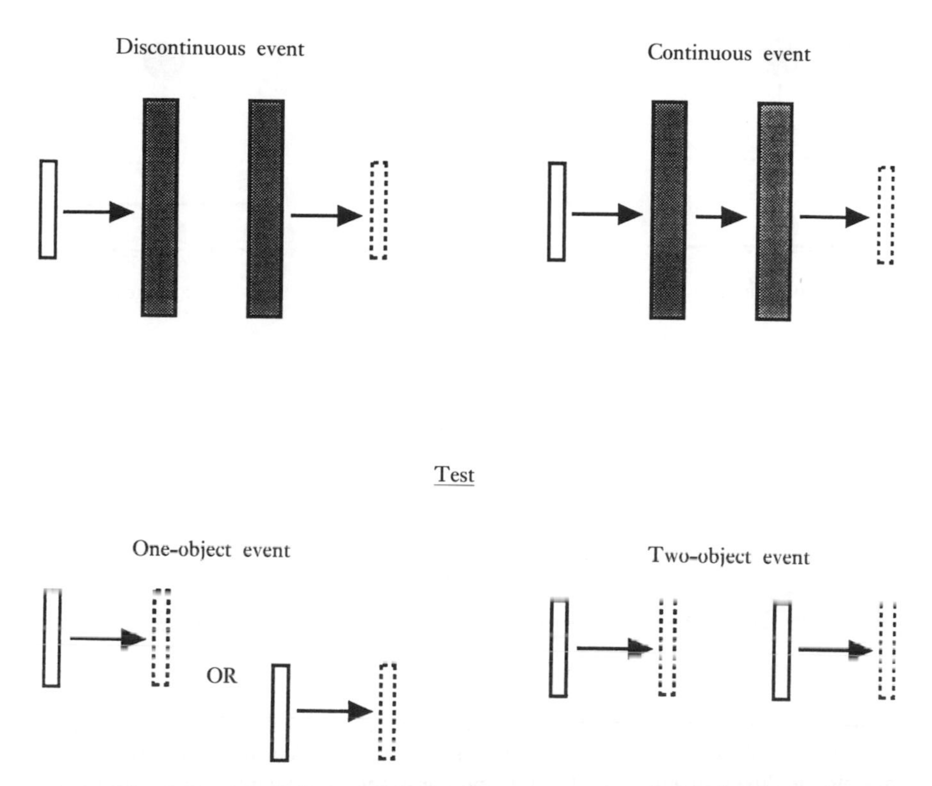

Figure 5.6 Schematic depiction of displays from an experiment on infants' apprehension of object identity. Arrows indicate the path of the object's motion from its initial position (solid lines) to its final position (broken lines). (After Spelke and Kestenbaum, 1986.)

habituated to the discontinuous motion showed a preference for the one-object event. This preference differed reliably from the preferences of infants habituated to continuous motion and from baseline. The experiment provides evidence that infants apprehended two continuously moving objects in the discontinuous event, in accord with the constraint that objects move only on connected paths. Research by Baillargeon and Graber (1987) supports the same conclusion.

2.2 Infants' inferences about hidden object motion The above experiments suggest that infants represent hidden objects and their motions, and that their representations accord with the continuity constraint. The next experiments extend these findings in two directions (Spelke et al., 1992; 1993b). First, they investigate whether infants are able to infer how a hidden, moving object continues to move and where it comes to rest. Second, they investigate whether infants' representations of the motions of hidden objects accord with a second physical constraint on object motion. The *solidity*

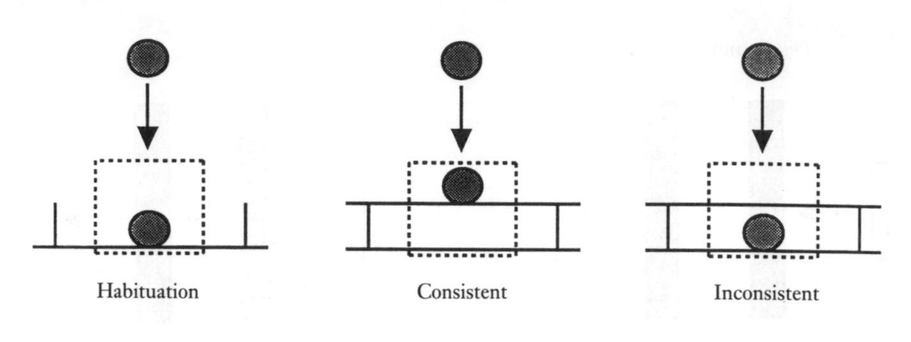

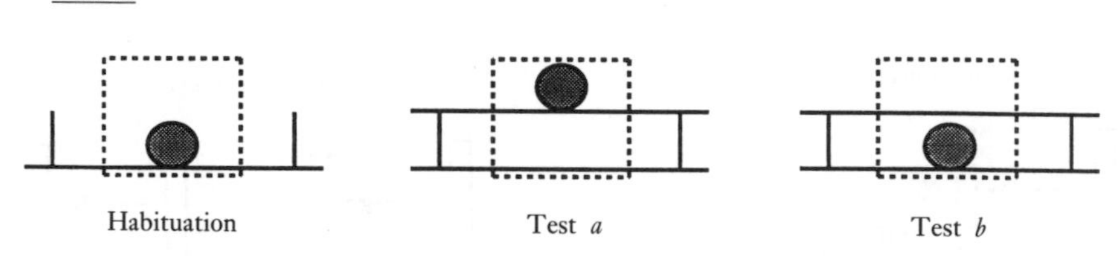

Figure 5.7 Schematic depiction of displays from an experiment on infants' inferences about hidden object motion. Broken lines indicate the position of the screen. Arrows indicate the direction and path of visible motion in the experimental condition. In the control condition, the ball was moved forward in depth to its final position. (After Spelke et al., 1992.)

constraint dictates that objects move only on non-intersecting paths, such that no parts of distinct objects ever coincide in space and time. If infants represent hidden object motion in accord with the continuity and solidity constraints, then they should infer that a hidden moving object will neither 'jump over' nor 'pass through' any obstacle in its path.[6]

These experiments used a modified preferential looking method so as to present young infants with a task devised by Piaget: an 'invisible displacement task'. In the first experiment (Spelke et al., 1992), four-month-old infants were first presented with an open stage with a brightly coloured floor (Figure 5.7). A screen was lowered to cover the lower half of the stage, including the floor, a ball was introduced above the screen and dropped behind it, and then the screen was raised to reveal the ball on the floor: an expected outcome for adults. Looking time to this event outcome was recorded, beginning with the raising of the screen. The event was repeated on a series of trials, until looking time declined by 50 per cent.

After this habituation sequence, a second, brightly coloured surface was introduced into the display above the floor, the screen was lowered to cover both surfaces and the ball was dropped as before. On alternating test trials, the screen was raised to reveal the ball at rest either in a new position on the upper surface or in its familiar position on the floor. Whereas the new position was consistent with all constraints on object motion, the familiar position was inconsistent with the continuity and solidity constraints. Given that the upper surface neither moved nor ruptured, the ball could only have reached the lower surface by 'jumping over' or 'passing through' the upper surface, in violation of the continuity or solidity constraints.

Looking times to the two test outcomes were recorded, beginning with the raising of the screen, and were compared to the looking times of infants in a control condition. In the control condition, infants were presented with the same outcome displays, preceded by events in which the ball was moved forward in depth to its final position, and then the screen was lowered and raised. As in the studies of Spelke et al. (1993a), therefore, infants in the two conditions viewed exactly the same displays throughout the time that looking was recorded.

If infants inferred that objects would move in accord with the continuity and solidity constraints, then the infants in the experimental condition were expected to look longer at the inconsistent event, relative to those in the control condition. The findings confirmed this prediction. The experiment thus provided evidence that four-month-old infants inferred that the hidden object would move on a connected path that intersected no other object, in accord with solidity and continuity constraints.

This finding complements the findings of research by Baillargeon (1986), using a different preferential looking method. It was replicated and extended by Spelke et al. (1992; 1993b) in four further studies. In these studies, reasoning in accord with the continuity and solidity constraints was tested with events in which an object fell through the air toward a surface with a gap, rolled on a horizontal surface in the frontal plane or rolled on a horizontal surface in depth. Experiments were conducted at four ages: two and a half months, four months, six months and ten months. At all these ages and for all the events, infants looked longer at event outcomes that were inconsistent with the continuity and solidity constraints.

Continuity and solidity are symmetrical constraints on object motion. The continuity constraint dictates that a single object travels on a connected path over space and time: its path can contain no gaps (Figure 5.8b). The solidity constraint dictates that distinct objects travel on separate paths over space and time: paths cannot intersect such that the objects coincide in space at any moment in time (Figure 5.8c). These two constraints, therefore, can be encompassed by a single *principle of continuity*: an object traces exactly one connected path over space and time. The above research provides evidence that this principle guides young infants' reasoning about hidden objects.

The continuity principle does not permit fully specific predictions about the motion or the final position of a hidden object. For example, an object that falls from view toward an obstacle might cease moving before reaching the obstacle. If the object does reach the obstacle, it might come to rest, rebound, displace the obstacle, break the obstacle or itself be broken. All these event outcomes are consistent with the continuity and solidity constraints:[7] adults distinguish between these outcomes by applying knowledge of other constraints on object motion and knowledge of the behaviour of objects of particular kinds. The next experiments have begun to investigate whether infants distinguish among some of these outcomes, and infer that objects will move in accord with two further physical constraints on object motion: gravity and inertia.

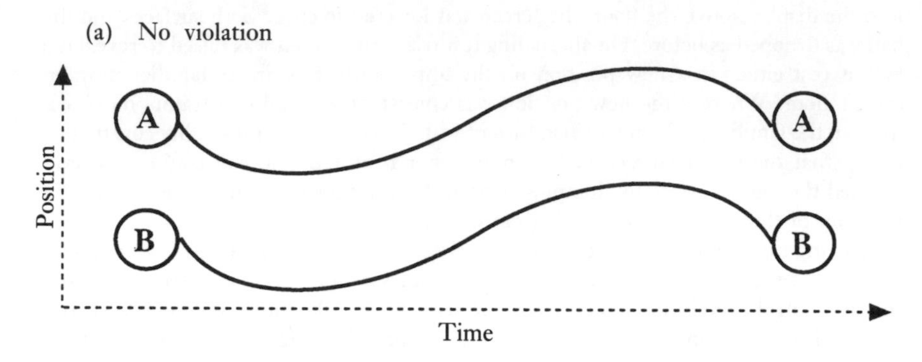

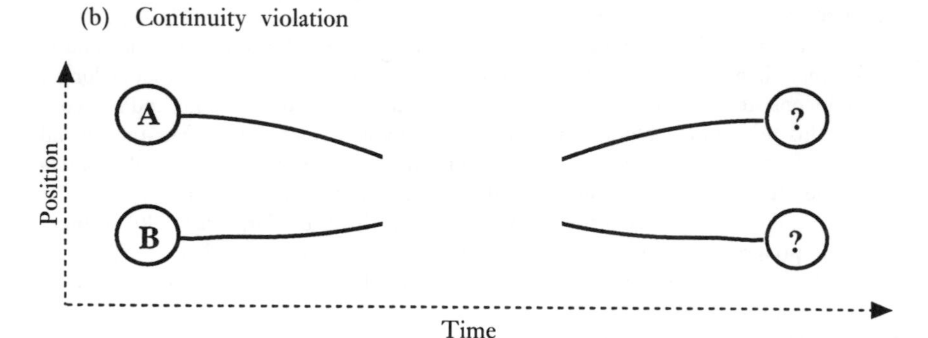

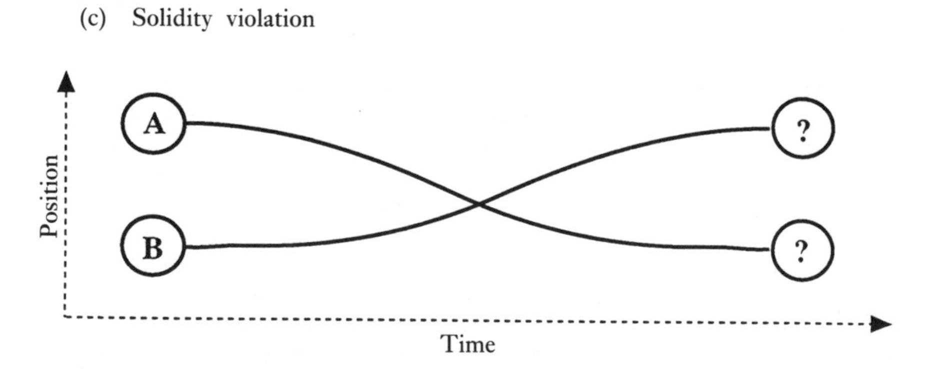

Figure 5.8 Schematic depiction of events that accord with, or violate, the continuity and solidity constraints. Solid lines indicate each object's path of motion, expressed as changes in its position over time. Each object traces (a) exactly one connected path over space and time, (b) no connected path over space and time or (c) two connected paths over space and time.

Experimental

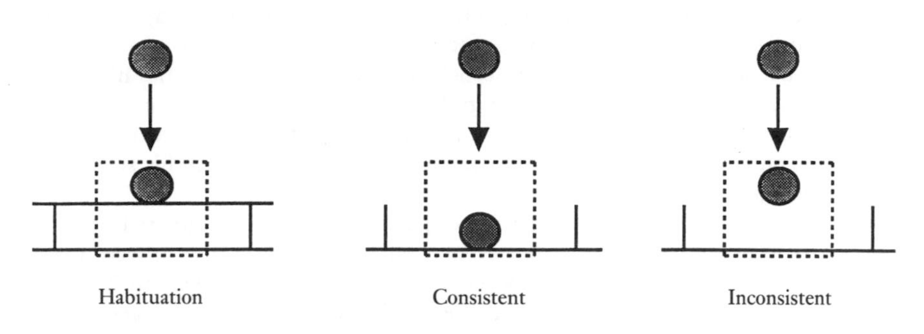

Control

Habituation Test *a* Test *b*

Figure 5.9 Schematic depiction of displays from an experiment on infants' inferences about hidden object motion. Arrows indicate the direction and path of visible motion in the experimental and the control condition. In the control condition, the hand held the ball throughout the event. (After Spelke et al., 1992.)

A number of experiments have tested whether infants infer that an object that moves from view will move downward to a supporting surface on a smooth path, in accord with the constraints of gravity and inertia (Spelke et al., 1992; 1993b; 1993c). One experiment by Spelke et al. (1992) serves as an example. This study used the same method and nearly the same events as the above studies of continuity and solidity (Figure 5.9). Four-month-old infants were habituated to a ball that fell behind a screen covering two surfaces and was revealed at rest on the first surface in its path. Then the upper surface was removed, the ball was dropped behind the screen as before, and it was revealed either in a new position on the lower surface or in its former position, now in midair. The latter position was inconsistent with gravity and with inertia: the ball appeared to stop falling in the absence of support and to change its motion in the absence of obstacles.

Looking time to these two outcomes was compared to the looking time of infants in a control condition, presented with a hand-held ball that was lowered to its resting position before the lowering and raising of the screen. Except for the hand, the infants in the control condition viewed the same outcome displays as those in the experimental condition, but with outcomes that were consistent with gravity and inertia. Infants in the experimental condition looked non-significantly longer at the *consistent* outcome. Their preferences differed reliably from the preferences of infants in the corresponding studies of continuity and solidity. These findings provide no evidence for sensitivity either to gravity or to inertia.

In further experiments using the invisible displacement method, sensitivity to gravity and inertia has been tested with other events and at other ages, ranging from three to twelve months. Infants under six months of age have shown no sign of sensitivity to either constraint in any situation tested. Sensitivity to both constraints appears to emerge at older ages, although it appears to emerge in a piecemeal fashion, with considerable variability across infants (Spelke et al., 1992; 1993b; 1993c). Comparisons across experiments have revealed that infants respond more strongly and consistently to the violations of the continuity and solidity constraints than to violations of the gravity and inertia constraints (Spelke et al., 1992; 1993b). These studies suggest that infants do not make fully specific inferences about the motion of hidden objects. They suggest, moreover, that sensitivity to continuity and solidity is more deeply rooted in human development than sensitivity to gravity and inertia. Sensitivity to gravity may not, however, be fully absent in early infancy (see Baillargeon, 1990; Baillargeon and Hanko-Summers, 1990; Needham and Baillargeon, 1991).

The next experiments investigate whether infants represent the motions of hidden objects in accord with one more physical constraint on objects' behaviour: no action at a distance. Experiments by Ball (1973), Borton (1979) and Baillargeon, DeVos and Black (in Baillargeon, 1992) provide evidence that they do. Infants reason about hidden interactions between objects in accord with the constraint that surfaces move together only on contact.

Ball's (1973) experiment may have been the first to use a preferential looking method to investigate young infants' inferences about hidden object motion. Infants ranging in age from nine to 122 weeks were familiarized with an event in which one object moved behind an occluding screen and, after an appropriate temporal delay, a second object emerged from the other side of the screen (Figure 5.10). Separate groups of infants then were presented with a fully visible display in which one of two events occurred. In one condition, the first object came to a halt on contact with the second object, and the second object immediately began to move in the same direction. In the second condition, the first object came to a halt before it reached the second object, and after a delay, the second object began to move. Looking times to the two event outcomes were recorded and compared to the looking times of infants in a no-habituation baseline condition.

Ball's analysis treated all the infants in each condition as a single group. He found that the infants in the experimental condition looked significantly longer at the display in which the objects were separated by a spatial and temporal gap, relative to baseline. Because Ball reported the data for each subject in an appendix to his paper, we were able to re-analyse his data separately for the youngest infants. The looking preference for the display in which the object motions were spatially and temporally separated differed significantly between the two conditions, among the subset of infants tested

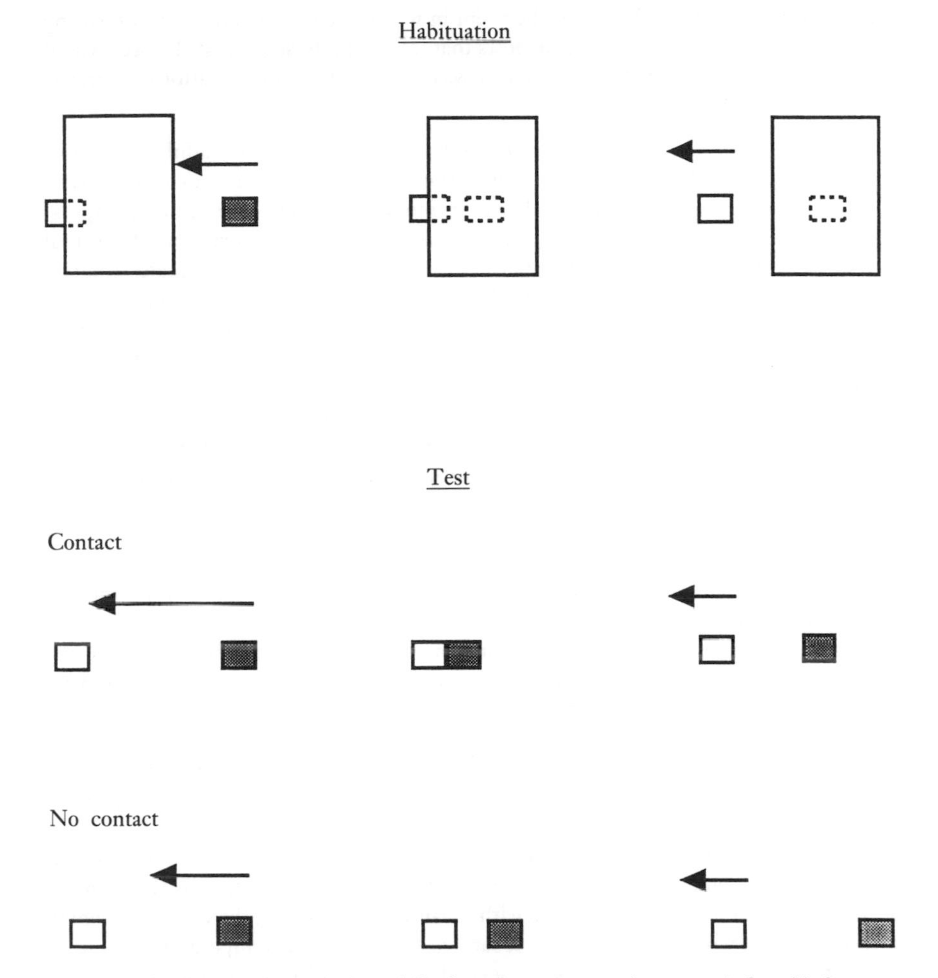

Figure 5.10 Schematic depiction of displays from an experiment on infants' inferences about hidden object motion. Each arrow indicates the direction and extent of motion of the object below it. (After Ball, 1973.)

at less than 28 weeks of age (Wilcoxon-Mann-Whitney $z = 2.48$, $p < .01$). Ball's experiment therefore provides evidence that by six and a half months of age, infants infer contact between the surfaces of the hidden objects, in accord with the constraint of no action at a distance.[8]

Two further experiments provide additional evidence for this ability (Baillargeon et al., 1991; in Baillargeon, 1992; Borton, 1976). In Baillargeon et al.'s study, six-month-old infants were shown a long narrow platform with a toy bear that either sat on the left end of the platform or sat next to the platform without touching it. The display was then hidden from view by a screen, a hand reached behind the screen and pulled the platform to the right and the bear moved rightward into view through an aperture in the screen. Infants looked longer at the event in which the bear had been shown sitting next to the platform. This finding suggests that infants understood that the bear would move together with the platform only if it was in contact with the platform's surface.

2.3 Infants' apprehension of causal relations between objects When adults view two objects that collide and immediately change their motion, we tend to experience each object as causing the other object's change in motion (Michotte, 1963). This impression accords with the constraint of action on contact. Adults' impression of a causal relation is removed, moreover, if a spatial gap or a temporal delay separates the two objects' motion (Michotte, 1954). The absence of an impression of causality between spatially or temporally separated objects accords with the constraint of no action at a distance. As noted above, these geometrical constraints can be captured by a single principle of contact.

We conclude this section by considering whether infants infer causal relations in accordance with this principle. Using preferential looking methods, Alan Leslie (1988; Leslie and Keeble, 1987) presented infants with animated film sequences in which two objects moved in succession (Figure 5.11). In one condition, the first object came into contact with the second object, and the second object immediately began to move in the same direction. The objects therefore moved together at the moment of contact. In other conditions, the motions of the two objects were separated in space or in time. To assess whether infants apprehended a causal relation in any of these events, Leslie familiarized separate groups of infants with one of the events and then presented the same event in reverse. He reasoned that in conditions where the familiarization event was perceived as causal, a reversal of that event would reverse not only the direction of the objects' motion, but also the causal relation between the objects. The infants who perceived a causal relation in the familiarization event therefore should recover their looking to the reversal of that event more than the infants who saw an event evoking no impression of a causal relation.

Leslie's experiments provided evidence that infants perceive causal relations as adults do. The infants presented with the objects that moved together at the moment of contact reacted to the event reversal with reliably longer looking than those presented with the objects whose motions were separated by a spatial or temporal gap.

In one respect, Leslie's findings go beyond the scope of this chapter: we offer no account of the mechanisms underlying impressions of causality in infants or adults (see Leslie, 1988; Michotte, 1954).[9] We believe it is significant, nevertheless, that infants' causal impressions accord with the contact principle. This finding converges with the findings from studies of infants' inferences about hidden object motion (Baillargeon et al., 1991; Ball, 1973; Borton, 1976). Infants may make sense both of visible and of hidden object motions in accord with a small set of general constraints on the behaviour of material bodies.

2.4 Principles guiding physical reasoning In summary, young infants appear to reason about the existence and the motions of both visible and hidden objects. Under certain conditions, infants represent the persistence of an object that is fully hidden, they trace object identity over successive encounters, they infer how a hidden, moving object continues to move and where it comes to rest, and they apprehend the causes of an object's motion. Early reasoning about object motion appears to accord with four

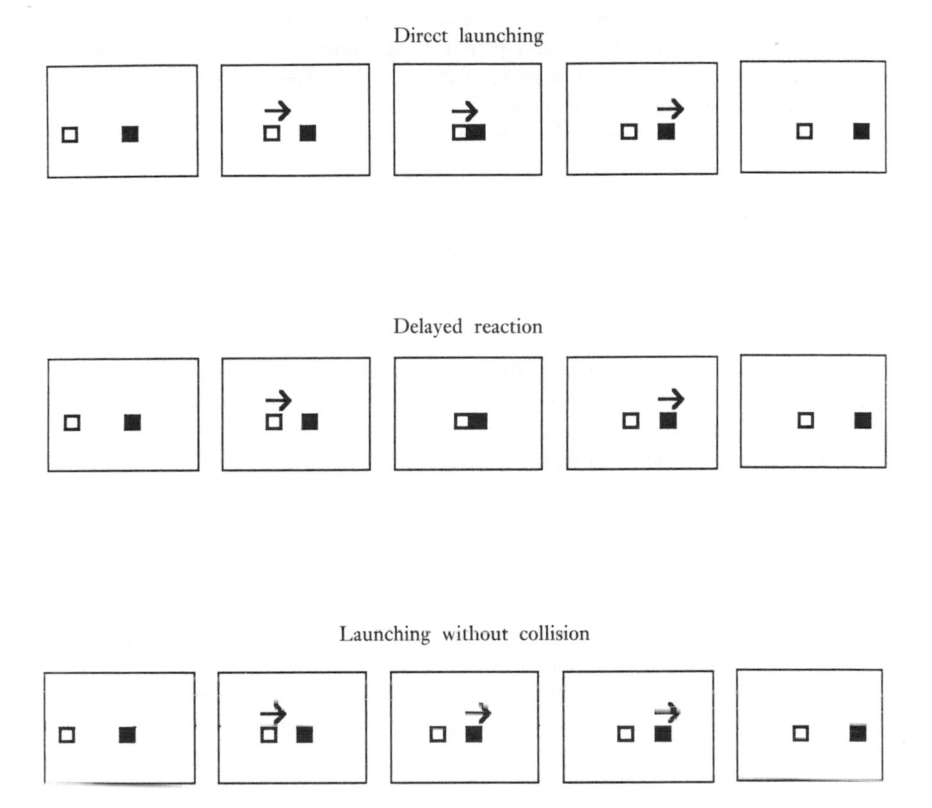

Figure 5.11 Schematic depiction of displays from an experiment on infants' impressions of causal relations between object motions. Arrows indicate the presence and direction of motion of the object(s) below. (After Leslie and Keeble, 1987.)

constraints: continuity, solidity, action on contact and no action at a distance. These constraints, in turn, can be captured by the principle of continuity (an object moves on a unique, connected path) and the principle of contact (objects move together if and only if they are in contact). How do these principles relate to the principles guiding object perception?

3 Object perception and physical reasoning in infancy

Infants' abilities to reason about objects resemble their abilities to perceive objects. First, object perception and physical reasoning both depend on an analysis of how surfaces are arranged in space, and how surface arrangements change over time. Second, the relevant spatio-temporal arrangements apply in both cases to surfaces in the three-dimensional layout, not to images in the two-dimensional visual field. Third, processes of object perception appear to depend on amodal mechanisms. It seems very likely that the same is true for the process of physical reasoning, although to our knowledge, this issue has not been investigated directly. We suggest now that there is a stronger relation between object perception and physical reasoning: the principles that encompass object perception and physical reasoning are closely connected, and possibly identical. This relation suggests that a single system of knowledge guides perceiving and reasoning about objects.

Consider first the contact principle. One could propose distinct and relatively restricted principles to encompass infants' perception of the simultaneously moving ends of one centre-occluded object, infants' inferences about the hidden motions of two successively moving objects, infants' impression of a causal relation between two objects that make contact, and infants' impression of no causal relation between two objects that do not make contact. The contact principle could, however, figure in all these abilities. When two surfaces move together, either throughout an event or at one point in an event, this principle dictates that the surfaces are in contact. When two surfaces move independently, this principle dictates that the surfaces are not in contact. Conversely, when two surfaces come into contact, this principle dictates that their motions are not independent. When two surfaces are separated by a spatial or temporal gap, finally, this principle dictates that their motions are independent; neither motion therefore could cause the other.

We now turn to the relation between the cohesion principle, specifying the conditions for object unity over space, and the continuity principle, specifying the conditions for object persistence over time. These principles are not identical: non-solid substances such as sand and water behave in accord with the continuity principle (a pile of sand moves on a connected, unobstructed path) but not in accord with the cohesion principle (a sand pile does not retain either its connectedness or its boundaries as it moves in relation to other objects and substances). Nevertheless, the principles are related. The necessary and sufficient condition for object unity, over space and time, is that the (spatial) parts and (temporal) appearances of an object be connected. Moreover, cohesion implies continuity: only continuously movable bodies could behave in accord with the cohesion principle.

The relation between the cohesion principle and the continuity principle can be discerned if one focuses on the separate constraints of boundedness and solidity, on one hand, and cohesion and continuity, on the other. According to the boundedness constraint, no parts of two distinct objects are connected. That is, every path from a part of one object to a part of the other object contains one or more boundary points, where at least one of the objects ends. This condition cannot be satisfied if points on distinct objects occupy the same location: no boundary separates a point from itself. It follows that no parts of distinct objects ever coincide in space and time, and therefore that the paths of two distinct, persisting objects can never intersect. Boundedness can only be satisfied by bodies whose behaviour accords with the solidity constraint.

Consider next the constraints of cohesion and continuity. According to the cohesion constraint, the parts of a single object are linked by enduring connections: connections that persist as the object moves. This condition cannot be met if there are gaps in the object's path through space or time. Cohesion can only be satisfied by bodies whose behaviour accords with the continuity constraint.

These considerations suggest that there is a close relationship between object perception and physical reasoning. It is possible, indeed, that the same principles govern perception and reasoning. Infants may infer that a hidden object moves not only continuously but also cohesively; infants may infer that two hidden, moving objects not only do not intersect but also do not merge. Ongoing research supports both these possibilities (Carey, personal communication). During infancy, object perception and physical reasoning may depend on a single system of knowledge.[10]

If object perception and physical reasoning are based on a single system of knowledge, then it is fair to ask whether this knowledge is embedded in 'perceptual' or 'reasoning systems'. We question both characterizations.

One of us has suggested that apprehending objects does not depend on 'perceptual processes' akin to those underlying the apprehension of colour, distance or motion (Spelke, 1988). This suggestion was based on a consideration of three aspects of object perception: the nature of its input, its amodal character and its accordance with physical constraints on objects' behaviour. The suggestion contrasted with the common view that object representations are produced by input systems operating independently of reasoning (e.g. Fodor, 1983; see also Gibson, 1966; 1979; Koffka, 1935; Michotte, Thinès and Crabbé, 1964) and echoed Marr's (1982) proposal that in moving from the representation of surfaces to the representation of objects, one passes the limits of 'pure perception' (see footnote 1). We believe the findings of the present studies cast strong doubt on the view that object perception depends on an input system operating independently of physical reasoning. Indeed, perceiving and reasoning about objects appear to be inseparable processes. In this respect, the present findings support Marr's view.[11]

Nevertheless, representing hidden objects and inferring their motions do not appear to depend on 'reasoning processes' akin to the processes involved in calculating one's income tax or solving crossword puzzles. For adults, inferences in accord with the constraints of continuity and contact appear to be immediate, effortless and unconscious. In some cases, these inferences are unaffected by specific knowledge of the objects being viewed: they occur even when such knowledge indicates that they are inappropriate (Michotte, 1963). Prototypical reasoning processes, in contrast, tend to be slow, effortful, deliberate and responsive to knowledge from any source (Fodor, 1983). These considerations led Michotte to propose that humans *perceive* the motions of hidden objects and the causal relations among visible objects. The present findings support aspects of Michotte's view (see also Leslie, 1988). Contrary to Michotte, however, the findings suggest that object perception and physical reasoning are based on a system of knowledge of physical objects, not on general perceptual principles such as those proposed by the Gestalt psychologists.

The processes that underlie perceiving and reasoning about objects therefore appear to lie at the border of what is traditionally considered 'perception' and what is traditionally considered 'thought'. An appreciation of the unique position of these processes may shed light both on the nature and on the development of physical knowledge.

4 *Physical conceptions in infancy and adulthood*

The above experiments suggest answers to the four questions that began this chapter, as those questions apply to infants. There appears to be a general process of object perception in infancy, applicable to all material bodies. Infants perceive the unity and boundaries of objects by analyzing the spatial arrangements and motions of surfaces, in accord with the principles of cohesion and contact. There also appears to be a general process of physical reasoning in infancy. Infants reason about the behaviour of material bodies in accordance with the principles of contact and continuity. The close relation between the principles guiding object perception and those guiding physical reasoning suggests a connection between these psychological processes. In infancy, perceiving and reasoning about objects may be guided by one system of knowledge: a single conception of material bodies comprising, at least in part, the principles of cohesion, contact and continuity.

What might these findings with infants suggest about object perception, physical reasoning and physical knowledge in adults? It is conceivable that studies of infants reveal nothing about mature perception, reasoning or knowledge. Studies of infants could be uninformative for either of two reasons. First, mature processes of object perception and physical reasoning might depend on mechanisms that emerge after infancy. Second, mature processes of perception and reasoning may result from developmental processes that bring radical change to physical conceptions (Carey, 1991; Gopnik, 1988; Kuhn, 1977; Piaget, 1954).

Contrary to these possibilities, we suggest that the principles guiding perception and reasoning in infancy are central to the physical conceptions of adults. This suggestion arises from a consideration of how one learns spontaneously about the physical world. It is supported, we believe, by a variety of observations, both from psychology and from philosophy, concerning adults' physical conceptions.

Insofar as humans develop knowledge by learning, and learn about objects by observing their behaviour, we suggest that the system of physical knowledge found in infants will tend to be enriched, not overturned, by the further development of knowledge (Spelke, 1991). This suggestion hinges on the claim that a single system of knowledge underlies infants' physical reasoning and object perception. If the principles of cohesion and contact underlie object perception, then infants will single out objects whose behaviour accords with those principles. As children learn about the bodies they single out, their learning may well enrich their initial conceptions. Children may learn, for example, that material bodies tend to have simple shapes, to move smoothly and to fall when unsupported. Children will not learn, however, that the behaviour of material bodies violates the cohesion and contact principles, even if the world that surrounds them contains myriad violations of the principles. Such learning will not occur, because entities that violate the cohesion and contact principles will not be singled out as objects.[12] Spontaneous learning will tend to deepen, not dislodge, the initial conception of material bodies.

A number of observations in philosophy and in psychology appear to support the view that cohesion, contact and continuity are central to mature conceptions of material bodies. Consider mature, common-sense intuitions about object identity over change. Intuitions about the persistence of a material body are clearest when the body's behaviour accords with all of these constraints; intuitions about the non-persistence of a body are clearest when the body's behaviour violates all the constraints. In contrast, intuitions about physical identity appear to become least certain, and decisions most problematic, when the behaviour of a body accords with some constraints and violates others: for example, when two particles fuse or one object is disassembled and reassembled (see Hirsch, 1982).

Consider next the sortal concepts that children acquire: concepts such as 'chair' or 'tiger'. These concepts tend strongly, although not universally, to apply to bodies whose behaviour accords with the principles guiding object perception and physical reasoning. In English, there are sortal terms such as 'lock', 'cup', 'bolt' and 'bee', but not one term for 'lock-and-key', 'cup and saucer', 'nut-and-bolt' or 'bee colony', despite the functional coherence of the latter notions. The accordance of sortal concepts with the principles guiding object perception and physical reasoning may not be accidental. Initial conceptions of material bodies may underlie the acquisition of sortal concepts and sortal terms (see Markman, 1989; Mervis, 1987; Shipley and Shepperson, 1990; Soja, Carey and Spelke, 1991).

In contrast, the constraints on material bodies that infants fail to recognize may be weaker and less central to the physical conceptions of adults. To be sure, adults recognize that objects tend to have simple and regular shapes, that objects tend to fall when unsupported and that objects tend to move smoothly. Nevertheless, adults quite readily perceive, reason about and learn sortal terms for objects that are irregular in shape (e.g. rocks), that can rest without apparent support (e.g. kites) and that can change direction abruptly and spontaneously (e.g. animals). Moreover, adults' sensitivity to physical constraints such as gravity and inertia is incomplete, at best (McCloskey, 1983; Proffitt and Gilden, 1989).

These observations suggest that the earliest developing conceptions of physical objects are the most central conceptions guiding mature object perception and physical reasoning. For adults, such conceptions are overlaid by a wealth of knowledge about the appearances and the behaviour of particular kinds of objects. Even this more specific and limited knowledge, however, reflects the core knowledge from which it grew. Both for adults and for children, cohesion, contact and continuity may be central to our understanding of the material world. Through studies of early development, psychologists may investigate the nature of this core knowledge.

ACKNOWLEDGEMENTS

We thank the editors of this volume for their insightful comments on an earlier version of this manuscript. Supported by a grant, to ESS, from the National Institutes of Health (HD23103) and by a graduate fellowship, to GAV, from the National Science Foundation.

NOTES

1 More specifically, Marr suggests that perceptual mechanisms recover from retinal images a description of the visible surface layout: the '2.5D sketch'. This description does not indicate where an object ends and the next begins. According to Marr, it is likely to mark 'the limits of what one might call pure perception – the recovery of surface information by purely data-driven processes without the need for particular hypotheses about the nature, use or function of the objects being viewed' (Marr, 1982, p. 269).
2 We intend our account of object perception and physical reasoning to apply to all material bodies, including machines, animals and persons. Unfortunately, few studies exist concerning infants' perceiving or reasoning about objects in the latter categories. The chapter therefore focuses on objects such as balls and blocks.
3 We do not know what infants perceive when visible objects undergo non-rigid motions that preserve their connectivity, such as uniform bending or rotation around a joint. Current research is investigating some of these situations.
4 The rings were connected by a long and highly flexible elastic band that in all likelihood was not detectable by the infants.
5 Although the present account of object perception is substantially in agreement with the account in Spelke (1990), the bi-directional cohesion principle now replaces Spelke's uni-directional principles of cohesion and boundedness, and the bi-directional contact principle

replaces and revises Spelke's (1990) uni-directional principles of no action at a distance and rigidity.

6 The terms in quotations should not be interpreted as in ordinary language. When an object hits an obstacle, it may rupture the obstacle and thus move beyond it. Although this event may be described as one object passing through another, it does not violate the solidity constraint, because the two objects never coincide in space at any time. Similarly, an object that hits an obstacle may alter its motion and bounce over it. Although this event may be described as one object jumping over another, it does not violate the continuity constraint, because each object exists continuously and moves on a connected path. It is significant, we believe, that the English language has no term for events that violate the continuity and solidity constraints, even though such events are ubiquitous (for example, in the motions of shadows). People do not ordinarily communicate about such events, we suggest, because they do not really apprehend or understand them.

7 Infants may rule out the last two outcomes if their inferences about hidden object motion are guided by the cohesion constraint, because events in which the object or obstacle breaks apart violate that constraint. Two recently completed experiments provide evidence in favour of this possibility (S. Carey, personal communication).

8 Because the surfaces moved together only at the moment of impact and were otherwise both spatially separated and independently moving, the cohesion principle dictates that the surfaces comprised two separate objects.

9 The contact principle formulated here – *objects move together if and only if they are in contact* – must be combined with other premises in order to yield the inference that one object has *caused* another object's motion.

10 In our usage, a 'system of knowledge' is a set of principles that characterize the entities in a domain and underlies inferences about those entities (see Chomsky, 1980). We do not imply that this knowledge is explicit or conscious in infants or adults.

11 Contrary to Marr, however, studies of infants suggest that both object perception and physical reasoning depend in part on sensitivity to *general* constraints on the behaviour of material bodies. They do not depend exclusively on 'particular hypotheses' about objects of particular kinds.

12 Shadows provide an interesting test case for this thesis. Both infants and young children detect shadows, and they may be predisposed to consider shadows as objects, subject to the cohesion and contact principles (Piaget, 1960; deVries, 1987; Rubenstein, 1991). With development, children discover that shadows do not behave in accordance with these constraints. This discovery does not, however, lead to a change in conceptions of material bodies. Rather, children come to view shadows as non-material (see Carey, 1991)

REFERENCES

Baillargeon, R. 1986: Representing the existence and the location of hidden objects: Object permanence in 6- and 8-month-old infants. *Cognition*, 23, 21–41.

Baillargeon, R. 1987a: Object permanence in 3.5- and 4.5-month-old infants. *Developmental Psychology*, 23, 655–64.

Baillargeon, R. 1987b: Young infants' reasoning about the physical and spatial material characteristics of a hidden object. *Cognitive Development*, 2, 179–200.

Baillargeon, R. 1990: The development of young infants' intuitions about support. Paper presented at the 7th International Conference on Infant Studies, Montreal, Canada.

Baillargeon, R. 1992: The object concept revisited: New directions in the investigation of infants' physical knowledge. In C. E. Granrud (ed.), *Visual Perception and Cognition in Infancy*. Carnegie-Mellon Symposia on Cognition, Volume 23. Hillsdale, NJ: Erlbaum.

Baillargeon, R. and Graber, M. 1987: Where's the rabbit? 5.5-month-old infants' representation of the height of a hidden object. *Cognitive Development*, 2, 375–92.

Baillargeon, R., Graber, M., DeVos, J. and Black, J. C. 1990: Why do young infants fail to search for hidden objects? *Cognition*, 24, 255–84.

Baillargeon, R. and Hanko-Summers, S. 1990: Is the top object adequately supported by the bottom object? Young infants' understanding of support relations. *Cognitive Development*, 5, 29–54.

Baillargeon, R., Spelke, E. S. and Wasserman, S. 1985: Object permanence in five-month-old infants. *Cognition*, 20, 191–208.

Ball, W. A. 1973: The perception of causality in the infant. Paper presented at the Society for Research in Child Development, Philadelphia, PA.

Bornstein, M. H. 1985: Habituation of attention as a measure of visual information processing in human infants: Summary, systematization, and synthesis. In G. Gottlieb and N. A. Krasnegor (eds), *Measurement of Audition and Vision in the First Year of Postnatal Life*. Norwood, NJ: Ablex.

Borton, R. W. 1979: The perception of causality in infants. Paper presented at the Society for Research in Child Development, San Francisco, CA.

Bower, T. G. R. 1967: The development of object permanence: Some studies of size constancy. *Perception and Psychophysics*, 2, 411–18.

Bruner, J. S. and Koslowski, B. 1972: Visually preadapted constituents of manipulatory action. *Perception*, 1, 3–14.

Carey, S. 1985: *Conceptual Change in Childhood*. Cambridge: Bradford/MIT.

Carey, S. 1991: Knowledge acquisition: Enrichment or conceptual change? In S. Carey and R. Gelman (eds), *Epigenesis of Mind: Studies in biology and cognition*. Hillsdale, NJ: Erlbaum.

Chomsky, N. 1980: *Rules and Representations*. New York: Columbia University Press.

Clifton, R. K., Rochat, P., Litovsky, R. and Perris, E. 1991: Representation guides infant reaching in the dark. *Journal of Experimental Psychology: Human Perception and Performance*, 17, 323–9.

Craton, L. and Yonas, A. 1990: The role of motion in infant perception of occlusion. In J. Enns (ed.), *The Development of Attention: Research and theory*. North-Holland, Elsevier Science Publishers, B.V.

deVries, R. 1987. Children's conceptions of shadow phenomena. *Genetic Psychology Monographs*, 112, 479–530.

Fantz, R. L. 1961: The origins of form perception. *Scientific American*, 204, 66–72.

Fischer, K. W. and Biddell, T. 1991: Constraining nativist inferences about cognitive capacities. In S. Carey and R. Gelman (eds), *Epigenesis of Mind: Studies in biology and cognition*. Hillsdale, NJ: Erlbaum.

Fodor, J. 1983: *The Modularity of Mind*. Cambridge, Mass: Bradford/MIT.

Gallistel, C. R. and Gelman, R. 1990: The what and how of counting. *Cognition*, 34, 197–9.

Gelman, S. 1989: Children's use of categories to guide biological inferences. *Human Development*, 32, 65–71.

Gibson, J. J. 1962: Observation on active touch. *Psychological Review*, 69, 447–91.

Gibson, J. J. 1966: *The Senses Considered as Perceptual Systems*. Boston: Houghton-Mifflin.

Gibson, J. J. 1979: *The Ecological Approach to Visual Perception*. Boston: Houghton-Mifflin.

Gopnik, A. 1988: Conceptual and semantic development as theory change. *Mind and Language*, 3, 197–216.

Halloun, I. A. and Hestenes, D. 1985: Common sense concepts about motion. *American Journal of Physics*, 53, 1056–65.

Held, R., Birch, E. and Gwiazda, J. 1980: Stereoacuity of human infants. *Proceedings of the National Academy of Science*, 77, 5572–4.

Hirsch, E. 1982: *The Concept of Identity*. New York: Oxford.

Hofsten, C. von and Spelke, E. S. 1985: Object perception and object-directed reaching in infancy. *Journal of Experimental Psychology: General*, 114, 198–212.

Johnson, M. H. 1990: Cortical maturation and the development of visual attention in early infancy. *Journal of Cognitive Neuroscience*, 2, 81–95.

Kaiser, M. K., Jonides, J. and Alexander, J. 1986: Intuitive reasoning about abstract and familiar physics problems. *Memory and Cognition*, 14, 308–12.

Keil, F. 1979: The development of the young child's ability to anticipate the outcomes of simple causal events. *Child Development*, 50, 455–62.

Kellman, P. J., Gleitman, H. and Spelke, E. S. 1987: Object and observer motion in the perception of objects by infants. *Journal of Experimental Psychology: Human Perception and Performance*, 12, 586–93.

Kellman, P. J. and Spelke, E. S. 1983: Perception of partly occluded objects in infancy. *Cognitive Psychology*, 15, 483–524.

Kellman, P. J., Spelke, E. S. and Short, K. R. 1986: Infant perception of object unity from translatory motion in depth and vertical translation. *Child Development*, 57, 72–86.

Kellman, P. J., Van de Walle, G. A., Hofsten, C. von and Condry, K. F. 1990: Perception of motion and stability during observer motion by pre-stereoscopic infants. Poster presented at the International Conference on Infant Studies, Montreal, Canada.

Kestenbaum, R., Termine, N. and Spelke, E. S. 1987: Perception of objects and object boundaries by three-month-old infants. *British Journal of Developmental Psychology*, 5, 367–83.

Koffka, K. 1935: *Principles of Gestalt Psychology*. New York: Harcourt, Brace and World.

Kuhn, T. S. 1977: A function for thought experiments. In his *The Essential Tension*. Chicago: University of Chicago Press.

Leslie, A. M. 1987: Pretense and representation: The origins of 'Theory of Mind'. *Psychological Review*, 94, 412–26.

Leslie, A. M. 1988: The necessity of illusion: Perception and thought in infancy. In L. Weiskrantz (ed.), *Thought and Language*. Oxford: Oxford University Press.

Leslie, A. M. 1991: Infants' understanding of invisible displacement. Paper presented at the Society for Research in Child Development, Seattle, WA.

Leslie, A. M. and Keeble, S. 1987: Do six-month-old infants perceive causality? *Cognition*, 25, 265–88.

Markman, E. M. 1989: *Categorization and Naming in Children*. Cambridge, Mass: Bradford/MIT.

Marr, D. 1982: *Vision*. San Francisco: Freeman.

McCloskey, M. 1983: Naive theories of motion. In D. Gentner and A. L. Stevens (eds), *Mental Models*. Hillsdale, NJ: Erlbaum.

Mervis, C. M. 1987: Child-basic object categories and lexical development. In U. Neisser (ed.), *Concepts and Conceptual Development: Ecological and intellectual factors in categorization*. Cambridge: Cambridge University Press.

Michotte, A. 1954: *The Perception of Causality*. New York: Basic Books.

Michotte, A., Thinès, G. and Crabbé, G. 1964: Les compléments amodaux des structures perceptives. *Studia Psychologica*. Louvain: Presses Universitaires de Louvain.

Needham, A. and Baillargeon, R. 1991: Reasoning about support in 3-month-old infants. Unpublished manuscript.

Piaget, J. 1954: *The Construction of Reality in the Child*. New York: Basic Books.

Piaget, J. 1960: *The Child's Conception of Physical Causality*. Paterson, NJ: Littlefield, Adams and Co.

Prather, P. and Spelke, E. S. 1982: Three-month-old infants' perception of adjacent and partly occluded objects. Paper presented at the International Conference of Infant Studies, Austin, TX.

Proffitt, D. R. and Gilden, D. L. 1989: Understanding natural dynamics. *Journal of Experimental Psychology: Human Perception and Performance*, 15, 384–93.

Rubenstein, J. 1991: Just me and my shadow: Infant conceptions of shadow phenomena. Unpublished honours thesis, Cornell University, Ithaca, NY.

Schmidt, H. 1985: The role of Gestalt principles in perceptual completion: A developmental approach. Unpublished doctoral dissertation, University of Pennsylvania, Philadelphia, PA.

Schmidt, H. and Spelke, E. S. 1984: Gestalt relations and object perception in infancy. Paper presented at the International Conference on Infant Studies, New York.

Schwartz, K. G. 1982: Perceptual knowledge of the human face in infancy. Unpublished doctoral dissertation, University of Pennsylvania, Philadelphia, PA.

Schwartz, M. and Day, R. H. 1979: Visual shape perception in early infancy. *Monographs of the Society for Research in Child Development*, 44 (Serial No. 182).

Shipley, E. F. and Shepperson, B. 1990: Countable entities: Developmental changes. *Cognition*, 34, 109–36.

Slater, A., Mattock, A. and Brown, E. 1990: Size constancy at birth: Newborn infants' responses to retinal and real size. *Journal of Experimental Child Psychology*, 49, 314–22.

Slater, A., Morison, V., Somers, M., Mattock, A., Brown, E. and Taylor, D. 1990: Newborn and older infants' perception of partly occluded objects. *Infant Behavior and Development*, 13, 33–49.

Soja, N. N., Carey, S. and Spelke, E. S. 1991: Ontological categories guide young children's inductions of word meanings: Object terms and substance terms. *Cognition*, 38, 179–211.

Spelke, E. S. 1985a: Perception of unity, persistence, and identity: Thoughts on infants' conceptions of objects. In J. Mehler and R. Fox (eds), *Neonate Cognition*. Hillsdale, NJ: Erlbaum.

Spelke, E. S. 1985b: Preferential looking methods as tools for the study of cognition in infancy. In G. Gottlieb and N. Krasnegor (eds), *Measurement of Audition and Vision in the First Year of Postnatal Life*. Norwood, NJ: Ablex.

Spelke, E. S. 1988: Where perceiving ends and thinking begins: The apprehension of objects in infancy. In A. Yonas (ed.), *Perceptual Development in Infancy*. Minnesota Symposium on Child Psychology (vol. 20). Hillsdale, NJ: Erlbaum.

Spelke, E. S. 1990: Principles of object perception. *Cognitive Science*, 14, 29–56.

Spelke, E. S. 1991: Physical knowledge in infancy: Reflections on Piaget's theory. In S. Carey and R. Gelman (eds), *Epigenesis of Mind: Studies in biology and cognition*. Hillsdale, NJ: Erlbaum.

Spelke, E. S. and Born, W. S. 1983: Visual perception of objects by three-month-old infants. Unpublished manuscript.

Spelke, E. S., Breinlinger, K., Jacobson, K. and Phillips, A. 1993a: Gestalt relations and object perception: A developmental study. Unpublished manuscript.

Spelke, E. S., Breinlinger, K. Macomber, J. and Jacobson, K. (1992): Origins of knowledge. *Psychological Review*, 99, 605–32.

Spelke, E. S., Hofsten, C. von and Kestenbaum, R. 1989: Object perception and object-directed reaching in infancy: Interaction of spatial and kinetic information for object boundaries. *Developmental Psychology*, 25, 185–96.

Spelke, E. S., Katz, G., Purcell, S. E., Ehrlich, S. M. and Breinlinger, K. 1993b: Early knowledge of object motion: Continuity, solidity, and inertia. Unpublished manuscript.

Spelke, E. S. and Kestenbaum, R. 1986: Les origines du concept d'objet. *Psychologie Française*, 31, 67–72.

Spelke, E. S., Simmons, A., Breinlinger, K., Jacobson, K. and Macomber, J. 1993c: Developing knowledge of gravity I: Infants' sensitivity to hidden object motion. Unpublished manuscript.

Streri, A. 1987: Tactile discrimination of shape and intermodal transfer in 2- to 3-month-old infants. *British Journal of Developmental Psychology*, 5, 213–20.

Streri, A. 1990: *Voir, atteindre, toucher*. Paris: Presses Universitaires de France.

Streri, A. and Pêcheux, M. G. 1986: Cross-modal transfer of form in 5-month-old infants. *British Journal of Developmental Psychology*, 4, 161–7.

Streri, A. and Spelke, E. S. 1988: Haptic perception of objects in infancy. *Cognitive Psychology*, 20, 1–23.

Streri, A. and Spelke, E. S. 1989: Effects of motion and figural goodness on haptic object perception in infancy. *Child Development*, 60, 1111–25.

Streri, A., Spelke, E. S. and Rameix, E. 1993: Modality-specific and amodal aspects of object perception in infancy: The case of active touch. Unpublished manuscript.

Teller, D. Y. and Bornstein, M. H. 1987: Infant color vision and color perception. In P. Salapatek and L. Cohen (eds), *Handbook of Infant Perception, vol. 1, From sensation to perception*. Orlando, FL: Academic Press.

Wellman, H. M. 1990: *The Child's Theory of Mind*. Cambridge, Mass: Bradford/MIT.

Wiggins, D. 1980: *Sameness and Substance*. Cambridge, Mass: Harvard University Press.

6
Intuitive mechanics, psychological reality and the idea of a material object

Christopher Peacocke

I will be addressing three questions:

1. What is distinctive of the form and content of the principles of intuitive mechanics?
2. What is it for the principles of an intuitive mechanics to be psychologically real?
3. What is the relation between possession of an intuitive mechanics and possession of a conception of a material world?

There are interrelations between all three of these issues, or so I shall argue. I approach the subject as a philosopher, primarily concerned with what are in a broad sense conceptual aspects of the issues. In saying that, I do not mean in the least to endorse any controversial doctrine alleging the irrelevance of psychology to philosophy, or of philosophy to psychology. I mean only that questions of what has to be the case for a principle of an intuitive theory to be psychologically real, or questions about the appropriate goals of a psychological theory, cannot themselves be settled just by the discovery of additional empirical truths. Even more so does this apply to such a constitutive question as 'What is necessarily involved in having a conception of a material world?' All these are essentially philosophical questions; though, of course, it is certainly not only philosophers who are qualified to address them.

I

I will first consider the relation between spatial reasoning and mechanical reasoning. It will be convenient to have an example in play. Imagine lifting a dinner plate from the kitchen counter, and placing it in a rack in the dishwasher. You have owned the plate for many years, let us suppose, and know how heavy it is. You employ this knowledge of its weight in adjusting the force you exert in lifting the plate. You also have to locate a place for the plate in the dishwasher. You need to find a place which not only fits the shape and size of the plate, but one in which it can be inserted without knocking anything fragile *en route*, and without breaking anything in its final location.

In this and in a thousand more everyday examples, spatial reasoning and mechanical reasoning are unavoidably intertwined. An intuitive mechanics will be involved in any case of spatial reasoning in which an initial mental representation with a spatial content is transformed into a second representation with a second spatial content, and in which the transformation of contents involves some representation of the action of one material body on another. In effect, an intuitive mechanics will be involved in any spatial reasoning which is not purely geometrical. It is a condition of adequacy on any psychological theory of human intuitive mechanics that it give an account of how mechanical reasoning is intertwined with distinctively spatial reasoning. This condition of adequacy constrains both the form and the content of the principles of intuitive mechanics. I will take the constraints on content first.

When we formulate a principle of a mechanics, we commonly need to refer to distance, and to temporal derivatives of distance. The distances are taken as given in some particular unit, feet or metres. Any principle formulated in this way is unsuitable for giving the content of the principles of an intuitive mechanics, if that mechanics is to interact with spatial perception to produce action. It is unsuitable, because when we perceive distances, we do not perceive them as having a particular magnitude in feet or in metres, not even if we do in fact know how long a foot or a metre looks. Suppose you want to know how hard to push the book which is resting on the table, so that it will slide from one end to the other and then stop. A principle with an informational content that relates the required effort to distance in feet will be of no help to you, since although there is certainly a sense in which you see how long the table is, your perception does not supply the length of the table in feet. A principle of mechanics which uses a particular unit of distance cannot by itself engage with the kind of spatial information that perception supplies. As long as distances are specified in some particular units, this point continues to apply, whether the mechanical principles are quite general, like those of a Newtonian mechanics or a medieval 'impetus' mechanics, or whether they are much more specific, and concern the mechanical behaviour of some prototypical types of familar objects.

This point applies not only to spatial magnitudes, but to mechanical magnitudes too. Knowing the weight of the dinner plate you place in the dishwasher is not knowing its weight in pounds and ounces, even if you know how heavy a one-pound weight and a one-ounce weight are. Knowing how heavy the plate is seems to be based on a link with action. Such knowledge of its weight is knowing what force has to be exerted to lift it. Here the thought about force is also unit-free. It seems that the ability to think about force in that way is distinctively related to the ability to exert it. That would certainly explain why it is that you can in ordinary circumstances learn how heavy something is, on this way of thinking of weight, by trying to lift the object, and succeeding. In any case, the way of thinking about weight and force which is relevant to moving and manipulating objects in the environment is also unit-free. So again, any formulation of principles of intuitive mechanics which refers to weights and forces in any particular kind of unit will be unable to engage in the required way with action if it is also supposed that the subject has to have some grasp of those units. Knowing that a force of 15 pounds is needed to get the book along the table does not yet specify, in the way relevant to acting, how hard you have to push.

A first step in adequately describing unit-free perception and thought about spatial and mechanical magnitudes is to use a notion of magnitudes which is itself unit-free. We can employ an ontology of distances of such a kind that there is just one distance which has both the measure of one inch and equally has the measure of 2.54 centimetres. Such an approach is in effect already adopted by those theories which give the spatial content of perceptual experience in part by specifying which ways of filling out the space around the perceiver are consistent with the experience's being veridical. In such theories, the way the experience represents the world as being is given in part by the unit-free distances themselves between the various objects and features perceived. Similarly, we could introduce an ontology of unit-free forces and weights, an ontology which includes a certain unit-free weight which is measured both by one kilogram and equally by 2.2 pounds weight. We would also have to supply statements of the individuation of the appropriate ways of thinking of these magnitudes, where these ways of thinking have distinctive connections with action. But though these are all moves in the right direction, a problem remains. The problem is that in abandoning, with good reason, the unit-involving magnitudes, we seem at first blush to have abandoned something which in another respect we needed.

Within a certain range, and to a certain degree of precision, we have everyday knowledge of some of the relations which hold between weight, force, distance and time. In some cases we have misinformation about their relations. But whether our information is true or false, the question arises of what the informational (or misinformational) content *is*. The usual way of stating a relation between two or more physical magnitudes in a theory makes use of the assignment of numerical values to the magnitudes, as in the statement that force is the numerical product of mass and acceleration, or that the initial force needed to make an object move across a given kind of surface is proportional to the distance it is to travel. We proceed by specifying the numerical value of one magnitude as a certain mathematical function of the numerical values of one or more other magnitudes. But now we have abandoned the assignment of numerical values for the purposes of specifying the content of certain mental states, how are we to capture the content of states which involve a general relation holding between many values of two or more magnitudes?

There are other ways of expressing relations between physical magnitudes once we have bleached out the units, but they do not seem to help solve our particular problem. It is true, for example, that unit-free magnitudes can still stand in comparative relations. For unit-free magnitudes, we can continue to make sense of such relations as 'distance d_1 is greater than distance d_2', or 'force f_1 is less than force f_2'. But any simple statement using only these comparative relations does not seem adequate to capturing the content of our intuitive mechanical knowledge. For instance, the statement 'Movement across a horizontal surface requires a greater force if the object is to travel a greater distance' is much less specific than our intuitive knowledge; it is compatible with any shape of increasing function.

With an ontology of unit-free magnitudes, it is also true that we can continue to speak of what Russell would call the function-in-extension of the relation thought to hold between certain magnitudes. The problem with the function-in-extension as a means of capturing the content of intuitive mechanical information or misinformation is its lack of generality. A function-in-extension relating, say, force and distance for a given type of situation is essentially a list of cases. There may indeed be some cases of entirely specific mechanical knowledge, such as your knowledge of how hard you have to push the gear lever in your car to change from second to third gear. But it is very hard to believe that all cases are like this. Our intuitive mechanics generates expectations about new situations we have never encountered before, and in which the physical magnitudes about which we have expectations may take values that we have never experienced before in those situations. The mechanical information or misinformation we exploit in such cases must have a more general character than can be given by the list of cases captured in a function-in-extension.[1]

So, on the one hand we both need the specificity of the relation between magnitudes, and the generality captured by the finite statement, which is supplied by a numerical formulation of a principle of mechanics. On the other, we have to reconcile this with the non-numerical character of magnitudes as perceived. I think we can see a way out of this dilemma, and also make some progress on the issue of psychological reality, if we reflect on the consequences of a thinker's representing a particular relation between two magnitudes in his or her intuitive mechanics. That the thinker is representing one relation rather than another will be reflected in which pairs of mental representations, with spatial content, he treats as specifying the way the world around him will develop, or empirically could develop, or hypothetically would develop, depending on the particular task of spatial reasoning in which he is engaged. One of these two mental representations may be perceptual, or they may both be the products of imagination. If one of them is perceptual, it may give a represention of the world around the perceiver at either an earlier or a later time than that represented by the other spatial representation. The perceptual representation will represent an earlier time if the thinker is engaged in imaginative prediction of how the world will develop. It will represent a later time if the thinker is engaged in imaginative spatial and mechanical reasoning about what earlier state of affairs could have led to the presently perceived environment. For enthusiastic readers of the philosophical literature, the spatial contents of these representations of imagination or perception will more particularly be what in earlier work I called the scenario and protopropositional content (Peacocke, 1992a; 1992b, ch. 3).

We can now suggest what it is for a particular relation between physical magnitudes to be represented in the thinker's intuitive mechanics. To fix ideas, let us suppose we are concerned with the non-numerical analogue of the principle that force is the product of mass and acceleration, and that the question arises of what would have to be the case for this principle to be psychologically real. I suggest that for it to be a psychologically real principle of a thinker's intuitive mechanics is for the following condition to be met:

> there is a uniform explanation of the fact that the thinker's accepted triples of particular non-numerical magnitudes of force, mass and acceleration, as given in his mental representations of a spatial array at any particular time, all have the property that their numerical analogues are such that the first is the product of the second and the third.

Note the order of the quantifiers here: there is an explanation which, for each triple of values, explains why it has this property. If, by contrast, there were a partition of the accepted triples, with a different explanation for each range of triples, then the displayed criterion would not be met. In such a case, the most that would be psychologically real is not the unrestricted law, but rather a set of principles, each one dealing with one of the ranges of accepted triples. The distinction here is that between possessing a general rule and possessing some form of list of individual cases, or kinds of case – a 'look-up table'. The distinction has various consequences. For instance, when a general rule is psychologically real, damage of the state which realizes it will remove the basis of all the accepted values covered by the general rule. By contrast, the basis of the information given in one line of a look-up table can be damaged without damaging the states underlying the information given in the other lines. The general distinction is familiar from discussions of the psychological reality of grammars (Davies, 1981, ch. 4; Evans, 1985; Fodor, 1985; Peacocke, 1986). The distinction is also fully applicable if we concentrate on the state of a subject at a given time. In the case of intuitive mechanics, for instance, there will be counterfactuals about what other values of the magnitudes would be accepted at a given time. We can distinguish the case in which these accepted values have a uniform explanation – a 'common cause' as the literature on the psychological reality of grammars has it – from that in which they do not.

There will be an analogous version of this criterion when, as will be the case for any finite being, the mentally represented unit-free magnitudes do not have sharp numerical equivalents. There can still be a uniform explanation of why a certain somewhat fuzzier relation holds between somewhat fuzzier magnitudes. Note also that under this criterion, if we have one principle formulated in one set of units, say metric units, and a second strictly equivalent principle formulated in imperial units, the non-numerical analogue of the first principle is a psychologically real principle of intuitive mechanics if and only if the non-numerical analogue of the second is. The criterion filters out differences of units, while preserving some psychological significance for the relation between magnitudes which the principle states. The unit-involving principle merely determines a proposed relation between unit-free magnitudes.

We were led into this discussion by way of considering what the content of a general principle of intuitive mechanics might be, once we acknowledge the need for unit-free magnitudes. We now have the resources for answering that question. Let us continue with the example of the principle that force is the produce of mass and acceleration, since the answer to be given generalizes to other principles. Let ϕ, μ and α be variables over non-numerical values of force, mass and acceleration. Let f, m and a be variables over numerical values of these magnitudes, in a given system of units (metric, let us say); these values are numbers pure and simple, rather than the 'impure numbers' introduced in some discussion of measurement. Let $\mathrm{n}(x)$ be that function which maps the non-numerical value of a magnitude to its numerical metric value. Let us now fix on that relation R such that

$$\mathrm{R}\phi\mu\alpha \quad \text{iff} \quad \mathrm{n}(\phi) = \mathrm{n}(\mu)\ .\ \mathrm{n}(\alpha).$$

The unit-free content of the general principle in question is then this: force, mass and acceleration are related by the relation R. The way in which the relation R has been introduced should be regarded as a means of reference-fixing for a relation: it is not a way of introducing a concept, or a way of thinking – if it were, it would be dependent upon the particular system of units used in its introduction. It is the relation itself, and not a way of thinking of it, which enters the content of the principle of intuitive mechanics.

The criterion of psychological reality which I have put forward suggests a way in which we can steer a middle course between two more extreme views about the sense in which an explicitly formulated psychological theory may be psychologically real. At one extreme is the view that the only sense in which a physical theory may be used in characterizing someone's informational state is that his or her physical predictions, retrodictions, hypothetical reasoning and explanations are in agreement with those which can be derived from the explicitly formulated theory. This extreme view is committed to holding that attribution of a particular physical theory to a subject cannot have any explanatory power. Even if only one physical theory were consistent with the subject's predictions and spatial reasoning, it would remain the case that on this extreme view, mention of that theory functions only in giving a compact description of the competence.

At another extreme is the view that in saying a particular physical theory is psychologically real, one is committed to holding that the principles of the theory are explicitly represented in a language-like medium, a language of thought in Fodor's sense. The view that there is such a commitment is held by Jack Yates and others. In one paper, Yates and his co-authors aim to set up a contrast between accounts which emphasize simulation, and those which appeal to intuitive physical theories. They write: 'As opposed to a language-like set of general propositions, [our] view relies on the basic metaphor of what is often called, in everyday parlance, an image (with the understanding that much more than visual information is included, and that a mental object is not anything like an unstructured photograph)' (Yates et al., 1988, p. 253).

Yates is making a false contrast if the discussion of the present chapter is along the right lines. I have been emphasizing the role of intuitive mechanics in mediating transitions between mental states with spatial contents. These mental states are prime candidates for realization in mental representations with a distinctively imagistic character. If pressed on the nature of this imagistic character, I would follow Michael Tye's exposition of the notion of what makes a representation quasi-pictorial, an exposition which develops naturally from reflection on the work of Shepard, Cooper and Kosslyn (Tye, 1988). But however the notion of an imagistic representation is to be elucidated, the important point for present purposes is that it is very far from obvious that the only way a principle of a mechanics can be psychologically real, even when mediating transitions between states involving imagistic representations, is by being explicitly formulated in a language of thought. In fact the criterion of psychological reality I endorsed already suggests other possibilities.

We can give a highly oversimplified example to illustrate the point in the abstract. Suppose we are concerned with Newton's first law, the principle that everything continues in a state of rest or uniform motion in a straight line in the absence of forces acting on it. One way in which this principle could be psychologically real for a computational device in mediating transitions between states with spatial contents is as follows. In building up a representation of how the world around it will be at a later time, the device first checks whether in its representation of an earlier time, any forces are represented as acting on the represented objects. Where no forces are represented as acting on an object already in motion, the device constructs, in a quasi-pictorial imagistic medium, a representation of a straight line connecting a suitable pair of the object's earlier positions. It then extends this imagistic representation of a line in constructing a representation of the object's position at the later time. For an organism incorporating a device which works this way, Newton's first law would be psychologically real; but it need not involve any sentential representation of the law in a language of thought. It does not need a word in a language of thought for the property of being a straight line: it just needs to operate in the way described.[2]

Principles of intuitive mechanics which are psychologically real but not explicitly represented in naturally evolved systems provide one kind of instance of a category whose importance Daniel Dennett has long emphasized. This is the category of regularities 'intermediate between the regularities of the planets and other objects "obeying" the laws of physics and the regularities of rule-following (that is, rule-*consulting*) systems. These intermediate regularities are those which are preserved under selection pressure. . .' (Dennett, 1991, p. 43). I would add only that within Dennett's intermediate class of such regularities in behaviour and thought, we can still distinguish between those which result from a psychologically real theory not written out in a language of thought, and those which do not result from any psychologically real theory at all. I have been emphasizing that some principles of intuitive mechanics may fall in the former sub-class of Dennett's category.

I am not claiming that all principles of a psychologically real intuitive mechanics receive a non-linguistic mental representation. There are other principles which seem ripe for mental representation in a linguistic medium. One such is the principle employed by the six-year-olds investigated by Annette Karmiloff-Smith, the principle that blocks balance in the middle (Karmiloff-Smith 1984; 1988). When a principle of an intuitive mechanics is actually expressed in an external language by its possessor, the arguments for some internal representation in a language of thought are further strengthened. The broadly philosophical question then arises of whether we can give some criterion for distinguishing the cases in which there is some presumption of a non-linguistic mental representation of a principle of an intuitive mechanics. One conjecture which seems plausible is as follows. Let us suppose that indeed the mental representations underlying states of perception and imagination with spatial content are of an imagistic, quasi-pictorial character. It is then those principles of intuitive mechanics whose fundamental role is to assist in the generation of one such representation from another which are the most plausible candidates for non-linguistic realization. This is because a linguistic representation of the principles would involve a detour in generating one spatial representation from another. Information in the first spatial representation would have to be converted into some linguistic form; this linguistic form would then be transformed, or made the starting point of an inference, to yield another linguistic form; and this would then have to be converted back to generate a spatial representation in the imagistic medium. Such an unnecessary detour is avoided when there is a mechanism for generating the second spatial representation from the first directly. It is also to be expected that when a principle of intuitive mechanics is non-linguistically represented, a subject may be quite bad at articulating its content in his or her public, external language. For there will be no underlying linguistic mental representations which can be linked with his or her mental representation of expressions in the external language in which the principle in question could be articulated.

Principles of intuitive mechanics which are mentally represented in linguistic form must of course be capable of interacting in various ways with the representations of a spatial content. The child who believes that blocks balance in the middle will rule out some imagined spatial representations as possible, and accept others as correct predictions. The required interaction of principles is possible as long as the spatial content, and its corresponding mental representation, has suitable structure. At the level of content, we can acknowledge a level, a level which I like to call that of proto-propositional content, at which the subject's perception is characterized as representing some line as straight, some shape as symmetrical, two lines as equal in length and so forth. This is a level which goes beyond simply specifying which ways of filling in the space around the perceiver are consistent with the veridicality of the experience, since two experiences may have the same characterization at that basic level, but differ at the level of proto-propositional content. As long as we have some mental representation of the proto-propositional content overlaying an imagistic representation, we will have something with which linguistically represented principles of an intuitive mechanics can interact in the required fashion.

What on the present view is the status of derivations carried out in an explicitly formulated mechanics which is put forward as psychologically real? A derivation in the explicitly formulated theory cannot be said to correspond to a possible sequence of transitions between states involving the corresponding formulae in the language of thought, since we have been asserting that there need not be any such formulae in the case of some of the axioms of the mechanical theory. But it should not be concluded that derivations in the explicitly formulated theory are of no psychological significance. At the very minimum, a derivation of a sentence in the explicit, linguistically formulated theory will give information about what is true in any model of the theory, in the logician's sense of 'model'. It will therefore give information about what the subject will predict, or retrodict, or be prepared to offer as explanations, provided of course that the inferential apparatus of the theory corresponds to operations on the represented contents that the subject's computational mechanisms perform. We know from discussions of the psychological reality of a semantic or a syntactic theory that a theory can state the information which is drawn upon in some psychological mechanism without specifying precisely which algorithm the mechanism employs (Peacocke, 1986). In such linguistic cases, the theory tells us something about the results of an arbitrary algorithm which draws in suitable ways upon the semantic or syntactic information. For a range of explanatory purposes, it does not matter what the algorithm is, as long as it draws upon a certain body of information. Similarly, derivations in an explicitly formulated mechanics supply information about the spatial and mechanical contents which anyone for whom the theory is psychologically real will be prepared to accept as predictions and explanations. The derivations will also show more. They will show which principles of an intuitive mechanics are crucial for a subject's acceptance of a particular spatial or mechanical content.

II

In the second part of this chapter, I want to discuss another link in which intuitive mechanics is a partner. In a nutshell, my argument will be as follows:

> We experience objects specifically as material objects. Part of what is involved in having such experiences is that perceptual representations serve as input to an intuitive mechanics which employs the notion of force. This involvement is in turn a consequence of a general principle governing the ascription of content to mental representations, together with very general philosophical considerations about what it is for an object to be a material object.

I will start immediately on the elaboration and defence of these claims.

Probably only the starting point of that argument is sufficiently uncontroversial to count as a datum. The starting point is that requirements on matter occupying space are part of the correctness conditions for ordinary visual experience. An experience produced in perception of what is in fact a perfect hologram of a boulder in front of you does not represent the world as it really is. This is true even if the boulder-hologram is visually indistinguishable from a real boulder from a wide range of angles, and for a wide range of different types of perceiver. The experience is still not veridical even if we suppose the setup modified so that when the subject extends a hand against the apparent boulder, he or she has a sensation of resistance. This subject would equally still be subject to an illusion if some superscientist had a device that produces in the subject just those sensations of force and resistance which would be produced by a real boulder there – an incorporation of a form of internal 'force feedback' without external forces really acting on the subject's hand. (The superscientist would also, of course, have to ensure that internal forces cause the extended hand to stop on the apparent surface of the apparent boulder.) The correctness condition for a visual experience as of a boulder will always be incompletely stated until we add the requirement that there be a *material object* at the relevant location. This is why it is the case that in genuine tactile perception of a material object, the subject has to experience the resistance exerted by the material object itself.

When I say that the perceptual representations of a subject who experiences the world as made up of material objects must serve as input to an intuitive mechanics, I do not mean that this mechanics must make correct predictions about the mechanical behaviour of the perceived objects, nor that the subject must be in a position to know the weight, or forces necessary to move an object, just on the basis of those perceptions. All I mean is that when a subject perceives an object as a material object, and takes the experience at face value, then he or she will have some conception of the magnitude of force, and will take it that force is necessary to change the state of rest or of certain kinds of motion of the perceived object. This can still be true of the subject, even if his or her intuitive mechanics contains systematic errors and gaps.

It will be convenient to call an *inertial principle* any principle to the effect that objects in a certain class are such that changes in their state of rest or of certain kinds of motion require the action of a specified type of force. Any inertial principle for the class of perceived objects provides a further example of a principle that can be psychologically real yet only implicitly represented. The particular inertial principle in question can be implicit in the way in which the representations delivered by a perceptual system are employed by the subject's intuitive mechanics.

Let us now prescind for a few paragraphs from all issues in psychology and the philosophy of mind, and consider a question which is in itself one of descriptive metaphysics. What is it for an object to be a material object, for it to be of a material composition? I suggest that for something to be a quantity of matter is for changes in its state of motion to be explicable by the mechanical forces acting on it, and for its changes of motion to exert such forces. This is not meant to be an eliminative definition. It is a thesis about a property, rather than a thesis about a concept or a way of thinking. Indeed, it is plausible that it would be circular if it were intended as a definition of a concept or way of thinking of a property. There are forces other than mechanical forces, such as electromagnetic forces. If we are asked which are the specifically mechanical forces, we would probably have to answer 'those which produce changes of motion in quantities of matter'. The point of the claim about the property of being a quantity of matter is rather this: what makes something a material object is its relations to a certain kind of physical magnitude in the world, that of mechanical force. Note that the thesis respects a distinctive property of matter – that it has an identity over time. To say of a quantity of matter that *its* changes of motion are explained by mechanical forces already presupposes identity over time for the quantity of matter.

This account of what distinguishes the material character of an object is to be distinguished from the impenetrability discussed by Locke, which he often called 'solidity'. For matter to be impenetrable is for it to be impossible for two quantities of matter to occupy exactly the same place at the same time. Recently some philosophers, notably Johnston and Kripke, have made a good case that such impenetrability is not a metaphysically necessary property of matter (Johnston, 1987; Kripke, unpublished lectures). On their view, there is a good account to be given of what could distinguish between there being two material particles, each of mass *m*, at a given location, and there merely being one particle of mass *2m* at the same place. But the link between matter and force could still be in place even if impenetrability fails. In this respect, the position for which I am arguing can be described as being even more Leibnizian than Leibniz. Leibniz wrote:

> Matter. . . resists motion by a certain *natural inertia*, as Kepler has well called it, so that it is not indifferent to motion and rest. . . but it needs in order to move an active force proportionate to its size. Wherefore I make the very notion of. . . mass. . . consist in this very passive force of resistance (involving impenetrability and something more). (Leibniz, 1951)

Leibniz was clear that such a requirement on matter was additional to impenetrability; I have been noting – and here I differ from Leibniz – that it does not even seem to require impenetrability.

The point of this brief excursion into metaphysics is that there is a principle which connects metaphysical distinctions with mental phenomena. The principle is this:

> If an account of what is necessarily involved in something's having a certain property makes reference to some substantial condition which must be met by things which have it, a thinker's mental representations of that property must be suitably sensitive to the existence of this substantial condition.

The particular property in which we are interested is that of being a material object, but this principle is quite general in its application. The intuitive argument for the general principle is that if it were false, it seems impossible to answer the question of what makes it the case that it is the given property, rather than some other, that the thinker is mentally representing. It is a task of some interest to spell out in various different kinds of case exactly what being 'suitably sensitive' here requires. It does seem clear that knowledge of a very detailed condition for having the property in question is not necessary. As Hilary Putnam long ago emphasized, early nineteenth century scientists had beliefs with contents of the form that acids have such-and-such characteristics, without knowing the real sub-atomic basis of acidity. What they did have was an unspecific characterization: acids are those substances with the microphysical properties which explain their engaging in reactions of a certain specified sort. One can have beliefs whose content is given by using the word 'acid' in oblique position whilst being unsure, or indeed mistaken, as to what those underlying microphysical properties are.

The principle linking metaphysical distinctions with mental representation, together with what I have said about what it is to be a material object, jointly imply this: that a thinker who mentally represents objects as material must be suitably sensitive to the fact that certain motions of material objects exert, and are caused by, mechanical forces. That the thinker takes objects he or she perceives to be subject to an intuitive mechanics is certainly one form of such a required sensitivity. I further put forward the conjecture that it is not only sufficient; it is also necessary. If a subject is to mentally represent objects as *material* objects – and not merely to have the existentially quantified thought that there is some physical magnitude meeting such-and-such conditions which the objects possess – possession of an intuitive mechanics mentioning forces is essential. Once again, it is consistent with these points that the intuitive mechanics is wrong or incomplete, just as the nineteenth-century chemists' theories of acidity may have been wrong or incomplete.

It is important to acknowledge that the magnitude of force recognized in an intuitive mechanics is a primary quality in the classical sense, if that sense is properly expounded. A primary quality is a quality of a general kind such that necessarily any material object has some quality of that general kind. If what I have been saying is right, then necessarily, for any material object, mechanical force is required to alter certain of its motions. And as with any other primary quality, it is a matter of considerable complexity and delicacy to give a proper philosophical account of the relations between thought about the physical magnitude of force and conscious mental states.

There is such a thing as the sensation of force. It is experienced when, for instance, a heavy book is resting on your lap and pressing downwards. But it is not necessary ever to have experienced such sensations in order to mentally represent the magnitude of force and to use it in an intuitive mechanics. We can conceive of beings whose afferent nerves are congenitally damaged so that they are incapable of such sensations of force. This is no obvious conceptual or *a priori* bar to their operating with an intuitive mechanics, adjusting their expectations of the movement of objects to their beliefs about their mass and the forces acting on them, and adjusting their own actions on material objects likewise. Whether a particular being incapable of passive sensations of force has an intuitive mechanics is an empirical question.

Reflection on this case brings out the fact that there are actually at least two kinds of conscious states which are a systematic function of forces acting in certain relations to your body. There is not only the bodily sensation of pressure. There is the state of consciously exerting a greater or lesser force with one of your own limbs. This is a conscious state which can be present even when the relevant afferent nerves are not functioning. That is what makes possible the example of the preceding paragraph. Yet this conscious state, too, does not seem to be essential to having a conception of force. There is equally apparently no conceptual or *a priori* bar to a conscious subjects with no limbs and incapable of force-producing actions thinking about force as long as they are capable of the experience of forces acting on them. It seems that either sensation or action may each in principle provide routes to acquisition of a conception of force (if it is acquired).

It is never a sufficient account of possession of a conception of force that the subject is willing to form certain beliefs, or representations, involving it on the basis of his or her sensations, if those sensations are ones he or she could have without already having an intuitive mechanics. The same applies to sensitivity to conscious exertions. To have the conception of a certain force acting on one, or of oneself as exerting a certain force, is to have a conception on which it makes sense, and can be true, that an inanimate object is subject to forces of exactly the same kind. A mere sensitivity to sensations enjoyable without possession of an intuitive mechanics would never suffice to meet that requirement. If such a sensitivity to sensations were all that is involved in having a conception of force, conceiving of forces no one experiences would be none too easy a thing to do: you would have to conceive of something felt by no one on the basis of sensations you *do* feel (cp. Wittgenstein, 1958, section 302). It is no solution for this theorist simply to supplement his account with a reference to whatever is the primary quality ground of these sensations. It is force itself which is their primary quality ground, and one of our original problems was precisely what is involved in thinking of that magnitude in the way we do. When we are in possession of an intuitive mechanics, we do not think of force in a way which would allow it to be an as yet unidentified ground of certain sensations; any more than we think of particular shapes as potentially unidentified causes of features of visual experience. In each case, conscious experience makes available to us the physical magnitude and the physical shape itself, respectively, presented as such.

To make these points is not to deny that someone capable of experiences of force may have available ways of thinking of that magnitude which are unavailable to someone who does not have experiences of the same general kind. There will be such ways of thinking available; just as someone capable of perceiving a particular kind of shape will have available a distinctive way of thinking of such shapes. What these points do imply is that possession of the distinctive ways of thinking of force made available by certain kinds of conscious experience does not consist in a sensitivity to sensations free of any internal connection with an intuitive mechanics. This is again parallel to the corresponding point about the perception of shape and the ways of thinking of particular shapes that it makes available. The relevant experiences of shape which make available a distinctive way of thinking of a particular shape are ones which involve a conception of the object with that shape as having a location in the objective world, and as being perceivable from different standpoints in that world, together with all that that involves. It is not just a matter of sensitivity to uninterpreted sensations.

The claim that possession of an intuitive mechanics is constitutively required for objects to be represented as material objects in no way entails that someone who has an intuitive mechanics always uses it in his or her predictions and explanations of the motions of objects. The results of Lynn Cooper and her co-workers establish that it is often a kinematics rather than a dynamics which is used by human subjects in the prediction of the trajectory of an object. As Cooper and Munger note (chapter four), it can often be computationally expensive to calculate forces and force-related magnitudes. We would expect that in a wide range of circumstances, the extra expense will not justify the small increase in accuracy of a dynamics over an approximately-correct kinematics. These points and empirical findings are fully consistent with the main claim of the second part of this chapter, which requires only that if an object is represented as material, then in some circumstances, some representation of force involving it must be employed in the subject's psychological economy. It is not required to be used in all circumstances in which it could be used. The phenomenon of efficient short-cuts in reasoning and thought we should expect to be as commonplace at the level of non-conceptual content and representation as it is at the level of conceptual content and thought. It would in general be fallacious to move from the absence of any mention of a property in certain widespread forms of reasoning or transition involving the mental representation of *F*'s, to the conclusion that *F*'s can be mentally represented as such without any mental representation of that property in any circumstances.

To give a counter-example to the constitutive claim I am making, it would rather be necessary to describe a case in which there are no circumstances in which the subject attributes a magnitude of force to objects in his or her environment. Consider a case in which (say) a creature is in fact surrounded by holograms, which come into 'contact' with one another, and whose movements and regular interactions on 'contact' are predicted by the subject by means of an intuitive kinematics. (We can imagine an external controller of the subject's environment who has designed this arrangement.) This subject is unsurprised when these objects exert no force on him when they they 'touch' his surface, and does not expect to have to exert any force to 'move' one of them out of his way. I have no objection to saying that this subject conceives of these things in his environment as objects. But if he in no way represents them as exerting or being subject to forces, I do not think that he conceives of them as material objects. It would, of course, be a quite different case if the subject conceives of the objects as being *very lightweight*. Then he will indeed be conceiving of them as material objects; but he will then also think of them as having some mass, and exerting some forces, albeit of a small magnitude.

I have been arguing for a constitutive link between the capacity to perceive objects as material objects and possession of an intuitive mechanics. There is a competing approach, one which would state that perceptions are as of material objects because their predominant distal causes are in fact material objects. Such a proposal would be in the spirit of what have come to be called 'causal covariance' theories of perceptual content (Cummins, 1989, chs. 4–6). But it seems to me that it is not sufficient for perceptions to be as of material objects that their predominant distal causes are material objects. We can imagine an underwater creature whose perceptions of underwater objects are caused by the presence of what are in fact material objects. Suppose that this creature never comes into contact with the objects it perceives, and that it never experiences them as coming into contact with one another. (We could tell a story about some chemical emissions by these objects which cause them to move away from one another if they approach too close.) Such a creature may have no expectations one way or the other about what would happen were one of these objects to come into contact with another, or with the perceiver itself. It would not be surprised if they apparently passed through each other; it would not expect the motion of a relatively large object on collision with a smaller to cause the smaller to move or deform; and so forth. In short, its expectations would not be disappointed were the objects holograms rather than material objects. In these circumstances, it seems that the perceptions of this subject do not represent their objects as material objects. What is lacking, according to me, is any connection between the perception and an intuitive mechanics possessed by the subject. If my earlier remarks are right, it would not change the situation if we altered the case in order to imagine this subject with an intuitive kinematics as well.

Conversely, regular causation of perceptions by distal material objects does not seem to me necessary for the perceptions to be as of material objects, provided that the perceptions are input to an intuitive mechanics. If some superscientist of the sort we considered earlier were able to surround a subject frequently with realistic holograms, and produce illusions of force and action upon them when the subject tried to interact with them, this scientist could be producing an illusion of material objects – rather than completely veridical perception of holograms!

Causal covariance accounts of perceptual content are clearly externalist accounts, in that they make the content of perceptual states depend constitutively, and not just causally, on the perceiver's relations to his or her environment. If we reject pure causal covariance accounts, are we thereby committed to rejecting externalism? Certainly not: an account of perceptual content may equally be externalist in virtue of what it says about the consequences, and not just the antecedents, of perceptual states. Possession of an intuitive mechanics affects what forces you exert on objects around you, the directions in which you exert the forces, which objects you move towards or away from, and quite generally what you attempt to do. These are all relational facts about you and your environment, relational facts whose explanation makes reference to your intuitive mechanics. Of course, there is no one-one correlation between possession of a certain principle of an intuitive mechanics and any particular relational fact about you. Attribution of an intuitive mechanics has its explanatory consequences only in combination with a batch of auxiliary hypotheses: in this respect, it is like any other theoretical claim. I conclude by suggesting that the case of intuitive mechanics ought to encourage us to undertake a more general investigation into the ways in which intuitive theories and conceptions on the output side of perception help to make possible the distinctive explanations given by a content-involving psychology.

ACKNOWLEDGEMENTS

I have been helped by the comments of John Campbell, Tony Marcel, Gabriel Segal, Elizabeth Spelke, Michael Tye, the editors and an anonymous referee.

NOTES

1 This point about generality also applies against a more sophisticated use of the comparative relations which hold between unit-free magnitudes. We know from the theory of measurement that if we choose the right set of relations, specified without use of numbers, and a suitable set of statements involving them, the numerically specified distance between two objects will be uniquely determined (up to linear transformation). So someone might suggest that to accept the particular relation between physical magnitudes embodied in a certain statement which mentions numbers is for the following condition to obtain. It is for the subject to accept sets of contents not involving numbers, for each of the magnitudes in question in the statement, and for these contents each to determine statements of distance, mass (or whatever), unique up to linear transformation, and which stand in the relation given in the statement. This account involves all kinds of questionable suppositions: but in any case, it still lacks the appropriate generality. What it offers specifically in answer to our question is still no more than the function-in-extension as determined by the numerical statement of the relation between magnitudes. The additional theory it offers, building on ideas from measurement theory, of what it is to mentally represent a certain magnitude, does not address the issue of generality. It is also questionable whether this approach can accommodate the application of a principle of intuitive mechanics to a new case. If which principle is being accepted is not determined until all singular qualitative statements involving the magnitude are fixed, the possibility of application of an already accepted principle to an object newly thought about is undermined.

2 To say that a principle is psychologically real is not to say that some sub-personal system *consults* some representation of it. A principle of inference can be psychologically real within a system which has a language of thought; this does not require that the system contain a representation of the inference rule (and indeed it cannot always do so, on pain of a familiar regress).

REFERENCES

Cummins, R. 1989: *Meaning and Mental Representation*. Cambridge, Mass: MIT.
Davies, M. 1981: *Meaning, Quantification, Necessity*. London: Routledge.
Dennett, D. 1991: Real Patterns. *Journal of Philosophy*, LXXXVIII, 27–51.
Evans, G. 1985: Semantic theory and tacit knowledge. In his *Collected Papers*. Clarendon Press: Oxford, 322–42.
Fodor, J. 1985: Some notes on what linguistics is about. In J. Katz (ed.), *The Philosophy of Linguistics*. Oxford: Oxford University Press, 146–60.
Johnston, M. 1987: Is there a problem about persistence? *Proceedings of the Aristotelian Society*, Supplementary Volume LXI, 107–35.
Karmiloff-Smith, A. 1984: Children's problem solving. In M. Lamb, A. Brown and B. Rogoff (eds), *Advances in Developmental Psychology III*. New Jersey: Erlbaum, 39–90.
Karmiloff-Smith, A. 1988: The child is a theoretician, not an inductivist. *Mind and Language*, 3, 183–95.
Leibniz, 1951: On Nature in Itself (Essay, 1698). In P. Wiener (tr. & ed.), *Leibniz Selections*. Scribners: New York, 137–56.
Peacocke, C. 1986: Explanation in computational psychology: Language, perception and level 1.5. *Mind and Language*, 1, 101–23.
Peacocke, C. 1992a: Scenarios, concepts and perception. In T. Crane (ed.), *The Contents of Perception*. Cambridge: Cambridge University Press, 105–35.
Peacocke, C. 1992b: *A Study of Concepts*. Cambridge, Mass: MIT.
Tye, M. 1988: The picture theory of mental images. *Philosophical Review*, XCVII, 497–520.
Wittgenstein, L. 1958: *Philosophical Investigations*, tr. G. E. M. Anscombe. Oxford: Basil Blackwell.
Yates, J. et al., 1988: Are conceptions of motion based on a naive theory or on prototypes? *Cognition*, 29, 251–75.

Part III

Spatial representation in the sensory modalities

Introduction: Spatial representation in the sensory modalities

Naomi Eilan

The following chapters cover a wide range of questions. They are less closely related to each other than are chapters in others parts, in that each is concerned with one or more of the very many different problems raised by the issue of spatial representation in the modalities. Nevertheless, there is one meta-question that they can all be said to be either directly concerned with or to have bearings on, and this is whether there is any sense at all in which the perception of, and thought about, objects, space and spatial properties is modality specific.

Thus, both chapters seven and eight suggest that there are ways in which touch and body sense respectively do exhibit some modality-specific features. For chapters nine and ten, concerned with questions of cross-modal integration and matching, a central issue is whether perceptions are modality-specific and, if so, in what way. Finally, chapter eleven is concerned with the kinds of errors it argues are prevalent in contemporary writing on perceptual contents, including claims to the effect that these contents are either amodal or modality-specific. Given this variety of claims, one immediate question to arise when comparing them is what are the senses of 'modality specificity' under attack, defence or mere consideration. This is the question we will focus on in these introductory remarks.

These different senses of modality specificity bring out both the complexity of the problem of spatial representation in the modalities, and its importance to a general account of spatial perception and its relation to spatial thought. However here, perhaps even more than elsewhere, it should be stressed that our central question is not an explicit concern of the following chapters, but one which arises mainly from their comparison.

Molyneux's question has been the focus of much psychological and philosophical discussion on the nature of spatial representation in the modalities, since it was taken up and publicized by Locke in his *Essay Concerning Human Understanding*. Although only three of the chapters are explicitly concerned with variants of the question, it serves as a useful peg for examining the relation between different modality-specificity claims that might be made. We begin, in section I, with the question in its original formulation, and then go on to list the various notions of modality-specificity touched on in Part III and which might or might not be relevant to answering the question. In section II we give a brief summary of the main claims in each of the five chapters, and in section III we inquire about the relation among some of the senses of modality-specificity touched on and the bearings they have on Molyneux's question.

I

1 Molyneux's question

> Suppose a Man born blind, and now adult, and taught by his touch to distinguish between a Cube and a Sphere of the same metal, and nighly of the same bigness, so as to tell, when he felt one and t'other, which is the Cube, which the Sphere. Suppose then the Cube and Sphere placed on a Table, and the Blind Man to be made to see: quaere, Whether by his sight, before he touched them, he could now distinguish, and tell, which is the Globe, which the Cube? (Locke, 1975, II, ix, 8, p. 146).

As we noted, this question has provided the framework for much philosophical and psychological discussion of the nature of spatial representation in the modalities.[1] Its format has been extended to properties other than shape, indeed to all spatial and physical properties and entities, and to other sensory modalities and to action. Very broadly, the general understanding of the question, in any of these various extensions, has been that someone who holds a modality-specific view of spatial representation will tend to deliver a negative answer to Molyneux's question; and that someone who holds an amodal view will tend to give a positive answer. However, there is wide variation in understanding of what modality specificity and amodality come to, and consequently a great variety in the bases for delivering positive and negative answers, a variety which suggests that answers based on one notion of modality specificity may not carry over to questions based on another.

2 Varieties of modality specificity

In what follows we will list some of the different kinds of modality specificity claims that are touched on, implicitly or explicitly, in Part III. We will return to how these various claims bear on Molyneux's question, and to the general issue of what their bearing on the question tells us about the relation among them.

(1) The properties and objects that are experienced by the various modalities and that are referred to by judgements based immediately on the deliverances of the modalities, are themselves modality-specific, in the following sense. The essential features of the objects and properties experienced in each modality are determined by subjective, modality-specific characteristics of the way they are experienced. The most obvious form of such a claim is to say that the essential features of the objects and properties experienced in each modality are determined by the character of modality-specific sensations. On this view, the shapes experienced in touch are essentially distinct from the shapes experienced in sight. There is no one property of, say, being a sphere or a cube, experienced by both modalities and no common referent to shape judgements based on these experiences. The same holds for the space and objects experienced and judged about. (This radical modality specificity claim was made, most famously, by

Berkeley (Berkeley, 1975, CXXVII, ff.) It is tantamount to claiming for spatial properties what many say holds for secondary qualities such as colour, that is, denying the distinction between primary and secondary qualities, and with it the usual understanding that the former are in some sense mind-independent.)

(2) The content of token perceptions in each one of the various modalities are modality-specific in the sense elaborated in the Introduction to Part IV, under the heading of the 'separate coding view'. On this notion, the codes are distinct just in case for each modality there are modality-specific principles for grouping token perceptions in that modality into types, where these principles bring out the psychologically relevant similarities and differences between these token perceptions. (For more on this notion of separate coding, see the Introduction to Part IV and chapter 13.) This notion of modality specificity is consistent with the claim that these distinctly classified perceptions in each modality are caused by and, indeed, on some versions and in some cases, refer to the same amodal properties and objects in the external world. That is, claims of modality specificity in this sense can be consistent with denying modality specificity in sense 1.

An example of such a separate coding claim would be that vision and touch use distinct, non-overlapping criteria for classifying shapes as the same, similar or different; or that hearing and vision use distinct, non-overlapping frames of reference for classifying places as the same or different (on this notion of 'frame of reference', see the Introduction to Part I). Another example might be that spatial contents are extrinsic to the character of experience proper in all the modalities, and that the type to which token experiences in each modality belong is determined by the kind of modality-specific sensations the experiences consist in.

(3) The concepts used in judgements based immediately on perceptions in each modality are distinct. On this notion of modality specificity, the concept 'cube', say, is ambiguous depending on whether it is used in haptically or visually-based judgements. There is no one concept common to judgements in both modalities. This claim might be based on claims of modality specificity in sense 1 and/or 2 above, but the relation between these three notions of modality specificity is somewhat complex, and depends on general views about the relation between concepts, reference and perceptual content.

(4) There are differences among the modalities with respect to which spatial and physical properties they inform us about. Thus, uncontroversially, we do not believe that hearing informs human subjects about shape. More controversial are claims comparing vision and touch in this respect, in particular claims about which spatial properties touch informs us about, and which physical or material properties vision informs us about. To claim that there is modality specificity in this respect is, of course, consistent with denying modality specificity in senses 1–3, for one might claim that with respect to those properties for which there is overlap, there is no modality specificity in any of those three senses.

(5) There are differences among the modalities with respect to which features are exploited for purposes of recognizing object-types, particular objects and properties. Thus, the claim might be, that for purposes of recognizing cats, say, texture and

temperature are typically used in touch, whereas shape is typically exploited in vision. This is not the same as saying that the concept 'cat' means something different in judgements based on the deliverances of each modality, that is, modality specificity in this sense does not entail modality specificity in sense 3 above. Nor would such a claim of modality specificity entail modality specificity in perceptual codes, that is, in sense 2 above. Thus, the claim might be that mass distribution is often a diagnostic cue for recognizing shapes in touch, while various aspects of surface structure the cue used in vision. But this need not mean that touch and vision use different classifications of three-dimensional shapes as the same, different or similar.

(6) Collateral information stored about objects and properties identified through the various modalities is distinct. Thus, the claim might be that more information about material properties of object-types is stored on the basis of touch, and information about colour stored on the basis of visual encounters. Again, this is not to say that the concept 'cat' is ambiguous, for the information is, precisely, collateral information, that is, information that is additional to whatever it is we think of as the essential properties of cats.

(7) Claims that there are differences in collateral information, in turn, give rise to the possibility, but only the possibility, of claiming functional disassociability in the information acquired through the various modalities, in the sense currently much discussed in neuropsychology under the heading 'modality specificity'. Thus, it turns out that for some kinds of concepts, such as animal concepts, particular lesions lead to the loss of the ability to recognize visually presented instances of particular kinds of animals, coupled with the retention of verbally accessible general information about them (McCarthy, personal communication); others lead to the converse effects (McCarthy and Warrington, 1988). How such effects should be explained is, of course, a highly topical issue in neuropsychology. But claiming that a similar effect is to be found with respect to information acquired though different sensory modalities would involve showing that subjects may, for example, lose the ability to access information typically acquired through touch, without losing the ability to access information acquired through vision, or vice versa. Again, this would not show that we have two concepts of, say, 'cat', a visual and a haptic one, but, at most, that the information that goes into grasp of a unitary concept is non-unitarily stored and accessed. Or more generally, that cognitive architecture does not mirror conceptual structure.

(8) The processes underpinning spatial perception in the various modalities are distinct, where such distinctness can include differences in information drawn on for various computations, in algorithms used at various stages, in whether processes are sequential or parallel, in whether, indeed, information pick-up is direct or involves computations, and so forth. Claiming such differences is in principle consistent with denying modality specificity in all the senses indicated above.

(9) There are differences of kind with respect to the vehicles of spatial representation in the modalities. That is, there are differences among the modalities of the kind that go into the distinction between pictures and sentences, or between sentences in distinct languages. Among differences sometimes cited with respect to touch and vision we find claims to the effect that the vehicles of haptic shape perception are, characteristically,

such that the units of representation are sequentially ordered, unlike the vehicle of visual shape perception, which are in some sense simultaneous. Claims about vehicles may or may not be said to constrain the content of perceptions, that is, may or may not be said to result in modality specificity in sense 2. But, in principle, it is certainly possible to make claims about distinctness in the type of vehicle while denying modality specificity in any of the above senses.

(10) There are phenomenological differences among the modalities with respect to the way spatial and physical properties are experienced that go beyond, and are distinct from, any of the above notions of modality specificity. Claiming that there are such differences is in principle consistent with denying modality specificity in any of the above senses, though whether in fact it is depends, of course, on how exactly these phenomenological differences are explained.

II

(a) The puzzle Klatzky and Lederman are concerned with is this. On the one hand, experiments they have done with their co-workers suggest that we are remarkably quick and accurate at recognizing familiar object-types by touch alone. On the other hand, there is a body of research which suggests that haptic shape perception is very poor. Their question is: Should we take the latter results to suggest that in touch, unlike vision, shape is not diagnostic of object category?

The authors' response to this question falls into two parts. First, with respect to haptic shape perception, they argue that the research used for claiming that the haptic system is poor at shape identification is based mainly on the results of tests on the ability to identify raised contours. However, on the basis of experiments they have conducted, they argue that there are systematic and distinct procedures we use in haptic exploration, depending on the kind of information we wish to extract (shape, texture, weight and so forth), and that contour tracing is not the way we normally begin shape exploration. In seeking information about shape we use what they call 'enclosure and lifting' to provide a rough idea of global shape, with contour tracing resorted to only in order to fill in additional details. For this reason alone, we should be cautious about claims based on contour tracing. They go on to argue that contour tracing used on its own imposes processing demands (of temporal integration) that our natural way of finding out about shape by-passes. (They supplement this claim with experiments designed to show that when analogous processing demands are placed on vision, the results are almost as poor as those for haptic exploration that relies on contour tracing alone.)

With respect to whether shape is, in fact, used in haptic object-type recognition, they suggest that a rich assortment of experiments they have conducted show that despite the fact that the haptic system is good at shape perception, material properties (such as weight, temperature and texture) are more salient in haptic perception than in vision, and used for sorting unfamiliar objects into types. This, in turn, suggests, they argue, that such properties may be used more often than they are in vision for recognizing familiar object-types. Moreover, they suggest that this shows up in the 'cognitive representations' of familiar object-types. For example, when asked how one would identify a hammer by touch subjects are more inclined to mention weight distribution than they are when asked how they would identify it by sight. It is such phenomena that lead Klatzky to suggest that what they call the 'cognitive representation' of object-types is likely to depend on the modality in which it is encoded.

(b) Martin is concerned with the puzzle of what he calls 'body sense' the way in which each of us is aware of his or her own body 'from the inside', through bodily sensation, kinaesthesia and our sense of balance. He suggests that the reason for grouping together these ways of acquiring information about our bodies is that they underpin a kind of awareness of the object our body is which is distinctive in that the proper and sole object of awareness in this case is our own body. There is, in bodily awareness, no identifying of one's body as one object among others experienced, as there is when one becomes aware of one's own body through touch or sight. But this fact raises the puzzle Martin is concerned with: if, in bodily awareness, we do not identify our bodies as one among others experienced, then in what sense is there any awareness of our body as an object?

The puzzle is explored primarily through the problem of the felt location of one's bodily sensations. As Martin interprets it, the sense in which the sole object of awareness in bodily sensation is our own body is that wherever a sensation is felt to be located, one thereby feels one's body to extend to that place. In this sense, feeling a sensation in one's body does not rest on identifying one's body as one object among others, for wherever one feels the sensation there one feels one's body. But this alone would not yield a sense of one's body as an object within space. For that one needs some contrast between what is and what is not one's body. This, Martin argues, is provided by our sense of the boundaries of our bodies, and this sense of boundary,with its concomitant sense of a space that extends beyond the felt body, is intrinsic to the phenomenology of bodily awareness. Martin then goes on to explore what exactly this kind of awareness of our boundaries involves.

Martin then discusses the relation between bodily awareness, touch and vision. He argues with respect to touch that there is a kind of mutual interdependence between bodily awareness and touch which means that some aspects of the phenomenology of bodily awareness carry over to the phenomenology of haptic awareness of objects and space. However, he suggests that, in contrast, there are fundamental differences between the spatial phenomenology of vision and body sense, both with respect to the role of the notion of a boundary, and with respect to the relation between awareness of objects and awareness of the space in which they are located. He also suggests that behind philosophical denials of there being any such thing as body sense there may well lie an illegitimate requirement to the effect that all awareness of objects and space conform to the visual model.

(c) Meltzoff is concerned with two interrelated questions. The first turns on variants of Molyneux's question applied to very young infants who have not yet had time or opportunity to form associations between the deliverances of the modalities, in a manner analogous to the way the blind person who regains sight will not have had the opportunity to form associations between the deliverances of sight and touch. His first question here is whether infants are able to match and integrate across modalities without the benefit of association, and he tests this for a wide range of properties and modalities, including feeling shapes and seeing them, imitating seen gestures, and matching heard sounds to seen and to produced lip movements. The general answer is that there is an impressive degree of cross-modal matching. His second question is: What does this tell us about the way infants code properties across the modalities? Here, he argues that his results support the claim that infants' perceptual coding is amodal, in contrast, for example, to Piaget's claim that infants begin life with distinct, modality-specific 'universes' which they gradually learn to co-ordinate through association (Piaget, 1954, p. 13).

Meltzoff is also concerned with the problem of the ontogenetic and constitutive core of self awareness. As he notes, on a variety of developmental theories, such awareness is supposed to emerge suddenly, when babies are approximately 18 months old and begin to recognize themselves in mirrors and in photographs. This awareness is, moreover, often supposed to be connected with the emergence of various linguistic achievements. In contrast, Meltzoff argues that infants' and babies' awareness of their own bodies and movements 'from the inside' is more fundamental, both constitutively and ontogenetically. He goes on to describe series of experiments he has conducted to test and further articulate this claim, experiments that, he argues, show that 14-month-old babies can recognize that they are being imitated. He argues that the results show that these babies have a primitive grip on the likeness of others to themselves, and goes on to suggest the principles, psychological and physical, that may be involved in the kind of self awareness that goes into registering these kind of similarities between oneself and others. (This second issue is discussed in the Introduction to Part IV and will barely be touched on here. See also the last section of the General Introduction.)

(d) Eilan suggests that part of the enduring interest in Molyneux's question lies in the idea that there is immediate connection between answers to it and explanations of what it is for perceptions to be perceptions as of an external world. Very broadly, it is generally supposed that if the answer to the question and to its various extensions should be 'Not', on the grounds that our spatial perceptions are modality-specifiic, in some sense, this poses a serious threat to our access through perception to the world out there.

Her aim is to examine whether this is indeed the case. She does so by focusing on a variant of Molyneux's question with respect to place representation. She considers imaginary subjects who are unlike us in that their perceptions of places are modality-specific in a variety of ways which she spells out, some more radical than others, all of which would result in delivering negative answers to the variant of Molyneux's question she constructs. Her question is: Which, if any, accounts of such modality specificity would threaten which, if any, ingredients in explanations of what makes perceptions perceptions as of places in the external world?

If there is some deep connection between how we explain the relations between spatial perceptions in the modalities, and how we explain what makes perceptions perceptions as of places out there, one would expect examination of the exact nature of this link to illuminate particular aspects of what goes into making perceptions perceptions as of an external world. Eilan suggests that one such aspect is how exactly we spell out, motivate and justify the connectivity requirement, the intuitive idea that there is some link between representations counting as representations as of places out there, and their representing places as spatially related to each other. She argues that only accounts which constitutively link objective place representation to self-conscious self-location are such as to generate an explanation of connectivity which actually rules out modality specificity in place representation. Weaker accounts of what it is for a subject's perceptions to be perceptions as of external places do not, she argues. She goes on to consider the implications of her arguments to the way we explain the mapping abilities of animals and developing children.

(e) Millikan is concerned with illustrating abstract varieties of what she regards as a fundamental and pervasive error in the way in which perception is discussed in much contemporary work. The basic form of the error she describes involves two complementary moves. The first consists in what she calls 'content internalizing', projecting properties of the content of the perceptual state or act onto its vehicle. One example of such an error is thinking that the vehicle of a perceptual representing of temporal properties must itself have these very temporal properties; as in thinking that the perceptual representing of succession must involve successive representing elements. The second move is one she labels 'content externalizing', projecting properties of the vehicle onto the contents. A temporal example of this error would involve claiming that if the vehicles of representation are sequentially ordered this suffices for the perceptual representation as of succession, that is, for the perceptual state to have succession in its contents.

The effect of making these two moves, Millikan suggests, is to create an illusion of having produced an explanation of mental representing. The resultant picture is one in which the mind is moved by the contents represented in virtue of the vehicle bearing precisely the properties which the content represents. But in adopting this mythology we simply evade the central and difficult project of explaining the mechanics of representing, how representational contents can and do move the mind.

Millikan draws out subtle and pervasive examples of this form of error in contemporary writing on perception, including ones that involve internalizing and externalizing constancy, determinacy and sameness and difference. Many of the examples occur within the framework of responses to Molyneux's question, in a wide range of different kinds of argument for both amodality and modality specificity.

III

There is a general tendency in many appeals to Molyneux's question to assume that by focusing on one notion only of modality specificity one can secure a definitive positive or negative answer to the question, which at the same time yields a general explanation of the relation between the modalities and a general account of perceptual content as either amodal or modality-specific. This tendency rests on any of four tacit assumptions. First, that arguments with respect to the perceptual experience of one particular spatial or physical property carry over to the experience of all other properties. Second, that arguing for or against modality specificity with respect to one or more modalities carries over to all other modalities. Third, that the answers given to Molyneux's question relative to one notion of modality specificity will automatically carry over to the question as asked relative to a different notion of modality specificity. Fourth, that only those notions of modality specificity that would result in delivering a negative answer to Molyneux's question are relevant to explaining the phenomenology of spatial perception in the various modalities.

It seems to us that the chapters in Part III provide the materials for questioning all four assumptions and, more generally, the grounds for making these assumptions. We will not attempt to match all the notions of modality specificity raised in each chapter with the list of possible modality specificity claims we gave in section II, nor will we list all of those that might be relevant to delivering conflicting answers to Molyneux's question. Rather, we end with a few illustrations of how the following chapters serve to raise doubts about these four assumptions.

(1) A variant of Molyneux's question suggested by Klatzky and Lederman is this. Imagine a person born blind who, without tracing their contours, can tell which of two objects is the hammer and which the screwdriver. Suppose she suddenly regains her sight: will she be able to tell which of two visually presented objects is the hammer and which the screwdriver?

If Klatzky and Lederman are right about typical cues used for recognizing object kinds in touch – material properties, local shape fragments and rough global shape, in contrast to the pervasive reliance of contour in vision – it might be that the answer is no. Or it might vary from subject to subject and from object-kind to object-kind. But if the answer is no, based on the possibility of modality specificity in sense 5 above, that is, modality specificity with respect to cues used in recognition, this does not entail modality specificity in any of the other senses mentioned earlier, except, perhaps, for the sense in which there might be differences in collateral information stored, i.e. modality specificity in sense 6 above (which is, roughly, how it seems we should interpret Klatzky's and Lederman's appeal to the idea that 'cognitive representations' might depend on modality of input). So a negative answer to Molyneux's question may well leave questions about the nature of our concepts, perceptual codes or the referents of these completely untouched, contrary to many philosophers' assumptions.

(2) Klatzky's and Lederman's initial concern is with the claim that haptic perception is very poor at shape detection, which is to make a claim about modality specificity in haptics in sense 4 above, a claim about the kind of properties about which a modality informs us. Their suggestion is, in effect, that the notion of modality specificity relevant to explaining the findings on which this claim is based is modality specificity with respect to the processes involved in such tasks, that is, modality specificity in sense 8 above, rather than modality specificity in sense 4. Thus, they suggest that these findings should be explained by the fact that the ways of acquiring information about shape relied on for these findings, especially contour tracing, are typically extended in time, and therefore impose processing and memory demands which affect performance.

Although it is not their concern in their chapter, it is worth mentioning in this connection the extraordinary idea that the blind have no conception of space at all, and that time stands for them as space stands for the sighted.[2] This kind of claim might be explained as a case of transferring properties of the process onto the vehicle of representation and then externalizing these properties of the vehicle onto the content represented, in the manner described by Millikan, in such a way that we get the claim that what is represented is temporal succession or duration, rather than spatial properties. As a result of this move, modality specificity is claimed for touch in senses 2 and 3 above, i.e. both with respect to perceptual code and to concepts. What Klatzky's and Lederman's remarks bring home is that this kind of externalizing move can be aided and abetted by the empirical consequence of focusing on unsuitable stimuli, of the kind that do in fact deliver poor spatial perception. One question that arises is why have both philosophers and psychologists focused, in experiment and thought experiment about touch, on two dimensional stimuli? It may well be that this focus is the consequence, at least in part, of transferring to the external stimuli used, properties tacitly assumed to be true of the image in vision, namely that it is a two-dimensional spatial array, contemplated by the mind's eye, an assumption which is, on the one hand, a classical example of the consequences of the kind of internalizing and externalizing moves Millikan describes; and, on the other, an illustration of Martin's suggestion that various features held to be true of vision are often assumed to be true of other modalities, in particular touch and body sense.

(3) The point of focusing on a person born blind in Molyneux's original question is that it is supposed to rule out the possibility that using the same terms in response to tactile and visual inputs is a consequence of pervasive acquired associations among inputs that are modality-specific in all or some of senses 1–3. Meltzoff's appeal to very young infants is intended to rule out the possibility of acquired association as the basis for matching, from which we may conclude that the basis for this capacity must be innate. His particular suggestion is that this also serves to rule out modality specificity with respect to perceptual code, i.e. in sense 2 above.

It might be held that innateness does nothing to rule out modality specificity with respect to perceptual code, on the grounds that anything acquired can in principle be innate, and there is nothing that rules out the possibility of hard-wired associations among different perceptual codes. In principle, this answer is indeed an option. But in practice there are two challenges that must be met if this form of response is to hold. First, what are the consequences of the hard-wired association supposed to be? They had better not be that the inputs from both modalities are treated as identical in all relevant psychological respects. The onus of proof, once presented with evidence for innate matching capacities, is on the separate coding adherent to show some ways in which inputs from the modalities are psychologically distinct in a way that rules out the possibility of a common perceptual code.

Second, what is the explanation of the hard-wiring supposed to be? In some cases, it is plausible that it will turn on the advantage of associating inputs caused by different kinds of stimuli: it is arguable that the example discussed by Meltzoff, of innate matchings of sounds to mouth movements, is one such case. But in the case of shape or more generally spatial representation, it is hard to escape the idea that the point of such hard-wired matching is to connect inputs that are caused by what are in fact the same properties or objects. If this is the explanation, then it is at least arguable that with respect to perceptual code, such explanation is partially determinant of the content of the perceptual code, in which case we are at least close to having a common code in play.

To say that innateness is at least relevant to arguments for a common code is not, of course, to say that all experiments that show innate matching are equally persuasive in this respect. Thus, in Meltzoff's experiments designed to show matching of shapes, the response that indicates matching is one of preferential looking. In the imitation experiments, on the other hand, the response involves the production of spatial behaviour and, Meltzoff argues, the correction of that behaviour to achieve exact match. It is arguable that the production of spatial responses that match the spatial character of the perceptual stimulation in the other modality provides stronger reasons for opting for a common code view than does the production of preferential looking, in which there is no correlation between the spatial structure of the response and the spatial character of the stimulus in the other modality. (For reasons for thinking so, see chapter 13, the

Introduction to Part IV, which discusses Meltzoff's imitation results in this light, and chapter nine.)

(4) One substantial question raised by Martin is this. There is nothing in his description of the distinctive phenomenological character of bodily awareness that would rule out Meltzoff's imitation results, assuming infants are aware of their bodies in the way described by Martin. That is, to the extent that Meltzoff's results tell for a common code between the spatial representation of our bodies and of visually perceived bodies, Martin's insistence on a distinctive phenomenology for bodily awareness does not tell against there being such a common code. So his notion of modality specificity is consistent with rejection of modality specificity in sense 2. But he suggests a stronger point, namely that the spatial phenomenology of perceptual experiences cannot be captured wholly in terms of their representational content, and that modality specificity comes in on this non-representational level. That is, he gestures at modality specificity in sense 10. He does not develop this intriguing suggestion here, but even without such development, a minimal suggestion we may extract from it is that insistence on modality specificity with respect to the intrinsic phenomenology of perceptual experience is, on this notion, consistent with delivering a positive answer to Molyneux's question.

Though Martin's appeal to the distinctive character of bodily awareness is consistent with admitting a common code, to what extent does describing one's awareness of one's body as spatial awareness actually depend on a common code of any kind, even if the nature of this awareness cannot be exhaustively accounted for in terms of such a code? Even if Martin's claims are consistent with and, even more strongly, actually require a common code, is this the kind of code needed for explaining imitation? Thus, as the Introduction to Part IV notes, the kind of spatial code implicated in successful reaching for places in the environment in response to perceptual input arguably requires less in terms of the representation of one's own body than the kind of spatial representation of one's own body needed for explaining imitation; to which one might add that the latter is in turn perhaps less developed, in some sense, than the kind needed for recognizing that one is being imitated. Martin's phenomenological characterization of bodily awareness is arguably intended to capture the least reflective and most basic way we have of being aware of our own bodies. A question of great interest, raised by Martin and Meltzoff is, then, how we should build up degrees of objectivity in our awareness of our own bodies, both informationally and phenomenologically. (See the Introduction to Part IV, and the General Introduction, for more on this issue.)

(5) One of the general problems highlighted by Martin is the relation between a phenomenologically-based individuation of the modalities and one based on the idea of different information channels. As he says of body sense, it rests on what would count as different kinds of information channels: nevertheless, there is reason to group these together on phenomenological grounds, as they underpin a unitary and distinctive kind of spatial awareness. Similar differences between these two approaches to the modalities, phenomenological and information-based, arguably show up with respect to all the modalities. Thus, if we begin with a phenomenological description of visual experience, we will assume constancy in how things are perceived as being, and we will assume various symmetry effects, where the first arguably depends on proprioceptive correction for eye and head movements and the second on sensitivity to gravity, two information channels that underpin bodily awareness.

The information channels, and their content, underpinning perceptual experience are certainly relevant to explaining the phenomenology of perceptual experiences in the various modalities. A point we may extract from Martin's chapter is the suggestion that they are not, however, sufficient for yielding the phenomenological individuation of the modalities. One of the central issues, raised in very different ways by Martin and Millikan, is what are the tools for describing the phenomenology of perceptual experience and, in particular, what is the relation between claims made about their informational content, their vehicles and their phenomenological character.

It is arguable that the four assumptions we described at the outset of this section, assumptions that lead to the collapsing into each other of the various notions of modality specificity, rest in many cases on slipping into many of the internalizing and externalizing moves brought out by Millikan. That is, in addition to the examples she actually discusses, many of them, as we noted, culled from arguments for amodality and modality specificity of perceptual contents, there are many others. These would involve various illegitimate internalizing and externalizing moves from and to claims about one of the kinds of modality specificity argued for or against in Part III, to and from other kinds mentioned in other chapters. As Millikan says, the result of many such simplifying moves is to produce illusions of having explained difficult problems in the nature of perceptual representation.The following chapters do much to counter such illusions, by opening up and distinguishing the various notions of modality specificity relevant to explaining the nature of spatial perception in the modalities.

NOTES

1 For a historical review of philosophical and psychological debates about Molyneux's question see Morgan (1977). For a recent and highly influential philosophical discussion of the question, which draws largely on psychological work and which informs the setting up of issues here, see Evans (1985).

2 The most recent exponent of this view is M. von Senden (1960), see especially pp. 285–6. For discussion of this view, as one possible motivation for delivering a negative answer to Molyneux's question, see Evans (1985), pp. 367–70. For comments on some particular arguments Evans raises against this view see chapter 11 of this volume.

REFERENCES

Berkeley, G. 1975: *New Theory of Vision*. M. A. Ayers (ed.), London: Dent.

Evans, G. 1985: Molyneux's question. In his *Collected Papers*. Oxford: Oxford University Press, 364–99.

Locke, J. 1975: *An Essay Concerning Human Understanding*. P. H. Nidditch (ed.), Oxford : Clarendon Press.

McCarthy, R. A. and Warrington, E. K. 1988: Evidence for modality specific meaning systems in the brain. *Nature*, 334, 428–30.

Morgan, M. J. 1977: *Molyneux's Question*. Cambridge: Cambridge University Press.

Piaget, J. 1954: *The Construction of Reality in the Child*. New York: Basic Books.

von Senden, M. 1960: *Space and Sight*. P. Heath (tr.), London: Methuen.

7
Spatial and nonspatial avenues to object recognition by the human haptic system

Roberta L. Klatzky and Susan J. Lederman

Introduction

When we think about our ability to learn and represent spatial information, we generally think of the visual modality. Vision simultaneously conveys the relative positions of multiple features (e.g. landmarks, object vertices) within a coherent frame of reference, which is provided by the field of view. We should not forget, however, that touch is also a spatial modality. Kinaesthetic information from muscles, tendons and joints tells us about the positions and movement of the limbs in space and hence provides us with the ability to construct a spatial map of objects and surfaces that we encounter. The distribution of pressure on the skin tells us about the local spatial properties of objects. The system that provides and processes cutaneous and kinaesthetic information is called the haptic system (Loomis and Lederman, 1986). This chapter concerns the capabilities of the haptic system with respect to a specific problem in spatial information-processing, that of object recognition. The general issue we address is: What is the importance of spatial information about an object, particularly its shape, size and part structure, to haptic identification?

If one considers object recognition in the visual modality, such information is undoubtedly of primary importance. A critical step in visual recognition appears to be the extraction of primitive volumes or geons from viewpoint-independent properties of an image (Biederman, 1987). This constitutes a form of image parsing. Relations among geons are then determined, with two to three geons and their arrangement being sufficient to determine an object's identity. People can recognize a two-dimensional visual depiction of a common object that is presented as briefly as 100 milliseconds.

The usual criterion for object recognition is naming. The name that is provided is, for most objects, at what Rosch called the basic level of categorization (e.g. Rosch, Mervis, Gray, Johnson and Boyes-Braem, 1976). At this level, objects within a common category tend to share common features, and objects from contrasting categories tend not to share features. In particular, objects at the basic level tend to have a common shape, so much so that an averaged shape prototype is recognizable. On the basis of attribute listings by subjects who were asked to describe objects at the basic level, Tversky and Hemenway (1984) suggested that it is an object's part structure that is particularly diagnostic of its basic-level identity. Below that, or at what is called the subordinate level, non-part attributes tended to emerge.

Work on visual object recognition, then, makes the fundamental points that (a) an object's structure – the layout of edges, vertices, surfaces and parts in space – is strongly diagnostic of its identity at the level most commonly used in naming; (b) access to structural information, particularly about shape, underlies the process of visual object identification. Together, these results suggest that object identification is an important paradigm with which to assess the spatial processing capabilities of a sensory modality.

In this chapter, we consider these arguments in the context of a sensory modality other than vision, that of the haptic system. As noted above, the general concern is with the importance of information about an object's spatial characteristics when categorizing or identifying it by touch. We address the issue as follows. First, we describe evidence that the haptic system is highly effective at recognizing objects. Only in this case could we go on to consider the importance of spatial information for recognition. However, we also describe circumstances that seem to undermine haptic object recognition, namely, those in which the object is portrayed as a raised two-dimensional depiction. The inadequacy of touch in this situation is important for understanding how the haptic system normally performs object recognition. Finally, we discuss how object properties other than size and shape, those related to an object's material, may contribute to the identification and cognitive representation of objects.

1 Haptic identification of real common objects

Level of accuracy

In 1985, we published a finding that was considered somewhat surprising (Klatzky, Lederman and Metzger, 1985). The finding was quite simple – given the task of naming each of 100 common objects, using touch alone, people performed with almost 100 per cent accuracy. Moreover, their response times were typically within two seconds, as shown in Figure 7.1. These findings were surprising simply because the haptic perceptual system was generally considered to be a relatively ineffective one – for reasons we will discuss further below.

More recently, we followed up these results with a somewhat more focused identification task (Lederman and Klatzky, 1990). Subjects were asked a question about a common object that they explored haptically. The question took the form of a true – false item; for example, 'Is this tableware further a plate?' or 'Is this plate further a paper plate?' Subjects simply answered yes or no. As is indicated by our examples, the question targeted an identification response either at the basic level (plate) or a more subordinate level (paper plate). We again found that accuracy was very high – an average of 95 per cent at the basic level and 86 per cent at the subordinate level.

Attributes predictive of an object's name under haptic exploration

These studies establish that haptic object recognition is very good, which, given our initial discussion about the critical role of an object's structure in recognition, might suggest that the haptic system is also excellent at encoding spatial information. But

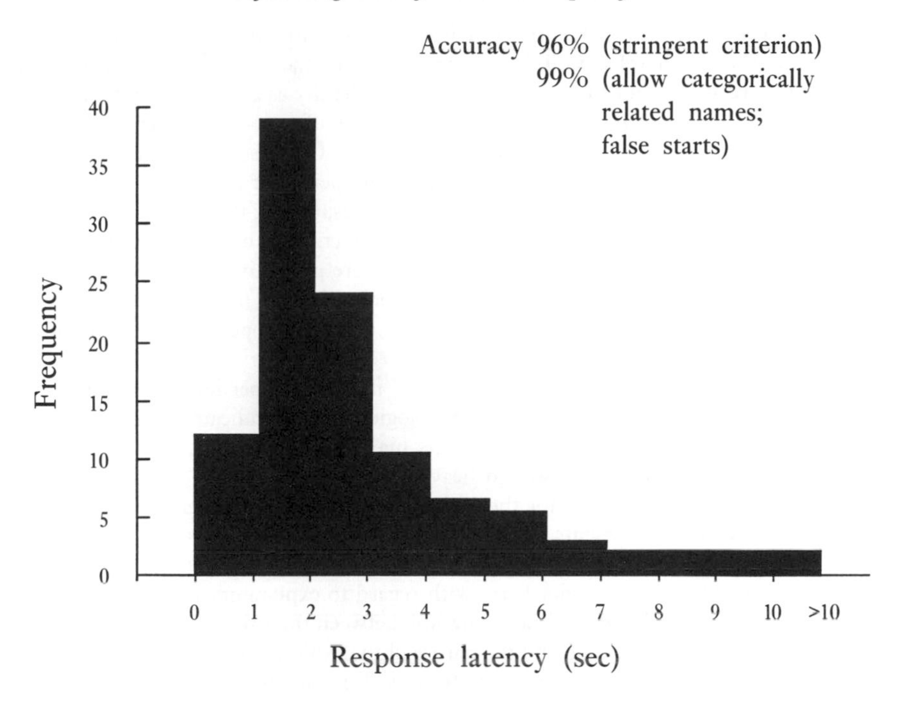

Figure 7.1 Performance in haptic object identification task of Klatzky, Lederman and Metzger (1985); each bar shows the number of observed responses completed with a particular response latency. Also shown is the overall accuracy.

several caveats are in order. First, most studies of the defining cues for object categorization have not considered the modality of touch. Do an object's shape and part structure remain critical for determining its basic-level category, when the determination is made haptically? If not, we should not use categorization proficiency to infer successful processing of structural information.

To answer this question, we asked a set of respondents to think about identifying a set of objects by touch alone (Lederman and Klatzky, 1990, Experiment 1). For each object category, they selected the attribute(s) predictive of category membership from a closed list: hardness, texture, weight, temperature, size, shape, part (defined for the subjects as a section of the object independent of any of its perceptual attributes such as shape), and motion of a part. An attribute from the list was assigned a score reflecting its frequency of listing by subjects, weighted by serial position in the list. The results confirmed the relative importance of shape for basic-level categorization. It was the most highly weighted attribute at the basic level, and its weighting dropped substantially when categories were defined at the subordinate level. Size was also found to be important at the basic level. These results give us little reason, it seems, to question the capability of the haptic system for spatial processing. At least they indicate that structural information is important for something that the system does well – determining the common name of a touched object.

On the other hand, our results suggested that properties other than shape, size and part structure were used to identify touched objects, particularly at the subordinate level but at the basic level as well. For example, the hardness, weight and texture of a paper plate are all relevant to its subordinate-level identity as a paper plate. Objects in our study had been initially selected so that a variety of properties other than shape might be predictive of their subordinate-level identity. Our results from the attribute-listing task confirmed the importance of material properties at this level. At the basic level, we allowed free variation of predictive properties; that is, the object pool was not selected so as to bias any particular property in predicting an object's identity. As we noted above, we found that structural properties were predictive of an object's basic-level category; however, we also found a substantial weighting for a material property, texture, at the basic level. Thus, it seems that material properties can be used to determine category membership at both levels.

There are some other indications that objects' material properties are used in identification by touch. For one thing, phenomenological reports indicate reliance on these properties (see Klatzky et al., 1985). Subjects may say, for example, 'I suspected the object was a hammer because it was so heavy'.

Yet another type of evidence for the importance of material properties comes from the exploratory hand movements that people use. Essentially, these movements suggest that people are seeking information about an object's substance as well as its structure.

A bit of explanation is in order here, with regard to exploratory movements of the hand. We have found that there is a strong link between the nature of the information that a haptic explorer is seeking, and the nature of the exploration itself (Lederman and Klatzky, 1987; Klatzky, Lederman and Reed, 1987). For example, when extracting information about surface texture, people move their hand so as to induce 'lateral motion', or shear, between the skin and the surface. Lateral motion is what we call an 'exploratory procedure' for texture. The principal exploratory procedures we have studied and the associated object properties are shown in Figure 7.2. The procedure that is associated with a given property tends to be used spontaneously when information about that property is sought; it also tends to be optimal, in terms of the accuracy or speed with which the property is encoded (Lederman and Klatzky, 1987). A procedure may also be sufficient to extract additional properties, but generally it does so at a less than optimal level.

With this in mind, let us return to what hand movements can tell us about the properties used to identify an object. In our study of yes/no categorization, we found that whether subjects were identifying objects at the basic or subordinate level, their exploratory movements occurred in a particular sequence (Lederman and Klatzky, 1990, Experiment 2). They began the sequence of exploration by grasping and lifting the object – or using our labels for exploratory procedures, they were executing Enclosure and Unsupported Holding. These procedures would give them coarse information about an object's structural and material properties. Enclosure, in particular, we have found to be sufficient to perform above-chance discriminations among objects on the basis of size, global shape, apparent temperature, roughness, hardness and weight. Beyond these two preliminary modes of exploration, subjects went on to use others, including procedures specialized for material properties such as texture and hardness. It would be a mistake, then, to conclude that categorization even at the basic level occurred entirely on the basis of an object's spatial properties.

This might lead us to question whether the observation of highly accurate object recognition is sufficient evidence for a strong contribution from spatial information. If

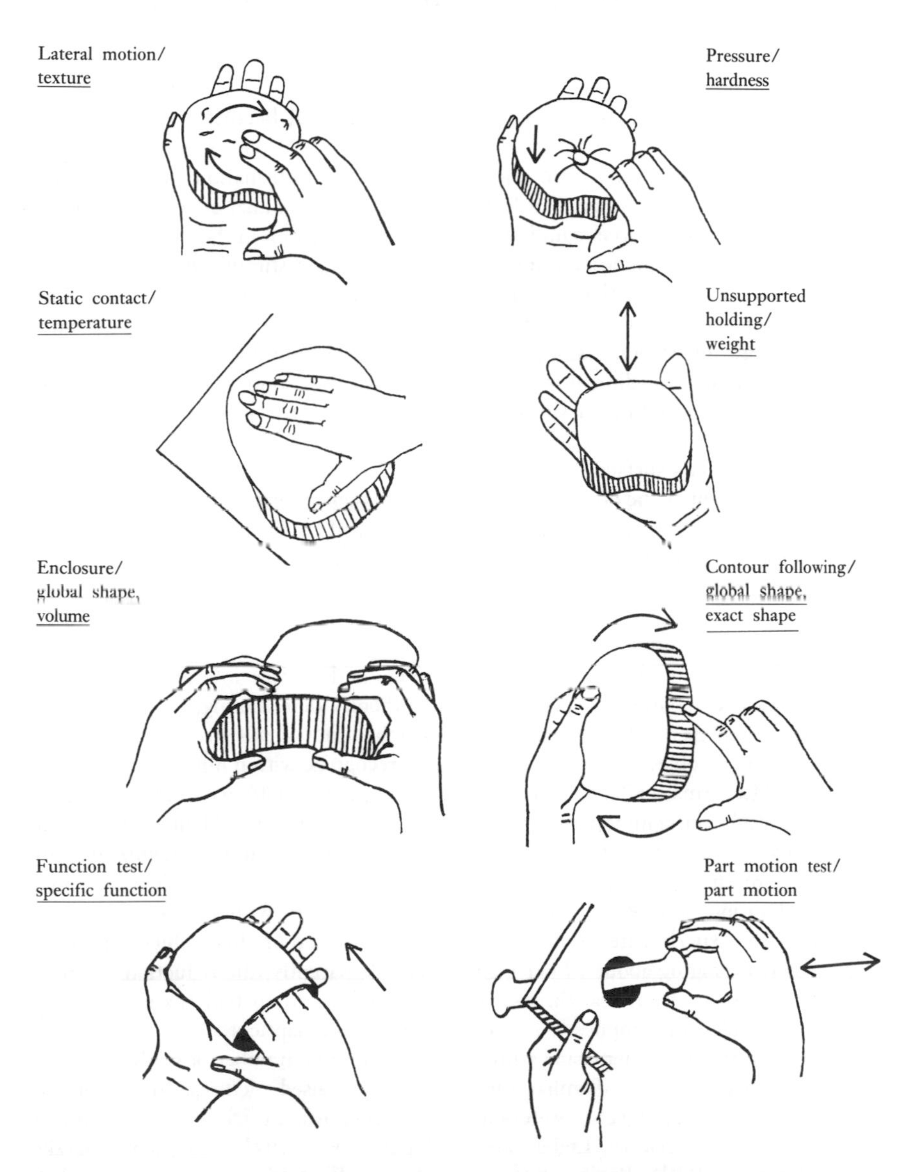

Figure 7.2 Haptic exploratory procedures and associated object properties. (From Lederman, 1991; adapted from Lederman and Klatzky, 1987. Copyright Academic Press Inc.)

the contribution of material properties is substantial enough, it might compensate for deficiencies in haptic encoding of structure. Subjects could then perform well in an object identification task, without underlying precision in spatial processing. This leads us to the issue of how well the haptic system encodes spatial layout information, which we consider next.

2 Haptic processing of 2-D spatial layout

Much of the research on haptic encoding of spatial information has used two-dimensional stimuli, either unfamiliar forms or depictions of familiar objects. The stimuli may be raised outlines or free-standing forms without variation in the third dimension. In this section we present results indicating that with such stimuli, the haptic system decidely does not excel relative to vision in discrimination or identification tasks. This relatively poor performance can be attributed in large part to the substantial demands on memory and integration resulting from sequential encoding of contour.

When we look at a two-dimensional projection from a three-dimensional object, certain contours of the object form a silhouette or envelope in the projection, and others become lines internal to the envelope. Also internal to the envelope may be visible signals of the object's surface properties, such as texture gradients, reflectance and so on. In order to identify the object, we must translate these visual cues into the primitive volumes and surfaces that originally projected them. The visual system, as we have noted, handles this problem with ease. The haptic system, however, does not.

Level of accuracy

Confronted with a raised drawing, in which edge and surface information becomes tangible lines and textures, people have great difficulty determining, by touch alone, what object it represents. We have commonly watched individuals struggle for a matter of minutes with an object that they could visually recognize within milliseconds. In one experiment (Lederman, Klatzky, Chataway and Summers, 1990), we simply replicated our earlier study on common-object recognition, but now using highly prototypical two-dimensional displays of familiar objects, which were readily recognizeable by vision. The mean percentage of correct responses to the raised-line drawings, under haptic exploration, was 34 per cent; the average time taken (with a two-minute limit imposed by the experimenter) was 91 seconds. A population of blind subjects did even more poorly, averaging about 10 per cent correct. Apparently, the reduction in information when we go from real three-dimensional objects to spatially extended, two-dimensional projections imposes considerable cost to the haptic system.

Difficulty with two-dimensional stimuli can be seen in a number of studies investigating performance with unfamiliar forms as well as raised pictures. In one of our studies, we used planar objects – wafer-like stimuli of constant thickness with a number of projecting parts (Klatzky, Lederman and Balakrishnan, 1991; Lederman, Klatzky and Balakrishnan, 1991). Typical items can be seen in Figure 7.3. Subjects were asked to perform a number of different tasks, including recognition memory for the whole stimulus or one of its projections (a projection was essentially an angle) and remembering the position of one of the projections relative to the others. The stimulus was stabilized by a supporting rod so that one or both hands could be used freely, and there was no

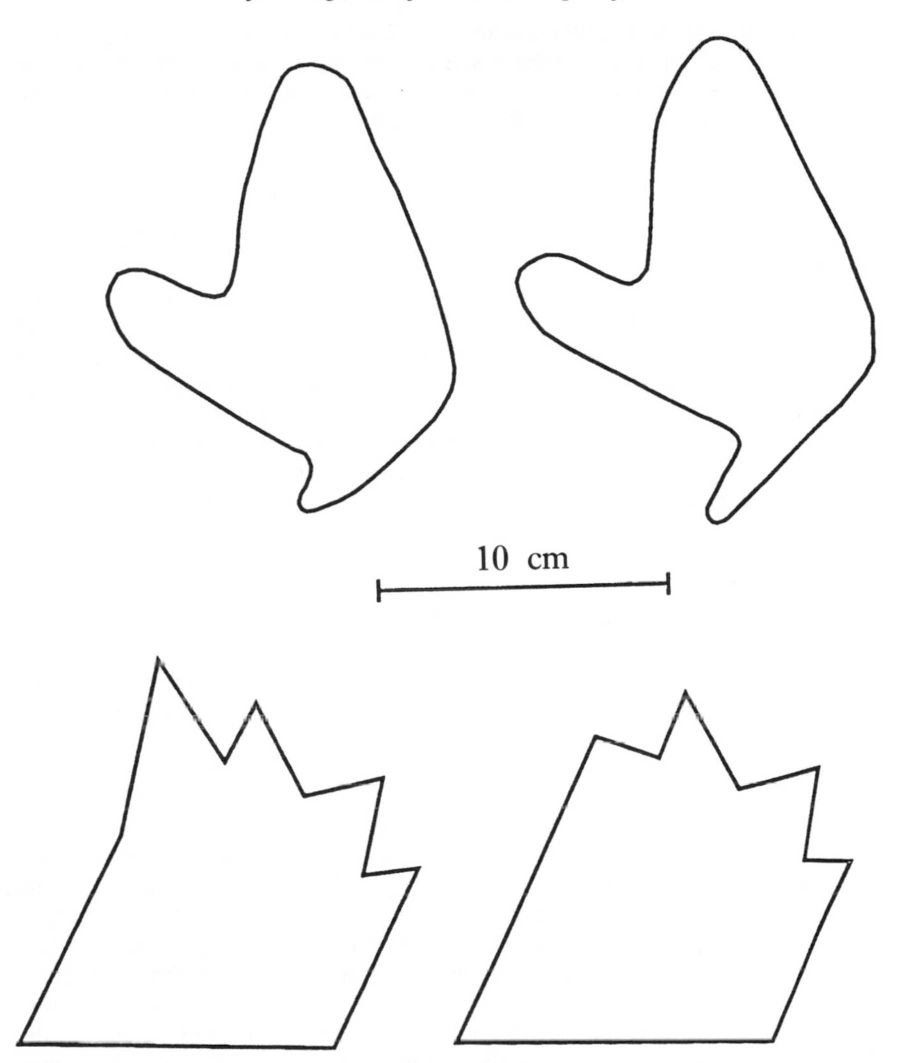

Figure 7.3 Samples of two-dimensional forms used in comparison task. Upper row: curved, low-complexity stimulus (left) and a changed version (right). Bottom row: pointed, high-complexity stimulus (left) and a changed version (right). (Adapted from Klatzky, Lederman, and Balakrishnan, 1991. Copyright Cambridge University Press.)

imposed limit on the time spent exploring it. Still, subjects performed only at about the 75 per cent level when recognizing a single projection or remembering its position relative to others, and they were no better than chance at a recognition test on the whole stimulus that immediately followed its exploration.

2-D displays are encoded sequentially

Why should people perform so badly in tests that require haptic encoding of two-dimensional contour? One potential answer to this question can be found in the nature of their exploration. Recall that subjects who are trying to recognize real objects tend to grasp and lift them, after which they execute exploratory movements that are useful in extracting texture, hardness and other specific properties. Subjects who are confronted with a two-dimensional object or a three-dimensional one without variation in the third dimension are reduced to a much smaller repertoire of exploratory movements. They can follow along edges with their finger(s), an exploratory procedure that we call 'contour following'. If the object projects sufficiently into the third dimension, subjects can attempt to enclose the whole or its parts in one or both hands. However, enclosure of even a moderately complex region is likely to give no more than an impression of curvature or sharpness rather than explicit contour. Contour following can access the entire object envelope, but it is of necessity extended over time and hence imposes demands on memory and spatio – temporal integration processes.

With Jack Loomis (Loomis, Klatzky and Lederman, 1991), we investigated the effects of encoding a contour sequentially. We asked subjects to recognize simple line drawings of objects (most of them used in our previous study with raised pictures) that they explored either haptically or visually. In the visual case, sequential exploration was imposed by means of a stationary visual aperture that allowed only part of the object to be seen at any one time on a computer monitor. The segment of object contour that could be seen within the visual aperture was actually controlled by movement of the subject's fingers (holding an electronic pen) over a digitizing pad. In essence, the object was spatially mapped by the computer onto the pad. As the subject moved the fingers over some part of the pad, the corresponding portion of the object was portrayed within the aperture. Further, the aperture was the same size as a typical fingertip, and resolution of the screen was matched to the subject's haptic two-point threshold. Thus, the subjects in the visual and haptic conditions had the same amount of information available at any one time, and they received the same kinaesthetic information, but the subjects in the visual condition saw the contour rather than feeling it. Under these conditions, the visual and haptic groups were virtually identical with respect to both accuracy (which was about 40 per cent for these pictures) and response time (an average of about 90 seconds). (Performance with vision was somewhat better if the aperture moved around the stimulus rather than having the stimulus appear to move behind the aperture, but it was still well below the level attained with full-field viewing.)

With respect to our central question about the importance of spatial information to object identification, experiments with two-dimensional stimuli make the important point that the effectiveness of the haptic system for encoding spatial information is limited. Limitations arise, in particular, when spatial information is encoded over time, by sequential exploration of sections of contour. It seems that sequential encoding imposes demands that undermine the success of object recognition. Given our efforts to match the amount of information available to the two senses, these demands are likely to be at a level that transcends the sensory modality. In particular, contours obtained from both haptic and visual sequential sampling may require storage in a spatial short-term memory and a process of integration to build an object depiction. The load on memory may mean that people retrieve relatively noisy spatial information, which would by itself limit their ability to integrate. In addition, whatever the fidelity of the retrieved information, there may be limits on the ability to integrate it.

One might argue that studies with two-dimensional displays underestimate the spatial competency of the haptic system, because they largely restrict exploration to being sequential; three-dimensional common objects, after all, can be explored simultaneously with enclosure. The adequacy of this argument is not entirely clear. First, we note that sequential exploration is not an atypical means of encoding contour information haptically even with three-dimensional objects. Contour following often occurred subsequent to the initial grasp-and-lift stage in our study with yes/no categorization of objects. And in our study with wafer-like objects, subjects who were to match relatively small single projections used contour following as the predominant exploratory procedure, even though enclosure was possible. Further, the simultaneous nature of enclosure does not necessarily render it more accurate than contour following. In our study of the associations between exploratory procedures and object properties, we found that contour following was the only means of exploration that produced above-chance discrimination of exact shape. However, we have not compared contour following to enclosure with stimuli that are differentiated on the basis of locally available (but still precise) shape information, such as the curvature of an ellipsoid. With certain types of stimuli, enclosure might fare better. Moreover, it might produce a level of precision that would allow for identifying common objects, particularly at the basic level, a point we will discuss further below.

Consequences of impoverished encoding: Heuristic estimates replace spatial computation

The sequential manner by which a two-dimensional display is encoded haptically appears to provide a representation that is imprecise, in that the locations of points or segments are known only within some range of uncertainty. It may be disorganized, so that the relative positions of distinct segments are uncertain. Under these circumstances, retrieval of information from the representation would become inaccurate, if it could be attempted at all. In this section we consider some consequences of having inadequate spatial information.

Consider in particular how someone might retrieve information about the distance between two points in a spatial representation that was encoded haptically; for example, the distance between the head and the tail of a dog presented as a raised line drawing. If the dog were represented as a coherent image, the distance between the targeted features could be determined by some process analogous to visual inspection or scanning. Given an estimate of image scale relative to the original drawing, the original distance could then be computed. But if the coherent image is lacking, what can be done?

We asked subjects to perform a similar task (Lederman, Klatzky and Barber, 1985). They felt a raised curved pathway with the fingertip, after which they were asked to indicate the distance between its endpoints. Our results suggested that subjects found it very difficult to estimate this distance along a line that they had never felt directly.

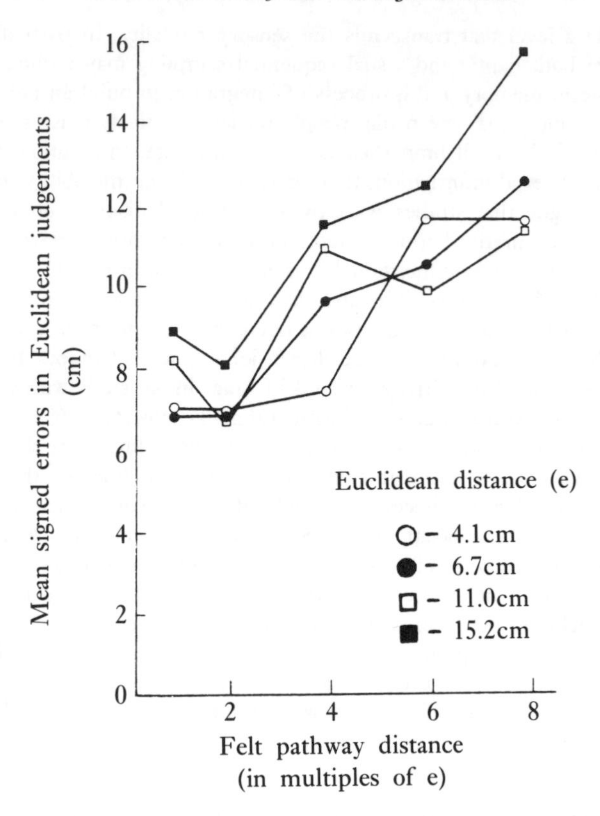

Figure 7.4 Signed error (observed minus actual value) in judgements of the straight-line (Euclidean) distance between two points after tracing a pathway between them. Data are shown for several values of Euclidean distance and pathway distance. (Adapted from Lederman, Klatzky and Barber, 1985. Copyright American Psychological Association.)

Instead, they seemed to make use of something we might call a spatial heuristic – a rule of thumb instead of a computation based on a spatial representation. Their heuristic was a sensible one – they simply used the pathway length to estimate the straight-line distance between endpoints. Because the pathway was followed by the hand, its length could presumably be encoded by some means other than a visual image; for example, from kinaesthetic or temporal cues. As a result of their using pathway length in this way, subjects' error on a given straight-line length increased directly with the pathway length, as shown in Figure 7.4. Under some circumstances, they overestimated the straight-line length by a factor of 2 : 1. We call this phenomenon 'length distortion'.

Length distortion appears to be a relatively high-level strategy, in that it can be observed quite generally under circumstances where encoding of spatial layout is likely to be deficient. We have found length distortion of pathways that are explored on foot without vision (Lederman, Klatzky, Collins and Wardell, 1987). We have also induced length distortion with a visual display (Balakrishnan, Klatzky, Loomis and Lederman,

1989). Subjects viewed a moving dot that traced out a pathway over time, after which they indicated the distance between pathway endpoints. An attentional load was imposed by having subjects monitor the dots in the pathway; if one of the dots was larger than the others, the subject was to press a button. Under these circumstances, estimates of the distance between pathway endpoints increased with the pathway length. The magnitude of length distortion in the visual display (measured by the estimation error as a percentage of the correct response) was much less with visual displays than with haptic exploration, but the pattern was similar, particularly with slow presentation rates and relatively complex pathways. These conditions would promote uncertainty in spatial encoding.

3 Reasons for differential performance with objects and 2-D displays

To summarize the preceding sections, we first documented that the haptic system produced accurate and rapid object recognition, and further, that haptic recognition at the basic level appears, as does visual recognition, to demand discrimination among objects with respect to shape and part structure. However, we note that material – particularly texture – is also important to haptic recognition at the basic level, and it becomes more important at the subordinate level. We next provided evidence that the haptic system is rather restricted with respect to precise shape discriminations. How, then, can haptic recognition of three-dimensional objects be so successful, and – returning to the original question raised in this chapter – what is the contribution of spatial information?

There are at least two general approaches to resolving the discrepancy between the importance of spatial information to object recognition, on the one hand, and the apparently impoverished spatial capabilities of the haptic system, on the other. One possibility is that the highly precise spatial discriminations that tend to be required in tasks with two-dimensional stimuli are unnecessary for real object recognition, and that the level of precision that is available under haptic exploratory procedures is sufficient. In this respect, experiments with two-dimensional forms may be ecologically unrepresentative.

In a recent experiment with Jack Loomis, Hiromi Wake and Nao Fujita, we assessed the contribution of structural information to naming of real objects at the basic level, particularly information obtained from enclosure. Subjects felt an object with a thick glove, which substantially reduced cues to its material. The object was also stabilized on a platform so that its weight could not be determined by lifting, and any moving parts were fixed. While some conditions allowed the subject to feel the object without impediment other than the glove, under other conditions the fingers were splinted so that they could not flex and enclose the object. We found that the unsplinted glove condition allowed for highly accurate performance and was faster and more accurate than the splinted fingers. It remains to be seen whether gloved exploration is sufficient for categorization at subordinate levels, but it appears at least that the combination of enclosure and contour following provides structural information that can be quite valuable to identification. It also seems that the coarse apprehension of three-dimensional shape that is possible with grasping an object makes a contribution that is not entirely redundant with information from contour following.

A second potential basis for the successful identification observed with real objects is that they offer nonspatial cues to their identities. Above, we offered evidence that an object's material properties are potentially very informative, particularly when categories are distinguished at the subordinate level. Although we have indicated that structural information alone can produce accurate identification, this certainly does not mean that material properties are not used when they are available. In an extension of the experiment just described, we used gloves that stopped before the fingertip, so that the contribution of material as well as structure could be assessed. The addition of material considerably facilitated performance, particularly with respect to speed. This result is consistent with our finding that the exploratory procedures associated with material properties tend to have much shorter spontaneous durations than does contour following (Lederman and Klatzky, 1987).

The use of material properties may constitute a fundamental difference between categorization by the visual and haptic modalities. Reliance on material under haptic exploration could result from both the limited ability to encode information about shape and the ready access to relatively high-quality information about material. While vision, too, allows access to material properties, the rapid naming observed with line drawings that fail to preserve these properties suggests that material is not an important cue to visual object identification. The accessibility of at least some material properties to vision is also much less than to touch (see Klatzky, Lederman and Matula, 1991). For an extreme example, the thermal properties of an object cannot be directly sensed by vision unless the object is heated or cooled, giving rise to cues such as condensation or colour change. Of course, material properties can be inferred after an object is identified, but this precludes their use in identification.

Together, material and structural information, even at an imprecise level, might be sufficient to haptically identify at least some common objects with reasonable speed. Consider, in particular, what happens when someone grasps and lifts an object. This was the first stage observed in our yes/no identification task, and it is an activity that potentially incorporates several exploratory procedures. It would permit the individual to quickly and simultaneously extract at least crude information about both material and structural attributes, which might be used to converge on the appropriate category. In a recent experiment testing this possibility, we asked subjects to identify objects at the basic and subordinate levels, using the same yes/no task as previously but now restricting exploration to grasping and lifting. This form of contact produced above-chance performance at both levels of categorization, confirming the power of knowledge about material and structure together (Klatzky and Lederman, 1992).

4 Contributions of spatial properties to the cognitive representation of a seen and felt object

In this section, we turn from the perceptual task of object identification to the cognitive representation of objects that occurs after their properties have been apprehended by vision or touch. We have suggested that the importance of material properties in identification differentiates haptics from vision. Does the same apply at the cognitive level? Obviously a differential emphasis on representation of spatial properties, depending on whether an object is seen or touched, is likely if vision can encode an object's structure but touch cannot. But assuming it were possible to determine the contours of an object by touch at least as accurately as was necessary for between-category discriminations, we can then ask whether the representation of object categories would be invariant across different modalities of encoding. In other words, if perceptual fidelity were equated or at least were very high for both modalities, would cognitive representation be equated as well?

We have addressed this issue in studies using an object sorting paradigm (Klatzky, Lederman and Reed, 1987) in which the objects are constructed from factorial combinations of properties. In one such study, for example, we crossed three levels each of size (all within the span of the hand), shape, texture and hardness, to make a set of 81 objects. The shape factor varied only planar contour and held constant the third dimension, so that the objects were like thin wafers in the shape of an oval, an hourglass or a clover (three lobes). Most importantly, these levels were all highly discriminable by touch, as were those of the other properties.

We asked subjects to form their own categories among these objects, so that we could see which properties they used. Because the distributions of property values were uniform and the properties were uncorrelated, there was no intrinsic categorical structure to dictate the subjects' responses. The partitioning of the objects should therefore reflect which properties the subjects find most salient.

More specifically, the subjects were asked to freely sort the stimuli into sets that 'were similar', with the definition of similarity and the perceptual conditions varying between groups. From the pattern of sorting, we derived a measure to determine how they weighted the dimensions of variation among the objects. The measure increased with segregation of objects along a dimension and decreased with aggregation. For example, someone who sorted by shape (e.g. placing all ovals together, all hourglass shapes together, and all clover shapes together) would, of necessity, have to aggregate objects that varied in texture, size and hardness. This person would have a high measure for the shape dimension and zero for the others.

The results of this study revealed that perceptual modality and instructions considerably affected the relative weighting of object properties. Subjects who could see the objects as well as touch them judged similarity primarily by shape, but they also tended to use texture. Subjects who could not see the objects, but who were told to think of how they would look if they could be seen, sorted almost exclusively by shape. Subjects who sorted only by touch, and who were not biased to think of visual appearance, showed higher weightings for texture and hardness than those in the other groups. In short, vision or visual bias increased the weighting of spatial properties, whereas other haptic conditions led to a relatively high weighting for material properties.

Recently, we conducted a similar study with fully three-dimensional objects of two different materials (which in turn altered weight and thermal variations; Summers, 1991; Summers, Lederman and Klatzky, in preparation). In this case, the shape dimension had two values; the object was either a cube or a sphere. With such a dramatic difference in values, the shape dimension became highly weighted by all the groups, but it was again most highly weighted by those with vision or visual-imagery bias. In contrast, subjects who sorted haptically without imagery bias showed higher weightings for material properties than did the other groups.

We have suggested that the dimensional weightings follow from the relative efficiency of hand-movement patterns. That is, material properties can be extracted relatively easily and locally from an object by a haptic explorer. Following contours takes more time and effort, and the most contour following is observed when subjects are instructed to form visual images of the objects. Without such a bias, then, material properties are likely to be more cognitively salient under haptic exploration.

The representation of many common objects is derived from both vision and touch. Hence, such a representation will store information about a variety of properties – the object's size, shape, apparent temperature, roughness and so on. When we think about an object, we may weight those properties in some way, so that some become relatively salient. The work just described suggests that the salience or weighting of a property differs, depending on the modality of exploration. When we see an object, we may think more about its spatial properties. When we feel an object, we may think more about its material properties. Of course, these patterns of salience are also likely to depend on the task at hand and our past experiences with the object as well. The relative importance of material properties under haptic exploration in this sorting task does not preclude the importance of spatial properties for the task of object identification. (Indeed, we have found size to be more important in categorization than in sorting; see Reed, Lederman and Klatzky, 1990.) But our present point is that it would be a mistake to think that the cognitive representation of an object will be independent of sensory modality, even if encoding precision can be equated.

Conclusions

We began by suggesting that touch is a spatial modality, raising the issue of the importance of spatial information in haptic object identification. Despite the fact that a lower level of spatial precision is possible with the haptic than with the visual modality, we have presented evidence that the information about an object's structure extracted through haptic exploration is of substantial value in determining its identity. This does not mean, however, that vision and touch rely equally on spatial properties; material properties also make an important contribution to object recognition by touch. Further, the salience of material properties seems to characterize the cognitive representation of objects that are apprehended through haptic exploration.

REFERENCES

Balakrishnan, J. D., Klatzky, R. L., Loomis, J. and Lederman, S. J. 1989: Length distortion of temporally extended visual displays: Similarity to haptic spatial perception. *Perception and Psychophysics*, 46, 387–94.

Biederman, I. 1987: Recognition-by-components: A theory of human image understanding. *Psychological Review*, 94, 115–47.

Klatzky, R. L. and Lederman, S. J. 1992: Stages of manual exploration in haptic object identification. *Perception and Psychophysics*, 52, 661–70.

Klatzky, R. L., Lederman, S. J. and Balakrishnan, J. D. 1991: Task-driven extraction of object contour by human haptics: I. *Robotica*, 9, 43–51.

Klatzky, R. L., Lederman, S. J. and Matula, D. 1991: Imagined haptic exploration in judgments of object properties. *Journal of Experimental Psychology: Human Learning, Memory and Cognition*, 17, 314–22.

Klatzky, R. L., Lederman, S. J. and Metzger, V. 1985: Identifying objects by touch: An 'expert system'. *Perception and Psychophysics*, 37, 299–302.

Klatzky, R. L., Lederman, S. J. and Reed, C. K. 1987: There's more to touch than meets the eye. *Journal of Experimental Psychology: General*, 116, 356–69.

Lederman, S. J. 1991: Skin and touch. In R. Dulbecco (ed.), *Encyclopedia of Human Biology*. San Diego: Academic Press.

Lederman, S. J. and Klatzky, R. L. 1987: Hand movements: A window into haptic object recognition. *Cognitive Psychology*, 19, 342–68.

Lederman, S. J. and Klatzky, R. L. 1990: Haptic object classification: Knowledge driven exploration. *Cognitive Psychology*, 22, 421–59.

Lederman, S. J., Klatzky, R. L. and Balakrishnan, J. D. 1991: Task-driven extraction of object contour by human haptics: II. *Robotica*, 9, 179–88.

Lederman, S. J., Klatzky, R. L. and Barber, P. 1985: Spatial- and movement-based heuristics for encoding pattern information through touch. *Journal of Experimental Psychology: General*, 114, 33–49.

Lederman, S. J., Klatzky, R. L., Chataway, C. and Summers, C. 1990: Visual mediation and the haptic recognition of two-dimensional pictures of common objects. *Perception and Psychophysics*, 47, 54–64.

Lederman, S. J., Klatzky, R. L., Collins, A. and Wardell, J. 1987: Exploring environments by hand or foot: Time-based heuristics for encoding distance in movement space. *Journal of Experimental Psychology: Learning, Memory and Cognition*, 13, 606–14.

Loomis, J. M., Klatzky, R. L. and Lederman, S. J. 1991: Similarity of tactual and visual picture recognition with limited field of view. *Perception*, 20, 167–77.

Loomis, J. M. and Lederman, S. J. 1986: Tactual perception. In K. Boff, L. Kaufman and J. Thomas (eds), *Handbook of Perception and Human Performance*. New York: Wiley.

Reed, C. K., Lederman, S. J. and Klatzky, R. L. 1990: Haptic integration of planar size with hardness, texture and planar contour. *Canadian Journal of Psychology*, 44, 522–45.

Rosch, E., Mervis, C., Gray, W., Johnson, D. and Boyes-Braem, P. 1976: Basic objects in natural categories. *Cognitive Psychology*, 8, 382–439.

Summers, C. 1991: Haptic exploration with and without vision: Property encoding and object representation. Ph.D. thesis, Queen's University.

Tversky, B. and Hemenway, K. 1984: Objects, parts, and categories. *Journal of Experimental Psychology: General*, 113(2), 169–93.

8
Sense modalities and spatial properties

Michael Martin

In this chapter I want to pursue some issues concerning the character of the awareness each of us has of their own body. In part, such awareness arises from the five senses, but it is not exhausted by that: one is also aware of one's body through bodily sensation, kinaesthesia and one's sense of balance. One might group together these latter experiences as comprising a body sense or bodily awareness. While neurophysiologists and psychologists have long been happy with the idea that there is such awareness, it has tended to be ignored or treated with some circumspection by philosophers.[1]

In particular, I shall be concerned to highlight some of the peculiarities of bodily awareness in contrast to the five senses. In such experience there is no identifying one's body as one among those experienced: the sole object of awareness is one's body. Nevertheless, one has a sense of one's body as a bounded object located within a space which can contain other objects. In what follows I shall set out to elaborate and defend these claims against certain objections.

My concern here is not really restricted just to the issue of bodily awareness itself. On one theory, the sense of touch is partly constituted by one's awareness of one's own body. This may seem obvious only of cutaneous touch, where contact with or pressure to the skin can elicit sensations. But it is no less appropriate to active touch and haptic perception: movement over objects and a grasping exploration of them involves the movement of and stimulation of parts of one's body. This is not only to claim that information about one's body is exploited in tactual cognition – the same is true, after all, of vision – but that an awareness of one's body is constitutive of an awareness of the objects of touch. Where one feels an object pressing against one's skin, there is also an awareness of one's skin as it feels pressed against. Where there is an awareness of the shape of an object as one traces over it, there is also a sense of the movement of one's body and a sense of the contact between an object and the part of one's body when one explores an object. This is not to say that one does attend to how one's body feels whenever one feels an object of touch – on the contrary, one's attention is normally directed at the object touched. Rather, as with aspect-shifts, one can switch one's attention between the object touched and how one's body feels. On the basis of this, the claim is that the awareness of one's body is a necessary accompaniment for, and partly constitutive of, one's sense of touch. In the light of this, I am interested in the question of how the features that I claim belong to bodily awareness provide for a sense of touch.[2]

In the body of the chapter, I shall be concerned to elaborate further the claims I have made above about the nature of bodily awareness and how it may be connected to the sense of touch. In the final part of the chapter, though, I shall briefly turn back to the question of how such awareness might contrast with other forms of perception, in particular sight. I suspect that philosophers' scepticism about one's awareness of one's body derives from the inappropriateness of a visual model of spatial experience when applied to one's awareness of one's body. With this thought I shall then leave open the question whether one needs different accounts of the varieties of spatial experience.

I

A starting point for any discussion of bodily awareness is the denial made by many philosophers that there is any such awareness of the body. They do not mean to deny that there are sensations, but that such sensations are an awareness of anything in the objective world. Instead, bodily sensations are to be viewed as examples of purely subjective states of mind. For instance, on one view of perception there is a sharp contrast to be made between sensation and perception (this dichotomy is most closely associated with Thomas Reid, 1983, ch. 5). Sensations are purely subjective mental events, whose phenomenological character is independent of the objective, physical world. These contrast with perceptions which are of the objective order. Sensations are to be thought of as mere parts, or alternatively causes, of perceptions. When applied to the sense of touch, the sensational component of perception is often taken to be bodily sensation, and these are then treated as purely subjective states of mind. And indeed, independently of this particular model of perception, there has often been a tendency to suppose that bodily sensations are purely 'inner' or subjective states of mind. While in the case of perception a distinction is drawn between an experience and what it is of, it is suggested that no such distinction can be drawn for a sensation such as pain. On this view, an experience of pain is not of anything, in the way that a visual experience of a red ball is of a ball (McGinn, 1982, p. 8). On this assumption, if there were any sense of one's body, then it too should have a purely subjective, sensational component, out of which the objective perception would arise. The sensational core of this awareness would itself lack any intrinsic bodily or spatial character, being simply the sensations internal to the mind.

It is this view of bodily sensation, and the relation between bodily sensation and other forms of awareness of oneself, such as kinaesthesia, that I suggest should be rejected. For the thought that bodily sensations such as feelings of pressure, of pain, of warmth and cold are purely subjective is not itself at all obvious. Typically such sensations are felt to be located in certain parts of one's body; and how they are felt to be indicates to one something about that part of the body. Furthermore, such sensations can misinform or mislead one as to the state of one's body: when one has a referred pain, it feels as if one part of one's body is unwell, when in fact another part is. So one might naturally suppose that these sensations are in fact an awareness of something objective: one's body which is a part of the physical world.[3]

The view that sensation is purely subjective can only accommodate the idea that sensation does inform one about one's body, if it can show that felt location of sensation, and the way in which it indicates something about the state of one's body, are somehow additional to its purely sensational core. I doubt whether this claim is true for either the location of sensations or what they tell one about the state of one's body; but the point is most easily made with respect to location alone.

If one denies that the location of a sensation is intrinsic to the sensation itself, one needs to give some account of how it comes to be associated with the sensation. One venerable view is that non-spatial qualities of sensations are correlated with parts of the body; one either learns to associate the sensation with the location or one knows innately that the non-spatial quality of the sensation indicates something about a certain bodily location. If the view is to be plausible one needs to assume that the variation in non-spatial qualities of sensations as one feels them should support such a correlation. But that seems not to be the case. Consider any twinge that one feels in one's left hand, and now imagine a qualitatively identical twinge in the right hand: need anything else be different about how the sensation feels other than being either in the right or left hand? Or again, consider the case of feeling a series of sensations at different, but relatively close positions on the back: need there be any qualitative difference between them other than their apparent position?

It is implausible, then, that the connection between sensation and bodily location is forged by some learnt association of intrinsic character of the sensation and a part of the body. But could the connection be made in some other way, without compromising the idea that these sensations are in themselves purely subjective? Anscombe has argued that this is so. Being highly suspicious of the idea that there is any such thing as kinaesthesia, or even a genuine feeling of the location of sensation, Anscombe suggested that there is simply 'knowledge without observation' of the position of one's limbs; and instead of a felt location for sensations, there is simply a disposition to act towards whichever part of one's body in which the sensation is said to be located (Anscombe, 1957, pp. 13–14, 49–51; 1962). On this view, sensations will still have a purely subjective phenomenological character, their location being something purely extrinsic to how they feel.

But Anscombe's position is surely no more satisfactory than the idea that there is a learnt association; for brute dispositions to respond fail to accommodate the complex relations between experiences and our responses to them, which we find present in the case of bodily sensation just as we do with perceptual experience. It is no accident that I reach to a particular position in space when I wish to grasp the jug that I see: an adequate account of visual experience must explain how the experience matches the action, such that that action appears appropriate to how things look to one. The same rationale applies to bodily sensation. When I place my hand on a hot stove, there is just no question for me which part of my body I have to move away. But at the same time, one must not forget that however automatic the response, it also appears appropriate to the agent in the light of where the pain feels to be. The sensation not only causes behaviour, but appears to rationalize it. An appeal on Anscombe's part to no more than brute dispositions would fail to accommodate that point. For one can be disposed to respond in some way to a stimulus whether or not that response appears to be appropriate to one: both basic reflexes and conditioned responses reflect that fact; it is not as if one's knee jerk appears to one to be the right thing to do in response to a hammer tap, it simply happens. While Anscombe may indeed be right to suggest that we look to the relation between experiences and behaviour to provide an account of the former, she is wrong in supposing that a dispositional connection will alone explain what it is for sensations to have a location.[4]

In the light of this, I suggest that we should take seriously the felt location of sensation, and in addition the sense one has of the position of one's limbs. Sensations are not, then, purely subjective events internal to the mind; they are experiences of one's body, itself a part of the objective world. The location of a sensation should be treated as on a par with the visual perception of location of an object: both relate to the physical world as experienced, and not to some attribute of the experiencing of it. But does this amount to supposing, with Armstrong and Pitcher, that sensations are a kind of perceptual or quasi-perceptual awareness of one's body?

One may quite rightly be sceptical of the idea of a body sense analogous to the five senses. In this case there is no obvious discrete organ of awareness. One cannot group together the various examples of bodily awareness, such as bodily sensations, kinaesthesia, one's sense of balance, as constituting a sense modality in virtue of all providing information about one's body. For it is equally true that one can perceive one's body via the five senses; and sight and touch are just as important as kinaesthesia with respect to providing information about the body for a body image or schema. This would suggest that there are simply diverse ways of gaining information about one's body.

It may be correct to draw as a conclusion from this that one's awareness of one's body does not deserve the epithet of sense modality. But I suggest that it would be wrong to suppose that there is no interesting unified phenomenon which links sensation and kinaesthesia and contrasts them with senses such as vision and hearing. In sight or touch, one encounters one's body as an object of perception only as one among many other possible objects of perception. But this is not so with bodily awareness, with sensation, kinaesthesia or sense of balance. In these cases there are no objects of which one can be aware other than one's own body. Sensations which feel to have a location feel to be located within one's own body, and not anyone else's; when one feels a pair of legs to be crossed, one feels that it is one's own legs that are crossed, and not another's; one is aware of whether oneself is upright through vestibular sense but not whether anyone else is (Evans, 1982, ch. 7, § 7.3). Bodily awareness is such because its proper and sole object is one's own body and not any other occupant of the objective world.

This feature bodily awareness shares with introspection: one does not single out one's own mind from among the various minds one introspects. Sydney Shoemaker has argued that the absence of identification of oneself tells against the view that introspection is a form of self-perception (Shoemaker, 1986, pp. 108–13). One might, therefore, appeal to the same consideration to deny that bodily awareness is a genuine form of body perception. Even if this is so, one cannot deny that there are genuine experiences, all of which share this feature of informing one solely of one's body. This may not be perceptual awareness, but it is still awareness.

The claim that the body is the sole object of awareness has two parts to it. The first is that one does not in such experience encounter one's own body as one among others which one also feels; there is no identifying a body as one's own through such awareness. The second part goes on to affirm that nevertheless one is aware of one's body as one's body, as an object in a world which can contain other objects. The first part alone would be consistent with the thought that one felt sensations to have a location but not to be located within one's body as opposed to somewhere else; or that one felt limbs to be disposed in a certain way, but did not feel them to be one's own limbs. As I shall describe below, it seems to me that this is a genuine possibility, but is nevertheless not true of us. This raises the question of how it can be true that we have a sense of our bodies as our bodies, when we don't identify them as such.

When applied to the example of sensation, one can interpret the claim that one is only aware of one's own body, and aware of it as one's body, as the claim that when a sensation feels to have a location that location feels to be internal to one's body: that is, that a sensation is an awareness of it as happening in something which is felt to be a part of one's body. This would appear to rule out the possibility of feeling a sensation which has a location but which is not felt to be internal to one's body, or is even felt to be external to it. Brian O'Shaughnessy endorses this suggestion when he writes: 'I think it all but impossible to comprehend a claim concerning sensation position that detaches it from actual or seeming limb, e.g. "A pain to the right of my shoulder and not even in a seeming body part"' (Shaughnessy, 1980, vol. 1, p. 162). Rather than admitting to the limitations of his own intellect, O'Shaughnessy is here claiming that there is a necessary connection between the felt location of sensation and its apparent location within one's body. What connection exactly should one suppose this to be? It is simply false to claim that all sensations must feel to have a location, since some fail to do so. Nor can it be right to claim that if sensations feel to have a location, then they must feel to be within the actual limits of one's body. That claim obviously goes contrary to the incidence of phantom limb sensations. At best, then, the claim must be that there is a necessary coincidence between felt location of sensation and apparent location of parts of the body: wherever a sensation feels to be located, one's body appears to extend to at least that point in space.

Even this claim is open to both further interpretation and counter-example. On one interpretation, one might suppose that the apparent extent and limits of the body were determined independently of the apparent location of sensation. The alleged connection would then amount to a constraint on where it was possible to feel sensations to be: if my body appears to have a certain extent, then I simply can't feel a sensation to be located anywhere beyond those limits. This suggestion is clearly open to counter-example. First, there are cases of projected sensation, as when one comes to feel a sensation of contact in a tool or some prosthetic device which one is manipulating – there is no reason to think that such tools have been previously determined to be an apparent part of one's body. Secondly, there are cases of extra-somatic sensation. There is, for instance, von Békésy's example of feeling a sensation to be located between two fingers (Békésy, 1967, pp. 220–26) and documented pathological cases of people having exosomesthetic sensation, locating sensations out in space or in others' bodies (Shapiro, Fink and Bender, 1952). These kinds of cases seem to support the idea that sensations can be felt to be positively outside the apparent limits of one's body.

However, O'Shaughnessy's claim can be given another interpretation, and on this the apparent counter-examples are much less clear-cut than might at first appear. One might reject the thought that the apparent extent of one's body is fixed entirely independently of where one might come to feel a sensation to be located. On the contrary, one might instead suppose that the apparent location of a sensation can determine the apparent extent of one's body, such that wherever one feels a sensation to be located, one thereby feels one's body to extend to at least that point. This would suggest that the apparent extent and shape of one's body could be rather plastic, subject to sudden alteration. This supposition is indeed supported by experiment on induced proprioceptive illusions (Lackner, 1988). One might then think of the occurrent sense of the limits of one's body as subject to alteration given the current apparent location of sensations within it.

This view accords well with examples of projected sensation, for it is commonly commented that in these cases there is also the phenomenon of the apparent extension of one's body along the tool manipulated, as if it becomes part of one. That appearance can be abolished by focusing attention instead on sensation within the fingers grasping the tool, but it is coincident with feeling at the tip of a scalpel or other device. Extending this observation to von Békésy's example – and he himself notes the connections between this case and that of projected sensation – one might suppose that while the sensation fails to have a location in either finger, it nevertheless feels to be located in some indeterminate extension of the body between the fingers. Similarly, case studies of exosomesthetic sensation suggest that the examples are consistent with the hypothesis that the body image has distorted to include the apparent location of the sensation, although they note certain significant differences from familiar cases of projected sensation.

But one might reasonably question whether one can attribute to sensations any character of being internal to the body, if O'Shaughnessy's necessary connection had to hold. For its consequence is that wherever a sensation is felt to be located will be felt to be a part of one's body. But then there can hardly be a contrast in how things feel between sensations being internal to one's body and sensations being external, since that is what has been ruled out – the location of a sensation might actually be external to the body, but it won't be felt as such.

One can indeed imagine a creature for which it would be true to say that it felt sensations to have a location without thereby feeling them to have some bodily location. Suppose we have a creature which has sensations, but has no sense of the contrast between itself and the rest of the world. Would its sensations be just like ours? For instance, suppose we have a kind of jellyfish living in currents good enough to move it towards food and away from harm. The jellyfish lacks all sense of its boundaries with the rest of the world, and has little time for detecting predators. Its main concern might rather be with events taking place beneath its skin – perhaps it needs to monitor the rate of certain processes depending on how far they are from its centre. At a stretch one can imagine the creature having sensations which indicate what processes are taking place at what rate at certain locations under its skin. Although it has sensations which inform it about its body, it is doubtful whether we should think of it as sensing its body as its body. In this case there does not seem to be a useful contrast to be made for the creature itself between its body and the rest of the world. For instance, were it to have a rogue sensation which felt to be at a point which in fact lies outside its skin, we would not be inclined to say that it felt something external.

But while the suggestion that sensations might simply be felt to have a location without thereby being felt to be located within one's body is appropriate to this imaginary case, I suggest that it does not apply to us in the same way. In contrast to the case of the jellyfish, we have a sense of our limits and boundaries with the rest of the world, some sense of the contrast between what is oneself and what is other. This is not merely the claim that we have some conception of the difference, it is surely also true that the contrast applies as equally to the sense that we have of ourselves through bodily sensation and kinaesthesia. The question remains, then, how this can be, if all that one feels, feels to be a part of one's body?

The thought that one's sensations might each have some positive quality of feeling to be internal to one's body is in tension with O'Shaughnessy's claim. For if a sensation is only felt to be internal to the body because it has some positive quality, then it should be conceivable that the sensation might lack that quality and thereby fail to be felt as internal to one's body. In that case, it should be conceivable that some of one's sensations should be felt to be located in one's body, while others would be felt to be located but not internal to one's body, directly contrary to O'Shaughnessy's contention. Instead, one needs to look for some structural feature of feeling sensations which would apply to all or none at once.

That one feels one's sensations to be internal to one's body might also be seen as one feeling one's sensations to be located within one's boundaries. The question would then become what sense could one have of one's boundaries, given that all one's sensations which are felt to have a location are felt to fall within those boundaries. That is, why should it feel to one that one does have boundaries, rather than that one's body might simply extend to encompass the whole world, or the world apparently shrink to the limits of one's body? One answer one might give to this is just that in addition to having some sense of the extent of one's body, one also has some sense of the world extending beyond those limits. That is, wherever one does feel a sensation to be located, one may also have a sense that the world must extend beyond that point, the world extending beyond one's limits being composed of regions of space which one couldn't at this time be feeling a sensation to be located in. However plastic one's body schema, however much it seems to extend, one would still have a sense of one's body being bounded and limited if there was also a sense of the world extending beyond those limits.

One might suggest, therefore, that it is here that one can locate the difference between our sense of ourselves and that of the jellyfish. Unlike the jellyfish, we have a sense of ourselves as being bounded and limited objects within a larger space which can contain other objects. A sense of our limits brings with it a sense of the world extending beyond those limits. The important contrast is not between different qualities that sensations might have, but between places where one does feel sensations to be located and places where one simply cannot feel them to be.

The same claim extends to other aspects of bodily awareness. In the case of kinaesthesia one is aware only of parts of one's own body, and of no one else's. There is just no question of whose legs it is that are crossed: since one can feel the position of the limbs, they must be one's own. As with the location of sensation, one might suggest that feeling the position or movement of a body part is sufficient for that to feel to be a part of one's body. But one is also aware of one's body as in a space which extends beyond it and contains it. One way of illustrating this is to consider one's sense of the relative position of parts of one's body. If one extends one's arms out in front of one, one has a sense of the position of both hands, and their positions in space relative to each other. No part of one's body occupies the region of space lying between the two hands; and it does not feel to one as if any part of one's body is there. One does not have, therefore, in position sense any awareness of what occupies that region of space, if indeed anything does. Nevertheless, one does feel one's hands to be separated across that region of space. In this way regions of space which extend beyond what one feels at a time through bodily awareness enter into the character of how one feels things to be through bodily awareness. What one feels in this way is felt to be located in a larger space which one does not also feel. Since one is aware of nothing but one's body, it does not have to be identified as such within experience; there are no other objects of awareness to contrast it with. But since one is aware of it as in a world which contains many other objects, one nevertheless has a sense of it as one's body in contrast to other objects, things which one doesn't feel.

The model suggested by the five senses is that one can be aware of a number of objects at a time, and that perceptual awareness of these objects enables one to single one out from among the others. The multiplicity in bodily awareness is only an awareness of the parts of one's body; the only object of such awareness is one's body itself. But the lack of contrast between one's body and other objects does not mean that it appears to one as if one's body is all there is; rather, there is the sense of one's body being only one object within a space which extends beyond it. It is important, therefore, not only to stress that there is genuine awareness of one's body, as philosophers such as Armstrong and Pitcher do, but also to stress how different in structure such awareness is from other senses.

At the outset, I suggested that an account of one's awareness of one's body also has a role to play in a theory of touch. I suggest that it is the above claims about the sense of one's body as bounded within a larger space which shows how touch and bodily awareness might be interdependent, and how one's sense of one's body becomes an important part of the sense of touch.

On the above account, one's sense of one's body is a sense of it as a bounded object within a larger space which one doesn't also feel in the same way. One is aware of one's boundaries and movement through space. Suppose some object in fact comes into contact with one's body, such that it impedes the movement of one's body through that region of space. One may be aware that one cannot move one's body through that region, something impenetrable is there. In being aware of one's skin as a boundary of one's body, one has some sense of space extending beyond that boundary. In having some sense that one cannot move through that region of space immediately beyond one's skin, one has some sense that it is occupied. So being aware of one's body and where it can and cannot move can also provide information about what does or does not occupy space around one. This suggests that one can view one's body as a kind of template against which one measures other objects in the world. Just as one has a sense of things as belonging to one's body through bodily awareness, in having a sense of the world extending beyond those limits, one has some sense of objects as being external, as existing outside those limits, even when they press against them.

In the space of tactual objects, one's body is just one object among others,[5] these various objects impeding the movement of each other. But in order that one can come to be tactually aware of objects within this space, one's own body must play a special role within the space: it is that with which one feels other objects. Touch is dependent on one's awareness of one's body in as much as one's body is used to measure other objects. This is not to claim that bodily awareness is prior to or independent of the sense of touch. For awareness of one's body as one's body involves a sense of its being a bounded object within a larger space, and that just is to locate it within a space of tactual objects. The two kinds of awareness are rather interdependent.

II

In the preceding discussion I've argued that one is aware of one's body in a way that one is aware of no other object, and that one is aware of it nevertheless as one's body. I've argued that this requires that one should have a sense of one's body as being a bounded and limited object within a larger space, and that this can be exploited in the sense of touch. What is central to this account is the idea of a felt boundary between oneself and the rest of the world, where there is a contrast between what is felt and what is not and cannot be felt.

I have already noted that bodily awareness in taking just one object of awareness contrasts with the five senses, which can take an indefinite number of objects. I shall now argue that the sense in which one feels oneself to be bounded and limited suggests further ways in which bodily awareness differs from certain other kinds of spatial experience, in particular vision. For no boundary or limit to visual experience plays a role in such experience analogous to that of the apparent limits of one's body in bodily awareness and touch. That is not to deny that there is a limit or boundary phenomenologically present in visual experience. For the field of vision necessarily imposes a limit on what one is visually receptive to at a time. The important difference is not that there are no limits to the field of vision, but that they do not play the same role in vision that they play in one's sense of one's body. That is to say that there is no apparent contrast between objects within the field of vision and objects which fall outside of it. Rather, both regions of space and objects lying outside the field of vision seem irrelevant to how things within it appear to be. The contrast between being currently visible and not being visible is not a mark of any property of objects themselves, in the way that being felt through bodily awareness identifies what is felt as a part, or apparently a part, of one's body.

In addition to this, there is in normal visual experience some sense of not only being aware of objects in space, but being aware of a region of space they occupy. The field of vision can be thought of as delimiting a cone of physical space which is experienced at a time. One can certainly see not only the presence of visual phenomena, but also their absence. Think, for instance, of the case of seeing a ring-shaped mint head-on. One experiences not only the white parts of the mint, but also the hole in the middle and the area around its outer edge. In order to see the mint as a ring shape, one needs to distinguish the figure from the ground, but the ground here need be no more than the empty space around the object.

Indeed, one can have some sense of the space that objects are located in, even when that space is not itself completely illuminated, or is partly obscured. Shining a torch across a room illuminates in turn different parts of the room and objects within it, but it does not necessarily alter the field of vision; rather, there is the sense of different parts of the scene being lit up. In the same way, the rearrangement of objects so as to reveal what was once obscured seems to be a rearrangement of objects within a space experienced, rather than an alteration in that space.

The idea that visual experience is not only spatial experience but experience of space is an undercurrent of much philosophizing about experience. It seems to be echoed in Kant's idea of space as the form of outer sense, there to be experienced whether or not any objects are present.[6] In a rather different way, the idea is enshrined in sense-datum theories of vision. On such accounts, the experience of a space is taken to be a necessary condition of spatial experience. For, according to such a theory, perception of physical objects is mediated by an awareness of an array of purely mental colour-patches which occupy a non-physical space internal to the mind. Since such mental entities as sense data, and the space that they occupy, are normally thought to depend for their existence on a subject's awareness of them, the space occupied by such mental mosaics will itself exist only as experienced.[7]

It is doubtful whether the idea that the experience of space is central to spatial experience holds for vision in general. It seems implausible to suppose that a subject who can identify the shapes of seen objects, but fails to locate them, experiences them all within a region of space which is itself experienced. But perhaps the clearest example of spatial experience which this model is inappropriate for is bodily awareness.

The field of vision delimits a region of space at a time within which one may visually experience objects; no object outside of the field can be experienced without its either moving within the field, or the field itself being altered by moving one's gaze. The parallel field of receptivity for awareness of one's body can only be the apparent extent of one's body itself. For, as stressed earlier, one is only aware of what appears to be part of one's body through such experience. But now the region of space apparently occupied by one's body cannot be an analogue in bodily experience of the visual field. It does not seem correct to say that what one experiences through bodily awareness is a region of space and objects located within it. Rather, what one experiences are apparent parts of one's body, and happenings within it. Such events are not organized in terms of location in space independent of one's body, but are organized in terms of location within the body. In this way, where one might think of visual space as the organizing form of visual experience, with visual phenomena located at points within it, the body appears to play that role in bodily awareness: what is felt is organized in terms of the apparent arrangement of one's body.

But one can't treat the body itself as if it were the space within which one experienced things to be happening. For, as has been stressed above, although one is only aware of one's body through bodily awareness, it does not appear to be the only thing existing; it is not as if one's body were the complete world of awareness. One has no awareness of unoccupied regions of space, since any region of space of which one is aware feels to be occupied by some part of one's body. Nevertheless, one's body is felt to occupy a space which extends beyond it, since it is felt to be bounded. Hence, unoccupied regions of space are relevant to how one's body is felt to be, although they are not themselves objects of awareness. That is to say that the area of space which one's body would have to occupy in order to possess all of the spatial properties it appears to have is necessarily larger than any area of space which one could be aware of through bodily awareness. In this case experience of the spatial properties of an object cannot coincide with an experience of a region of space within which the object can appear to possess those properties.

What is the significance of this contrast between bodily awareness and sight? Well, one might question whether the task of explaining how bodily awareness has the spatial content it does, as of an object bounded and limited within a larger space, can fit in the same way the task of explaining visual experience, an experience, which in the normal case at least, involves the awareness of objects as located within a space itself experienced. One way of expressing the worry is to question whether the differences here are merely ones to do with the spatial information that sight and tactuo–kinaesthetic experience can both provide.

Consider for instance the contrasts between a case of visually experiencing a certain spatial array and having a tactuo–kinaesthetic experience of such an array. Exploiting an example of Evans's (1985),[8] one might consider the case of viewing four points arranged in a square. In the visual case, one is aware of four points of light in space, at some indeterminate distance, and of nothing else. A close analogue of this in the case of bodily awareness might be the following: a rock climber on a sheer cliff face might move her hands and feet into four cracks on the face which happen to be arranged in a square. When moving in to them she might realize that they are arranged in a square, and come to be aware that her appendages are thereby also so arranged. So in both cases, the subject is aware of four points arranged in a square.

If we imagine that the climber is not in contact with any other part of the cliff face, then all that she is aware of are the four cracks arranged in a square. Similarly, in the visual case, one can imagine that all that is visible are the four points of light, and nothing else. Both experiences are of a square arrangement of points; is there any difference in content between the two experiences? If the suggestions above about a visual experience of space are correct, the answer will have to be yes. Even though the viewer can only see the four points of light, she has in addition some sense of the space between the points; the experience is not only of the four points, but also the space which contains them all and through which they stand in spatial relations. This is not so in the tactuo–kinaesthetic case; there the climber can feel the four holes and is aware that they stand in certain spatial relations, but she has no sense of the space between them: she is not in contact with that part of space, and cannot tell whether anything is there or not. In the visual case there is a region of space experienced as well as objects experienced; in the tactuo–kinaesthetic case there are simply the objects experienced as having certain spatial properties.

Is this difference simply a difference in the informational content of the two experiences? Certainly with respect to the four points in each case and their spatial arrangement, neither experience provides more information (disregarding, that is, the irrelevant information about the distance of the cracks from one's body). But one might think that there is an informational difference nevertheless, as the story has been told. In the visual case, if the region around the four points of light is itself dark, then the viewer will indeed have information lacking in the other case: she will know that there are no points of light in that area. In contrast, the climber is not aware whether there are any cracks between the four cracks she is holding on to, or indeed whether there is a gaping hole in the rock face at that point. But is there only an experience of space if the visual experience is more informationally replete in this way? Suppose that the area around the four points of light is not dark, but obscured, perhaps by fog. The viewer will now no longer be in a position to tell whether there are any lights or anything else in the area between the four points, so the informational advantage she possessed over the climber is now lost. Nevertheless, it does not seem obvious that the notion of a visual field or visual space is no longer applicable to the visual experience. If this response is correct, then one can't explain the difference between experience of the spatial properties of objects without the experience of space and such experience with experience of space in purely informational terms.

I have emphasized above not only the idea that there is a genuine awareness of one's body, even if it does not amount to a sense modality of the body, but also the ways in which it is different from certain models of experiential awareness. The fact that it does not easily fit into certain conceptions of perception may help explain why philosophers have been so wary of developing an account of it. What may yet remain, once we take on board the fact that there is such awareness, is a suspicion that there are different kinds of experience, even different kinds of spatial experience. But such a suspicion will only be borne out once one has a developed account of perception, body sense and spatial experience. The discussion in this chapter is only an indication of directions in which such accounts might develop.

ACKNOWLEDGEMENTS

A rather different version of this chapter was presented in seminars in Oxford and Cambridge. I am grateful to the participants of those seminars, and particularly so to Tony Marcel, Brian O'Shaughnessy and Peter Sullivan for their comments. I am also grateful to the editors for their comments and the opportunity to revise this chapter at a late stage.

NOTES

1 The most notable exceptions among philosophers are David Armstrong (1962) who developed a perceptual model of bodily awareness, and Brian O'Shaughnessy, who has developed in various writings a very subtle account of these matters.
2 I argue for this view of touch in Martin (1992); the source of this view of touch can be found in O'Shaughnessy's seminal paper (1989).
3 The claim that bodily sensations are bodily perceptions has been advanced by David Armstrong (1962) and George Pitcher (1970). However, unlike these two authors, I remain neutral here on the question of whether all aspects of the phenomenological character of sensations can be treated as aspects of how one's body appears to be.
4 This does not address Anscombe's more startling claim that there is no such thing as kinaesthesia. That suggestion looks much less plausible, once bodily sensations are admitted to have a genuine spatial character. For a vigorous attack on Anscombe's claim, see C. B. Martin (1971).
5 Cf. O'Shaughnessy's (1989) claim that 'in touch a body investigates bodies as one body amongst others', p. 38.
6 Viz. the Metaphysical Exposition of the concept of space in the Transcendental Aesthetic (Kant, 1929), in particular A24 (B38–9). One might also note that in discussion of whether 'phenomenal' space is Euclidean, a debate arising from Kant, the phenomenal space in question is normally taken to be a visual space. See, for instance, Strawson (1966, Part V) and Hopkins (1973).
7 However, this view cannot be universally attributed to all variations of sense-datum theories: Jackson (1977) locates visual sense-data in physical space.
8 'To have visual experience of four points of light arranged in a square amounts to no more than being in a complex informational state which embodies information about the egocentric location of those lights. . .' (Evans, 1985, p. 392). The contrast I am here interested in making does not depend, however, on Evans's view that the informational content of an experience can be reduced to information that the subject has about the location of objects in egocentric or behavioural space.

REFERENCES

Anscombe, G. E. M. 1957: *Intention*. Oxford: Basil Blackwell.
Anscombe, G. E. M. 1962: On sensations of position. *Analysis*, 22, 55–8.

Armstrong, D. M. 1962: *Bodily Sensations*. London: Routledge and Kegan Paul.
Békésy, G. von, 1967: *Sensory Inhibition*. New Jersey: Princeton University Press.
Evans, G. 1982: *The Varieties of Reference*. Oxford: Clarendon Press.
Evans, G. 1985: Molyneux's question. In his *Collected Papers*. Oxford: Clarendon Press, 364–400.
Hopkins, J. 1973: Visual geometry. *Philosophical Review*, 82, 3–34.
Jackson, F. 1977: *Perception*. Cambridge: Cambridge University Press.
Kant, I. 1929: *The Critique of Pure Reason*, tr. N. Kemp Smith. London: Macmillan.
Lackner, J. R. 1988: some proprioceptive influences on the perceptual representation of body shape and orientation. *Brain*, 281–97.
McGinn, C. 1982: *The Character of Mind*. Oxford: Oxford University Press.
Martin, C. B. 1971: Knowledge without observation. *Canadian Journal of Philosophy*, 1, 15–24.
Martin, M. 1992: Sight and Touch. In T. Crane (ed.), *The Contents of Experience*. Cambridge: University Press.
O'Shaughnessy, B. 1980: *The Will*, 2 vols. Cambridge: Cambridge University Press.
O'Shaughnessy, B. 1989: The sense of touch. *Australasian Journal of Philosophy*, 67, 37–58.
Pitcher, G. 1970: Pain perception. *Philosophical Review*, 79, 368–93.
Reid, T. 1983: *Inquiry and Essays*. Indianapolis: Hackett.
Shapiro, M. F., Fink, M. and Bender, M. B. 1952: Exosomesthesia or displacement of cutaneous sensation into extrapersonal space. *AMA Archives of Neurology and Psychiatry*, 68, 481–90.
Shoemaker, S. 1986: Introspection and the self. Midwest *Studies in Philosophy*, X, 101–20.
Strawson, P. F. 1966: *The Bounds of Sense*. London: Methuen.

9
Molyneux's babies: Cross-modal perception, imitation and the mind of the preverbal infant

Andrew N. Meltzoff

Gareth Evans (1985) contemplated novel variants of the Molyneux problem to illuminate certain questions in epistemology. I and others have used variants of the Molyneux problem to address questions about the nature of the infant mind. When we study infants instead of blind adults, the questions cannot be verbal; we must infer infant perception and thought through their actions.

Imagine that the blind man is presented with a sphere to explore by touch. A sphere and cube are placed before him and sight is bestowed. In this modified example we do not ask the man to use a verbal label to designate which object he has just touched; we simply ask that he point towards or even to look longer at that object. The subject's response in this instance is nonverbal, but it raises most of the critical issues of the original case.

I will discuss cases in which we draw inferences about the classical epistemological problems of the Molyneux example from patterns of actions such as these. Our subjects were not sight-recovered patients, but young infants. There are parallels between the newly-sighted man and an infant. Like a blind man, a newborn infant has not visually inspected objects and has not had a chance to associate visual and tactual experiences of the same object. The original Molyneux problem stimulated many innovative studies of the blind. Had Molyneux posed his question using a newborn infant fresh from the womb, this might well have stimulated careful research with infants a century or two earlier than it became popular. Accelerated growth in the field of genetic psychology in turn may have aided philosophers. The information garnered from sight-recovered patients and cited in the philosophical literature is notoriously variable, as might be expected given this rare, and neurologically rather bizarre, population. An intact, normally-developing brain may be more useful as a touchstone for theorizing.

Focusing on newborn infants and not blind adults raises a whole family of new Molyneux-like questions. In fact, infants turn out to have far more sophisticated abilities to co-ordinate information from different modalities than we would ever have expected. While Molyneux wondered whether information about an object derived from touch and vision might be related prior to associative experience, it did not occur to him to ask whether proprioceptive information and visual information concerning one's own body movements in space, or speech sounds and mouth movements, might be intrinsically related prior to such experience. Yet the behaviour of actual infants suggests that such cross-modal connections come into play quite early in development, and perhaps are available at birth. I will argue that the behaviour of young babies raises puzzles analogous to the ones raised by the Molyneux problem. One of the deepest and most intriguing of these is the problem of imitation.

1 The Imitation Problem

Imitation involves the form of an action in behavioural space, not the shape of an external object, a sphere or cube. Molyneux might have posed the following Imitation Problem:

> Suppose a blind man can perform simple body movements, such as mouth opening and closing; he can identify the movements when he produces them and can produce them on demand. Suppose then that an actor is placed before the blind man and the blind man is made to see. The actor silently opens his mouth. Can the newly-sighted man, without being allowed to touch the actor, imitate the actor's gesture by opening his own mouth?

Locke, Molyneux, Berkeley and others who answered the original query negatively would also answer the Imitation Problem negatively: there is no way for the subject to know that in order to produce a certain visual spectacle he must move his body in a particular way. There is no immediate equivalence between the mouth-opening-as-seen and the mouth-opening-as-done.

The Imitation Problem has not, to my knowledge, ever been tested with newly-sighted patients. However, we do know that one-year-olds can imitate facial gestures. Unfortunately, this does not provide us with the data relevant to the Imitation Problem. Although infants cannot now see their faces while they are imitating, they may have seen themselves in the past in mirrors. Mirrors are a tool for making the invisible face visible. Mirrors provide a tutorial in 'connecting' action-as-seen with actions-as-done. Mirror-experienced one-year-olds are not like newly-sighted men. Fortunately, developmental psychologists may easily locate mirror-naive infants. In fact, the relevant population is one that is naive to this whole set of potential mediators, not just mirrors. At about one year of age infants begin to reach out and touch their mother's mouth and then touch their own, thus providing tactual comparisons. But with appropriately young and inexperienced infants, we can test the Imitation Problem directly. Such infants can produce the relevant motor movements with their own faces, but have never seen them. They can see the relevant acts of others, but have never felt them. Will they be able to imitate? If so, what does this tell us about the organization of the infant mind?

2 Imitation in infancy: The phenomena and initial suggestions about supramodal perception

Meltzoff and Moore (1977) discovered that 12- to 21-day-old infants, who by all reports are mirror-naive, could successfully imitate a variety of facial acts (Figure 9.1).

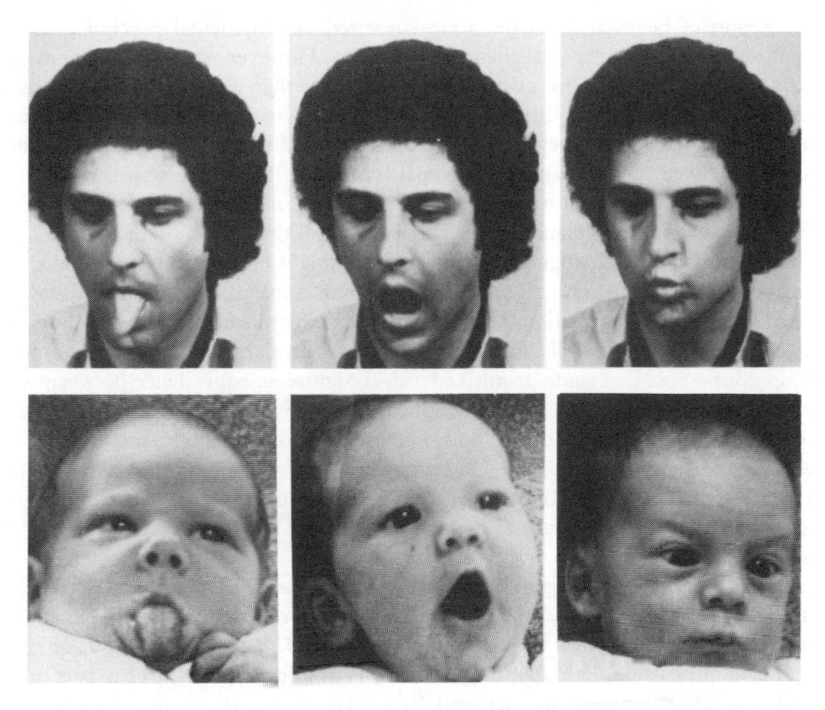

Figure 9.1 Photographs of 12- to 21-day-old infants imitating facial gestures shown to them by an adult.

We showed imitation of four body actions: lip protrusion, mouth opening, tongue protrusion and sequential finger movement. These particular gestures helped evaluate the specificity of the imitative response. Infants responded differently to two movements of the same body part (mouth opening v. lip protrusion) and also to two body parts producing the same general movement (lip protrusion v. tongue protrusion). This suggested that infants were matching particular *acts*, not just activating a certain region of their body (lips) or producing very generally-defined movements in space (protrusions).

We also found that young infants could imitate from memory. A pacifier was put in infants' mouths as they watched the display so that they could only observe the adult demonstration but not duplicate the gestures. After the infant observed the display, the adult assumed a passive-face pose and only then removed the pacifier. Infants were then given a 2.5-minute period in which to respond, during which the adult maintained this passive face regardless of the infant's response.[1] Even with this pacifier technique, the infants imitated the two displays. Moreover, infants did not produce exact matches early in the response period. The first responses of the infants were often with the correct body part but an approximation of the adult's act. Infants would move their tongues but not produce full tongue protrusions. Infants appeared to home in on the detailed match, gradually correcting their responses over successive efforts to more exactly correspond to the details of the display. The adult was sitting with a passive face all this time, thus the infant was comparing his or her motor performance against some sort of internal model or representation of what had been seen.

The report of neonatal facial imitation surprised developmental psychologists because it did not fit with classical theories of infancy. However, the basic phenomenon of early imitation has now been demonstrated in more than a dozen studies, by independent investigators using different designs; apparently early imitation is a cross-cultural phenomenon: positive results have been reported in the USA, Britain, France, Switzerland, Sweden, Israel and rural Nepal (see Meltzoff, 1990a, for a review). The fact that neonates will duplicate certain basic acts performed by an adult is now well-established, though the explanation for this fact is still unclear.

There are several psychological mechanisms that might underlie this behaviour. The hypothesis I favour is that imitation is based on infants' capacity to register equivalences between the body transformations they see and the body transformations they only feel themselves make. On this account early imitation involves a kind of cross-modal matching. Infants can, at some primitive level, recognize an equivalence between what they see and what they do. We might imagine that there is something like an 'act space' or very primitive and foundational 'body scheme' that allows the infant to unify the perceptual and action systems into one framework. In this view, although the infants' own facial gestures are invisible to them, they are not unperceived, for even unseen body movements can be monitored by proprioception (O'Shaughnessy, 1980, and ch. 13 of this volume).

Learning theorists could argue that all this is unnecessary. The subjects were 12 to 21 days old. Perhaps they had been trained to imitate during the first weeks of life. Infants could be conditioned to poke out their tongues to a ringing sound, or to an adult tongue protrusion. Perhaps the conditioning of a few oral gestures is part of the natural interaction between mother and baby. To resolve the point, Meltzoff and Moore (1983) tested 40 newborns in a hospital setting. The average age of the sample was 32 hours old. The youngest infant was only 42 *minutes* old. The results showed that the newborns imitated both of the gestures shown to them, mouth opening and tongue protrusion. One further study showed that newborns also imitated a non-oral gesture, head movement (Meltzoff and Moore, 1989). We can infer that the capacity to imitate certain gestures is innate. In line with some of the ideas advanced by Bower (1982), I would like to argue that imitation is just one manifestation of a larger capacity, a supramodal perceptual system, that can be tapped by other tasks. The next study pursued this point.

3 Asking infants Molyneux's question: Tactual–visual object perception

Bryant showed that six- to 12-month-old infants could recognize, by sight, an object that they had previously explored by touch alone (Bryant, Jones, Claxton and Perkins, 1972). This does not address the classic issue, however, because infants in the second half-year of life regularly reach out and tactually explore objects that they see; they also bring objects that are in their hands before their eyes for visual inspection. Through such simultaneous bi-modal exploration of objects, infants may have learned to associate particular tactual impressions with particular visual sensations.

Meltzoff and Borton (1979) designed a test to evaluate tactual–visual cross-modal perception in much younger infants, before such learning experiences were likely. The average age of the subjects was 29 days at the time of test.[2] How can we induce these

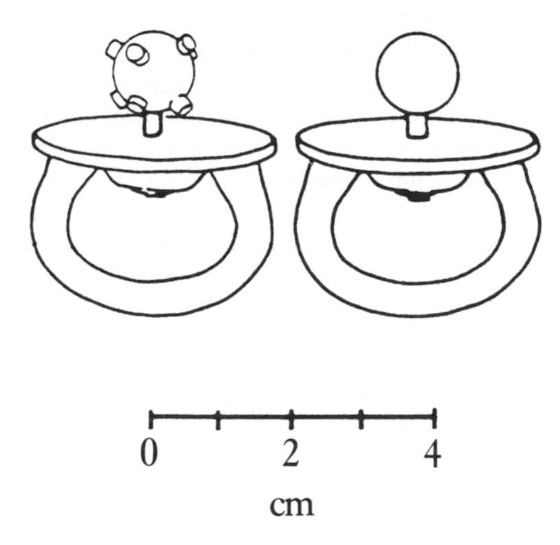

Figure 9.2 Shapes used to pose Molyneux's question to preverbal infants.

young babies to explore a shape by touch; and then how can they indicate to us which shape they felt?

Pacifiers were modified so that mouth sized shapes could be mounted on them (Figure 9.2). The tactual shapes used in the test were a small sphere and a sphere with nubs. The pacifier was cupped in the experimenter's hand and slipped into the infants' mouths without them seeing it. Most infants were quite happy to suck on the object, rolling it around on the tongue and furrowing their brows, as if the tactual exploration of the novel object was of some interest. They were allowed to feel the object for 90 seconds; it was then slipped out of the mouth, unseen. Each infant was randomly assigned to an experimental condition, half the infants were given one shape to explore and half the other shape.

All infants were then shown two visual objects, side by side, that were the same shape as the two tactual objects. We measured infants looking to both objects during a fixed response period. If infants can relate shapes-as-felt to shapes-as-seen, then the object they looked at should vary as a function of tactual condition. The hypothesis was that infants would systematically look longer at the shape they had felt. The results showed just that: of the 32 infants tested, 24 fixated the shape matching the tactual object longer than the non-matching shape, which differs significantly from the 16 v. 16 split expected by chance alone. The mean percentage of total fixation time directed to the matching shape was 71.8 per cent, which was also significantly different from chance.

The infants in this study were not literal newborns, but one-month-olds. On purely logical grounds we cannot exclude the possibility that they may have learned (what for them are arbitrary) associations or connections between visual features and tactual features in these initial days of life, associations that then were used in the experiment. However, what we know about the actual behaviour of young infants argues against such a notion. Such associative connections are said to be formed when an object was simultaneously seen and touched and hence the two sense impressions are connected in the observer's mind. It is well established that infants this young do not engage in simultaneous visual and tactual explorations of objects. At this early age, they do not yet use their fingers to explore objects while they visually inspect them. The empirical literature shows that this sort of co-ordinated bi-modal exploration only begins to occur at about three to six months of age. It is true that infants suck objects during the first month, but it is, after all, impossible for them to look at the objects *while* they suck them. Nor do infants of this age visually inspect objects before they are inserted into the mouth. During feeding, infants' commonest visual impressions are those of the mother's face (Spitz, 1965). Based on pure association, most infants should come to believe that seen faces feel like nipples! There must be prior constraints, psychological biases based on spatial properties, on the impressions that can and cannot be 'associated'.

I have suggested that newborns, without associative experience, register the same information about the shape of the object even if it is picked up through two different modalities, touch and vision. Having perceived the form through one perceptual mode, they are familiar with it when it is presented to them in the new mode. In the present experiment, the neonates may have been particularly interested in the visual instantiation of the form, because it provided modality-specific information, such as colour, that was not available through touch. Hence, they looked significantly longer at the matched than the mismatched shape. The neonates in this experiment seem to act in a way that is compatible with Evans's (1985) hypothetical philosopher, 'V'.

4 Speech by ear, eye and mouth

Are material objects unique in being registered through more than one modality? Speech is typically considered to be an auditory phenomenon. The sounds of speech are, of course, auditory. They are not seen. But articulatory acts, the causes of speech sounds, can be seen. In this sense, speech is not uniquely auditory, but a polymodal phenomenon. That speech can be seen, at least in adults, is demonstrated by the fact that we can 'read' a person's lips and grasp what was 'said', even when there is no sound. At what age and by what mechanism does the human perceiver apprehend the correspondence between auditory speech and visual speech, between audition and articulation?

4.1 Relations between seen and heard speech: A perception task

Kuhl and Meltzoff (1982; 1984) presented four-month-olds with an infant-tailored lip-reading problem. We tested whether infants recognized that an /a/ vowel sound (as in 'pop') corresponded to one articulatory gesture and that an /i/ sound (as in 'peep') corresponded to another articulatory gesture. The infants were shown a film of two faces articulating the vowels: one face was articulating the /a/ vowel and the other the /i/ vowel. The two faces were life-sized and in colour. The faces were filmed and edited so that they would articulate in perfect temporal synchrony with one another. The auditory vowel sounds, either the /a/ or the /i/, were presented from a loudspeaker placed midway between the two faces. Each infant heard only one of the sounds (played repeatedly for a two-minute test period), but had the visual choice of two faces.

This set-up allowed us to rule out two possible bases that infants might use to detect an auditory–visual match: The central placement of the loudspeaker ruled out any spatial cues; the temporal synchrony between the two faces ruled out temporal cues. The only way infants could solve this problem was by recognizing that the auditory /a/ corresponded to the articulation involving a wide-open mouth and the auditory /i/ to the articulation involving narrowed lips with the corners pulled back.

We posed this problem to 18- to 20-week-old infants. We reasoned that if infants could detect the correspondence between auditory speech and visual speech, they would look longer at the face that produced movements appropriate to the sound they heard. The hypothesis was strongly supported: infants listened intently to the sound and looked back and forth between the two faces, settling on the particular face that matched the sound they heard. Subsequent studies in our laboratory using other vowel sounds, /i/ and /u/, extended these basic effects. The experiments suggest that by 18 weeks of age infants recognize that /a/ sounds go with mouths that are open wide, /i/ sounds with mouths that have retracted lips and /u/ sounds with mouths whose lips are protruded and pursed.[3]

4.2 Relations between audition and articulation: A production task

The foregoing studies probed the infant's knowledge of auditory–articulatory links in a speech perception task. A more important, but deeply related skill, is the link between audition and articulation in speech production.

Humans around the world do not sound the same; the sound pattern of English is vastly different from German or Japanese. Early auditory experience is critical to the development of a particular phonology: growing up in a particular language environment indelibly influences the 'accent' one uses, even in adults who may not have been exposed to the original language for decades. At what age does auditory input begin to influence the sounds people make? Do young infants mimic the speech sounds they hear, adjusting their unseen articulators to match the auditory model?

The infants in our speech study provided a convenient way of addressing this question. Recall that half the infants were randomly assigned to hear the /a/ vowel and half the /i/ vowel. The formant frequencies of /i/ are spread widely apart, while /a/'s formants are close together in frequency.[4] The formants of the infants' vocalizations were measured by computer and the relevant values of the infants' formants were calculated. The results supported the imitation hypothesis. Infants hearing /i/ produced sounds that were more /i/-like, while infants hearing /a/ produced sounds that were significantly more /a/-like (Kuhl and Meltzoff, 1988).

On the basis of this work, we have suggested that in 18-week-olds the representation of speech is not limited to its auditory properties. Rather, speech representations, like the body transformations in facial imitation, are probably organized in a way that is not exclusively auditory, motor or visual, but instead is supramodal. This internal representation is such that an auditory signal can influence behaviour in two other modes. The data show that an auditory signal influences where infants look, causing them to look at a silent moving mouth that is phonetically equivalent to the sound they hear. The auditory signal also influences what infants say, causing them to move their mouths in a way that will result in an event that is equivalent to the one they hear. It seems likely that both these phenomena, cross-modal perception and vocal imitation, are linked by some common representation of speech.

5 Imitation and the roots of the notion of self

The next line of studies concerns the development of the notion of self in the preverbal infant, a topic intrinsically tied to imitation, cross-modal functioning and the co-ordination of perception and action. It is difficult to design studies to address the preverbal notion of self. In fact there is only one established experimental paradigm for examining self in infants: the mirror self-recognition paradigm first used with infants by Amsterdam (1972) and adopted by many since then. The procedure is simple. It involves putting rouge on the infant's forehead without its knowledge (usually as part of wiping its face). Then the child is put in front of a mirror. Infants older than about 18 months look at themselves in the mirror and then reach up and rub the mark on their foreheads. Infants younger than about 18 months make no such attempts, although they will rub off marks that are put on visible parts of their bodies, such as hands or arms. The inference that has been drawn by some psychologists (e.g. Kagan, 1984) is that at about 18 months of age a sense of self suddenly emerges which is linked to the emergence of language. Compatible with this, it is said, are the related findings that children give no indication of recognizing photographs of themselves before about 18 months.

It is clear that the mirror or photographic self-recognition tests only assess a narrow aspect of self: the recognition of visual features. A prior, developmentally more fundamental aspect of self concerns one's own movements and body postures. You may need mirror experience to learn that your face does not normally have a red mark on it or that your eyes are green. However, if the arguments about facial imitation are sound, you don't need visual experience in order to know what your own unseen body movements would look like. Visual instantiations of your own body movements can be directly related to the movements that are felt. In short, self-recognition based on static featural information is quite different from self-recognition based on spatio–temporal movement patterns, and I believe the latter provides the ontogenetic foundation for the former.

How can we begin to investigate infants' ability to recognize that seen human movements are 'like me' (or, 'like the movements that are felt', or even 'like those that are intended' – distinctions to which I shall return)? Several approaches are possible; we chose one in which an adult experimenter acted as a kind of 'social mirror' to the infant, reflecting back everything the baby did. We wanted to know if infants could recognize this self–other correspondence despite the absence of featural identity. We tested infants at 14 months of age, an age at which infants fail the mirror self-recognition test.

Three converging experiments were conducted. The first investigated whether or not infants at this age showed any interest in their own behaviour being reflected back to them by another person. The infants sat at a table, across from two adults who sat side by side. All three participants were provided with replicas of the same toys. Everything the infant (X) did with his toy was directly mimicked by one of the adults (X′). If X banged the toy three times on the table, X′ banged his toy three times on the table. It was as if X′ was tethered to the infant, a puppet that was under X′s control. The second adult (Y) was not so tethered. This adult sat passively, holding the toy loosely in her hands on the table top.

We thought that if infants could detect that their actions were being matched, they would prefer to look at X′ and also smile at him more. We also thought that infants would tend to test the relationship between the self and the imitating other by experimenting with it. For example, infants might modulate their acts by performing sudden and unexpected movements to check if the X′ was still shadowing them. Adults do this when they unexpectedly catch sight of themselves in a store video camera; they wave their arms or make a sudden movement to check whether the image on the screen follows suit.

Scorers watched a videotaped record of the experiment and noted which side the infant looked at and all instances of smiling and testing behaviour. The results showed that infants had a clear-cut preference for X′ over Y. Infants looked significantly longer at X′, there were more smiles directed toward X′ and infants directed more test behaviour at X′.

There are several alternative interpretations of these findings. One is that infants can recognize the self–other equivalence that is involved when an adult imitates them. Alternatively, infants may simply be attracted to any adult who actively manipulates a toy, without invoking any detection of action equivalence. Such a simple interpretation does not explain why infants would direct more 'testing' behaviour towards the imitating adult, but perhaps such behaviour is displayed to any active adult, whether or not the adult is mimicking the baby.

In a follow-up study, the general procedure was similar to the first study, but the control experimenter Y did not remain passive. Instead, this adult actively manipulated the toys. Furthermore, we wanted the adult not only to be active, but to do 'baby-like' things with the toys so that no preference for the imitating experimenter could be based solely on a differentiation of adult versus infantile actions. This was achieved by using a yoked control procedure. There were two TV monitors situated behind the infants and in view of the adults. One monitor displayed the actions of the current infant, live. The other monitor displayed the video record of the immediately preceding infant.

The job of each adult was to mimic one of the infants on TV. Both adults performed in perfectly infantile ways, but only one matched the perceiving infant. Could the infants recognize which adult was acting like they were and which was acting like another baby? The results again showed that infants looked longer at X′, the person who imitated *them*, smiled more often at him and, most importantly, directed more testing behaviour toward him.

These findings constrain the possible interpretations. The demonstrated effects cannot be explained as simple reactions to activity, for both experimenters were active. Nor can they be explained by saying that the infants recognized a generic class of baby-like actions, for both experimenters were copying the acts of babies. It would seem that the subjects are recognizing the relationship between the actions of the self and the actions of the imitating other.

How did the babies detect this relationship? Very broadly speaking, two kinds of information are available. The first is purely temporal contingency information. According to this alternative the infant need only detect that whenever he does *a* the adult does *b*. The infant need not detect that *a* and *b* are in fact structurally equivalent, only that they are temporally linked. A second alternative is that the infant can do more than recognize the temporal contingency. In particular, the infant may be able to recognize that the actions of the self and other have the same form: they are structurally equivalent.

To differentiate these alternatives, we used a design similar to the previous two.

However, in this study the purely temporal aspects of the contingency were controlled by having both experimenters act at the same time. This was achieved by having three predetermined pairs of target actions. Both experimenters sat passively until the infant performed one of the target actions on this list. If and only if the infant exhibited one of these target actions, both experimenters began to act. The imitating adult performed the infant's act, and the control adult performed the other behaviour that was paired with it from the predetermined target list.[5]

For example, whenever an infant shook a toy, the imitating adult also shook his toy, carefully shadowing the infant. The behaviour of the other adult was also under complete temporal control of the infant, but this adult performed a different type of action. Whenever the infant shook his toy, the control adult would slide his matched toy, also carefully shadowing the speed and duration of the infant's act. If the infant began waving his toy, both adults stopped acting in unison, because waving was not one of the 'target acts' to which they were programmed to respond. This design achieves the goal of having both the adults' actions contingent on the infant's. What differentiates the two experimenters is not the purely temporal relations with the acting subject, but the structure of their actions *vis-à-vis* the subject.

The results showed that the infants looked, smiled, and most importantly, directed more testing behaviour at the matching actor. Thus, even with temporal contingency information controlled, infants can recognize the structural equivalence between the acts they see others perform and the acts they do themselves. In that sense they have already begun to elaborate a notion of self. This sense of self consists of a kind of extended 'body scheme'; a system of body movements, postures and acts.

6 Implications for genetic psychology

These findings and other recent work with young infants (Bower, 1982; Butterworth, 1981; ch. 5 of this volume) alter the classical story of early psychogenesis. In the classical view, infants from birth to at least several months old have separate 'heterogeneous spaces' (Piaget and Inhelder, 1969, p. 15), a tactile space and a visual space and an auditory space that are then gradually co-ordinated as 'the child begins to grasp what he sees, to bring before his eyes the objects he touches, in short to co-ordinate his visual universe with the tactile universe' (Piaget, 1954, p. 13). The project for genetic psychologists was to trace how an infant starting from such a deficient initial state developed into the mature adult. However, such development may never need to occur, because the initial state is not as limited as we supposed.

It seems likely that the young infant is not limited to registering isolated bits of sense data, such as tactual impressions, retinal images and acoustic frequencies. There is probably no time in development in which infants are restricted to modality-specific fragments, sense scraps that are connected through empirical correlations. Instead, infants may represent the world more abstractly, in terms of objects and events that transcend a single sensory modality. These 'distal projections' are not the product of a long period of experiencing sense-data → sense-data correlations. More likely, the psychological world of the human newborn is populated by objects and events that can be accessed by more than one modality. When a young baby brings a round rattle before his eyes, he is probably not engaged in discovering what visual sensation is associated with this particular tactual impression; he already knows that. Instead, he is fascinated by the additional modality-specific features (the rich colours, visual sheen and shadows that could not have been known by touch alone) of the abstract form that he already apprehended through touch.

This picture of the infant's world follows from the results of the the tactual–visual experiment in which we put an unseen shape in the infants' mouths. These infants, 29 days old at the time of the test, were too young to have had many experiences associating shapes-as-felt with shapes-as-seen. Nevertheless, the results showed that infants who were given a shape to feel would systematically seek out the matching shape by eye. The mouth cannot see, the eye cannot touch, yet information picked up by one modality directs the other. It is also important that the test was designed to tap memory and representation. The tactual object was not felt at the time that the visual objects were seen: rather, the shape was removed from the infant's mouth, and only at that point was the visual choice presented. The results demonstrate that young infants can relate a visual perception to the memory of the information that was picked up through touch.

The studies of neonatal imitation push the story a bit further. In this case the infants tested were truly newly sighted. The youngest infant in the study was just 42 minutes old. We can say with assurance that the capacity to imitate certain facial acts is truly an innate aspect of the human mind. When the newly sighted infant sees certain human gestures he or she can immediately mimic these acts. Such facial imitation entails cross-modal functioning: the infant can see the adult's actions, but he cannot see his own face; indeed has never seen his own face in his entire life. There is some primordial connection between our own acts and the acts we see others perform.

Two details about the findings of innate imitation are particularly noteworthy. First, as in the cross-modal case, memory and representation are involved. It is conceivable that infants might have been restricted to imitating only if the to-be-matched target was in sight at the time the infant action was performed. Imitation might be the result of a kind of 'perceptual-motor resonance'. But recall that we inserted a pacifier in the infant's mouth during the adult demonstration, and it was only after the adult had stopped gesturing and assumed a passive-face pose that the pacifier was removed. In this situation, the infant needed to bring his own *unseen* body movements into accord with a currently *unseen* target act; not a task for an organism confined to a here-and-now world of raw sense impressions.

Second, the imitative response does not appear to be a mindless reflexive reaction. Simple reflexive acts don't bridge temporal gaps. Consider an analogy: suppose an infant's startle reflex was inhibited (by tight swaddling) and a loud sound was suddenly presented. A few seconds after the sound ceased, the infant is unwrapped. Would the infant show a delayed startle reaction once freed to do so? Certainly not. Reflexes do not work like that: they are triggered by the presence of the stimulus, not by a memory of the stimulus. In contrast, infants can imitate after a delay and may be particularly motivated to respond once the adult has stopped gesturing, a point I will return to below.[6] Moreover, instead of imitation bursting forth in a stereotypic, fully-formed manner, the infants correct the response over successive attempts so that they more and more closely approximate the adult target. Simple, mindless reflexes do not involve correction or a homing in on a target.

In my view, early imitation is an intentional act, in the minimal sense that it involves goal-directed matching of the target. The infant is trying to correct his or her motor performance, which may not be accurate for all sorts of reasons, so that it more accurately matches what he intends. Thus conceived, the newborn encodes the adult's act in neither exclusively visual nor exclusively motor terms. Instead, the newborn's representation is a modality-free description of the body transformation. This internal representation is the 'model' that directs the infant's actions and against which he can match his motor performance. Thus, infants compare the proprioceptive information from their own unseen body movements to their representation of the visually perceived model and sharpen their match over successive efforts. Similarly, imitative responses would not need to be 'tripped', 'released' or 'fired-off' in the presence of the model, but might be initiated from the infant's memory of what the adult had done.

This returns us to the observation that infants seem prompted to imitate once the adult has stopped. When the gesturing stops the infants are confronted with a mismatch between their current perception of the adult and their stored representation. The infant may generate a matching response in order to reinstate the absent event, to make it perceptually present again. Thus, the disappearance of the adult gesture gives the infant a cognitive problem to work through, the conflict between the world-as-represented and the here-and-now world present to the visual system. This mismatch or disequilibrium between perception and representation motivates the infant to act, and so to imitate (Meltzoff, Kuhl and Moore, 1991; Meltzoff and Moore, 1992).

Some of these notions also can be applied to the cross-modal speech effects. In classical theory, infant speech is considered an acoustic event in the province of the 'sense of hearing', processed along the eighth nerve. The new research indicates that as early as 18 weeks of age, speech is not purely a matter of hearing. Infants probably access the same underlying phonetic representation whether they see, hear or motorically produce speech. The distal entity of interest is not sense-specific, but rather a supramodal phonetic unit.

An important caveat is that the infants in the speech studies, unlike those in the imitation studies or the cross-modal object studies, were 18- to 20-weeks old. Moreover, unlike the infants in the other studies, these infants do have some experience with matches between auditory and articulatory events. They watch adults talk, and they babble and hear the results of their babbling. Accordingly, three ontogenetic accounts may be offered, with the third being of special interest.

First, the infants may simply have learned which articulatory gestures go with which sounds by simultaneously watching and listening to adults. This might reduce to associative learning. Second, the hypothesized supramodal phonetic units might be innately specified, inasmuch as all the world's languages draw from a pool of only about 100 phonemes. If so, infants should succeed on a cross-modal test using foreign-language phonetic contrasts that were never heard or seen in the infant's particular culture. Similarly, newborns might also demonstrate cross-modal auditory–visual perception. However, there is also an intriguing third alternative, namely, that the infants' self-produced babbling experience may play an important role. This interpretation is interesting because it ties together several of the phenomena discussed in this chapter.

In the babbling account, infants are conceived of as carefully monitoring their own vocal play during cooing and babbling. They 'feel' their articulatory movements through proprioception and can perceive the consequences of these articulatory efforts through audition. Thought of in this manner, the seemingly aimless vocal play of young babies is actually a way of practising the basic act of speaking, practising the production of phonetic units at will.

How could this babbling experience help infants in the cross-modal situation? It could help only if infants can relate the speech acts they see the adult perform in the experiment to the auditory–articulatory events they produced themselves during babbling. This is a cross-modal generalization. The research indicates that infants may well be able to do this. Infants' ability to imitate visual gestures demonstrates that they can relate mouth movements they see to their own mouth movements. The mouth-opening movement in Meltzoff and Moore's imitation experiments is similar to the mouth opening used to produce /a/ in Kuhl and Meltzoff's cross-modal speech case. So there is a foothold on the articulatory side – infants may relate their own unseen speech mouth movements to those they see the adult perform. There is also a similar foothold on the auditory side. Kuhl's (1979; 1983) speech categorization work indicates that infants can recognize the equivalence between the vowels uttered across talkers, including those produced by children and adults, despite the differences in the actual frequencies of the sounds that are caused by the differences in the size of the vocal tract. It therefore is reasonable to suppose that the infants in our speech study can recognize the auditory equivalences between the vowels they hear in the experiment and their own previous vocalizations. To summarize: the knowledge gained during their own babbling may contribute to infants' ability to recognize the auditory–visual correspondences for speech when seen and heard on another's body.

The infants' use of cross-modal capacities as leverage in grasping self–other relations seems an avenue worth pursuing, for there may be both constitutive and ontogenetic connections (Meltzoff, 1990b). This point came more clearly into focus in the tests involving the adult imitation of the infant. We arranged a situation in which infants were presented with two adults, one who was pre-programmed to imitate the child and the other who systematically mismatched the infant's behaviour. The results showed that the infants acted in very special ways toward the particular other who matched the self. The infants devoted most of their visual attention to the imitating adult, smiled at him more, and also directed more of what we called 'testing' behaviour towards him.

I suggest that the infant recognizes the adult as acting 'like me', that the self–other correspondence is evident to the child. A closely related notion is that the infants perceived the causal relationship with the imitating other. The imitating adult is seen by the infant as being 'more under my control'. This sense of causality might follow from the fact that the imitating adult acts in a way that is not only temporally contingent on the infant (both adults do this equally), but also that the patterns of behaviour are spatially matched.[7] The point is, however, that infants' intense interest in the imitating other ultimately derives from the infants' perception of the self. It is the cross-modal spatio–temporal correspondence between the pattern 'out there' and the pattern of self action 'here' that gives the imitating adult special psychological salience.

Indeed, it is probably the infants' own finely-tuned perceptual mechanisms that also allow them to differentiate the imitating adult as clearly 'non-self'. There are at least two pieces of information that can be used: the adult is seen to be located in a different spatial location than the one in which the self's actions are perceived to be (both visually and proprioceptively); and moreover, the imitating adult, no matter how practiced a mime, does not provide a perfect match.[8] The results of the study show that infants at this age spend virtually all their time watching the movements of the non-selves rather than watching the movements of the self; for example, they are not very interested in the movements of their own arms as they shake the toy (again indicating there is not total confusion, non-differentiation). Yet between the two non-selves, infants prefer the one that is more like the self. This attention to 'non-self that is none the less like self' may not be an altogether bad recipe for psychological development.[9]

The imitation game provides a kind of tutorial in the world of the child. It seems possible that children this age differentiate physical and psychological causality. Physical causality in the ordinary world of middle-sized objects has both spatial and temporal characteristics, there is 'physical contact' between the cause and effect. In the imitation game the infant 'causes' the adult to move in a particular way, but there is no physical contact between baby and adult. The child may interpret the perception of cause and effect without physical contact as psychological control or even communication. Such an ascription might be natural for the child when the agent is self and the recipient is another like-me human. If so, then the imitation game provides a situation in which to explore the parameters of psychological contact and communication. Just as hitting objects and watching them bump provides opportunities for exercising and enriching the child's 'naive physics', the imitation game provides opportunities for the exercise and development of the child's 'naive psychology'.

I want to return to the earliest phase of infancy. Young infants are known to be fascinated with other human beings; no toy can compete with people during the early phases of life; people are the infant's favourite playthings. We may inquire why even the youngest of infants are so fascinated by human beings. A simple learning view might be that there is nothing intrinsically special about other people to young babies; the human figure begins to command attention as it becomes associated with primary pleasures like food, warmth and comfort. A traditional nativist view might hold that there are built-in preferences for certain visual properties of human beings; perhaps the face gestalt is innately attractive.

In contrast, I would like to suggest that infants find human beings interesting because they have a primitive ability to recognize that the distinctively human movements they see are like the movements that they feel themselves make. It is not only, perhaps not even primarily, the features of the human form – these lips, these eyes – that attract attention and give humans special meaning. Rather, it is the fact that the spatio-temporal patterns of human body movements are, in a sense, 'familiar', in some very primitive way reminding babies of themselves. Of all the things in the newborn's visual field, it is only other human beings that will have this fascinating trait; neither inanimate objects, nor even other animals, will match in quite the same way. It is plausible that the infants' own perception of the self, coupled with a capacity for recognizing cross-modal similarities, may lead them to feel a primordial kinship with their fellow human beings.

ACKNOWLEDGEMENTS

For the title and philosophical guidance I am especially grateful to Alison Gopnik; for help in putting questions to babies and interpreting their reactions I owe an enormous debt to Patricia Kuhl and M. Keith Moore. Naomi Eilan and Bill Brewer helped me to sharpen my thoughts on many of the problems discussed here. This work was supported by a grant from NIH (HD22514).

NOTES

1 Infants' sucking reflex took precedence over any tendency to imitate. They did not tend to open their mouths and let the pacifier drop out during the mouth display; nor did they tend to push the pacifier away with their tongues during the tongue display. It was only after the pacifier was removed that the response was inaugurated, sometimes after a long period of motor inactivity coupled with careful inspection of the experimenter. Thus, the technique was effective in disrupting imitation when the target was perceptually present.

2 Infants spend most of their first month sleeping; they are in an awake and alert state less than five hours a day, with their hands swaddled much of the time.

3 Precisely what aspect of the auditory signal is needed? In further studies Kuhl, Williams and Meltzoff (1991) systematically dissected the speech signal into elementary parts. The principal findings showed that the cross-modal performance was not supported if parts of the acoustic signal that comprise speech ('distinctive features'; Jakobson, Fant and Halle, 1969) were provided as the auditory stimulus instead of the whole phonetic unit. The broader theoretical inference is that infants' cross-modal perception of speech does not originate through a process that progresses from 'simple parts' to 'wholes', in which infants initially relate faces and voices on the basis of a simple acoustic feature, and then gradually build up a connection between the two that involves, on the auditory side, an identifiably whole speech stimulus. These findings thus provide an instance in which young infants are responsive to the wholes, the phonetic unit *per se*, rather than to isolated components (the distinctive features of speech).

4 Formant frequencies are an acoustic property of speech pertaining to the frequencies where energy is concentrated.

5 The three pairs of actions were: (a) shake = slide, (b) pound = poke and (c) touch mouth with toy = touch non-oral region on the head, neck or shoulders. These pairs were chosen from an extensive video review which showed that these were six common 'action schemes' of infants this age, and that the acts within each pair were similar.

6 We have performed studies in which the adult demonstrated the gesture and then assumed a passive-face pose without using a pacifier. The results show that many infants will watch the display with fascination and only begin to inaugurate the matching response after the adult has stopped.

7 Two other alternatives also bear mention, and I thank Naomi Eilan, Bill Brewer, James Russell and others of the Cambridge Spatial Representation Workshop for highlighting them for me. (1) The imitating adult might also be of special interest because he is fulfilling the infant's action intentions; whenever the infant wills there to be a toy-in-the-mouth, this occurs (actually two such events occur, one for the infant and one for the imitating adult). (2) Infants might prefer looking at behavioural synchrony, two people doing the same thing, without regard to one of the actors being the self. However, the second alternative has difficulty explaining the infants' testing behaviour, in which infants suddenly deviate from the behavioural synchrony.

8 Mirror reflections could well be puzzling to young infants because this duality is not so easily resolved. In terms of spatial location the image is clearly non-self; yet it moves *completely* under the self's control, a characteristic of self. The resolution is to infer a reflecting surface and virtual image of the self.

9 The imitation game is a form of preverbal communication between adult and infant. It plays an important role in early enculturation, because a social mirror (unlike a physical mirror) is both selective and interpretive in its reflections. Parents, as social mirrors, provide 'creative reflections' to their infants – reflections that capture aspects of the infant's activity, but then go on beyond it to read in intentions and goals to that behaviour. The infant waves an object, but the parent interprets this as waving in order to shake, and therefore waves intensely enough to shake the toy and produce a sound, which in turn leads the infant beyond his or her initial starting point. Similarly, actions that are potentially meaningful in the culture, will

be reflected back more often than others (Bruner, 1975; 1983). Social communication via the imitation game begins a long time before verbal communication.

REFERENCES

Amsterdam, B. 1972: Mirror self-image reactions before age two. *Developmental Psychobiology*, 5, 297–305.

Bower, T. G. R. 1982: *Development in infancy*, (2nd ed.). San Francisco: W. H. Freeman.

Bruner, J. S. 1975: From communication to language – a psychological perspective. *Cognition*, 3, 255–87.

Bruner, J. S. 1983: *Child's Talk: Learning to use language*. New York: Norton.

Bryant, P. E., Jones, P., Claxton, V. and Perkins, G. M. 1972: Recognition of shapes across modalities by infants. *Nature*, 240, 303–4.

Butterworth, G. 1981: *Infancy and Epistemology: An evaluation of Piaget's theory*. Brighton: Harvester Press.

Evans, G. 1985: Molyneux's question. In his *Collected Papers*. Oxford: Clarendon Press, 364–99.

Jakobson, R., Fant, C. G. M. and Halle, M. 1969: *Preliminaries to Speech Analysis: The distinctive features and their correlates*. Cambridge, Mass: MIT.

Kagan, J. 1984: *The Nature of the Child*. New York: Basic Books.

Kuhl, P. K. 1979: Speech perception in early infancy: Perceptual constancy for spectrally dissimilar vowel categories. *Journal of the Acoustical Society of America*, 66, 1668–79.

Kuhl, P. K. 1983: Perception of auditory equivalence classes for speech in early infancy. *Infant Behavior & Development*, 6, 263–85.

Kuhl, P. K. and Meltzoff, A. N. 1982: The bimodal perception of speech in infancy. *Science*, 218, 1138–41.

Kuhl, P. K. and Meltzoff, A. N. 1984: The intermodal representation of speech in infants. *Infant Behavior & Development*, 7, 361–81.

Kuhl, P. K. and Meltzoff, A. N. 1988: Speech as an intermodal object of perception. In A. Yonas (ed.), *Perceptual Development in Infancy: The Minnesota Symposia on Child Psychology*. Hillsdale, NJ: Erlbaum, vol. 20, 235–66.

Kuhl, P. K., Williams, K. M. and Meltzoff, A. N. 1991: Cross-modal speech perception in adults and infants using nonspeech auditory stimuli. *Journal of Experimental Psychology: Human Perception and Performance*, 17, 829–40.

Meltzoff, A. N. 1990a: Towards a developmental cognitive science: The implications of cross-modal matching and imitation for the development of representation and memory in infancy. In A. Diamond (ed.), *The Development and Neural Bases of Higher Cognitive Functions. Annals of the New York Academy of Sciences*, 608, 1–31.

Meltzoff, A. N. 1990b: Foundations for developing a concept of self: The role of imitation in relating self to other and the value of social mirroring, social modeling, and self practice in infancy. In D. Cicchetti and M. Beehgly (eds), *The Self in Transition: Infancy to childhood*. Chicago: University of Chicago Press, 139–64.

Meltzoff, A. N. and Borton, R. W. 1979: Intermodal matching by human neonates. *Nature*, 282, 403–4.

Meltzoff, A. N., Kuhl, P. K. and Moore, M. K. 1991: Perception, representation, and the control of action in newborns and young infants: Toward a new synthesis. In M. J. Weiss and P. R. Zelazo (eds), *Newborn Attention: Biological constraints and the influence of experience*. Norwood, NJ: Ablex, 377–441.

Meltzoff, A. N. and Moore, M. K. 1977: Imitation of facial and manual gestures by human neonates. *Science*, 198, 75–8.

Meltzoff, A. N. and Moore, M. K. 1983: Newborn infants imitate adult facial gestures. *Child Development*, 54, 702–9.

Meltzoff, A. N. and Moore, M. K. 1989: Imitation in newborn infants: Exploring the range of gestures imitated and the underlying mechanisms. *Developmental Psychology*, 25, 954–62.

Meltzoff, A. N. and Moore, M. K. 1992: Early imitation within a functional framework: The importance of person identity, movement, and development. *Infant Behavior and Development*, 15, 479–505.

O'Shaughnessy, B. 1980: *The Will: A dual aspect theory*. Cambridge: Cambridge University Press.

Piaget, J. 1954: *The Construction of Reality in the Child*. New York: Basic Books.

Piaget, J. and Inhelder, B. 1969: *The Psychology of the Child*. New York: Basic Books.

Spitz, R. A. 1965: *The First Year of Life*. New York: International Universities Press.

10
Molyneux's question and the idea of an external world

Naomi Eilan

Introduction

> Suppose a Man born blind, and now adult, and taught by his touch to distinguish between a Cube and a Sphere of the same metal, and nighly of the same bigness, so as to tell, when he felt one and t'other, which is the Cube, which the Sphere. Suppose then the Cube and the Sphere placed on a Table, and the Blind Man to be made to see: Quaere, Whether by his sight, before he touch'd them, he could now distinguish, and tell, which is the Globe, which the Cube? (Locke, 1975, II, ix, 8, p. 146)

The basic format of Molyneux's question can be and has been extended to spatial properties other than shape, such as location, length and so forth; and to modalities other than vision and touch, such as hearing and proprioception, and to action. Very loosely, negative answers to the question are said to involve the idea that spatial perceptions are modality-specific; positive answers, the idea that they are amodal. Its enduring interest lies in the idea that the kind of issues it raises about the relations among the sensory modalities, whether they are amodal or modality-specific, have fundamental implications for the way we explain the nature of spatial representation in general.

In particular, a recurring theme in discussions of Molyneux's question is the idea that there is an immediate connection between, on the one hand, answers to it and, on the other, explanations of what it is for perceptions to be perceptions as of an external world. Very broadly, it is generally supposed that if the answer to the question and to its various extensions should be 'Not', on the grounds that our spatial perceptions are modality-specific, in some sense, this poses a serious threat to our access through perception to the world out there.

Now it is certainly true that one common reason for being drawn toward a negative answer begins with the thought that what touch delivers when one is feeling a cube, say, is a succession of tactile sensations. It is these sensations, the thought continues, that are the basis for post-perceptual judgements that the shape of the object one is feeling is cubic. Such sensations are then thought of as constitutive of the modality, something that vision could not deliver, and in this strong sense modality-specific. So the basis for judging an object cubic which is used in touch is absent in vision. The claim is then that a newly sighted subject would have to learn to associate the tactile sensations with the new visual input before extending his or her tactually-based concept of 'cubic' to the deliverances of vision. On this kind of account, the answer to the question should be 'Not'.

A negative answer based on such reasoning does, indeed, often involve the idea that what perceptions proper deliver is a collection of sensations, mind-dependent items, rather than immediate information about the external world. And it is to this kind of reason for delivering a negative answer to an application of Molyneux's question to infants that Meltzoff alludes in his arguments for a positive answer, when he suggests that such negative answers are based on the idea that infants are initially 'restricted to modality-specific fragments, sense scraps that are collected through empirical correlations' (chapter nine).

But is it the case that all negative answers to Molyneux's question necessarily involve such explicit (or more subtle) denial of our immediate access through perception to the world out there? And if it is, why should it be so? What in general is the connection between explanations of what is involved in perceptions representing a world out there, and the way we explain the relation between the spatial contents of perceptions in the various modalities? This is the issue I want to begin to examine in this chapter. Before starting, a few points about the style of inquiry I will be pursuing, about the particular questions I will be asking and, also, about the kinds of questions I will not be engaged with.

(1) It seems to me that there are massive amounts of empirical evidence which suggest that our perceptions are amodal in their spatial content in such a way as to deliver a resounding 'yes' to important variants of Molyneux's question.[1] The issue I will be pursuing is not whether the answer to these variants should, for humans or animals, be yes or no, but rather that of whether and how, in arguing for the amodality of our spatial perceptions, we can and should appeal to what is required for perceptions to be as of an external world. For that reason, in order to avoid confusion with empirical questions, I will be describing imaginary subjects who are unlike us, in that their perceptions are modality-specific in various ways which would result in negative answers to the variants of Molyneux's question I will be considering. I will be asking: Suppose their perceptions are modality-specific in this way, would such subjects' grip through perception on the idea of an external world be threatened in any way, and if so, in what way? Towards the end of this chapter, I will return to the implications of this discussion for the kinds of arguments appropriate for claiming that our own spatial perceptions are amodal, and for the role of these arguments in explaining our grip through perception on the external world.

(2) My general aim will be to examine which, if any, accounts of modality specificity would threaten which, if any, explanations of what makes perceptions perceptions as of a world out there. The issues that must be considered when we are examining the relation between perceptions being as of a world out there, and the way spatial properties are represented in the various modalities, vary greatly, depending, first, on which spatial properties and, second, on which modalities we are considering. They also vary depending on which notions of modality specificity are being considered, and on which

ingredients in our conception of an external world are in question. It is, to my mind, a project of no small interest to consider separately questions introduced by different permutations on possible positions along these four distinct dimensions of variation. However, if only for reasons of space, the issues we will be considering here will be restricted along the following lines.

I will focus exclusively on a variant of Molyneux's question with respect to *place* representation; and will consider mainly the case of two modalities, *vision* and *touch*. Within this context, I will be considering the cases of imaginary subjects whose place representations are modality-specific in a variety of basic ways, all of which would yield negative answers to the variant of Molyneux's question on place representation I will go on to formulate. The exact nature of the various kinds of modality-specific place representations that will be examined will be spelled out as we go along. Our question will be: Which, if any, ingredients in explanations of what makes perceptions perceptions as of external places would be threatened by which, if any, of these different notions of modality specificity in place representation?

There are very many accounts that one might give of what makes perceptions perceptions as of places in the external world. These various accounts can be seen as yielding a spectrum of kinds of place representation ranging from those that deliver a weak or impoverished notion of places out there, to those that yield a fully objective and detached notion of a world out there. I will be focusing on the robust end of the spectrum. It is certainly consistent with everything I will be saying that quite different kinds of arguments and substantive claims apply to the weak end of the spectrum. (Indeed, I believe there are independent arguments for amodality on the weak end of the spectrum that do not apply to the kind of place representations I will be considering).

The subjects I will be imagining are ones whose perceptions count as representations as of external places partly in virtue of the fact that they serve as input to cognitive maps of their environment. Perceptions will count as modality-specific just in case the cognitive maps to which they are input are modality-specific. These maps can be defined, initially, as representations of places in the environment which are such that (a) all the places they represent are represented as spatially connected to each other, and, in virtue of this, they enable the subject to have some kind of grip on the connectedness of places represented on the map; and, (b) they enable re-identification of places perceived in the past, from whatever spatial direction they are re-encountered.

In addition to assuming that maps provide for a grip on the connectivity of places they represent, and for their re-identification, I will be assuming that any explanation of what makes cognitive maps representations as of external places will link up, in some way, with explanations of what gives the subject a grip on the idea that the places represented are places in the physical world, rather than points in a mathematical space. In particular, I will be assuming that explanations of how a cognitive map provides a subject with a grip on the connectedness of the places represented on the map, and with the capacity for re-identifying places represented on the map, will link up in some way with explanations of a subject's grip on the idea that places out there are, essentially, places in and through which movement, change and more generally causal processes occur.

Finally: traditionally, at least in philosophical discussions, the idea of an external world is connected with the notion of 'mind independence'. In particular, the question of what is involved in having a grip on the idea of an external world is often formulated as a question about what is involved in having the idea of places or objects existing unperceived. Without, initially, going into any details of what having such a grip might involve, I will be assuming that any explanation of the nature of the link between perceptions and cognitive maps is intended, *inter alia*, to provide the subject with some kind of grip on the idea of places existing independently of the subject's current perceptual experiences.

These general assumptions leave much unspecified as to how exactly cognitive maps provide subjects with a grip on the idea of places out there, and how exactly the link with the map makes perceptions perceptions as of places out there. Such lack of specificity is intentional. If there is some deep connection between the modality and the externality issues one would expect examination of the exact nature of this link to illuminate particular aspects of what goes into making perceptions perceptions as of an external world. As we shall see, the problem in accounts of externality spotlighted by the way we will be examining this link is how exactly we spell out, motivate and justify the connectivity requirement, the intuitive idea that there is some link between representations counting as representations as of places out there, and their representing places as spatially related to each other. (For a fuller account of ingredients that go into our conception of an external world, which I shall later be implicitly drawing on, see the General Introduction.)

(3) Before turning, finally, to formulating the variant of Molyneux's question we will be concerned with, a last caveat. One issue often raised in connection with Molyneux's question is this. Suppose it is claimed that perceptions do not represent a world out there but, rather, consist in bundles of sensations. Does such a view inevitably lead either to some form of idealism, i.e. to the view that there is no mind-independent world out there for perceptions to inform us about, or to radical scepticism, i.e. to the claim that there are no grounds to believe that there is a mind-independent world out there? If such consequences do hold, then the implications of a negative answer to Molyneux's question, on the grounds that perceptions consist in modality-specific sensations, are radical indeed.[2]

Although this issue is, of course, closely related to the question we shall be concerned with, it is distinct. Our concern will be only with the relation between accounts of the immediate contents of perceptual experiences and Molyneux's question. Does delivering a negative answer to the question mean that we cannot think of perceptions as including, as part of their immediate content, representations as of a world out there? We will not be concerned with the implications of denying that perceptions are as of a world out there for the issues of idealism and scepticism.

1 Place representation in the modalities

Imagine creatures who are born blind and who, in addition to navigating by dead reckoning, build up cognitive maps of their environment of the kind that allow them to re-identify landmarks through which they move and to represent the spatial relations among them. Imagine too that their perception of landmarks in the environment is restricted to the deliverances of haptic exploration of these landmarks and, more generally, that they lack the capacity to see, hear, smell or taste anything. Suppose now that during their sleep they suddenly acquire the capacity to see. When they wake up and look around will they immediately recognize the places they see? And will new landmarks they see automatically be added to their previously acquired cognitive map, such that visually and haptically perceived places are immediately spatially related to each other? This is the first, rough formulation of the variant on Molyneux's question that will concern us here.

There are several ways of conceiving how they represent places before they acquire the capacity to see and when visually perceiving their environment, which would yield negative answers to these questions. In what follows I want to consider a few of these ways. Our question is: Which if any of these accounts put pressure on our subjects' perceptions counting as perceptions as of places out there?

To begin, it will help to have before us the idea of different frames of reference for individuating places, in one of the senses of 'frame of reference' mentioned in the Introduction to Part I. On this notion of frames of reference, commonly used in both physics and everyday talk, a frame of reference consists in the objects or object parts relative to which places are individuated. Frames used in place individuations are distinct just in case the object or object parts they consist in are distinct. For example, if places are classified as the same or different relative to their spatial relations to a table in the room, this will involve the use of a different frame of reference from that involved in classifying places as the same or different relative to their spatial relations to a chair in the room. On this notion, if we want to know whether places referred to are the same or different we must use a single frame of reference for individuating them. So, if I want to know whether a place individuated relative to my nose, and described as being five feet from my nose, in such and such a direction, is the same as a place individuated relative to the table in the room, and is described as being seven feet away from it, in such and such a direction, I will have to be able to map one frame of reference onto the other, either directly or through mapping each one onto a third frame of reference, relative to which I can ask the question of whether the places referred to are the same or not.

The negative answers we will be considering to our variant of Molyneux's question begin with this simple notion of frames of reference. In some form or other they all involve the claim that the frames of reference that give the contents of our subjects' place representations prior to acquiring the capacity to see are distinct from those that give the contents of their visually-based maps. As we shall see, in the course of our discussion we will have to bring in other ways of explaining the content of place representations, which go beyond appeal to this notion of frame of reference. In the next section, however, we will be concerned with claims that appeal only to the notion of frames of reference just specified.

2 Radical incommensurability

One reason for delivering a negative answer to the question about our imaginary subjects would be to claim that their cognitive maps before acquiring the capacity to see are *radically incommensurable* with those used in their visually-based maps, in the sense that there is no way in principle of mapping the frames of reference used in one map onto those of the other in order to tell whether the places represented are the same or connected. The claim is not that the subjects cannot do it, for reasons of psychological inadequacy, but, rather, that it is not possible for anyone to do it because of the nature of the frames. If we conceive of their place representations in the modalities as radically incommensurable in this way it simply does not make sense to speak of our subjects identifying places across modalities, or of relating places perceived in one modality onto the cognitive map employed in representing places in the other modality.

Is this way of conceiving of their place representations in the modalities consistent with the idea that their perceptions inform them about a mind-independent, external physical world? An initial difficulty turns on understanding how, on the notion of frames of reference that we are employing, place representations *could* be radically incommensurable. If one wants to know whether a place individuated relative to one object or object part is the same as a place individuated relative to another object or object part, all that is needed is a specification of the spatial relations among the objects or object parts that constitute the original frames of reference (where such specification may be relative to one of these frames, or relative to a third frame of reference, that uses other objects). Such specification will always provide a way of aligning place individuations on different frames of reference.

Blocking the possibility of such alignment would require denying that the objects that constitute the frames of reference used in the modalities are all spatially related to each other. There are various ways in which this might be done. The first would be to say that we can make sense of the idea of physical objects in the external world, that is, physical objects in the world out there, rather than fictional objects represented in novels or pictures, which are such that there are no spatial/physical relations of any kind between them; and that it is these kinds of objects that constitute the frames of reference used in the modalities. A second way would be to say that the objects that constitute the frame of reference used in the modalities should be thought of as mathematical points in distinct, unrelated, mathematical spaces. A third way would be to say that the objects that constitute the frames of reference used in the various modalities are modality-specific, mind-dependent sensations, which need not be thought of as spatially related precisely because they are mind-dependent entities and, therefore, need not be thought of as having a spatial or physical location at all.

All three ideas would yield a notion of radically incommensurable frames of reference. And all three threaten some fundamental ingredient(s) in our notion of what it is to have access to the external world through perception. In the case of the second and third way of blocking the possibility of aligning the frames of reference this is perhaps obvious enough. To adopt the third way of blocking the possibility of aligning the frames of reference used in the modalities, in which the entities that provide the incommensurable frames of reference are modality-specific, unlocated sensations, is to say that the immediate contents of perceptions are exhausted by appeal to the idea of 'sensational spaces', spaces that are akin to fictional spaces in that the objects that constitute the frame are mind-dependent items which have no location in the mind-independent world. If this is the position, the immediate world made available by perception is not external, at least in being mind-dependent.

As to the idea that the incommensurable spaces to which we have immediate access in the modalities should be thought of as mathematical spaces: certainly there is no difficulty in thinking of subjects having knowledge of such unconnectable spaces; and there is no need to think of mathematical spaces as mind-dependent, simply because they are thus unconnectable. Nevertheless, even if we think of mathematical spaces realistically, the idea of mathematical spaces falls far short of what we mean by external spaces, at least in that they are not physical spaces, spaces in which movement and physical change occur.

Only the first way of blocking the commensurability of the frames of reference that give the content of place perceptions in the modalities holds initial promise of such incommensurable place representations none the less being representations of the world out there, for the objects that constitute the incommensurable frames of reference are said to be spatially unrelatable, physical objects. However, such initial promise is short lived. For appealing to this notion of spatially unrelatable physical objects to explain the sense in which the frames of reference used in representing places in the modalities could be radically incommensurable would have very bizarre consequences. If the object-parts that provide the frame of reference are distinct parts of the subject's body, this would require splitting the subject's body between spatially unrelatable worlds. The situation is not much improved if the spatially unrelatable objects that constitute the frames of reference are said to be other objects, which the subjects perceive. We may assume that physical objects become relevant to explaining the frame of reference intrinsic to he contents of perceptions only if subjects have either perceived them in the past or are currently perceiving them. Perception of an object requires a causal link between the perceived object and the subject's embodied brain. If the perceived objects that constitute the frames of reference in each modality are spatially unrelated, so too must be the embodied brain-parts responsive to objects in each modality. For, to introduce a single embodied brain that is causally related to the objects that constitute the frames of reference exploited in each modality, just is to introduce a causal/spatial link between these objects. So, here too, maintaining the incommensurability of the frames would require splitting the embodied-brain between spatially unrelatable worlds.

Causal access through perception to the external objects that constitute the frames of reference that give the contents of perceptions in each modality requires, then, the connectivity of these places, on pain of losing the link between perceptions in the two modalities, on the one hand, and a single body, on the other. We shall return to the role of appeals to the unity of the perceiver in explanations of why representing places objectively requires a grip on the connectivity of places. But for the moment we may say, in summary, that the idea that the answer to our variant of Molyneux's question might be 'no' on grounds of the radical incommensurability of the frames of reference exploited in each modality, raises prima facie insurmountable difficulties for conceiving of the perceptual experiences of our imaginary subjects in those modalities as representing an external world, at least for the variety of reasons and in the variety of senses we have briefly examined.

3 *De facto* absence of commensuration

Let us now turn to the other, less radical reasons one might have for delivering a negative answer to our variant of Molyneux's question about places. On these, the claim is the objects that provide the frames of reference used in each modality by our subjects are distinct but spatially related objects in the external world; that is, the frames are distinct but not radically incommensurable. However, the commensurability of these frames is not exploited by, or manifest in, the contents of the subject's perceptions in the various modalities because the frames that give the contents of the maps linked to each modality are not, in fact, related to each other by the subject. The consequence of this is that places perceived in one modality are never recognized in another modality, and the spatial relations among places perceived by each modality are never intrinsic to the contents of place representations in either modality.

The following example will give an initial, informal feel of the phenomenology of the kind of experiences that one might appeal to in elaborating such an account. Imagine sitting in a highly contorted yoga position and hearing a recurring pattern of several sounds, each sound type at a different distance from you, such that you learn to recognize the places of each type of sound. Suppose that over the same period you touch with one of your hands several surfaces that differ in texture, such that you are able to match these different textures to these various locations. Both the sounds and the felt surfaces seem to be out there, and that they seem to be out there is intrinsic to the phenomenology of the experiences in each modality. However, it may be that because of the contortion, although you can locate the felt textures relative to the palm of one hand, say, and the sounds relative to your head, you do not know where your head is relative to your hand, so your experiences in each modality give you no clue as to the spatial relation between the places of the sounds and the felt surfaces.

On one kind of description of this situation, an exhaustive account of the spatial contents of the experiences in each modality would appeal to different frames of reference in explaining what is involved in each case in perceiving the sounds and felt surfaces as out there. These frames are, of course, not radically incommensurable. The important point about this example, on the account we are considering, is that the experiences themselves do not represent the locations of the sound and the felt object *as* related to each other; the commensurability of the frames is, in some sense, extrinsic to the contents of the experiences in each modality.

Certainly, reflecting on the experiences, one does think of them as related, because one knows one's hand is related to one's head, even if one does not know, at the moment, what this relation is. But the spatial relatedness of hand and head are extrinsic to the phenomenology of the initial, unreflective experience of locating felt textures and sounds respectively. The crucial claim in this account of the phenomenology of such experiences, from our perspective, is that their outward facingness or externality is said to owe nothing to, and be wholly independent of, the subject's ability to reflect on the relatedness of the frames that give the contents of its experiences in each modality.

Generalizing from this informal description of the example, consider the following story about our imaginary subjects. Their spatial experiences, once they have regained their sight, are always of this kind: their experiences in each modality, prior to any reflection about them, are such that the relatedness of the places perceived by each modality, in virtue of the relatedness of the object or object parts that constitute the frames of reference used in each modality, is extrinsic to the phenomenology of their experiences in both modalities. Nevertheless, each modality does represent the landmarks perceived as being out there, and the fact that it does so is wholly independent of the subjects' ability, if they possess it, to reflect on the relatedness of the places perceived in each modality. Our question is: Is such a story coherent? (Perhaps this is the kind of description of infants' experiences that Piaget had in mind when he describes their perceptions in the modalities as initially representing heterogeneous spaces, and of infants gradually learning to co-ordinate their visual and tactual universes. See Piaget and Inhelder, 1969, p. 15; Piaget, 1954, p. 13).

4 Egocentric versus objective place representation

In order to make progress here we must bring in an intuitive distinction often drawn in discussing place representation, the distinction between engaged/egocentric ways of representing spatial properties, and disengaged/objective/absolute/allocentric ways of representing spatial properties.[3] There are very many ways of explaining what this intuitive distinction involves. Although the issues we will be raising can be brought out on most of these accounts, for our immediate purposes the most succinct and illuminating is to be found in Campbell's account (chapter three). For the question we are considering is, in effect, whether and how the *de facto* connectedness of places represented in the various modalities must be manifest in the contents of perceptions in each of those modalities for the perceptions in those modalities to be perceptions as of an external world; and Campbell's account of the distinction between engaged and disengaged representations of spatial properties gives a primary role to two ways of grasping the connectedness of places.

We should bear in mind two ingredients in Campbell's distinction between egocentric and objective ways of representing the connectedness of places. The first is the insistence on a link between explanations of what it is to have a grip on the connectedness of places in general, and what it is to have a grip on places as physical places (as opposed to points in mathematical space). These come together, on Campbell's account, in the suggestion that in explaining what it is for a subject's representation of the connectedness of places to count as a representation of the connectedness of physical places, we must explain what it is for a subject to have a grip on the idea that places are causally connected, a grip on the idea that how things are at one place may depend causally on how they are at any other represented place.

The second ingredient in his account is the one that is important for our purposes. He suggests that we should explain what makes a way of representing places egocentric is that subjects' grip on the causal or physical connectedness of places is exhausted by their grip on the pragmatic implications of such connectedness for their own perceptions or actions; what is to be seen next, how to get from *a* to *c* through *b*, and so forth. Detached or objective ways of representing the causal connectedness of places in one's environment are, in contrast, on his account, ways of grasping such connectedness that are, in some sense, independent of the subject's engagement, through action and perception, with the world it represents. Campbell goes on to suggest that this will involve grasping such connectedness in virtue of grasping the intuitive physics one employs in representing the movement of objects in the space one represents. (For less condensed explanations of the distinction, see ch. 3, section 5). The important idea to keep in mind for our immediate purposes is the characterization of egocentric representations of the connectedness of places as representations whose contents are exhausted in terms of pragmatic implications for the subject in its interaction with the environment.

The distinction between different ways of grasping the connectedness of places in one's environment, however it is explained, cuts across the distinction between the different frames of reference, as so far introduced, in the following sense. It may be correct to explain both an animal's or infant's navigational ability in a room, say, and an adult human's verbal description of locations in that room, by appeal to spatial representations that individuate places relative to what is in fact, say, a table in the room. But the first may not be, and the second may be, disengaged. To the extent that we think of engagement or lack of it as intrinsic to the content of spatial representations, intrinsic to the way the external world is represented, this means that specifying the landmarks relative to which places are individuated can only be a partial specification of the content of the spatial representation; what matters, in addition, is how those landmarks contribute to grasping the connectedness of the space. On Campbell's account of the egocentric/objective distinction, this is a question of whether they figure only as landmarks in a system geared exclusively to navigating among them, or whether they figure in a representation of space that is based on thinking of the space as one in which objects move.

Returning to our variant of Molyneux's question, the problem of whether different perceptual modalities can be said to represent the external environment and be *de facto* uncommensurated turns, to a large extent, on the issue of whether we are dealing with engaged or disengaged representations of spatial connectedness within each modality, and on how exactly we explain what the distinction is between these two ways of representing the external environment. The remainder of the chapter will be devoted to spelling out why this is so, and what this tells us about the way we should understand the intuitive idea that the capacity to represent places out there is linked in some way to the capacity of having a grip on their connectedness.

5 Egocentric place representation and cross-modal connectivity

Let us return to our imaginary subjects and suppose, first, that their grip on the physical or causal connectedness of places in their environment, prior to acquiring the capacity to see, is exhausted by their navigational abilities, and is therefore, on Campbell's account, wholly engaged. That is, their grip on the connectivity of places is exhausted by the fact that they can navigate from any arbitrary place they are at to any other place represented on their map, on the basis, let us suppose, of computations of the kind proposed by O'Keefe (see chapters two and three). Let us also assume that they lack any reflective abilities, any conception of themselves as objects, any ability to reflect on their own interaction with the environment they perceive. They are excellent navigators, and what grip they have on the notion of places out there, whatever that comes to, derives wholly from this ability.

Second, let us suppose that their environment and perceptual capacities are such that touch and sight are, in fact, sensitive to different kinds of objects in the environment, and that there is no overlap in the objects each modality responds to. Alternatively, we may suppose that each modality responds to different features of the same object-kinds, texture in one, colour in the other, say, so that the landmarks used in each modality are different located features, rather than different objects. On this supposition, the maps our subjects will build up when they acquire the capacity to see take as input only visually perceived landmarks, and are only triggered into use, for the purposes of orientation and returning to represented places previously seen, by such visually perceived landmarks. The objects or features that constitute the frames of reference that give the contents of the maps in each modality are, then, distinct, and the frames of reference used in each modality are, in this sense, modality-specific.

Third, we must imagine, too, that when they acquire the capacity to see and begin to build up maps on the basis of visual input, their old maps are retained, but are only put to use on the basis of haptic stimulation. This is perhaps easiest to picture if we think of their haptically-based maps being used only during the night and their visual ones only during the day. The important point is that what we must imagine is that the deliverances of each modality are functionally disassociated from each other in the following sense. For reasons of cognitive architecture, there is no mapping of information about places acquired through one modality onto frames of reference employed in the other. An exhaustive account of the content of our subjects' experiences in each modality will, therefore, appeal only to these modality-specific frames of reference. So when these subjects acquire the capacity to see, the frame of reference used prior to the acquisition of this capacity is irrelevant to describing the content of their place representations in vision.

Finally, we must remember that the question of whether the places represented by the two maps are the same or not just does not arise for the subjects we are now considering. They do not, prior to acquiring the capacity to see, possess the power of reflection, of asking whether places are the same or not. They just navigate. And when they acquire the capacity to see they immediately get on with the business of getting their bearings on the basis of visual input and building up new maps, maps that individuate places relative to a different frame of reference, without asking any questions about whether the places they see are the same as the places through which they have navigated when blind. During the day they use one map; during the night they use another, and they no more ask questions about the relations among the places represented by each map than does a rat, transported anaesthetised backwards and forwards between two different environments, about the relation between places in these two environments.

Here it seems we do have a story where it seems right to say that the maps that are constitutively linked to the experiences in each modality, and therefore partially determine their content, are uncommensurated, and for reasons of cognitive architecture cannot be commensurated by the subject. Does such functional segregation among the maps linked to perceptions in each modality pose any kind of threat to our subjects' perceptions being perceptions as of external places?

Prior to the acquisition of vision these creatures were able to navigate from any arbitrary place they were at to any other place represented on their cognitive map. On the basis of this ability many would argue that it is plausible to ascribe to their perceptions contents that represent places in their external environment. They re-identify places over their own movements through them, thus giving them some kind of grip on the idea of places as the stable framework through which movement occurs, a basic ingredient in the idea of a physical world. They also have a grip of some kind on the connectivity of places, a grip captured in the fact of their navigational abilities. Whether or not one is convinced, as I am, that this gives them some kind of grip on the idea of places out there, for our purposes the important point is this. *If* we are willing to say that their perceptions do represent places in the external environment prior to acquiring the capacity to see, surely introducing another cognitive map, which enables them to navigate from any arbitrary place they are at to any place represented on that second map, cannot deprive them of whatever grip they had prior to acquiring the capacity to see. Or so goes the rough intuition.

The issue underlying this intuitive argument is this. If we begin, as we did, by simply helping ourselves to the idea that a subject's perceptions will count as perceptions as of an external world at least partly by being constitutively linked to a cognitive map which is such that all the places it represents are represented as connected to each other, there is nothing here to rule out the possibility that a subject, in representing places as out there, may employ more than one such map, where on each map, represented places are represented as connected to each other. Nor is there anything to rule out the possibility that perceptions in different modalities will count as perceptions as of external places by being linked to different maps. There is nothing in what we have said so far about connectivity that amounts to the requirement that for perceptions to be perceptions as of external places they must be linked to a *single* map in which *all* the places represented by the subject must be represented as related to each other.

Another way of putting the point is this. Connectivity initially gets into accounts of what is involved in place representation via the thought that places are essentially spatially related to each other, or, at least, the places in the subject's environment are, and that if a subject is to have some kind of grip on the very notion of places he or she must have some kind of grip on such essential connectivity. The notion of a cognitive map can then be introduced as that which provides the subject with some such grip by meeting the following requirement. Each place it represents must be represented as related to every other place it represents, in such a way as to provide the subject with a grip on the connectivity of those places represented on the map. This in itself does not introduce the requirement for a single map; it does not introduce the requirement that all places represented by the subject be represented as related to each other on a single map. And so long as we do not have the latter requirement there is no reason to deny that our imaginary subjects' modality-specific maps are such as to invest the perceptions in each modality with place-representing content.

Now there are two immediate counter claims one might make in response to the above intuitive argument. The first is that for perceptions to count as perceptions as of external places subjects must at least have the wherewithal to understand that the places perceived by each modality are in fact related to each other, even if the relations among these places are not manifest in the contents of the perceptions in each modality (as in fact happens to us in strange situations such as the yoga example cited earlier). The second is to require that for perceptions in each modality to count as perceptions as of external places the subject must be able to recognize places across modalities, and to have experiences in each modality which make manifest the relations to places perceived in the other.

Whichever intuitive line one takes here, the question is what justifies it? Which ingredient in our account of what makes representations representations as of external places brings in the idea of a single map, or at the very least, the idea that a subject must grasp that the places it represents are all related to each other? Intuitively, such requirements stem from ideas about what makes place representations detached or objective. For it would appear that underlying the claim that all places represented by the subjects must be represented as related or relatable to each other, there is the thought that unless subjects can do this they are leaving out important spatial facts about the places in their environment, and that their conception of places is therefore incomplete, and in this sense, at least, less than objective.

Now, the first question we must address is whether we can spell out with any more precision exactly what might be the link between explanations of what makes representations of places objective, and the global connectivity requirement, as I shall call it. The second question is: Where does the account we give with respect to objective place representations leave the problem of whether we can make sense of engaged representations of places that are modality-specific in the way we have described? In the next section we examine the first question in some detail, and then later come back to the second.

6 Objective place representation and cross-modal connectivity

Let us, then, formulate our last two variants of Molyneux's question with respect to places. Suppose we want to credit our subjects with perceptions in each modality the content of which are constitutively linked to objective place representations. Our first variant is: Is there an account of what makes place representations objective that actually requires that our subjects be credited with grasp of the idea that all places they represent are related to each other? The second variant, which is closer to our original question, is this. Does such an account, if one is found, actually require the use of a single, amodal map which is such that places perceived in the various modalities are immediately related to each other? Let us begin with the first.

On one account of what an objective or disengaged representation of one's environment consists in, we should think of it as a representation from no point of view, one that takes all the facts into account, including one's own interaction with the world, in such a way as to end up with a representation which owes nothing to one's first personal engagement with the world. Metaphors of 'a God's-eye view', or 'the view from nowhere', or a representation 'from no point of view' are often used to convey the notion of objectivity in question here.[4] The main ingredient in this account of objectivity, from our perspective, is the fact that the relations between the representer and the world represented are bleached out of the contents of the representation. If we think of normal subjects employing such a representation, the fact that they are in the world they represent and engaged with it plays no part in the contents of the fully objective description of the world.

Now suppose we think of perceptions as achieving objective content in virtue of being linked to spatial maps, conceived of as objective along these lines. Could this be used to motivate the requirement that the subject think of all the places it represents as related to each other? An immediate difficulty is this. There is nothing in this notion of an objective representation which rules out having objective representations of two unrelated spatial worlds. If there are such worlds, a complete description from no point of view would require representing them as such. And this would be precisely a case in which one would be employing two unrelatable (in fact, radically incommensurable) maps. Objectivity understood in this way does not, then, *a priori* require thinking of the places one represents as related to each other.

Are there ways of explaining what it is to have an objective or detached conception of places which do actually link the account of what makes the representation objective with the requirement that a subject's place representation can only count as objective if all places the subject represents are explicitly thought of as related to each other?

One thing that does require conceiving of places as being spatially related is thinking of them as all perceived or perceivable by the same person. As we remarked earlier when discussing radical incommensurability, if the same person is causally related through perception to two sets of places, these sets must themselves be causally related. But for this to work in motivating the requirement that objective place representations must be such that they are constitutively linked to grasp of the connectivity of all places represented by the subject, the thought that they are, were or could be perceived by the same person must be built into an explanation of what makes the representation objective.

Now, on one account, the idea of existence unperceived or, rather, now perceived and now unperceived, is the core notion that subjects must grasp in grasping the idea of a mind-independent world, a world as it is anyway, independently of whether and how it is experienced. Gareth Evans has suggested that the role of spatial concepts in giving subjects a grip on this idea turns, in its most basic form, on the use of these concepts in a primitive theory of perception, a theory which explains how and why a state of affairs might obtain even when there is not experience of it obtaining (Evans, 1985, pp. 261–2). In order to see a particular object one must be correctly located relative to it, there must be nothing in the way and so forth. Grasping such basic ingredients in our primitive theory of perception involves grasping the idea that the existence of the object and one's experience of it can be prised apart. Experience requires more than the existence of the object; it requires, in addition, that these perception-enabling conditions be met. In grasping this one gives the object a life of its own, a life independent of one's experiences. Spatial concepts, on this account, provide for the idea of a world as it is any way by providing one with a grip on these perception-enabling conditions, the meeting of which is necessary for perceptions to occur. Correlatively, the most basic use of these concepts, as they occur in giving the contents of perceptions, is always linked to the subject's thoughts about its actual and possible current, past and future locations.

In other places Evans has drawn on this notion of objectivity to explain the sense in which cognitive maps provide for an objective grip on places in one's environment. On his notion of a cognitive map, what makes such maps objective representations of the environment is that they license reasoning of the kind that provides for ongoing self-conscious self location on the basis of one's perceptions (see, for example, Evans, 1982, pp. 162–3; pp. 222–3). More generally, the notion of objectivity to which he links the capacity for employing cognitive maps is described as follows:

> Any thinker who has the idea of an objective spatial world – an idea of a world of objects and phenomena which can be perceived but which are not dependent on being perceived for their existence – must be able to think of his perception of the world as being simultaneously due to his position in the world and to the condition of the world at that position. The very idea of a perceivable objective world brings with it the idea of the subject being *in* the world with the course of his perceptions due to his changing position in the world and to the more or less stable way the world is. The idea that there is an objective world and the idea that the subject is somewhere in the world cannot be separated, and where he is given by what he can perceive (Evans, 1982, p. 222).

If we explain the objectivity of spatial representation along such lines it does seem that we have the materials for explicitly linking our account of what makes the representation objective with the requirement that a subject think of places, at least initially and most primitively, as related to each other. For on this account, what makes the representation of places objective is explicitly linked to the subject's conception of itself as a possible perceiver of the places represented on the map, as possibly located at them. And this just is to introduce the requirement that places be thought of as spatially related to each other.

Suppose we hold that something along these lines, if further developed, could be used to motivate the requirement that the capacity for objective place representations requires a subject to grasp that all the places it represents are related to each other. Let us now turn to our second variant of Molyneux's question. Does the way in which the idea of connectivity comes into grasp of a primitive theory of perception, actually require that a subject's perceptions in each modality be linked to a single map which provides for re-identification across modalities, and which is such as to make the relations among places represented in each modality explicit in the contents of the perceptions in each modality?

Let us, for the last time, return to our imaginary subjects. Suppose that before they acquire the capacity to see they build up a reflective conception of the places in their environment, one built on grasp of a primitive theory of perception. Suppose now that, overnight, they acquire the capacity to see. Could not the following story be told of them? When they wake up and have visual experiences as of places around them, they recognize none of the places, but set about building new reflective maps of their environment, in the way one would on waking up in a new place to which one has been transported anaesthetized. They do think of the places as connected but this is wholly on the basis of their grip on their own causal continuity over both kinds of experiences, a grip that underwrites their first-person self-locating thoughts when presented with their first visual experiences. That is, their grip on the connectivity of places rests wholly on their unproblematic use of 'I' thoughts in their perceptions, and is consistent with their not being able non-inferentially to identify objects and places across the modalities. On such a story, the answer to our question would be no, despite the subjects' grip on the connectedness of all places in their environment.

Is there anything wrong with such a story? I end this section with an objection to its coherence that takes the story about our imaginary subjects on its own terms. In the final section we will bring these imaginings down to earth, to a comparison with the way our sensory modalities and mapping abilities actually operate.

When transported anaesthetized to a new place, one's belief that the places one is now perceiving are connected to earlier places does not rest wholly on one's grip on one's own causal continuity between past and present places. For one thing, one has had experiences of unreflectively re-identifying places over movement relative to them, and has had unreflective, non-inferential experiences of the connectedness of places over time and over one's own movement relative to them. In the transportation case, one is assuming that the kind of identity and connectedness of places over travel which one normally does experience holds in this case without experience, a thought that immediately leads to the search for explanations of how this has happened, explanations that will, again, begin with appeal to familiar cases of moving without being aware of it, as in falling out of bed when asleep and so forth. Secondly, and relatedly, one's thought that the places are connected is underwritten by the thought that one could travel between them and unreflectively experience how exactly the places one is now in are connected with places one was in before being transported. That is, the belief that they are connected does not end there: there is the further belief that the exact nature of the connection could be discovered by observation over time.

In our imaginary modality analogue, however, the subjects cannot avail themselves of either of these ways of underwriting the thought that the places perceived through one modality are connected to those perceived through another. The burden of supporting the belief in the connectivity of places across modalities rests wholly on the subjects' awareness of their own causal continuity across experiences in the two modalities. Thus, as the story has been told, their belief is not supported by appeal to experiences of the identity and connectivity of places across modalities: the subjects never have unreflective awareness of the connectedness of places across modalities, and never non-inferentially recognize landmarks across the modalities. Connectedness and identity are never perceived. In addition, there can be no such thing as finding out by experience how exactly the places are connected, whether they are identical, adjacent or whatever, for the experiences in each modality are wholly neutral as to whether and how the places in one modality are connected to those in the other. These subjects can never get further than believing that the places must be connected, given their own causal continuity. They can never discover by experience what exactly the connection is.

This makes their belief that their experiences in both modalities are experiences as of connected places out there extremely tenuous. An equally plausible thought for them would be that the experiences they have when they acquire their capacity to see are dreams or hallucinations or imaginings, rather than experiences as of a world out there; or, alternatively, that their previous haptically-based experiences were hallucinations or the like. There is nothing in their awareness of their own causal continuity through both sets of experiences that makes this supposition any less plausible than the supposition that in both modalities they are perceiving connected places out there, though they can never establish by observation what exactly the connection between them is. The hallucination explanation, that is, is at least as plausible from the inside as an explanation they might give themselves about the functional disassociation of the deliverances of their haptic and visual systems.

Taking the story on its own terms, then, the kind of objections we roughly sketched do lend support to the following claim. If subjects' objective conception of the world out there rests on their capacity to employ spatial concepts in the context of a primitive theory of perception, there are prima facie serious difficulties with supposing that the contents of spatial experiences in the modalities are *de facto* uncommensurated; that is, that the maps constitutively linked to their perceptions in each modality are functionally disassociated in the manner described above.

7 Explaining amodal mapping abilities

So far, the only explanation of what is involved in perceiving places as places out there which actually seems to rule out, or at least raise serious difficulties for, the idea of functionally disassociated modality-specific maps, is an account of objectivity in place representation of a kind that requires the capacity for self-conscious self-location be built into the contents of the spatial perceptions.

It may be that there are other ways of ruling out modality specificity of the kind we have been describing, though I cannot at present see what these might be. I want to end with a different question. Let us return to the original story we imagined when introducing the idea of *de facto* uncommensurated cognitive maps. We imagined subjects who lacked any reflective abilities and whose grip on the connectivity of places was exhausted by their navigational abilities. Such subjects could not be credited with objective representations of the kind we have been describing, could not be credited with grasp of a primitive theory of perception.

One response to this story might be to deny that these subjects should be credited with any grip on the notion of physical places out there. Place representations must be objective and reflective or nothing. The other response might be to suggest that engaged place representation is consistent with modality specificity, unlike objective place representation. Without underestimating the complexity and indeed depth of issues that are involved in explaining the sense in which engaged place representations are, nevertheless, representations as of places out there, it seems to me that the second line must be the correct one to take. The question I want to end with is the following.

Suppose we do think we can make sense of the idea of wholly engaged place representation. If the line we have been pursuing is correct, there would appear to be nothing in our account of engaged ways of representing places that could motivate the requirement for a single map in which all the places represented by the subject are represented as connected to each other. This means that to the extent that we find creatures, the content of whose maps should be cashed solely in terms of, say, the navigational abilities they provide for, and who at the same time employ only one, functionally commensurated map across the modalities, such that they are able to navigate from any arbitrary place to any other, whatever the modality, this goes beyond what is actually required by the kind of grip on the connectedness of places provided by such maps. As there is every reason to suppose that there are, in fact, many creatures of precisely this kind (see chapter two) this raises the question of how we explain this ability. Must we conclude that there are no independent arguments which link the capacity for engaged grips on the idea of a world out there with the capacity to employ amodal frames of reference in place representation? Must we think of the evolutionary development of such amodal mapping abilities as a piece of fortuitous luck which, once developed, provided necessary conditions for the development of self- conscious grips on the idea of a world out there, but which had no independent evolutionary point?

Here we must come back to earth and recall two features that set our imaginary subjects apart from most of us; two features that let the various modality-specific stories we have sketched get off the ground in a way they would not for us. First, there is the fact that they are born blind and have time to build up a coherent conception of the world before vision sets in. Second, there is the fact that once they do acquire the capacity to see, their visually-based maps are employed only during the day and their haptic maps only during the night. Each modality has long stretches of time working on its own, as it were, and the transition between the employment of sight and touch is orderly and reliable.

Contrast this with the state of most subjects who are born into the world with all their sensory modalities functioning. They are simultaneously and unpredictably bombarded with stimulation from various modalities. Consider, first, what would be involved, in such conditions, in building up separate sound-, sight-, smell- and touch-based maps. For each modality there will be possibly long and certainly unpredictable gaps while stimulation from other modalities impinges and demands inclusion in other maps. Thus, suppose such a subject is set down in a new environment, and simultaneously hears a sound, smells a smell, touches an object and sees one. Suppose for the next ten minutes as it moves around it gets only visual stimulation. Subsequently, it touches an additional object. Some time later it hears another sound and so forth. In building up a touch-based map, say, it will not be able to use any of the visually acquired information about its environment during its movements. It will have had to have kept separate track of its movement relative to the first touched object, and similarly for stimulation from the other modalities. Each one must employ a separate movement-tracking device that ignores information from other modalities.

Secondly, consider what would be involved in using such maps for the purposes of navigation. Imagine our subject has, somehow, built up all these modality-specific maps. Suppose it gets hungry and simultaneously smells an object, sees another and touches a third, all of which are landmarks it recognizes, each one of which is associated with a different map. Which map will it use to get to food? If it compares the maps and opts for the closest food cache, the maps just are commensurated. To block such commensuration, and to avoid a completely random process of one map getting the better over another, as it were, we must imagine some mechanism for determining which map is used in such circumstances, for example a mechanism whereby vision always dominates, followed by touch, say, and so forth for the other modalities.

Doubtless it is possible to imagine a system that would work on the above lines. What is less easy to imagine is the evolutionary point of such a system, given the way the world is. The separate-mapping account is exceptionally complicated and makes huge demands on the computing capacities of the system. Moreover, there is only the possibility of modality-relative sensitivity to questions of where the closest food cache or hiding place might be, which severely undermines the utility of the system for the purposes of survival. The simplest and most effective solution, by a very long chalk, is one in which input from the various modalities is taken into a single, amodal navigation system. There are, then, strong independent arguments for postulating commensurated, amodal mapping abilities when we are concerned with the kind of grip on the external world that is exhausted by the ability to navigate. However, these arguments are of a very different kind from those we have been concerned with. They do not involve establishing any *a priori* link between what is involved in representing a world out there and the use of a unified–amodal spatial map. Rather, these are arguments from the simplicity and effectiveness of the amodal mapping system relative to the modality-specific story, to the kind of mechanism it is most plausible to think would have evolved, given the way the world is.

Concluding remarks

We began with the intuition that there is some kind of connection between the answer we give to Molyneux's question and explanations we give of what it is for perceptions to inform us about a world out there. Focusing on the case of place representation, I have suggested that if there are any *a priori* connections here they will turn on a way of representing the world out there which is constitutively linked to self-conscious thought of oneself as a perceiver of and actor in one's environment; that is, to a grip on the idea of a world out there that stems from the capacity to employ a primitive theory of perception. If this line of reasoning is correct, it would suggest that when we are dealing with subjects, such as various animals, to whom we do not want to ascribe the capacity to grasp such a theory, explanations of why their mapping abilities are in fact amodal should not appeal to what is required for having a grip on the idea of an external world, but, rather, to arguments from simplicity of design. But with respect to human infants, explanations of their mapping abilities should face two ways. On the one hand, there will be evolutionary arguments for an inbuilt, amodal place-representing ability, which we would expect to underpin their wholly engaged ways of representing the environment around them. On the other hand, if we are interested in the connection between the modality issue and having a grip on the idea of a world out there, we should be looking to the relation between the development of their ability for thinking of themselves as an object, and their ability for employing detached or objective maps of their environment. That is, in terms of contributions to this volume, we should be thinking of the developmental connection between the kinds of issues raised by Meltzoff, in his account of the origins of self consciousness, and the kinds of issues raised by Pick in his account of the development of mapping abilities.

ACKNOWLEDGEMENTS

For comments on (somewhat distant) ancestors of this chapter I am indebted to Bill Brewer, John Campbell, Roz McCarthy, Tony Marcel and Mark Sainsbury. For comments on the current version I am very grateful to José Bermudez and Julian Pears.

NOTES

1 Much of the work relevant to Molyneux's question is to be found in infant research which, as Meltzoff suggests in chapter nine, easily lends itself to Molyneux-type questions. In addition to Meltzoff's work, see also chapter five. A small selection of the great quantity of work which strongly suggests amodality in spatial perception, of the kind that would yield positive answers to variants of Molyneux's question as applied to infants and children, is the following: Lipsitt and Rovee-Collier (1982), which is a collection of good and representative papers and debates about the problems of cross-modal integration in infants and young children; Carpenter and Eisenberg (1978); Marmor and Zabak (1978); Landau et al. (1984); Streri and Pecheux (1986); Streri (1987); Landau (1991).
2 Berkeley's reasons for devoting much of his *New Theory of Vision* to arguing for a negative answer to Molyneux's question were, of course, motivated by an interest in the link between negative answers and establishing his version of idealism. See Berkeley (1975).
3 On the general relation between this distinction and our earlier notion of frames of reference, see the introduction to Part I; and chapter three.
4 The idea that cognitive maps should be thought of as objective in being representations from no point of view is mooted by Evans (1982, p. 152); but he raises difficulties for conceiving of maps in this way (pp. 299–301, pp. 264–5). The phrase 'the view from nowhere' comes from the title by Nagel, a book in which he develops this notion of objectivity and the problems he believes it raises.

REFERENCES

Berkeley, G. 1965: *New Theory of Vision*. In D. M. Armstrong (ed.), *Berkeley's Philosophical Writings*. London: Collier Macmillan.

Carpenter, P. A. and Eisenberg, P. 1978: Mental rotation and the frame of reference in blind and sighted individuals. *Perception and Psychophysics*, 23, 117–24.

Evans, G. 1982: *The Varieties of Reference*. Oxford: Clarendon Press.

Evans, G. 1985: Things without the mind. In his *Collected Papers*. Oxford: Clarendon Press.

Landau, B. 1991: Spatial representation of objects in the young blind child. *Cognition*, 38, 145–78.

Landau, B., Spelke, E. and Gleitman, H. 1984: Spatial knowledge in a young blind child. *Cognition*, 16, 225–60.

Lipsitt, Lewis P. and Rovee-Collier, Carolyn (eds) 1982: *Advances in Infancy Research*, vol. 4. New Jersey: Ablex Publishing Corporation.

Locke, J. 1975: *An Essay Concerning Human Understanding*. P. H. Nidditch (ed.). Oxford: Clarendon Press.

Marmor, G. and Zabak, L. 1978: Mental rotation by the blind: Does mental rotation depend on visual imagery? *Journal of Experimental Psychology, Human Perception and Performance*, 2, 515–21.

Nagel, T. 1986: *The View From Nowhere*. New York: Oxford University Press.

Piaget, J. 1954: *The Construction of Reality in the Child*. New York: Basic Books.

Piaget, J. and Inhelder, B. 1969: *The Psychology of the Child*. New York: Basic Books.

Streri, A. 1987: Tactile discrimination of shape and intermodal transfer in 2-to-3-month-old infants. *British Journal of Developmental Psychology*, 5, 213–20.

Streri, A. and Pecheux, M. G. 1986: Cross-modal transfer of form in 5-month-old infants. *British Journal of Developmental Psychology*, 4, 161–7.

Streri, A. and Spelke, E. S. 1988: Haptic perception of objects in infancy. *Cognitive Psychology*, 20, 1–23.

11
Content and vehicle

Ruth Garrett Millikan

When my daughter Natasha was very small, she often arrived in my bed in the night. One morning she woke there, eyes big with excitement: 'Mommy, I saw your dream!' The dream was mine, it seems, because it was in my bed. A charming mingling of the intentional contents of a representation with attributes of the vehicle of representation – something that we philosophers try hard to avoid.

The possible forms of this confusion are numerous, however, and some are vanishingly subtle. Some, I believe, persist in the work of contemporary philosophers, philosophers explicitly aware of the danger. The purpose of this chapter is to illustrate certain very abstract varieties of this confusion as they appear in contemporary work on perception.[1]

1 The error to be excised

Let me begin by retrieving the unconscious roots of the neurosis to be excised. These roots are most easily traced through naive theories of perception. In tracing these roots through the early mind that tries to understand perception, I will be following a path well mapped already by others.

It all begins with an impulse to reify the claims of perception. I cannot say 'the impulse to reify what is perceived', because 'perceive' is a success verb; whatever I perceive is surely already real and needs no resurrection. Rather, what gets resurrected is something to serve as correspondent to another thing, itself unquestionably real but lacking a layman's name. This is the act or state that stands to perception as belief stands to knowledge. Call this act or state, without prejudice as to its nature or mode (nominal, verbal, adverbial; visual, auditory, tactual, etc.), a 'visaging'. The impulse, then, is to resurrect to correspond to a visaging something that exemplifies the properties that the world would have to have for the visaging to be veridical. Thus, primitive peoples take dreams to be true, though true of another realm.

That is the first ingredient in the confusion. The second is the impulse to take visaging as like having pictures in the mind and, simultaneously, to take pictures to be literally *like* what they picture. That ordinary pictures should at least seem to be literal likenesses of what they picture is understandable. Similarly, that perception involves some form of representation is an obvious idea. Perceptions, like representations, can be mistaken. Like pictures drawn with the purpose of showing how things are, visagings can misrepresent.

From these three ingredients, out falls the irresistible theory that visagings involve items appearing before the mind that *have* the properties that they represent. The properties claimed by visagings to characterize the world exist in 'objective reality' (Descartes), or they, or doubles of them, are true of sense data, or percepts, or phenomenal objects, or visual fields, etc. When and only when the world resembles the inner picture, then the visaging is veridical, showing how things really are.

Most philosophers nowadays have undergone analysis so that they explicitly understand the unconscious motivations behind this error, and why they are not supposed to make it. Gareth Evans calls this error 'the sense datum fallacy'. He continues:

> It might better be called 'the homunculus fallacy'. . . when one attempts to explain what is involved in a subject's being related to objects in the external world by appealing to the existence of an inner situation which recapitulates the essential features of the original situation to be explained. . . by introducing a relation between the subject and inner objects of essentially the same kind as the relation existing between the subject and outer objects (Evans, 1985, p. 397).

He thus suggests that the main error that results from this tempting move is the invoking of a regress. How will the inner eye then perceive the inner picture? In the same way that the outer eye does?

I would like to suggest another way of looking at the sense datum fallacy. For it is surely possible to give a perfectly coherent answer to the question of how the inner eye works that does not invoke a regress. The purpose of introducing inner representations was to account for error. But the inner eye should not have the problems the outer eye does of sometimes misperceiving what is there before it. So there would be no need to suppose that it must use additional still-more-inner representations in order to see. The inner eye or mind can be taken to understand the representations before it merely by reacting to them appropriately, by being guided by them for purposes of thought and action. Thus the regress can be avoided. The broader trouble with the sense datum fallacy, I suggest, is that it produces a facade of understanding that overlooks the need to give any account *at all* of the way the inner understander works, any *account* of the mechanics of inner representation, of what kind of reacting is comprehending. Having projected the visaged properties to the inside of the mind, the assumption is that there can be no problem about how they manage to move the mind so as to constitute its grasp of them *as* what is represented. Their mere reclining in the mind constitutes the mind's visaging of them. Call this the 'passive picture theory' of inner representation. What's wrong with it in the first instance is the passive part. And once you see that it must be the mind's reaction that constitutes understanding of an inner representation, you see that the picture part is also suspect. Why would a picture be needed to move the mind appropriately? At least, wouldn't something more abstractly isomorphic do as well?[2]

2 Internalizing and externalizing; demands for coherency and completeness; intermediaries

The passive picture theory produces projection of properties claimed in or by the visaging onto the inner vehicle of the visaging. Call this move 'content internalizing'. It also produces 'content externalizing', whereby properties of the vehicle of the visaging are taken, reciprocally, to show up in the visaging's content. The illusion is thus created that one both directly apprehends the nature of the vehicle of perception through the visaging and also the reverse, that one can argue from the nature of the vehicle of perception to what must be being visaged. One result is that it becomes problematic how genuine incoherences or contradictions could occur in the content of a visaging. Incoherences in content would have to correspond, *per impossibile*, to incoherences in the actual structure of the representation's vehicle. We can call this the *demand for coherency in content.* A sister result is that there could be no visaging that does not visage also all logically necessary or internal features of what is visaged. Taking what will later emerge as a central example, there could be no visaging of properties without a simultaneous visaging of their internal relations. Contents lacking or failing to claim logically necessary or internal features associated with their contents would have to correspond, *per impossibile*, to vehicles lacking logically necessary or internal features of themselves. We can call this the *demand for completeness in content.*

Internalizing and externalizing moves are enormously interesting, for these moves can survive the contemporary turn that explicitly denies the phenomenally given, substituting neural representations for phenomenal ones. Indeed, there are forms in which these moves can survive even the turning of inner representations into mere dispositions and capacities, or into the states that account for these. Elsewhere (Millikan, 1991), I argue that in the case of Fregean *Sinn* they have managed to survive also the turning from perception to cognition, a mode generally thought of as very unpicturelike. Because the confusions that I wish to discuss cut in this way across theories that postulate experienced and those that postulate non-experienced inner representations or other non-phenomenal states, I propose to ignore such distinctions entirely. Sense data, percepts, sensations, perceptual capacities or the neural states upon which these depend are none of them exempt from internalizing and externalizing moves. I will speak indiscriminately, then, of the postulation of 'intermediaries'. Let me emphasize that: I am counting as intermediaries even capacities and the states in which they are grounded, when these are understood to account for the intentional contents of mental episodes.

The error to be eradicated, then, certainly is not that of positing intermediaries. Postulation of intermediaries of some kind is essential to understanding perception and thought. The error is that of projecting, without argument, chosen properties of what is visaged or conceived onto these intermediaries, and vice versa. The error is equally that of taking this sharing of properties to constitute an *explanation* of mental representing. The passive picture theory causes the underlying nature of the vehicle of perception to disappear from (the theoretician's) view as an agent. The nature of the actual intermediaries for perception or thought, the actual mechanics of these, retires, leaving in its place a frictionless substitute that translates meaning directly into mental or physical action and vice versa. The resulting illusion is that intentional content, that which is visaged, moves the mind or body directly or, the reciprocal, that intentional contents are merely an image of these movements.

3 Internalizing and externalizing temporal relations

Our first example concerns temporal relations. No one supposes, nowadays, that visaging colours or shapes requires that any similarly coloured or shaped intermediaries should appear either before the mind or in the brain.[3] But have we assimilated the parallel truth about temporal visagings? In a truly fine essay, Daniel Dennett and Marcel Kinsbourne (1990) have recently spoken to the multitude of confusions about this that persist in the psychological and philosophical literature. But here are two leftovers worth examining.

In *Molyneux's Question*, Gareth Evans (1985) discusses the classic view that the blind cannot perceive space, this because the parts of an object can only be touched in *succession*, and because successive touchings could not yield a perception of the object's *simultaneous* spatial layout. Evans's counter is that one cannot argue 'from the successiveness of *sensation* to the successiveness of *perception*', and that there is no reason why 'the information contained in the sequence of stimulations' might not be 'integrated into, or interpreted in terms of, a unitary representation of the perceiver's surroundings' (1985, p. 368). Evans calls such representations 'simultaneous perceptual representations of the world' (1985, p. 369), thus expressing his basic agreement with the assumption behind the classic view, that a representation of simultaneity can only be accomplished by simultaneity among elements in the representation.[4] In similar vein, Evans answers with a confident but unargued 'yes' the question 'whether a man born deaf, and taught to apply the terms 'continuous' and 'pulsating' to stimulations made on his skin, would, on gaining his hearing and being presented with two tones, one continuous and the other pulsating, be able to apply the terms correctly' (1985, p. 372). The assumption behind Evans's confidence seems to be that continuousness and pulsatingness in whatever medium must be represented *by* continuousness and pulsatingness, hence will always be recognized again. Yet first Evans, and then I, have *just now* represented pulsatingness and continuousness to *you* without using the pulsatingness or continuousness of anything in order to do so. Evans's assumption illustrates first 'content internalizing', then 'content externalizing'.[5]

4 Internalizing and externalizing constancy

A second example concerns the perception of change v. constancy. Consider one of Christopher Peacocke's arguments (Peacocke, 1983) for the existence of an intermediary called 'sensation'. This argument, ironically, is presented in support of the view that the properties of sensation are *not* derivable as mere correlates of the intentional contents of perception. The argument concerns the 'switching of aspects' that occurs as one fixates on a necker cube (or a duck–rabbit). 'The successive experiences have different representational contents', Peacocke says, yet 'the successive experiences fall under the same type. . . – as Wittgenstein writes, "I *see* that it has not changed"' (Peacocke, 1983, p. 16). Peacocke's conclusion is that beneath the change in representational content lies a constancy in properties on the level of sensation. Now assuredly, 'that it has not changed' is something that I see, but that *what* has not changed? My visaging has as part of its content that the *world* has not changed – that is where the constancy lies. Peacocke has internalized this content to yield an intermediary, a sensation, that has not changed. Compare a man looking through a perfectly ordinary window who erroneously believes he is watching a 3D film. He quite automatically takes it that whenever he sees a change or a constancy, that is because the film screen image has changed or been constant. Analogously, Peacocke's effective assumption seems to be that a perception of constancy can only be accomplished via an inner intermediary that is itself constant.

This assumption, call it 'constancy internalizing', which both philosophers and psychologists routinely fall into, has pervasive and far reaching effects. It produces the illusion of constancy at an intermediary level not just as shifts in aspect occur, but more devastating, as shifts in attention occur, and over episodes of perceptual learning. Shifts of attention are, of course, routinely coincident with perception of constancy in the object perceived; indeed, coincident with perception of constancy in the very properties upon which attention focuses and then withdraws. And so for episodes of perceptual learning. Learning to perceive, for example, learning to distinguish major triads, or learning to see the microbes in the field of the microscope, is simultaneous with the perception that what is perceived is not itself changing or undergoing reorganization over the interval. When these constancies are internalized, the illusion is produced that there is a background intermediary corresponding to the whole detailed scene before or around one in perception, an intermediary that changes only when caused to change by changes in the world outside or by shifts in the perceiver's external relations to that world. This intermediary is traditionally labelled 'the sensory field', for example, 'the visual field'.

The constancy of the hypothesized sensory field may then be externalized again. If the intermediary that supposedly stays the same is projected to become a constant *content* for the visaging, we arrive at a backdrop of continuing content from which there emerges a varying foreground as learning or attention switches occur – perhaps as connections are made into conception. Peacocke calls such contents, which in the case of vision determine (densely grouped alternative sets of)[6] complete spatial configurations of objects or surfaces around one, 'scenarios' (Peacocke, 1987).[7]

Combined with this sort of subsequent externalizing move, internalizing of constancy threatens to produce contradiction. How can the intermediary of perception remain constant so as to account for the perception of constancy, yet change so as to account for changes in content over changes in attention or over learning? When contradiction threatens, distinguish levels. Peacocke distinguishes two levels of properties for his intermediaries, 'representational properties' and 'sensational properties', the first of which concern content, the latter of which do not, although 'experiences with a particular sensational property also have, in normal mature humans, a certain representational property' (1983, p. 25).[8]

What would it be to refuse to internalize constancy? Perhaps the perceptual–cognitive systems manufacture perceptual intermediaries piece by piece, only as one needs them, each expressing only a fragment of the content that would be available for expression given other needs. The question whether this is how it works may turn on empirical evidence, perhaps on neurophysiological evidence, rather than *a priori* argument.[9]

5 Importing determinacy

If some aspect of content is merely internalized and then externalized again, this will not result in any change of content. But if an aspect of content is internalized and then filled out so as to make consistent the hypothesis of its inner reality before it is externalized again, the result may be an apparent change in content. This change may introduce an ambiguity concerning the scope of the visaging operator. For example, any property or relation that is internalized from a visaging to an intermediary must then be filled out and made determinate. If the intermediary really has the visaged property or relation it must have it in determinate form. Thus Berkeley's arguments against abstract ideas. Internalized contents cannot be abstract. But when they are first made to be concrete and then externalized again, the result is a change of scope for the visaging operator. Using a familiar example, if my visaging claims that there exists a large number that is the number of speckles on a certain hen, then there must exist a certain large number that the visaging claims to be the number of speckles on the hen. That this inferential move is in error becomes clearly evident when one applies it to the visagings of imagination, where its result is that I should not be able to imagine a speckled hen without imagining that it has a certain definite number of speckles. Call the move that first introduces determinacy at the intermediary level, then externalizes it as part of the visaging's content, thus moving the scope brackets over, 'importing determinacy'.[10] This move illustrates the demand for content completeness (see section 2 above), the internal feature required for completeness in this case being determinacy.

6 Importing determinate relata

A significant form of determinacy importation imports determinate relata. Any internal relation between properties (such as *larger than, a fifth higher than*) that is internalized from a visaging to an intermediary must be provided with appropriate relata: real relations don't exist without relata. If the relation is internalized as it were 'whole', that is, if it is taken that the intermediary embodies that very visaged relation, then the intermediary must possess determinate relata for the relation to relate, say, determinate sizes, pitches. But this is true for more subtle forms of internalizing too, where it is not the very content itself that is taken to be internalized but rather an analogue. More precisely, the mathematical form of the transformation space around the content is internalized; a simple set of transformations (in the mathematician's sense) of the vehicle is taken to correspond one-to-one to a simple set of transformations of the represented, thus yielding in one stroke a whole representational system. In either case, determinate relata must be introduced at the level of the intermediary. Externalizing, it then appears to follow that the original visaging was of determinate relata. That is, from the fact that my visaging claims that there exist relata related by a certain relation, it is concluded that there exist relata that the visaging claims to be related by that certain relation.

An easy example of the importing of determinate relata is found in Evans' essay 'Molyneux's Question' (1985). Evans has B, who gives the question an affirmative answer, use the 'very familiar' argument that there could not be an experience of something rotating 'in the visual field' without there being 'four sides' to the visual field, 'a, b, c, d, which can be identified from occasion to occasion' (1985, p. 386). That is, the experience of rotation requires determinate directions for the rotation to occur from and to.

The importing of determinate relata is implicit in Peacocke's claim that a 'matching profile' can be described, for example, for the visual experience of the direction from oneself in which the end of a television aerial lies (Peacocke 1986; 1989a). This matching profile is the area within which the aerial might lie given that the experience one is having of its direction is veridical. It corresponds to a solid angle, with oneself at origin, and it is determined by seeing how far to one side of its actual location the aerial can be moved without one's noticing the difference. But that one can perceive a discrepancy within a given small magnitude between A and B would be evidence that one is discriminating the *absolute* directions of A and B within that small magnitude only if visaging a discrepancy required that one visage a direction or range of directions for A and a different direction or range for B. And this would be necessary only if visaging a discrepancy required that the intermediaries for the visagings of A and of B be discrepant, thus having different absolute values. To appreciate that something has gone wrong here, compare pitches. If I can tell there's a difference when two pitches are as little as 2hz apart, does it follow that I can visage pitches (read the following transparently) within 2hz? I don't have absolute pitch. So either I don't hear absolutely within any such narrow range, or else I can visage exactly the same content twice without knowing it – a possibility to which I will turn a bit later. Certainly, it is not clear that my visagings of pitch are in fact so accurate. Indeed, notice that there is no reason to think that there is even any natural information present in me representing the absolute values of the pitches I hear, for the phenomenon of adaptation is very deep-seated in the structure of the nervous system. Quoting Oliver Selfridge (unpublished), 'the range of stimuli that can be distinguished is greatly increased by the power of adaptation [of the nervous system], although the ability to signal absolute intensities is lost'.

In similar vein, Peacocke remarks on 'what you can learn about the size of the room by seeing it' that you cannot necessarily learn by measuring it (1989a, p. 299). But my absolute sense of distance is not too good. What I can learn is mostly relative, it seems to me, and will help me only if I know independently something about the sizes of other relata involved. Suppose that I wrongly perceive two items on opposite sides of the room as different in length. In fact they are just the same length. Does it really follow that one or the other of my distance perceptions was wrong? How then is it determined *which one* was wrong? (Or can one, perhaps, wrongly perceive that the contents of two perceptions are different?)

7 Importing internal relations: The demand for same–different transparency

Another scope distinction ambiguity resulting from internalizing and externalizing moves imports internal relations. Any relata that are projected from a visaging to an intermediary must be provided with all necessary internal relations. If an intermediary really embodies the relata (or analogues of them) it must also embody these relations. Externalizing, it follows that the visaging was also of these relations. Thus, from the fact that there exists an internal relation R that items A and B bear to one another and the fact that I visage A and B, it is concluded that I visage R. For example, I could not truly visage middle C and then orchestra A without visaging one as higher in pitch than the other, or visage a square and a triangle without visaging one as having more sides than the other. The demand here is for content completeness.

Sameness can, for certain purposes, be treated as a relation.[11] So treated it is of special interest because although there is only one kind of sameness relation on the level of intermediaries, two separable relations correspond to this on the level of content. A visaging can involve (1) separate visagings of what is in fact the same or (2) separate visagings taken as of the same. Call the first of these 'visaging of sames', the second 'visaging of same*ness*'. Either can occur without the other or they can occur together. Similar remarks go for difference. We must distinguish 'visaging differents' from 'visaging difference'. Because there are two kinds of visaging for sameness and two for difference, there are two kinds of internalizing and two kinds of externalizing moves for each. This makes possible the following relation importation moves.

One can (1) import sameness: first internalize sames, yielding sames in intermediaries hence sameness, then externalize this sameness. The result is that sames are always visaged *as* the same. Similarly, one can (2) import sames: first internalize sameness, yielding sameness in intermediaries hence sames, then externalize these sames, yielding that what is visaged as the same always is the same; (3) import difference, yielding that what is different is always visaged as different; (4) import differents, yielding that what is visaged as different always is different. Each of these moves yields to a demand for content completeness. I mention these importing moves for two reasons; first, to round out the discussion of what importing consists in, and second, because it will be important in the discussion that follows not to confuse sames with sameness nor differents with difference. Elsewhere (1991) I have argued that Frege makes some of these importing moves with his *Sinne*. Here I will discuss simpler examples of internalizing and externalizing and importing moves involving just sameness and difference.

8 Moves involving sameness and difference

To externalize sameness is to take it that intermediaries that are (tokens of) the same necessarily project this sameness into the content visaged. Similarly, if we externalize difference, intermediaries that are different must present their contents as being different. Pair these two moves and then introduce determinacy: any two intermediaries must be, determinately, either same or different. It follows that visagings of a pair of contents must always visage these contents either as same or as different. For example, if my visaging is of two coloured items, it must either be a visaging of them as the same in colour or else as different in colour. The result is a scope distinction ambiguity that imports determinate sameness-or-difference. From, it is not the case that I visage A and B to be different, we get that I visage that it is not the case that A and B are different, hence that I visage A and B to be the same. Consider the ambiguity of 'they all look the same in the dark' (roguishly said of women). Harder to spot: is it true that colours tend to look more and more alike as it grows darker? Our ordinary conceptions 'seems' and 'looks' and 'appears' vacillate on such questions. They have ingested part of the passive picture theory of perception.

If one externalizes sameness, one is obliged also to internalize difference: things visaged as different are not visaged as the same,[12] hence, by *modus tollens*, cannot correspond to intermediaries that are the same, hence must correspond to intermediaries that are different. The converse does not hold, however. If one internalizes the difference relation, then when intermediaries are the same they cannot correspond to a visaging of difference, but this does not entail that they correspond to a definite visaging of sameness. Rather, internalizing lack of visaged difference, that is, the failure to visage a difference, is the stronger internalizing move, equivalent to externalizing sameness. In parallel fashion, if one externalizes difference one is obliged to internalize sameness but not vice versa, and externalizing difference is equivalent to the stronger move of internalizing lack of perceived sameness. Most important, there is no logical connection between externalizing sameness and externalizing difference, nor between internalizing sameness and internalizing difference. Nelson Goodman attempted to exploit these dissociations between internalizing and externalizing differences and samenesses in defining identity for qualia.

Goodman began by calling attention to an apparent paradox concerning the non-transitivity of identity over appearances: one thing, A, can appear to be the same colour as a second thing, B, and the second appear the same as a third, C, yet A appear to be a different colour from C. The weakest premises from which the paradox results is conjunction of the following internalization moves. (1) Internalize constancy: if B is visaged not to change as it is compared first with A and then with C, it corresponds to an intermediary that remains the same over the comparisons. (2) Internalize the sameness relation: if A and B are visaged as being the same, their corresponding intermediaries are the same, and so for B and C. (3) Internalize the difference relation: if A and C are visaged as different, their corresponding intermediaries are different. Goodman calls his intermediaries or their relevant qualities 'qualia', and he does not, of course, analyse the paradox in this way. But he tries to avoid it, in effect, by externalizing sameness but not difference. Qualia α and β are identical just in case every quale γ that matches either α or β also matches the other (Goodman, 1966, p. 290), where 'to say that two qualia are so similar that they match is merely to say that on direct comparison they appear to be the same' (1966, pp. 272–3). Being very careful, it is not merely difference that is internalized here but lack of sameness; that is, sameness is externalized. The assumption is that qualia that are the same never fail to appear so on 'direct comparison', so that not appearing the same on direct comparison – not matching – can be a criterion of qualia difference.

If one is seriously attempting to understand the ontology of qualia, a weakness of Goodman's solution is the problematic status of 'on direct comparison', which nearly always invokes a counterfactual. (Surely an infinitesimally small proportion of the actual pairs of qualia particulars in the world in fact get directly compared.) Applying Goodman's criterion not epistemologically but ontologically, as telling what constitutes the fact of the matter of qualia sameness, perhaps it is immaterial whether we can know that two qualia would 'match if directly compared'. But the very essence of the matter would surely be to define what would *constitute* the fact of these qualia's being directly compared, in particular, what would constitute its being the *same* pair of qualia as the original pair that is being compared. We would have to explain how the identity of each quale across different comparisons – the sameness of a as compared first with b then with c, etc. – is defined. And we must do this in a manner that permits possible comparisons for *every* pair of qualia (including, it should be noticed, all the merely possible ones). Appeal to counterfactual perceptions of constancy might do part of the job, but surely not very much of it. How is the job to be done, for example, for qualia separated by years of time, or appearing to different persons?[13]

It is easy to produce paradox by combining internalizing of constancies with internalizing of visaged samenesses and differences. Suppose, for example, that between two identically coloured objects a coloured band is inserted, one that is subtly graded in colour from side to side. The effect may be that while it appears that nothing has been changing colour, still what started out looking like two samples of the same colour now look like samples of different colours.[14] Or suppose while you are watching, someone draws arrow ends on each of two parallel lines, turning them into Mueller–Lyer arrows. While appearing not to grow or to shrink, the lines will begin by appearing the same length and end by appearing different lengths. Again: those trees in the distance looked the same size until I noticed the men standing beside them. Now they appear to be quite different sizes, yet things appear not to have changed. If we internalize constancy, sameness and difference, such visagings would appear to be impossible. The demand for content coherency (see section 2 above) seems to rule them out.

An entirely explicit externalizing and internalizing of the sameness relation occurs in Peacocke's (1986) discussion of manners of perception. Using perception of distances as his example he writes 'if μ is the manner in which one distance is perceived and μ' is the manner in which a second distance is perceived by the same subject at the same time, and μ = μ', then the distances are experienced as the same by the subject (they match in Goodman's sense)' (1986, p. 5). Next, '. . . the same manner can enter the content of experiences in different sense modalities. You may hear a birdsong as coming from the same direction as that in which you see the top of a tree: we would omit part of how the experience represents the world as being were we to fail to mention this apparent identity' (1986, p. 6). That is, an apparent identity between directions necessarily indicates a sameness in the intermediaries through which these directions are presented, a sameness of manner, regardless of the difference between these intermediaries with regard to modality. Thus there are amodal manners of perception. Peacocke clearly intends these moves to be stipulative, defining what constitutes sameness of perceptual manner. But such stipulations do not come for free. That there is or could be any such sameness must be argued. For exactly the same reason as in the case of Goodman's qualia, the sameness relation for perceptual manners could not be wholly constituted via facts regarding merely 'apparent identities'. Either (1) the manners themselves or, if manners are being defined here as equivalence classes, (2) whatever those intermediary items are that make up these equivalence classes, must have prior not-merely-phenomenal identity. If the former, we need an argument that the appearance of identity can only result from a real identity in these (independently defined) manners. If the latter, we need an argument that producing-the-appearance-of-identity-when-compared is a necessarily reflexive relation (i.e. that sameness can be externalized here) and a transitive relation. And we need to know that each pair of intermediary items is invariant over contexts with regard to producing vs. failing to produce the appearance of identity. But most important, we need an argument that the equivalence classes so formed are of relevant theoretical interest. For example, is there a necessary similarity among the ways the various members of each class determine their identical references or extensions and a necessary dissimilarity among the ways same-reference pairs from different classes do this?

Another way of externalizing the sameness relation is suggested when Evans (1985) gives a tentative 'yes' answer to Molyneux's question.[15] His reasoning is that if perceptions of shape by sight and by touch produce parallel behavioural orientations in the space surrounding one, hence constitute perceptions of space for the same reason, then they are understood to be perceptions of the same. Because 'there is only one behavioural space' (Evans, 1985, p. 390) within which grasp of visual and felt shapes are manifest, there is no problem about identifying across these modalities. Again, relevant sameness in relevant intermediaries – the intermediaries here are dispositions or the states in which these are rooted – is externalized to yield a visaging of sameness.

But what, you may ask, would it be to recognize a sameness in spatial content across modalities if not, just, to react in the same way to the same content when presented through these modalities? Elsewhere I have defended the following sort of view of the act of recognizing sameness (Millikan, 1984, ch. 15; 1991). Suppose that, lacking a wide angle lens, you have taken a series of overlapping photographs of a panoramic view. Later, by attention to the overlaps where the same pattern shows up in each of two photographs, you are able to combine photographs so as to observe or infer relations among objects that had not been photographed together. In so doing, you recognize sameness of content in the overlaps. In like manner, every mediate inference pivots on a middle term of some kind, on the ability to recognize that two thought tokens share a content. For a simple example, consider the inference from John's being taller than Sam and Sam's being taller than Bill to John's being taller than Bill. This pivots on recognition of the sameness of the person, Sam, as represented in each of the premises. Again, only through recognizing the identity of an item currently perceived with an item perceived earlier can what was learned earlier be joined with what is perceived now to yield informed action. Recognizing incompatible attitudes, hence avoiding contradiction, also turns on a grasp of identity. In short, grasp of identity is the pivot on which every exercise of perception and/or thought that collects together information, effects its interaction, or applies it, must turn. Indeed, to identify contents of two visagings or ideas consists in no more and no less, I suggest, than that the larger visagings or thoughts-that in which these are embedded interact or combine, or that there is preparation for them to combine, in this manner.[16] Returning, then, to Evans's speculations on Molyneux's question, only in the act of combining spatial information obtained through one modality with information obtained through another to yield behaviour or thought guided by both modalities together, does a person grasp the identity of spatial aspects over different sensory modalities.

ACKNOWLEDGEMENTS

Excepting only a few changes and additions, this chapter is a reprint of portions of my 1991 paper. A version written prior to 1991, roughing out the parts reproduced here, was presented at the Workshop on Spatial Representation at King's College Cambridge in the spring of 1991.

I am much indebted to Christopher Peacocke's challenging work on perceptual content (Peacocke, 1986; 1987; 1989a; 1989b). Although we disagree on some quite fundamental points, without Professor Peacocke's help, I should never have thought to look in this way at the issues. My ungrateful choice of a couple of Peacocke's claims and arguments to use as negative examples reflects only that these were what happened to be on my desk at the time of writing, not that they are singular in any other way. I also owe Professor Peacocke a great debt for his help with earlier drafts of this essay and for his unfailing patience with my views. Helpful comments by Justin Broakes, Elizabeth Fricker, Timothy Williamson, James Hopkins, Michael Lockwood and Mark Sainsbury also contributed to this part of the essay.

NOTES

1 In Millikan (1991) I also illustrate these confusions as they appear in the Fregean tradition on conception or thought.
2 On numerous occasions I have taken the position that thinking and perception likely both involve inner representation and that representation involves abstract mappings by which representations are projected onto representeds. But this claim does not entail that any particular concrete properties and relations are shared by representation and represented. Nor is it likely to be open to merely philosophical demonstration which abstract mathematical relations are shared.
3 But recall, for example, this passage from Strawson's *Individuals*, ch. 2: 'Sounds. . . have no intrinsic spatial characteristics. . . [by contrast]. . . Evidently the visual field is necessarily extended at any moment, and its parts must exhibit spatial relations to each other' (Strawson, 1959, p. 65).
4 That this is what Evans intends comes out clearly in his discussion of 'simultaneous' v. 'serial' spatial concepts in Part Four of Evans (1980).
5 But see also McDowell's footnote in Evans (1985, p. 373) suggesting that Evans may later have rethought this issue.
6 This feature allows for indefiniteness or indeterminacy due to lack of perfect visual acuity.
7 I had much the same scenario in mind when I wrote Millikan (1984). There are passages there on perception that may be uninterpretable if one does not take this view – and with it another relative of Peacocke's views, namely, that perception involves some type of analogue intermediaries. What I claim here is that at least certain arguments for this don't go through.
8 Drawing the distinction between these two kinds of sensational properties is not always easy. See Peacocke (1983, pp. 24–6).
9 Would refusal to internalize constancy result in rejection of secondary qualities? For there to be secondary qualities, there must be sensations, or some other designated intermediaries, that certain external objects produce in us and that remain the same both before and after the development of these concepts. An easy argument for the existence of such intermediaries is that we suspect, by analogy with other conceptual changes, that things appear not to change as one acquires these concepts. Are there other arguments to show that there exist, for example, sensations of red prior to conceptions of red? (Arguments that there exist prior *perceptions* of red will not do, of course, since perceptions of red need something to be about.)
10 Using 'V:' as an 'it is visaged that' operator, the move is from *V:(En) n is large and a hen has n speckles* to *(En)n is large and V:a hen has n speckles*. There is nothing wrong with exporting the existence of a number, but the result here is also to import determinacy to within the scope of the visaging operator.
11 It is not in fact a relation, as I argue in Millikan (1984, ch. 12).
12 Though I am ignoring this complication here, the assumption that what appears different does not at the same time appear same is not trivial. See, for example, Crane (1988).
13 See Goodman's remarkable discussion (Goodman, 1966, p. 132 ff.). Goodman supposes that, as with perception of the colours of ordinary objects under uniform lighting conditions, one sees qualia sames as same but cannot discriminate all differences. At the same time he proposes a purely phenomenal definition of qualia sameness. Thus, although Goodman's intermediaries are introduced as having aspects not externalized to content, still Goodman affords these intermediaries no aspects not originally internalized from content. They are not independent vehicles operating beneath perception with any properties not derived from awareness. Goodman's qualia abide in ontological limbo between the realms of appearance and reality.
14 Alternatively, perhaps nothing appears to have been changing colour. Then the paradox

results only if we also import determinate sameness-or-difference (see four paragraphs above).

15 Molyneux's question was whether a man born blind could, upon suddenly regaining his sight, immediately recognize shapes by sight.

16 More accurately, acts of identifying are performed by systems that have not mere dispositions but proper functions to perform such acts. See Millikan (1984, ch. 15) for a fuller discussion of these acts.

REFERENCES

Crane, T. 1988: The waterfall illusion. *Analysis*, 48.3, 142–5.

Dennett, C. and Kinsbourne, M. 1990: Time and the observer: The where and when of consciousness in the brain. *Behavioral and Brain Sciences.*

Evans, G. 1980: Things without the Mind. In Zak van Stratten (ed.), *Philosophical Subjects: Essays presented to P. F. Strawson.* Oxford: Clarendon Press, 76–116; reprinted in Evans (1985).

Evans, G. 1982: *The Varieties of Reference.* Oxford: Clarendon Press.

Evans, G. 1985: Molyneux's question. In his *Collected Papers.* Oxford: Clarendon Press, 364–99.

Goodman, N. 1966: *The Structure of Appearance.* Second Edition. Cambridge, Mass: Bobbs–Merrill.

Millikan, R. G. 1984: *Language, Thought, and Other Biological Categories.* Cambridge: Bradford Books/MIT.

Millikan, R. G. 1991: Perceptual Content and Fregean myth. *Mind*, 100.1, 1–21.

Peacocke, C. 1983: *Sense and Content.* Oxford: Clarendon Press.

Peacocke, C. 1986: Analogue content. *Aristotelian Society Proceedings*, Supplementary vol. LX, 1–17.

Peacocke, C. 1987: Depiction. *The Philosophical Review*, XCVI, 383–411.

Peacocke, C. 1989a: Perceptual content. In J. Almog, J. Perry and H. Wettstein (eds), *Themes From Kaplan.* Oxford: Oxford University Press, 297–329.

Peacocke, C. 1989b: *Transcendental Arguments in the Theory of Content.* An inaugural lecture delivered before the University of Oxford on 16 May 1989. Oxford: Clarendon Press.

Selfridge, O. (unpublished): *Tracking and Trailing.*

Strawson, P. F. 1959: *Individuals.* London: Methuen & Co.

Part IV

Action

Introduction: Action

Bill Brewer

Two problems are discussed under the head of 'Action'. The first concerns the connection between the spatial representations involved in perception and action. How does perception control and co-ordinate basic spatial action? More precisely, what can we say about the relation between the ways in which spatial information is coded in perception and in the organization of spatial behaviour, given the nature of perceptuo–motor control? The second problem concerns the nature of action itself, as the directed *doing of something* by an *agent*, as opposed to certain bits of animals' bodies moving in various ways. What is the relation between our talk of a person's perceptions, beliefs, desires and so forth causally explaining their intentional actions, and our understanding of why certain bits of a person's body move in the ways they do, in terms of the course of afferent stimulation on the perceptual surfaces, through the brain, and across the motor units? What marks off certain events involving animals as their intentional actions, others merely as reflexes or mechanical movements in bits of their bodies?

I

In chapter nine, Meltzoff graphically introduces the first set of issues by posing a perceptuo–motor analogue of Molyneux's question for vision and touch. In a general and rather vague form, the question might be put as follows:

> In the absence of prior experience, for whatever reason, of any association between perception of where things are around one, or perception of the movements others make, on the one hand, and one's own spatial behaviour on the other, is one immediately able to act spatially appropriately in connection with the things one perceives, or to imitate others' gestures?

The basic issue here, in connection with any kind of task involving perceptually guided action, is whether the ways in which spatial properties of the environment are represented in perception and spatial properties of bodily movements are represented in the initiation and control of action are the same or different, where this question of sameness or difference applies to the *content*, rather than the vehicle, of the spatial representations in question. Do perception and action code spatial relations in the same way, using a shared type of spatial content, or does their integration require some correlation or computation between intrinsically arbitrarily related modes of spatial representation?

There is a general empiricist presumption, which is also reflected in much theoretical psychology of perceptuo–motor control, in favour of the *separate coding* position. Intuitively, the considerations determining successive patterns of stimulation on the various sense organs as similar or different are quite unlike those which group token bodily movements into types. The intuition is sustained both by naive introspection of the subjective difference between tactual sensations, visual images, basic acts of bodily will and the like, and also by reflection on the wide variety of the computational raw materials at the disposal of the tactual, visual and motor systems, for making the required spatial discriminations. It is this thought of the basic principles of similarity and difference relevant to spatial categorization as quite disparate between the perceptual sense modalities themselves, which motivates many negative answers to Molyneux's original question (see Part III). The more general picture, which extends to action as well, gives rise to the Piagetian idea of initially 'heterogeneous spaces' represented by the various sense modalities and the motor system, which are gradually brought into registration as the developing infant acquires constant experience of their association and correlation (Piaget, 1929).

Thus, the normal adult's capacity for perceptually controlled and co-ordinated spatial action is supposed to be the upshot of an associative integration of intrinsically unrelated modes of representation, which is built up over experience of perceiving and acting in the world. Perceptual representations and those immediately involved in the initiation and guidance of action bear quite distinct types of spatial content.

Both Meltzoff and Brewer argue against this traditional view, in favour of a *common coding* account, in which there is a way of representing spatial properties and relations which is shared between perception of the world around the subject and the initiation and control of his or her physical action. This is not to say that this is the *only* mode of spatial representation available to the perception and action systems, but rather to insist that there is *a* type of spatial content common to both. They each focus on tasks involving the integration of visual perception and spatial action – imitation and simple reaching and grasping respectively – although we shall shortly see grounds for regarding these as interestingly different in the sophistication of bodily awareness involved. Both authors aim to bring out important similarities in the organization of spatial information between the visual and motor systems, in support of the common coding view. (See Bower (1979) for a general account of this anti-Piagetian position.)

Meltzoff addresses the imitation version of Molyneux's question directly, in experiments with young infants. He argues that their capacity for spatial imitation of others' bodily movements at an extremely young age (around 32 hours), along with their rather older discriminating sensitivity to being imitated themselves, strongly suggest a common type of spatial content for the representation of others' movements in vision and their own movements in action. There is an intrinsic registration of spatial isomorphism between an actor's mouth opening, sequential finger movement, or whatever, as this is seen by the infants, and their representation of their own performance of the same actions. In other words, the spatial features of others' and one's own bodily movements are represented in the same way, by the visual and motor systems respectively. This explains the older infants' direction of more attention and testing behaviour towards the actors imitating them, in preference to controls performing spatially distinct but temporally matched actions. Meltzoff concludes that there is some primordial connection between our own acts and the acts we see others perform (p. 232).

In a similar vain, Brewer argues that the widespread advantage in reaction time and accuracy to respond to visual stimuli in cases where these share spatial characteristics with the required bodily response, over non-isomorphic stimulus–response pairings, points to a common type of spatial representational content employed in both vision and action. Indeed, such results are a manifestation of the more general spatial *structure* in visuo–motor control (see Peacocke, 1983, pp. 61–4). Successful instances of visually guided action are not the result of one-off, individual associations between arbitrarily related visual and motor codes. They are rather an instantiation of a far more powerful general ability immediately to register the spatial significance of vision for action, which is not simply the pointwise sum of particular isolated visuo–motor correlations. For the subject treats as more-or-less equivalent the very many proximally quite different patterns of visual stimulation available from things at a given egocentric location around him or her, and the equally many proximally varied limb movements with fixed consequences for the things at such a location. Similarly, developments and breakdowns in visually guided spatial action are not individual additions and subtractions of particular visuo–motor code associations, with no consequences for the capacity elsewhere. They have widespread productive and destructive ramifications for performance in suitably spatially related cases throughout the network of visuo–motor correlations. Yet the separate coding theorist's attempt to incorporate this structure depends upon precisely the similarity of spatial classification of visual input and motor output which the position denies. Again then, the conclusion is that visual perception and basic bodily action operate with a common type of content in the representation of spatial properties of the environment and of bodily actions.

A question naturally arises as to what the relation is between the two common coding hypotheses advanced by Brewer and Meltzoff. Are the common coding claims for simple reaching and grasping and for the whole range of imitation-based behaviour discussed by Meltzoff just two instances of a single more general fact about the relation between spatial vision and action; or does one involve more sophisticated spatial information than the other? Intuitively there is a sense in which a more substantial body-image is invoked in sensitivity to imitation than in visually guided reaching and grasping. Two further thoughts seem to confirm this intuition. Firstly, if the Bernsteinian account, which Brewer develops, of the bodily biodynamics subserving reaching and grasping is on the right lines, little to no knowledge of current bodily configuration is required for this kind of task. Perception simply parameterizes the appropriate muscle assembly, which unwinds into the required bodily movement as a simple mass-spring finds its stable equilibrium. Secondly, on the other hand, since infants' capacities for imitation-based behaviour also involve reciprocal appreciation of their being imitated by others, this does seem to involve an ongoing sensitivity to the current spatial configuration of their own body parts.

The point can be put in terms of O'Shaugnessy's notion of the short-term 'here-and-now' body-image (1980, pp. 241–8). This is not the relatively long-term picture of one's physical dimensions and flexibility, which changes only with limb loss, skin graft, gradual growth and so on, part of whose rationale is 'to constrain intentional projects within the bounds of bodily possibility' (p. 246). It is rather the continually changing image 'given by the description or drawing or model one would assemble in order to say how the body seems to one at a certain instant' (p. 240). The thought would then be that although it is not essential for Brewer's simple reaching and grasping, on the Bernsteinian account he develops, some such short-term body-image *is* required for the whole range of Meltzoff's imitation-based behaviour.

It is interesting to note that along with this increased requirement on bodily knowledge for imitation, goes grasp of a non-mechanical mode of influencing things. The infant comes to the idea of him or herself as sharing certain specific spatial properties with the imitator, together with the idea of having an apparently physically unmediated effect on the other's body. This seems particularly evident in the testing behaviour directed preferentially towards imitating actors over controls. Adults' conception of this non-mechanical influence is psychological: the imitator *sees* what I do, and *acts* in the same way, as I do when I imitate him or her. Infants' understanding need not be anything like so sophisticated. But perhaps this basic idea of a non-contact based way of systematically affecting the spatial behaviour of things in the environment is a foundation for our grasp of others as perceiving, thinking agents. Here we might hope to fill out this conception of others as subjects with a notion of the child's naive psychology, or theory of mind, parallel to the notion of an intuitive physics (see Part II), which shapes and sustains thought and perception of the things around us as objects.

All this is extremely sketchy, and hardly the most parsimonious account, but it opens up the possibility of an interesting partner to a famous Kantian thesis. In the 'Transcendental Deduction' (Kant, 1929, pp. 129–75), Kant argues that unity of consciousness requires experience of independent objects. In other words, the thesis is that the conception of oneself as a persisting subject of experience requires that some such experience be perception of objects around one. Our current interpretation of Meltzoff's work suggests that our grasp of the idea of ourselves as spatially extended objects has (in part) to do with our recognition of others around us as subjects of experience, thought and action. This issue of the origin, nature and content of our awareness of ourselves as spatially extended bodies, and its relation with our conception of ourselves and others as psychological subjects, will figure centrally in our forthcoming interdisciplinary research into the complex interconnections between intuitive physics and intuitive psychology generally, in our grasp of our place in the world (see the General Introduction).

II

Recall that the second set of issues in Part IV concerns the distinction between what we might call an organism's mere bodily responses to its states and the world around it on the one hand, and its purposive spatial actions on the other. Again there are substantial agreements between the authors who explicitly discuss this question: Brewer, and Dickinson and Balleine. They each insist that the distinction is to be made, in the first instance, in terms of the type of *explanation* which is to be given for the occurrence in question; and both argue that this distinction in kinds of explanation is ineliminable:

actions and responses are fundamentally different categories of events and any identification or reduction between them is unacceptable.

The type of explanation which marks off spatial *action* from mere response is *purposive* teleological explanation, in terms of the presence of a goal and an instrumental contingency between the behaviour and access to the goal. In other words, it is explanation of the kind to which we appeal in the standard 'belief–desire' accounts of folk psychology. Purely mechanistic, non-rational, associative learning theory, on the other hand, provides an extremely successful framework for the explanation of quite complex response behaviour. Dickinson and Balleine's question is whether this dual psychology of behaviour – 'an intentional, cognitive psychology for goal-directed actions and an associative, mechanistic psychology for elicited non-purposive responses' (p. 277) – is ultimately justified; or whether a suitably sophisticated unified associative theory might explain both actions and responses.

They consider what they call the *associative–cybernetic model*, an optimal integration of the explanatory materials available within the mechanistic psychology, which seems on the face of it to make room for behaviour meeting the two conditions implicit in the teleological criterion for action. If actions are to be causally explained by the agent's desire for a goal and belief in the behaviour as a means to acquiring that goal, then he or she is rationally bound to desist given evidence either for the falsity of the belief, or for the undesirability of the goal. In other words, *actions* ought to cease in omission and goal devaluation schedules. This is what distinguishes the behaviour from a mere reflex-like response, with respect to which such questions of rationality do not arise. Although the associative–cybernetic model provides an explanation for the action/response distinction in connection with basic omission and goal devaluation schedules, neither it, nor anything much like it, can accommodate a further, intuitively highly plausible and empirically confirmed distinction between responses and actions in the context of goal aversion and shifts in primary motivational states (see p. 285 ff).

The upshot of all this is an endorsement of the implicit dual psychology of behaviour. Rationally governed intentional action explanation is an essential, irreducible addition to associative mechanism in our understanding of animal behaviour.

Brewer also discusses this same issue, although from a rather different angle, and comes to very much the same conclusion. Given the common coding account of the interface between visual and motor systems, it is tempting to identify the folk psychological explanatory antecedents of spatial action with the brain events which realize this visuo–motor achievement. Thus, one comes to an account of spatial action as such and such a movement of the body, brought about by the motor cortex activity which constitutes an appropriate intention or act of will. On such a view, normative psychological explanation is simply superimposed upon the underlying causal web of mechanical interactions. But, Brewer argues, the resultant picture fails to do justice to three aspects of our intuitive conception of an agent as the deliberate author in control of his or her action. Firstly, it cannot give a satisfactory account of agent's knowledge: the knowledge we have of what we are up to, simply in virtue of intentionally doing it. Secondly, it cannot give a satisfactory account of practical knowledge or ability: our knowledge *how* to act in the ways we do. Thirdly, it cannot give a satisfactory account of the characteristic experience of bodily action: it misrepresents what it is like to be an agent. He goes on to offer a sketch of an alternative account of the relation between psychological and mechanical explanations of spatial actions and bodily responses respectively.

REFERENCES

Bower, T. G. R. 1979: *Human Development*. San Francisco: W.H. Freeman & Co.
Kant, I. 1929: *Critique of Pure Reason*, tr. N. Kemp Smith. London: Macmillan.
O'Shaughnessy, B. 1980: *The Will*, vol. 1. Cambridge: Cambridge University Press.
Peacocke, C. 1983: *Sense and Content*. Oxford: Clarendon Press.
Piaget, J. 1929: *The Child's Conception of the World*. London: Routledge and Kegan Paul.

12
Actions and responses: The dual psychology of behaviour

Anthony Dickinson and Bernard Balleine

The psychological analysis of behaviour is riven by two different models of the agent. Students of human action usually adopt a cognitive approach to behaviour, and indeed contemporary action theory is replete with accounts of agents planning various behavioural strategies on the basis of their knowledge of action–outcome relationships and selecting courses of action by some rational criterion after an evaluation of the alternative outcomes (e.g. Frese and Sabini, 1985). When, however, psychologists are reluctant to shoulder the burden of intentionality implicit in such cognitive accounts, they usually resort to associationism, namely the thesis that the mind consists of 'representational' elements linked by excitatory (or inhibitory) connections. In contrast to cognitive theory, with its appeal to the rational interactions between the contents of mental states, associationism is essentially mechanistic in nature in that mental processes, such as excitation or activation, gain their explanatory power by analogy to physical ones.

Although associative theory has undergone a renaissance in general psychology, the sophisticated architectures and learning algorithms of the new connectionism (Rumelhart and McClelland, 1986) have not typically been applied to the analysis of behaviour; rather, they have addressed psychological capacities, such as perception, concept learning and categorization, which have been traditionally thought to lie within the province of cognitive theory. The systematic associative analysis of behaviour has been largely left in the hands of students of animal learning, where it has been most profitably applied in the case of Pavlovian conditioning (Dickinson, 1980; Mackintosh, 1983), which is usually assumed to reflect the elicitation of non-purposive responses by conditioned and unconditioned stimuli via associative structures of varying complexity.

Thus, there is, at present, an implicit dual psychology of behaviour; an intentional, cognitive psychology for goal-directed actions and an associative, mechanistic psychology for elicited, non-purposive responses. Given this context, the issue that we wish to address is whether there are good behavioural grounds for maintaining this dual psychology.

1 Actions and responses

As a prerequisite for our analysis, we must make explicit what we mean by 'goal-directed' and specify criteria for determining whether or not a particular activity is in fact goal-directed. Moreover, these criteria must make no reference to the mediating psychology, for clearly if they do so the point at issue, the behavioural warrant for a dual psychology, will be prejudged. For present purposes the status of a particular activity will be determined by whether or not it will support a teleological explanation, an account that explains the occurrence of the activity by reference to its goal. If it will, we shall refer to the activity as an action, and if not as a response.

Although a number of attempts have been made to rescue teleological explanations from the problem of circularity, we shall focus on that offered by Taylor (1964; 1970). According to Taylor,

> An explanation is teleological if the events to be explained are accounted for in this way: if G is the goal 'for the sake of which' events are said to occur, B the events to be explained, and S the state of affairs obtaining prior to B, then B is explained by the fact that S was such that it required B for G to come about (Taylor, 1970).

Thus, in order to warrant a teleological explanation, any particular activity must meet two criteria: performance of the activity must be shown to depend upon the antecedent situation being both one in which a goal exists for the agent and one in which there is an instrumental contingency between performance of the action and access to this goal. If both these properties are necessary for performance, the activity is a goal-directed action. Other things being equal, performance in the absence of either of these properties identifies the activity as a response.

The application of these teleological criteria can be illustrated by considering a simple activity, the approach to a food source by a hungry animal that, at least in terms of its manifest properties, invites a purposive explanation. Any one who has a pet dog is aware of its remarkable sensitivity to cues that signal the imminent arrival of dinner; wherever a dog of our acquaintance, Rolly, is in the house, it is impossible to open the tin of dog food without him bounding into the kitchen, jumping up, barking and generally causing chaos. Clearly, such anticipatory approach is a good candidate for goal-directed action; Rolly's hunger ensures the existence of a goal, namely access to food, and the sound of the can opener constitutes a state of affairs under which approach is required for the goal.

But the question is: How should we find out whether or not Rolly's behaviour is in fact goal-directed in this particular case? A requirement of the teleological criteria is that if we arranged a situation in which approach at the sound of the can opener is not required for access to his dinner while maintaining all other features of the situation the same, Rolly should no longer reek havoc in the kitchen. And a clearcut way of breaking this relationship is to arrange the opposite contingency, whereby if Rolly enters the kitchen during the signal provided by the can opener, he will not get his dinner or, in other words, the goal will be omitted. As such an omission schedule arranges a contingency in which the absence of anticipatory approach is required for the goal, surely on any teleological account Rolly should refrain from interfering with the culinary endeavours on his behalf if his normal behaviour in this situation is directed towards gaining access to food.

On reflection, dog owners may now have doubts about the purposive nature of their pet's feeding behaviour, for we have had no success in controlling Rolly's behaviour by refusing to give him his dinner unless he remains outside the kitchen until the preparations are complete. Moreover, Rolly's failure to adapt to an omission contingency is not an exception, but rather a general feature of anticipatory approach to food sources by various animals. Within the conditioning laboratory, it is well known that if the delivery of food to a hungry rat is signalled by a stimulus, such as tone, the rat will come to approach the food source during the tone, just as Rolly bounds into the kitchen at the sound of the can opener. What Holland (1979) showed is that rats will actually acquire this approach response even under a schedule in which approach during the tone causes the omission of the food. It is important to realize the maladaptive nature of this behaviour; by acquiring and persisting in anticipatory approach, the animals actually lost a significant proportion of the food available to them.

There is no doubt that, by the teleological criteria, anticipatory approach to a food source by hungry rats is not goal-directed; it develops and persists under a state of affairs where approach is not required for access to the goal – in fact, it will do so even when just the opposite behaviour is required.The purposive status of this action could be rescued by claiming that the very occurrence of anticipatory approach indicates that Holland's rats must have 'believed' that this action was necessary for access to the goal. Such a strategy, however, undermines the empirical status of the teleological explanation. If the primary evidence for an antecedent instrumental belief is the performance of the very action that it explains, the teleological account is fatally circular. Clearly, the onus must lie with the teleologist to give a principled account of why the animal should come to believe that anticipatory approach is required for access to food when, in fact, exactly the opposite contingency holds.

Doubt about the goal-directed nature of this particular response does not imply, of course, that rats and dogs are incapable of purposive actions, for a rat cannot be trained to press a freely available lever, for instance, with a food reward on an omission schedule. In fact, having established such an action rats are very sensitive to changes in the contingency between lever pressing and access to a goal, at least in a situation in which the lever is freely available to them and there is no explicit signal of the availability of food. Not only will the animal refrain from pressing if food is now delivered independently of this action, they will stop even faster if pressing actually causes the omission of the food (e.g. Davis and Bitterman, 1971).

The second feature of a teleological explanation is the dependence of the action upon the existence of a goal, for if an activity could be shown to occur independently of any plausible goal, we should once again have grave doubt about its status as a goal-directed action. Within the conditioning laboratory, a standard way of determining the goal dependency of an action is to conduct a revaluation test. Such a test can be illustrated by an experiment performed in our laboratory some time ago. Adams (1982) trained hungry rats to press a lever for sugar pellets. Clearly, the obvious goal of this action was access to the palatable and nutritious sugar. To investigate whether this was so, Adams then attempted to devalue this event so that it would no longer act as a goal. This he did by a food aversion procedure.

It is well known that if the consumption of particular food is followed by gastric illness, animals (including ourselves) develop an aversion to that food so that it will no longer function as an effective goal, even when hungry. So, in the second stage of the experiment, one group of rats (devaluation group) were allowed to eat the sugar pellets immediately before the induction of gastric malaise by an injection of a mild toxin. Although food aversions can be established with a single pairing of consumption with illness when the food is novel, the treatment had to be repeated a number of times to suppress completely the consumption of the sugar pellets. This is because the pellets were already familiar at the start of aversion training as they had previously been used to establish lever pressing. It is important to note that the animals were not required to earn these pellets by pressing the lever during aversion training; in fact, the lever was not present at this stage and the sugar pellets were delivered independently of any instrumental action. A control group received exactly the same treatment except for the fact that access to the pellets and the induction of gastric illness occurred on different days so that no aversion was formed to the sugar.

If we assume that 'access to sugar' was the original goal of lever pressing and that, as a result of the aversion treatment, this event could no longer function as a goal for the devaluation group, a prediction of the teleological account is that these animals should refrain from pressing the lever in the future. This is just what Adams found; when subsequently given access to the lever again, the devaluation group pressed significantly less than the control group. It is most important to note that this test was conducted in 'extinction' during which sugar pellets were never presented. If Adams had presented the sugar pellets during testing, the reluctance of the devaluation group to press the lever could be explained simply in terms of the direct suppressive effect of presenting this aversive consequence. By testing in extinction, however, he was able to change just the status of the antecedent situation for lever pressing, in that the test situation was no longer one in which lever pressing was required for access to a goal.

Thus, application of the teleological criteria has allowed us to identify a goal-directed action, lever pressing, and distinguish it from a simple response, anticipatory approach to a food source, for the rat. It must be emphasized that identification of the status of any particular activity requires consideration of both criteria. Anticipatory approach, although failing the omission schedule test, can be susceptible to devaluation by food aversion conditioning (Holland and Straub, 1979). Furthermore, there are training schedules under which lever pressing is impervious to the effects of goal devaluation (Adams, 1982; Dickinson, Nicholas and Adams, 1983).

The possibility that the same activity may be teleologically determined in one situation but not in another raises an assumption that has been implicit in the discussion so far. By its very nature a teleological explanation relates an action to a particular goal, a relationship that is examined by the omission and goal-devaluation tests. However, failure to pass these tests with respect to a particular goal does not mean that the activities constituting the response may not themselves be goal-directed. Thus, for example, there are a variety of ways by which a rat may press a lever, each of which may be directed towards the goal of bringing about the depression of the lever even though the act of lever pressing itself may not be teleologically related to a further goal, such as gaining access to food. Consequently, the present analysis always determines the status of an activity with respect to the particular goal identified by the omission and goal-revaluation tests.

The factors that determine whether a particular activity or set of activities constitute an action or a response (with respect to a given goal) must lie with the psychology of the agent. The obvious candidate for purposive action is a cognitive or intentional theory developed from the basic 'belief–desire' explanations of folk psychology. To embrace such theories, however, is to endorse, at least implicitly, a dual psychology, one for actions and another for responses. A hallmark of an intentional account is the assumption that its process must conform to some criteria of rationality, namely that the action must in some sense be a rational consequence of the instrumental 'beliefs' of the agent and his or her 'desires' (Heyes and Dickinson, 1990; Dennett, 1987), and it is this feature of intentional accounts that rules them out as potential explanations for responses. The central feature of response acquisition under an omission schedule or persistence in the face of goal devaluation is its irrationality.

The alternative is to eschew an intentional psychology in favour of one couched in concepts that potentially are applicable to both classes of behaviour, namely those of associationism. Consequently, in the next section we have sketched a possible associative architecture that is both capable of actions and responses under the appropriate conditions. There is nothing intrinsically novel about any of the components of the model, which have been gleaned from various sources in an attempt to offer the most powerful associative explanation that is currently available; all that is new is their integration.

2 Associative–cybernetic model

A cartoon of the basic architecture of the model is illustrated in Figure 12.1. Before addressing this model in detail, however, we shall describe its functioning in metaphorical terms. The basic idea is that being in a particular situation makes the agent imagine performing a response, but not very vividly. If this response has been previously associated with a goal, the agent will, as a consequence of the associative properties of the model, also think of the goal. The cybernetic component reflects the fact that imagining the goal feeds back to enhance the response image until, in the spirit of the ideo–motor theory, it is sufficiently vivid to trigger the response as an overt action. This basic idea has been proposed in one form or another by a number of theorists (King, 1979; Miller, 1963; Mowrer, 1960; Sheffield, 1965); indeed, Sutton and Barto (1981) have simulated a connectionist version of this account to explain goal revaluation effects. The present model expands this basic idea to provide an integrated account of both actions and responses.

Actions have their origin in what we have called a stimulus–response (S–R) store. Here it is assumed that an array of stimulus elements is linked to an array of response elements that correspond to the agent's behavioural repertoire. Thus, for instance, the stimulus element that is activated by sensing the lever tends to produce lever pressing as an untrained (or pre-trained) response due to the fact that this stimulus element is connected by a link to the appropriate response element. The activation of coupled pairs of stimulus and response elements is mutually exclusive, so that only a single pair of stimulus and response elements can be active at any one time, presumably because the perceptual system providing the input to the store ensures that each stimulus element is activated by a unique stimulus situation.

Each response element in the store has an output to a corresponding element in the motor system whose activation causes performance of the relevant action. This motor element has a high, though fluctuating threshold, so that activation from the S–R store is not normally sufficient to fire it and produce the overt action (except, of course, in the case of innate responses, fixed action patterns, reflexes, etc). This means that when the untrained rat is initially confronted with the lever the activity in the response element will not normally be sufficient to excite the corresponding motor element. Due to the fluctuating threshold of the motor element, however, occasional lever presses will occur as spontaneous, untrained responses.

The third component of the model is an associative memory which consists of

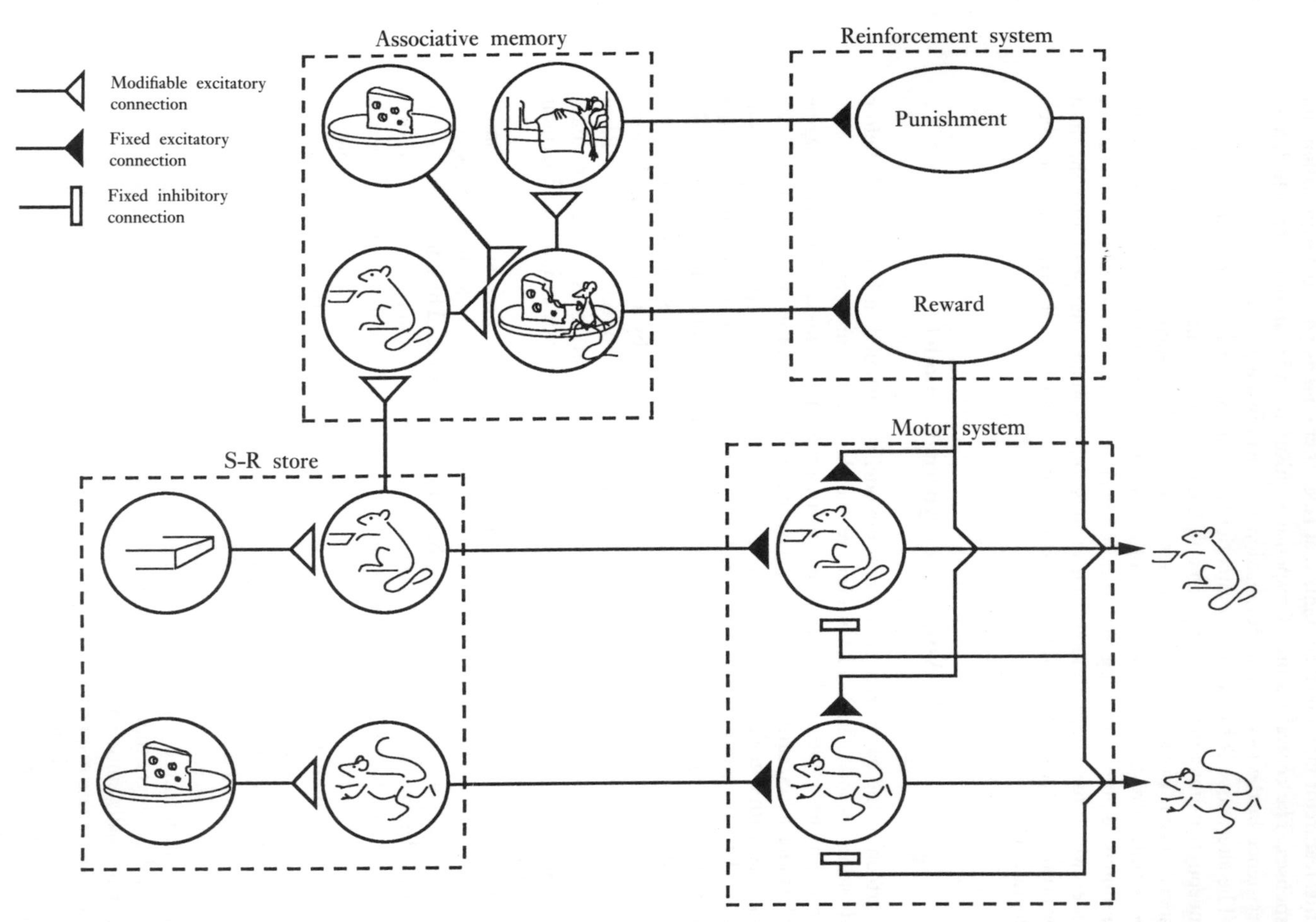

Figure 12.1 Cartoon of the associative-cybernetic model of the systems and associative structures mediating goal-directed actions and non-purposive responses. For details see text.

'representational' elements, each of which is directly activated by the occurrence of the relevant event. These elements are also interconnected by initially weak but modifiable links. The assumption here is that these connections are strengthened (and perhaps weakened) in accordance with one's favourite theory of associative learning (e.g. Mackintosh, 1975; Pearce and Hall, 1980; Rescorla and Wagner, 1972; Wagner, 1981); the particular learning algorithm adopted does not matter for present purposes as long as it exhibits the primary characteristics of the type of associative learning observed in conditioning. It is in this associative memory that the agent's knowledge of the relationships between events is stored. One important feature of an element 'representing' an action in this memory should be noted, however, namely that it is excited directly by activity in the corresponding response element in the S–R store, even in the absence of the overt performance of the action. It is for this reason that the response element for lever pressing in the S–R store is connected to the corresponding action element in the associative memory. Activation of an action element in the associative memory by a response element in the S–R store corresponds to the initial weak image of the action in the metaphorical account.

The final component, the reinforcement system, is designed to handle the cybernetic or feedback function. It is assumed that any element in the associative memory activated directly by an event of primary motivational or affective significance for the agent has a privileged status in that it has fixed connections to elements in the reinforcement system. The connection is to a reward element if the event is attractive, but to a punishment element if the event is aversive or noxious. The function of these reinforcement elements is to exert an influence on those of the motor system, a positive one in the case of the reward element and a negative one in the case of the punishment element.[1] It must be noted, however, that activity in the reward element is not by itself sufficient to fire those of the motor system. In order to activate a motor element the feedback influence from the reward element must sum with a temporally contiguous input from the S–R store. Thus, even under the influence of the reinforcement system, the particular motor element that is activated at any one time is determined by the input from the S–R store.

We are now in a position to trace the initial development of an action, such as lever pressing for food by a hungry rat. Sensing the lever will activate the appropriate stimulus element in the S–R store, and via the stimulus–response link, the response element and also the corresponding action element in the associative memory. Although this will not normally be sufficient to excite the motor element, as has been noted, activation will occur occasionally at a time when by chance the threshold of the motor element is low, thus producing the first overt lever press. As a consequence, food will be delivered exciting its element in the associative memory. This episode will bring about associative learning of the relationship between lever pressing and the delivery of food due to the strengthening of the connection between the elements for lever pressing and food access in the associative memory.

The consequence of this learning is that in future the motor element for lever pressing will receive two influences at any time when the corresponding response element in the S–R store is active, one a direct excitatory input from this element and the other from the concurrent activity in the reward element excited via the acquired connection in the associative memory. Thus, the likelihood of an overt lever press will progressively increase as the connection in associative memory is strengthened across the pairings of the lever press with the food reward, until this action becomes the dominant response in the situation. Sensitivity to the instrumental contingency arises from the fact that the appropriate connection in the associative memory will not be established unless there is a corresponding relationship between lever pressing and access to food. As a result, this action will neither be acquired nor sustained under an omission schedule.

Goal-devaluation effects arise from negative feedback from the reinforcement system to the motor elements. Conditioning a food aversion establishes a connection between the food element and one activated by gastric illness (represented in Figure 12.1 by a rat taken to its sick bed) in the associative memory. The aversiveness of such distress is produced by a fixed connection from its 'representational' element in the associative memory to the punishment element in the reinforcement system. This means that activation of the food element in the associative memory will now produce an inhibitory influence on the motor element for lever pressing, thus counteracting any excitatory influence on this element. As a result, the devaluation treatment will reduce the likelihood of the instrumental action of lever pressing in future.

The analysis so far shows how the model mediates goal-directed behaviour. The remaining requirement is that the system should distinguish such actions from activities that are intrinsically responses, such as anticipatory approach to a food source in the case of the rat. As can be seen in Figure 12.1, the only difference between this case and that of a potentially 'purposive' action, such as lever pressing, is that there is no element in the associative memory 'representing' anticipatory approach; rather, it is the stimulus element activated by sensing the signal for approach (sight of the food source itself represented by a slice of cheese in Figure 12.1) that enters into association with the food element and thus, via the reinforcement system, influences the motor elements. In effect, this account argues that responses are governed by the principles of Pavlovian or classical conditioning. Thus, it can be seen that the central distinction between actions and responses according to this model is that only the former are 'represented' in the associative memory and can enter directly into association with other events.

The application of this system to acquisition of anticipatory approach under an omission schedule is straightforward. Initially, sensing either a signal for or a source of food will fail to elicit approach because the relevant stimulus–response link in the S–R store cannot drive the motor element for approach by itself. Across successive pairings of the signal and access to food, however, a connection between their elements will be formed in the associative memory, producing an excitatory influence on the motor elements during the signal which will summate with that from the S–R store at the motor element for approach, thus producing the overt response. Of course, as soon as the rat starts consistently showing anticipatory approach under the influence from the associative memory (via the reinforcement system), the signal will no longer be paired with food on the omission contingency. This, in turn, will lead to a weakening of the signal–food connection, loss of the necessary excitatory influence from the associative memory and a consequent extinction of anticipatory approach. Of course, once the anticipatory approach response has extinguished, the signal will again be paired with food access, thus restarting the whole cycle of acquisition and extinction.

We have presented this model at considerable length because if one wishes to challenge a unified psychology of action, as we do, then an explicit and powerful target is required. The main problem with a unified approach is its failure to explain the dissociation of responses and actions that can be observed following goal revaluation, a dissociation that we shall illustrate by a personal anecdote concerning the origins of one of our food aversions.

3 Goal representations

AD's primary aversion is to water melons, an aversion acquired while on holiday in Sicily in his youth. At the end of a hot afternoon when thirsty, a companion suggested that they should go in search of water melons, of which AD had no previous experience. This they did and very refreshing the melons were. Unfortunately, being young and foolish, that evening AD drank far too much red wine with the inevitable consequences. He assumed, however, that once he had recuperated that would be the end of the matter; the only thing that he thought that he had learned from this experience was a respect for the local wine.

This was not so, however; once again in need of refreshment, he readily retraced the route to the water melon stall. But when confronted with the direct signal for approach, the sight of those sliced, juicy, rosy-red segments which had looked and were so refreshing on the previous occasion, his appetite abated. And indeed, when he managed to take a bite, he discovered that he had an aversion to water melon with the consequence that none has passed his lips since that day.

At first sight, this experience may appear to accord with the associative–cybernetic model, being just another example of a straightforward goal devaluation effect, such as that exhibited by Adams' rats. But a moment's reflection shows that this is not so. According to the model, AD should never have sought out the water melon vendor on the second occasion. Knowledge of the action necessary for access to water melons should have been 'represented' by a connection between the element for the action required to find the stall and the water melon element in his associative memory. In addition, the water melon element should have been connected to the one 'representing' the consequences of his over-indulgence, and hence should have been capable of exerting a negative influence on the motor elements via the reinforcement system. Consequently, according to this model, even contemplating finding the stall should have been sufficient to dissuade AD from this course of action, and yet he readily performed this goal-directed act in spite of the aversion. It was only after the discovery that he now had an aversion that AD subsequently refrained from pursuing the goal of water melons when thirsty.

The very fact of this discovery is also problematic for the unified psychology of the associative–cybernetic model. AD knew perfectly well that it was the wine that caused the illness and yet he formed an aversion to the water melons; in other words, his aversion was at variance with his cognition. The reason for this dissociation is obvious to students of conditioning. The key is that, as we have already noted, conditioning favours novel over familiar stimuli. Red wine was familiar to AD, who being usually moderate by nature, had no previous history of being routinely drunk after imbibing, whereas the water melon was novel and thus should have been favoured by aversion conditioning. Knowledge of the irrationality of his aversion has had no impact upon its strength, however, which remains to this day.

A final feature of this experience is also relevant. Although the presence of the aversion had no effect upon the initial instrumental act of seeking out the water melon vendor, this was not true of his anticipatory approach in the presence of direct signals of access to the goal. As soon as AD saw the slices of melon arrayed on the stall, his appetite abated. The associative–cybernetic model anticipates this direct effect on anticipatory approach because the feedback influence on this response is mediated by connection between the signal and food elements in the associative memory. What is problematic for this model, however, is the apparent dissociation between the readiness with which he sought out the vendor when again thirsty and his reluctant approach in the presence of direct signals for the goal. As both these activities are supposedly mediated by a common 'representation' of the goal in associative memory, the water melon element, the model anticipates no such dissociation between actions and responses.

In an attempt to establish whether this anecdotal dissociation has any reliability and generality, we (Balleine and Dickinson, 1991) gave thirsty rats an experience analogous to that of AD, except that we used a sugar solution rather than water melons as the goal. To parallel learning the route to the stall, we taught the rats to press a lever for the solution in a single session. Sickness was then induced after this session by an injection of a mild toxin, rather than by a bottle of red wine, for one group of rats. A second, control group was also made ill but after a delay sufficient to prevent any aversion being conditioned to the sugar solution. For reasons not of immediate relevance, but which will become apparent, both groups were then allowed to drink water in the test chamber on the next day in the absence of the lever, before being tested for their propensity to press the lever on the third day. The sugar solution was never presented on this test day, so that the animals did not have an opportunity to discover any aversion that they might have to the sugar.

If our rats' psychology is anything like that of AD, a specific pattern of performance should have been observed on the test day. First, just as he readily sought out the melon stall on the second occasion, the conditioning of the aversion in the group experiencing immediate illness should have had no impact on lever pressing during the test, as these rats had no opportunity to discover their aversion prior to the test. As the central panel of Figure 12.2 shows, this was so; these animals (IMM/H20) pressed just as frequently as the control group that had experienced the delayed illness (DEL/H20). That the animals receiving immediate illness did in fact have a strong aversion to the sugar solution is demonstrated by their performance on a subsequent reacquisition session in which lever pressing once again delivered the sugar solution. As can be seen in the right-hand panel of Figure 12.2, as soon as these animals (IMM/H20) started earning and thus making contact with the solution they stopped pressing, unlike the control rats (DEL/H20) which showed sustained performance throughout the reacquisition session.

The parallel between AD's instrumental action under a goal devaluation and that of our rats is exact. But what about their response of anticipatory approach? We measured this by recording their approach during the test to the source at which the sugar solution had been presented during training (in the jargon of conditioning this source is referred to as 'the magazine'). Just as AD was reluctant to approach the stall once it was in sight, so the rats' aversion was immediately manifest in their magazine approach. The left-hand panel of Figure 12.2 shows that animals with the aversion (IMM/H20) were reluctant to approach the magazine during the test session, relative to the control subjects (DEL/H20), although both groups were at the time pressing the lever with the same frequency. Thus, the dissociation between actions and responses which AD observed in his own behaviour was also manifest in that of our rats.

We argued that the failure of the aversion to be manifest directly in AD's instrumental actions arose from his ignorance of the aversion. If this is also true of our rats, giving them the chance to learn about their aversion should produce a very different pattern of lever pressing on test. Consequently, a second pair of groups (IMM/SUC and DEL/SUC) were given the same training as the previous animals except for the fact

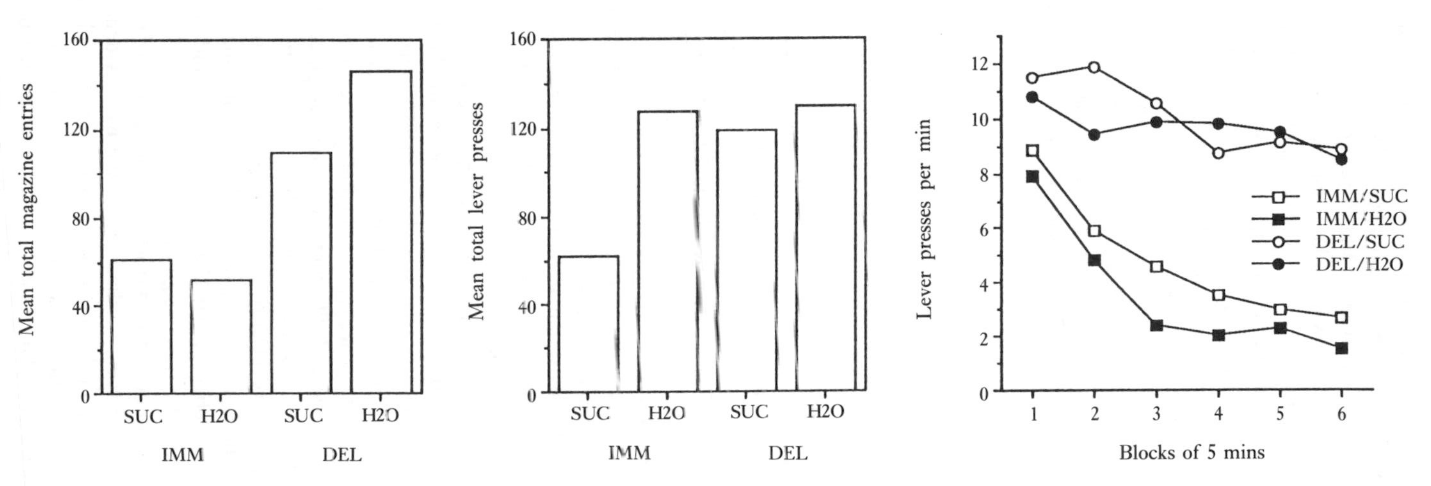

Figure 12.2 The left-hand and middle panels show the mean total magazine entries and lever presses, respectively, during the extinction test. The right-hand panel shows the mean rate of lever pressing during the reacquisition session. Illness was induced immediately after the instrumental training session in the IMM condition but after a delay in the DEL condition. All animals were exposed to either the sugar solution in condition SUC or water in condition H20 prior to the extinction test.

that they were allowed to contact the sugar solution rather than water on the day prior to the test for lever pressing. This experience should have allowed the animals with the aversion (IMM/SUC) to discover their aversion and therefore to refrain from pressing the lever on the test the next day. This is just what happened; these animals pressed significantly less than the other three groups on test (see middle panel of Figure 12.2).

This pattern of results poses considerable problems for any integrated theory of actions and responses, such as that espoused by the associative–cybernetic model. At the very least, the dissociation between magazine approach and lever pressing suggests they must be mediated by different goal 'representations', a difference that cannot be handled simply by positing separate goal elements in the associative memory. Such an account would have to give a principled account of why it is that only goal 'representations' mediating responses make direct contact with processes underlying aversion learning. At present, there is simply nothing in associative theory that independently motivates this distinction.

4 Basic desires

The dissociation follows naturally from a dual psychology, however, one for actions and the other for responses; indeed, it was predicted on the basis of the dual psychology implicit in the 'folk' interpretation of AD's experience. In general, the associative theory seems adequate for responses; the problem lies with actions. As far as we know, the only other explanatory candidate that we have for action is the intentional theory of belief–desire psychology. In one form or another, we interpret this as arguing that the antecedents of an instrumental action are a belief about the outcome of the action and a desire for that outcome, which operate through a process of practical inference to generate an intention to perform the action (e.g. Heyes and Dickinson, 1990). Such a dual psychology, the associative for responses and the intentional for actions, then allows us to make the required distinction between the 'representations' of the goals: in the associative case it is simply the activation of the appropriate elements in the agent's associative memory, whereas in the latter it is the content of a desire when participating in the practical inference for the action.

Given this characterization of goal 'representations', an interpretation of the outcome of our devaluation study is that intentional goals, or in other words the content of desires, have no direct contact with the processes that actually determine whether a particular stimulus or event is desirable, namely whether the stimulus, either directly or via associative links, activates the reward (or punishment) elements in the reinforcement system. Rather, the intentional goals of actions have to be acquired through self-observation or, more specifically, through acquiring beliefs about the attractive or aversive properties of stimuli and events through observing one's affective reactions to them. In other words, intentional goals are established by experiencing the psychological consequences of the interaction of our associative memory and reinforcement system when exposed to events of affective significance.

The outcomes of reinforcement devaluation studies by themselves, of course, do not supply a sufficient foundation for such a radical claim. We have evidence, however, that the claim is also true of the control of intentional goals by biologically determined motivational states, such as hunger. In the context of primary motivational states, this theory of desire predicts that agents should not know whether a particular stimulus, such as a specific food, is an intentional goal in a given motivational state, such as hunger, unless they have previously experienced their reactions to the stimulus in the relevant state.

This prediction has been tested in our laboratory (Balleine, 1992) by training rats that were not explicitly deprived of food to press a lever for food pellets (whose composition and taste differed from the animals' maintenance diet). Rats that are not deprived of food and therefore in a relatively low state of hunger will learn to press for food, although of course they do not perform vigorously. Performance of this action was then assessed in the absence of any reward (i.e. in extinction), when one group was deprived of food and therefore in a relatively high state of hunger (HIGH/0), while the other remained in the low state (LOW/0). As can be seen from the middle panel of Figure 12.3, the motivational state at the time of testing did not have any detectable effect on lever pressing; if anything the non-deprived animals tended to press more than the hungry ones, although this difference was not significant. The failure of the motivational state to affect performance was not due to the fact that the hunger state has no effect on performance. During a subsequent reacquisition test in which lever pressing once again produced the food pellets, the hungry animals (HIGH/0) pressed consistently more than the non-deprived rats (LOW/0). Performance during this reacquisition test is illustrated in the right-hand panel of Figure 12.3.

This pattern of results accords with the theory of desires based on the goal devaluation studies. A shift from a low to a high state of hunger did not appear to enhance our animals' desire for the food pellets in the absence of any prior experience with this incentive when hungry. Of course, if they received such an experience prior to testing, then we should expect the hungry animals to press more on test, a prediction tested in a second pair of groups. These groups (HIGH/PEL and LOW/PEL) received identical training and testing to the other groups except that prior to any instrumental training they were allowed to eat the reward pellets freely while hungry. Consequently, these animals had the opportunity to discover their reactions to the pellets when hungry. As the middle panel of Figure 12.3 shows, a very different pattern of results emerged when these groups were tested in extinction in either a high or low motivational state. Whereas the group tested in the low state (LOW/PEL) pressed at the same level as the previous pair during extinction, the animals that were hungry on test (HIGH/PEL) pressed significantly more than the remaining groups. Prior experience of consuming pellets while hungry had established for these animals a desire for this incentive when in a state of hunger, a desire manifest in their instrumental action.

Finally, and most importantly, it should be noted that the same dissociation between actions and responses that was observed following goal devaluation is also seen following motivational shifts. The left-hand panel of Figure 12.3 shows the magazine entries during the extinction test. For this response it is clear that the modulation of performance by the current motivational state does not depend upon prior experience with the incentive. Both groups of animals that were hungry on test (HIGH/0 and HIGH/PEL) showed elevated magazine entries relative to the non-deprived animals (LOW/0 and LOW/PEL). Moreover, in contrast to lever pressing, the degree of elevation was unaffected by whether or not the animals had previously consumed the pellets while hungry.

A parallel dissociation also occurs when animals are trained hungry, but not deprived prior to the extinction test (Balleine, 1992). Only rats that have previously experienced the pellets while in the low state of hunger show a reduced level of lever

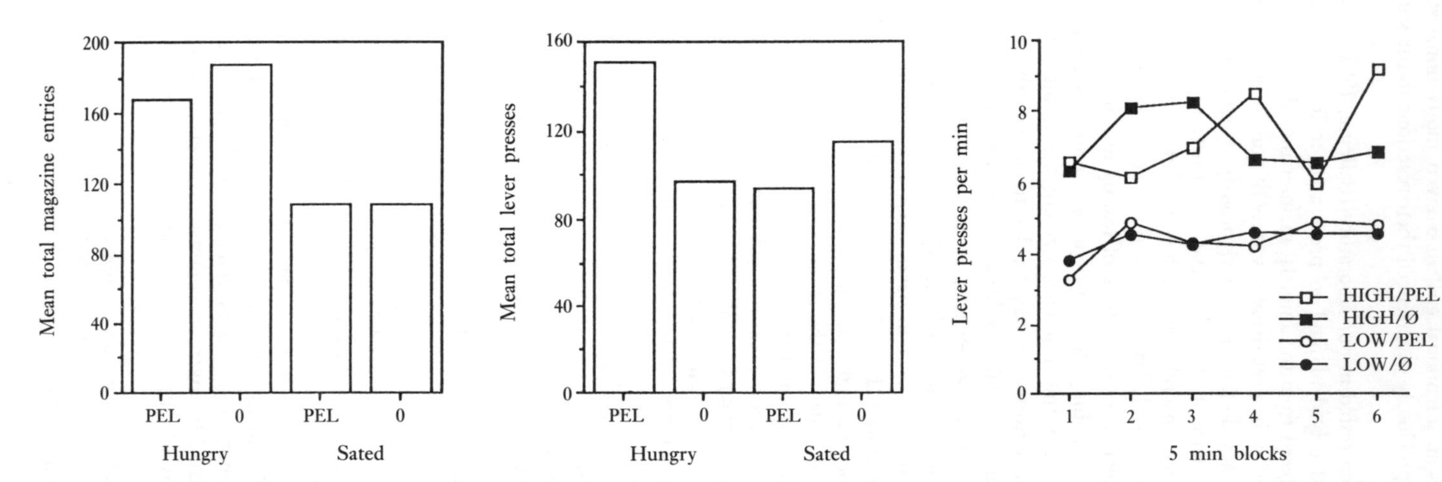

Figure 12.3 The left-hand and middle panels show the mean total magazine entries and lever presses, respectively, during the extinction test. The right-hand panel shows the mean rate of lever pressing during the reacquisition session. The animals were tested following either food deprivation in condition HIGH or no deprivation in condition LOW. Pre-exposure to the food pellets following food deprivation was given in condition PEL but not in condition 0.

pressing when tested in this state. By contrast, magazine entries are lower following a reduction in hunger irrespective of whether or not the animals have previously made contact with the incentive in the non-deprived state. This direct effect of primary motivational state on a response is in accord with the associative–cybernetic model if it is assumed that the ability of the element 'representing' the food pellets in the associative memory to activate the reward element of the reinforcement system is modulated by the agent's deprivational state. Such an assumption has long been the common explanation of motivational control of conditioned responses (e.g. Konorksi, 1948).

The dissociation between lever pressing and magazine entry under motivational shifts reinforces the conclusion that actions and responses are mediated by different 'representations' of the goal. Moreover, if we endorse an intentional account of actions, the present findings have important implications for the genesis of desires. What they suggest is that desires or intentional goals have no direct contact with the mechanism by which primary motivational states determine our responses and reactions to different stimuli. Rather, agents appear to have to learn about the value of particular commodities in a given motivational state through experience of their reactions to these stimuli while in that state. In other words, once again basic desires appear to be derived from beliefs based on self observation.[2]

Conclusions

We have argued that the status of a particular activity should be evaluated by two teleological criteria which can be assessed empirically by the omission and goal-devaluation procedures. Only if performance is sensitive to both the instrumental contingency and the current affective or motivational 'value' of the outcome should the activity be regarded as a goal-directed action; insensitivity to either of these factors indicates that the activity is a simple response.

The origin of the processes differentiating actions from responses must lie within the psychology of the agent, and an integrated, mechanistic psychology was offered in the form of the associative–cybernetic model. Within this model, actions are distinguished from responses in terms of whether or not an activity has a 'representational' element in the associative memory. The problem with the associative model lies with its assumption that actions and responses are mediated by common goal 'representations'. This is not so, however; actions and responses can be dissociated under goal revaluation brought about by both aversion training and shifts of primary motivational states.

It was for this reason that we argued for a dual psychology which retained the associative model for responses but attributed goal-directed action to intentional processes. We assume, moreover, that these two psychologies are interfaced by basic desires that have their origin in the agent's perception and consequent beliefs about his or her reactions to stimuli and events in different motivational states. These reactions, in turn, are mediated by the type of mechanistic processes identified by the associative model. Of course, our argument for a dual psychology is not one of principle, and it may well be the architecture and processes of the associative–cybernetic model or its brethren can be elaborated and developed to encompass the dissociation between actions and responses. All that we can do at present is to evaluate the evidence against the best integrated associative model available. Whether or not a mechanistic account of goal-directed action can in principle be given at the psychological level must remain an open question.

So, in the end, this long and tortuous line of argument and experimental evidence has yielded a conclusion that we could probably have found within our 'folk' psychology. Indeed, it turns out that we could have gone directly to that fount of all psychological wisdom, 'Peanuts'. In one episode, Snoopy is aroused from his repose by the sound of the can opener and bounds to his food bowl, just as Rolly does. During his brief journey he reflects upon the fact that as soon as he hears the opener his 'stomach knows that supper is coming' and is puzzled about how the ears could possibly 'tell the stomach'. Snoopy's puzzlement clearly manifests a dual psychology in that his gastric sensations, elicited associatively by the sound of the can opener, are the object of his intentional reflections. Snoopy is puzzled about the operation of the associative mechanism mediating his gastric responses in a way that he could never be, about why he rushes off to his food bowl when he hears the can opener. As an apparent goal-directed action, this would be just a rational consequence of the belief that approaching his food bowl when he hears the can opener will yield food and his desire for food at that time. Indeed, perhaps the only substantive point that this chapter adds to Peanuts' analysis is to suggest that, if Snoopy is anything like Rolly or our rats, anticipatory approach should probably not be an intentional action for Snoopy, but rather an associatively elicited response. So, in fact, Snoopy should be just as bemused about why he ends up at his food bowl when he hears the can opener as he is about his gastric sensations. But surely we can allow Peanuts this one anthropomorphic error.

ACKNOWLEDGEMENTS

The research reported in this chapter was supported by grants from the SERC and Commonwealth Scholarship Commission.

NOTES

1 This pattern of feedback from the reinforcement system is appropriate for appetitive responses that are directed towards the stimulus, such as 'approach'. The behavioural repertoire represented within the S–R store and motor system should also include defensive responses, such as 'withdrawal', for which the pattern of feedback is reversed. For such defensive responses, the punishment element exerts an excitatory influence on the respective motor elements and the reward element an inhibitory influence.
2 Of course, we (and possibly some other animals) can also acquire beliefs about the desirability of events on the basis of information from authoritative sources.

REFERENCES

Adams, C. D. 1982: Variations in the sensitivity of instrumental responding to reinforcer devaluation. *Quarterly Journal of Experimental Psychology*, 34B, 77–98.

Balleine, B. 1992: Instrumental performance following a shift in primary motivation depends on incentive learning. *Journal of Experimental Psychology: Animal Behavior Processes*, 18, 236–50.

Balleine, B. and Dickinson, A. 1991: Instrumental performance following reinforcer devaluation depends upon incentive learning. *Quarterly Journal of Experimental Psychology*, 43B, 279–96.

Davis, J. and Bitterman, M. E. 1971: Differential reinforcement of other behavior (DRO): A yoked-control comparison. *Journal of the Experimental Analysis of Behavior*, 15, 237–41.

Dennett, D. C. 1987: *The Intentional Stance*. Cambridge, Mass: MIT.

Dickinson, A. 1980: *Contemporary Animal Learning Theory*. Cambridge: Cambridge University Press.

Dickinson, A., Nicholas, D. J. and Adams, C. D. 1983: The effect of the instrumental contingency on susceptibility to reinforcer devaluation. *The Quarterly Journal of Experimental Psychology*, 35B, 35–51.

Frese, M. and Sabini, J. 1985: *Goal Directed Behavior: The concept of action in psychology*. Hillsdale, NJ: Lawrence Erlbaum Associates.

Heyes, C. and Dickinson, A. 1990: The intentionality of animal action. *Mind and Language*, 5, 87–104.

Holland, P. C. 1979: Differential effects of omission contingencies on various components of Pavlovian appetitive responding in rats. *Journal of Experimental Psychology: Animal Behavior Processes*, 5, 178–93.

Holland, P. C. and Straub, J. J. 1979: Differential effects of two ways of devaluing the unconditioned stimulus after Pavlovian appetitive conditioning. *Journal of Experimental Psychology: Animal Behavior Processes*, 5, 65–78.

King, D. L. 1979: *Conditioning: An image approach*. New York: Gardner.

Konorski, J. 1948: *Conditioned Reflex and Neuron Organization*. Cambridge: Cambridge University Press.

Mackintosh, N. J. 1975: A theory of attention: Variations in the associability of stimuli with reinforcement. *Psychological Review*, 82, 276–98.

Mackintosh, N. J. 1983: *Conditioning and Associative Learning*. Oxford: Clarendon Press.

Miller, N. E. 1963: Some reflections on the law of effect produce a new alternative to drive reduction. In M. R. Jones (ed.), *Nebraska symposium on motivation*. Lincoln: Nebraska University Press, 65–112.

Mowrer, O. H. 1960: *Learning Theory and Behavior*. New York: Wiley.

Pearce, J. M. and Hall, G. 1980: A model of Pavlovian learning: Variations in the effectiveness of conditioned but not of unconditioned stimuli. *Psychological Review*, 87, 532–52.

Rescorla, R. A. and Wagner, A. R. 1972: A theory of Pavlovian conditioning: Variations in the effectiveness of reinforcement and nonreinforcement. In A. H. Black and W. F. Prokasy (eds), *Classical Conditioning II: Current research and theory*. New York: Appleton–Century–Crofts, 64–99.

Rumelhart, D. E. and McClelland, J. L. 1986: *Parallel Distributed Processing*, vol. 1. Cambridge, Mass: MIT.

Sheffield, F. D. 1965: Relation between classical conditioning and instrumental learning. In W. F. Prokasy (ed.), *Classical Conditioning: A symposium*. New York: Appleton–Century–Crofts, 302–22.

Sutton, R. S. and Barto, A. G. 1981: An adaptive network that constructs and uses an internal model of its world. *Cognition and Brain Theory*, 4, 217–46.

Taylor, C. 1964: *The Explanation of Behaviour*. London: Routledge and Kegan Paul.

Taylor, C. 1970: The explanation of purposive behaviour. In R. Borger and F. Cioffi (eds), *Explanation in the Behavioural Sciences*, Cambridge: Cambridge University Press, 49–79.

Wagner, R. 1981: SOP: A model of automatic memory processing in animal behavior. In N. E. Spear and R. R. Miller (eds), *Information Processing in Animals: Memory mechanisms*, Hillsdale, N.J.: Lawrence Erlbaum Associates, 5–57.

13
The integration of spatial vision and action

Bill Brewer

Spatial vision and action are normally extremely well integrated. So much so that William James' suggestion of a mysterious, immediate psychophysical link between them has some intuitive plausibility (James, 1890, ch. 26).[1] But this intuition is notoriously difficult to accommodate theoretically. In this chapter I bring together two important psychological insights into the subpersonal organization of visuo–motor control with more general philosophical reflection on its personal level structure, to get a better understanding of how the spatial components of the visual and motor systems interact so smoothly.

Inevitably, fitting a philosophical theory of action onto a pre-prepared, although extremely incomplete, subpersonal framework, raises the fundamental question of how these stories are related. So most of the second half of the chapter (II) is an attempt to solve simultaneously for the right, general conception of agency – what is involved in being an agent and how to characterize the essential nature of bodily action – and for the correct account of the relation between subpersonal and personal level explanation in this area, on the basis of the partial framework proposed in the first half (I).

I

1 Separate coding

On a traditional and perfectly natural way of conceiving it psychologically, the way in which basic spatial action is immediately, reliably controlled and co-ordinated by visual perception in accord with needs and preferences can seem quite baffling. A pattern of retinal stimulation is processed and transformed by the early visual system, generating a sequence of 'sensory codes' which are systematically sensitive to, or correlated with, various successively 'higher order' properties of the retinal image. A central executive system takes a privileged member of this sequence as input, and computes the appropriate 'act code', given the subject's motivational state and environmental context, as its output. This is initiated by the motor system, which involves a similar series of processing stages, terminating in a pattern of activity across individual motor units which issues in the required bodily movement.

The difficulty with this conception stems from an implicit assumption that these two sets of codes are utterly incommensurable at every level of processing. This assumption characterizes what Wolfgang Prinz calls the *separate coding* approach to perception and action (Prinz, 1990).[2] Afferent codes are systematically correlated with various complex properties of the retinal image, whereas efferent codes are correlated with properties of the motor units. They stand for completely different kinds of things and bear very different types of representational content, with different frames of reference, different metrics and different functional roles. These intrinsically unrelated codes can only be brought together by the establishment of brute, arbitrary *production rules*. There is no way of reading the 'right' act code straight off the input sensory code. And the traditional psychological problem of visuo–motor control is often seen as that of understanding or modelling the mysterious operation of the central executive system which forges these links.

Consider a very simple task like reaching out to grasp a small object on the basis of its seen size, shape and location, to see how this approach works in connection with the spatial aspect of vision and action. On the afferent side, the relevant spatial properties of the object and relations between it and the subject are given by the results of various transformations on the spatial organization of the retinal image, which can be seen as extracting its higher order features. On the efferent side, the spatial properties of the required movement are carried by patterns of neural activity which are nomologically correlated with various properties of particular assignments of activation intensity to individual motor units. At this level of analysis, the connection between them is *quite arbitrary*. The problem is to see what has to go on to translate the upshot of the early visual system's operation on a given pattern of retinal stimulation from an attractive environmental object, into an appropriate instruction to the motor system to generate a temporally extended pattern of muscle activity, which will transport the hand to just the right place whilst forming an appropriate grip, and close it around the object successfully in preparation for the next stage of the task.

We can think of afferent and efferent coding as ways of grouping token patterns of retinal stimulation and motor unit response respectively into types, which bring out certain kinds of similarity and difference between them. And the definitive feature of separate coding is that the principles governing this typing by the visual and motor systems are completely different. The operations by which afferent tokens are treated as of the same, similar or different spatial types are very unlike those governing the classification of efferent types. Their respective dimensions of similarity and variation are quite unrelated. So the extraction of the spatial significance of visual input for motor control requires a complex translation between disparate types of spatial content. Very crudely, the relation between the classification of spatial relations by the visual and motor systems is akin to that between the visual classification of colours and the aural classification of sounds: in effect, arbitrary.

2 Objections

An initial cause for concern about this separate coding approach is the serious difficulty in specifying precisely the nature of the codes between which the central system establishes production rules. At *which* stage in the processing of perceptual input is the afferent code appropriate as input for the central translation and choice procedure; and what exactly is the content of the efferent code which emerges? In Marr's terms (Marr, 1982), is it the pattern of retinal stimulation itself, the primal sketch, 2.5D sketch or 3D model which is used to generate a range of possible or suitable motor programs? Similarly, is this range of act codes, from which the most appropriate is to be selected, given at the level of efferent nerve impulses, individual muscle contractions and extensions, muscle group configurations, forces to be exerted, displacements, velocities etc. of limb movements, perhaps involving a contingent sensitivity to certain closed-loop feedback information, or what? There seems to be no consensus here; nor has any particular selection produced any great success (Prinz, 1990, p. 170). Perhaps the controversy and failure of specificity over this issue is really just an unfortunate consequence of adopting an inappropriate framework for understanding visuo–motor integration.

More importantly, Prinz has advanced a whole barrage of experimental grounds for doubting the possibility of a complete and satisfactory account of visuo–motor control within this separate coding framework (Prinz, 1990). His comprehensive review of the evidence from choice reaction-time tasks highlights a well-known dichotomy in such tasks between those in which appropriate responses are to some extent spatio–temporally compatible or isomorphic with their environmental stimuli – *matching tasks* – and those in which the spatio–temporal aspects of the stimulus-response relation are arbitrary – *mapping tasks*.

Matching tasks are significantly easier and faster than mapping tasks, and prior specification of response alternatives is quite irrelevant for matching latencies but improves mapping performance. These results are extremely robust and emerge in almost every kind of reaction task. This is embarrassing for the separate coding theorist, according to whom all possible afferent-efferent assignments are on a par: isolated and quite arbitrary. The matching/mapping distinction should therefore be insignificant to choice RT performance.

It is tempting to appeal to past experience and learning to account for this general dichotomy in performance within the separate coding framework. But the move is either wildly implausible or self-defeating. As it stands, it is wildly implausible because production rules linking afferent and efferent codes are isolated individuals. Whatever mode of characterization is chosen of the incommensurable input and output codes, the prior establishment of production rules between them is only relevant in connection with the very same pairs of codes. And that there should have been precisely the right learning opportunities in all cases where the distinction between matching and mapping is manifest is extremely unlikely.

The obvious amendment is to insist that the value of past experience of visuo–motor co-ordination spreads, as it were, along shared dimensions of similarity. A previous production rule linking afferent code a with efferent code e will not only be of use in appropriately implementing e in response to a, but it will also ease the integration of similar visual input with relevantly similar motor output. So the learning will facilitate the move from afferent code a', which differs from a simply in respect r, to efferent code e', which differs from e in respect r'. This depends on the existence of a simple, if not trivial, function between r and r': the extent to which the current suggestion is effective is precisely the extent to which this relationship is close. Yet the whole point of separate coding is precisely to deny this. The principles for classifying token patterns of visual input and motor output activity as of the same, similar or different spatial types are quite distinct and unrelated. Hence the required function between afferent code and resultant efferent code variation is a computationally complex, discrete, lawless sequence.

Invoking an amendment along these lines to account for the quite general matching/ mapping dichotomy in visuo–motor control performance requires a smooth, simple dependence of the spatial types of efferent codes upon those of the afferent codes driving them. This is incompatible with the definitive assumption of separate coding. Therefore any appeal to learning and past experience which rests on this kind of generalization by similarity is self-defeating for the separate coding approach.

The point here is fairly straightforward, and can be made more generally. The way in which past experience and learning bear on present and future competence with respect to visuo–motor control is inconsistent with the atomism of separate coding. Past practice contributes massively holistically. Basic visually guided spatial abilities are *structured* in Christopher Peacocke's semi-technical sense: a person's complex ability to act successfully on objects in his or her immediate environment on the basis of seeing their location, size, shape, orientation and so on, is not merely the sum of its individual instances but, rather, 'these component abilities are themselves the result of interactions between members of a smaller set of more powerful abilities possessed by the agent' (Peacocke, 1983, p. 62).[3] In other words, mastery of a novel visuo–motor skill – reaching out to pick up a glass one sees on the table to the right in front of one, for example – is not simply achieved by a one-off linkage between the particular input sensory code and required output act code, at whichever incommensurable levels these are supposed to be characterized. Rather, it is the bringing together of a package of far more general abilities, whose recombination with others yields widespread further competence in a whole range of suitably related tasks – reaching out for other seen objects on the table, perhaps with the other hand, or pushing the glass off the table, for example.

Similarly, failures in the integration of spatial vision and action are not isolated or piecemeal. A person's inability to reach out to pick up the glass on the table to the right in front of her is due to her general failure to see anything in that area of egocentric space, to visually pick out glasses at all, or to get her arm moving to that, or even any peripersonal place, or perhaps due to a more global communication breakdown between the visual and motor systems. So any particular such failure has numerous related ramifications. It is not merely the destruction of a single, isolated and unconnected production rule.

Furthermore, and this is the crux, the separate coding theorist's attempt to mimic this spatial structure in visuo–motor abilities depends on the existence of a simple, or trivial identity function relating the visual system's mode of classification of token patterns of stimulation into spatial types with that by which the motor system treats token bodily responses as spatially of a type, similar or different. But this is precisely what the separate coding approach denies. Hence, the attempt is self-defeating.

3 *Common coding and spatial content*

Clearly, the relation between afferent and efferent codes is not at all arbitrary in the separate coding sense. Both the visual and motor systems are ultimately sensitive to the very same considerations in spatially typing token patterns of activity. Their respective dimensions of similarity and variation are the same, or at least very simply related. This is the basic thought behind Prinz's *common coding* approach to perception and action, in which afferent and efferent codes share the very same type of spatial content at some privileged level. So there is no need for any kind of homuncular translation and choice procedure. The problem of modelling the extremely complex computational activity of a central system in its construction of appropriate production rules linking intrinsically unrelated percept and act codes is finessed, by conceiving of appropriate pairs of such codes as sharing a single type of content for the representation of spatial properties and relations. The paradox of defining the undefinable operation of an executive system's generation of patterns of motor unit activity from patterns of retinal stimulation on the terms laid down by separate coding assumptions is revealed as an artifact of a misleading way of setting up the issue.

Of course, this common coding framework also has the virtue of predicting precisely the matching versus mapping dichotomy which is embarrassing to its traditional separate coding opponent. In the former type of task, but not in the latter, 'percept and act codes will partially coincide', as Prinz puts it (1990, p. 173). This explains both the superior processing speed as compared with mapping tasks, and the fact that latencies are unaffected by advance specification of alternatives: the spatial parameters of the response are read directly off the visual representation of the stimulus. No higher level translation or choice operation is required: representations of the visual and motor systems are fit to engage directly. So visuo–motor control has no need of anything like executive production rule construction.

Simply pointing to this alternative to the traditional separate coding approach to visuo–motor integration hardly amounts to a satisfactory positive account though. Firstly, more needs to be said about the nature and content of common visual and motor codes; and secondly, extended comment is required on the relation between the common coding idea and the personal level psychological explanation of visually controlled and co-ordinated spatial action.

The essence of Prinz's proposal is that a crucial component in visual processing codes an object as 'at *p*', say, in precisely the same way as the motor system classifies various bodily movements as 'targeted on *p*', or 'transporting the left hand to *p*', for example. That is to say, the dimensions and metrics with respect to which the visual registration of objects' locations at p and p' are treated as spatially similar and different, are the same as those by reference to which motor codes for various bodily movements to p and p' are compared. More generally, there is a privileged level of processing at which the places in a subject's environment, and the sizes and shapes of the things at them, are visually represented according to the very same principles of identification and discrimination as are employed in characterizing the spatial aspects of possible actions in connection with them. There is a single type of spatial content common to both systems. So no complex computation is required to pair an appropriately spatially isomorphic act code with an incoming visual code to achieve successful reaching and grasping, say. The spatial parameters of suitable efferent signals can simply be adopted straight from the afferent signal. But what more can be said about this privileged level of common coding?

A first point to make here, is that the non-arbitrariness of the spatial aspect of afferent–efferent code relations which lies at the heart of the common coding story, appears only to be available on the assumption that what is represented in common really is distal, spatial properties of external objects and bodily movements, rather than merely proximal correlates of these: that they are represented *as spatial*.[4] No one particular distribution of nerve impulses to motor units, so described, is intrinsically more appropriate than any other in response to a given pattern of retinal stimulation, described, say, in terms of its binocular distribution of light intensity. Similarly, the dimensions of variation in retinal stimuli which correlate with relevant spatial properties of environmental objects are quite unrelated to those of patterns of muscular activity whose variation is similarly correlated with appropriate distal spatial features of active bodily movements. The spatial significance of the typing of retinal images as the same, similar or different, *qua* patterns of light intensity is completely different from that of the classification of efferent nerve impulses *qua* patterns of motor unit activity. The same goes for any pair of sets of proximally typed, visual and motor spatial codes: they relate arbitrarily.

Arm, wrist and hand movements transporting the hand to *p* whilst forming a grip of size *s*, on the other hand, are *obviously* what is required to reach out and grasp an *s* sized object at *p*. The afferent–efferent code relation for reaching out to grasp a seen environmental object will be non-arbitrary, indeed quite trivial, in this way, just if the codes represent genuine spatial properties of distal objects and distal spatial properties of bodily movements *as such*. Only given this truly spatial content, can the model of common visuo–motor coding be realized.

Recall our identification of visual and motor coding with ways of grouping token patterns of retinal stimulation and motor unit response into spatial types, exhibiting the various similarities and differences between them. The point is that this coding will be *common*, the principles of typing isomorphic that is, just if the grouping is carried out according to distal spatial properties of seen objects and actions on them. Treating the tokens as more or less alike precisely on the basis of similarities and differences in such distal features, rather than in any of their proximal correlates, means that spatial variations in visual stimuli can be matched immediately with appropriate modifications in required motor responses: there is a single type of spatial content common to both visual and motor codes, so no need for any complex translation and choice. Thus common and distal spatial coding go hand in hand.

The primacy of this distal spatial code individuation for understanding the integration of vision and action, and therefore indirectly the common coding framework itself, receive independent support from Nicolai Bernstein's Principle of Equal Simplicity, which is closely connected with our earlier considerations of structure and generality in visuo–motor control: 'for every system which is capable of undertaking a set of different elementary processes of a given range, the lines of equal simplicity correspond to those directions in this range along which movement does not involve any change either in the structural principles or in the principles of operation of the system' (Bernstein, 1967, p. 52). In other words, one ought to conclude from the fact that proximally highly varied basic bodily movements sharing certain distal spatial features are similar in 'speed of completion, degree of accuracy, degree of variance and so on' (p. 52), that such spatial properties provide the explanatorily relevant taxonomy of the movements, and thus form the basis for the individuation of the motor codes controlling them.

Other things being equal, the principles for collecting token bodily movements into spatial types which best mirror the structural functioning of the motor system are those which best follow the contours of *some* such simplicity measure – in the sense that the closer the tokens score on overall simplicity, the more alike their relevant spatial types. Hence, these are the principles by which the motor codes driving the movements should be characterized and individuated. Intuitively and empirically, they concern the distal, genuinely *spatial* features of movements. So these are what enter into the contents of act codes.

Of course the details of the right measure of simplicity have to be argued. And these will surely include structural as well as purely temporal considerations – manner of variation from a paradigm exemplar, consistency and rhythm of execution and so on, are going to be at least as important as brute reaction time and performance time. But for now it is sufficient to press the intuitive idea that simplicity considerations of this general kind group token visually guided bodily movements as more or less alike according to their distal spatial properties rather than any of their proximal features. A large number of the more direct possible routes to reach out and touch a seen target with a given limb, although quite distinct in required patterns of motor unit activity or muscle innervation, would be grouped together. These would be closely related to similar movements to nearby targets with the same limb and to the same targets with different limbs and so on. Similarly, to extend an example of Bernstein's (1967, p. 54), the vast range of proximally quite unrelated movements involved in describing circles of varying sizes with an outstretched arm, at different speeds, carrying a range of weights, to one's front, side or above one's head, whilst standing, sitting or even lying down, 'are subjectively very much alike in terms of difficulty and objectively they display approximately the same amount of accuracy and variation' (p. 57), and their rhythmic structure is similar. Thus they are closely related in simplicity. Yet their unifying feature has nothing to do with their proximal patterns of muscle activity or innervation. It has to do rather with a shared distal spatial property. This then, provides their explanatorily relevant description, and hence the characterization of the motor codes guiding them. In other words, the spatial features of basic bodily acts are coded by the motor system in precisely these terms, rather than in terms of any of their proximal correlates. Motor codes represent distal spatial properties of bodily movements: their content is truly spatial.

To summarize the argument so far: the excellence of visuo–motor integration stems in a large part from the fact that visual and motor codes share a single type of spatial content; and this is genuinely spatial content, representing distal spatial properties of environmental objects and bodily movements as such.

4 *Biomechanics*

We can get further with developing the common coding insight by extending our detour via Bernstein's general approach to the problem of motor co-ordination. Our visually controlled and co-ordinated movements are those of a unified, harmonious machine. A person's individual movements must be counterbalanced by numerous others, so as to avoid their knocking her out of kilter in any way. Indeed, a voluntary arm movement to reach out and grasp an attractive object nearby say, is *preceded* by anticipatory, stabilizing muscle activity and movement in the legs and trunk (Belen'kii, Gurfinkel and Pal'tsev, 1967). In general with bodily movements, a whole range of prior and concurrent adjustment and compensation is extremely important for smooth, trouble-free performance (Pribram, 1971, p. 225; Turvey, 1977, pp. 230–5). This need is extended in response to practically inevitable, occasional, unexpected perturbations *en route*. The ideal compensation often requires adjustments in various elements of the motor apparatus other than those to which the perturbation is applied. This has been shown to be achieved (Abbs and Gracco, 1983). Similarly, the overall course and structure of such movements and counterbalancing compensations must be sensitive to the arrangement, extent and flexibility of the body. Its evolving configuration should not only preserve the agent's balance and stability throughout, but also maximize comfort and ease of performance. Balance must not come at the expense of awkward or even attempted impossible combinations of limb positions.

Bernstein's work is an attempt to make this mechanical–physiological achievement of unity, harmony and coherence intelligible, to understand how it is realized. Tuller, Turvey and Fitch capture the flavour of his approach as follows.

> The puzzle of the control and coordination of movement as seen by Nicolai Bernstein can be expressed succinctly: How can the many degrees of freedom of the body be regulated systematically in varying contexts by a minimally intelligent executive intervening minimally? A reasonable hypothesis is that nature solves this puzzle by keeping the degrees of freedom individually controlled at a minimum, and by using 'units' defined over the motor apparatus that automatically adjust to each other and to the changing field of external forces. (Tuller, Turvey and Fitch, 1982, p. 253)[5]

There are a massive number of motor units all over the body, which, along with the external forces acting on it, jointly determine its configuration and movement over time. Computing their required activity *individually*, in accord with current preferences in the perceived environment and as constrained by the various considerations of unity, harmony and general smooth running, is practically impossible in real time. Bernstein's thought here, is that we finesse the need for any such unwieldy central computation for each single motor unit, by operating with much larger, internally regulated, self-correcting basic functional units of motor activity. These *biodynamic structures* are complexes of linked and counter-balanced muscle groups spanning several joints, which are tuned and constrained to operate as a whole. On seeing an attractive object nearby, and in the absence of any competing, preferable plan of action, the relevant spatial parameters are simply transferred directly from the visual system to the appropriate biodynamic structure(s), which run(s) off the desired movement pattern for reaching out and grasping it. In this way, given a basic motivational state, and grasp of relevant wordly regularities, the muscle linkages are driven directly by visual perception of the environment. An important component in the successful achievement of simple spatial action in the world, is the possibility of the immediate parameterizing of bodily biodynamic structures by the visual system.

So Bernstein makes sense of the unified, balanced harmony of co-ordinated bodily movement by replacing the idea of an executive calculation of very many, intrinsically arbitrary, individual efferent assignments from afferent codes, with a picture of autonomous self-regulation on the basis of large scale functional constraints between many motor units, which automatically maintain the body's equilibrium during movement.

Now the intermuscular constraints which build biodynamic structures by reducing degrees of freedom amongst individual motor units produce mechanical systems with important properties in common with mass-spring balances, or perhaps more precisely, with mutually synchronizing limit-cycle oscillators. The 'entrainment', or tendency to global resonance, exhibited by connected systems of the latter, is surely important in understanding the automatic co-ordination of different biodynamic structures, which unifies and stabilizes bodily movement as a whole. More importantly for the current discussion though, is the 'equifinality' characteristic of all these mechanical systems, which underwrites our rationale for individuating act codes in terms of the distal spatial properties of the movements they bring about, regardless of the multiplicity of possible proximal realizations of these movements. For we have continually seen the importance of this distal individuation to the smoothness and success of visuo–motor integration. Common and distal spatial coding go together; and Bernstein's account of unity and harmony helps us start to see how these might be implemented.

A simple mass-spring system has a stable equilibrium point which depends on the elasticity of its spring and the mass of its load. When perturbed in any way, within limits, or released from a held position, it will return to this point. It is drawn to it, so to speak, from all directions, by any number of routes. Similarly, a suitably parametrized biodynamic structure is 'set' for a movement with the given distal spatial properties characteristic of its equilibrium, regardless of its starting configuration, particular route and minor perturbations *en route*. In executing this movement on any given occasion of course, it carries out a token movement equally describable in numerous different ways: in terms of individual motor unit activities, single muscle contractions and extensions, limb segment velocities and displacements and so on. But it performs a movement with those descriptions only because they capture what is more proximally required for a movement of the characteristic distal spatial type on that occasion: a movement taking the relevant body part to that egocentric location, say. In thus parametrizing physiological biodynamic structures therefore, the motor system is operating with codes with genuinely spatial content.

There is a substantial body of experimental evidence that Bernstein's picture captures significant features of our own motor co-ordination. For example, Kelso (1977) reports that blindfolded subjects, deprived of any position sense information from muscle spindles and joint and skin receptors, are very accurate at reproducing the final location of an initially actively moved finger when this is passively returned to different starting positions. In contrast, subjects have difficulty reproducing the distance covered by the initial movement from new starting positions. On the assumption that a voluntary finger movement to a freely chosen position is achieved by parametrizing a mass-spring-like biodynamic structure – setting its equilibrium – this is exactly the result one would predict. Similarly, Fel'dman (1966) shows that unexpected, last minute perturbations in the starting conditions of forearm movements – that is, alterations in initial 'values of the joint angle and its derivative' (p. 771) – have no significant effect on subjects' accuracy in actively setting a steady target joint angle.[6]

In this Bernsteinian context then, the common coding insight can be put as the idea that appropriate, distal visual representations immediately provide the spatial parameters for the motor system to distribute amongst the relevant biodynamic structures, to achieve the required goal movement.[7]

II

Personal level psychology

When we turn our attention to the personal level structure of visually guided spatial action – the structure reflected in explanations which make such behaviour intelligible as being more-or-less as it rationally ought to be, given our ascription of various folk psychological properties to the agent herself: perceptions, beliefs, desires, intentions and so on – we find striking similarities with this subpersonal picture. So it is tempting to propose some kind of identification between the central, common spatial coding system and the relevant aspects of the subject's mind. Such an identification is, I think, untenable; and we should explore the possibility of an alternative conception of the relation between subpersonal processing and the personal level.

1 Superimposition

Moving around in her environment, a person sees an attractive object nearby, just to the right on the table in front of her, and she intentionally reaches out there to grasp it. Implicit in this crude folk psychological account, is a parallel of the Prinz–Bernstein package of common, distal spatial coding facilitating the smooth integration of vision and action. We see things as at various places around us, egocentrically identified in precisely the way required immediately to satisfy our intentions to act with respect to them. Christopher Peacocke makes this explicit in formulating the following perfectly standard piece of practical reasoning.

[A person] forms the intention with this content:
  (1) I will move my arm in the direction of that tree.
He also knows from his perceptual experience that
  (2) That tree is in the direction *d* (identified egocentrically, using scenario content).
So he forms the intention with the content
  (3) I will move my arm in the direction *d*.

He can then carry out this intention without further practical reasoning. This description makes it clear that the connections between perception and action rest on two links. The first is the link between the perceptual demonstrative 'that tree' and the availability of the perceptually-based knowledge of (2) which contains that demonstrative way of thinking. The second is the link between the egocentric mode of identification of the directions and the subject's 'basic' actions. If either of these links does not hold for some other mode of presentation in place of 'that tree', such connections between perception and action will not hold, *ceteris paribus* (Peacocke, 1992, pp. 131–2).

That what are represented here are egocentrically specified directions – *direction d* – (and similarly in related cases, distances in those directions), illustrates the distal, genuine spatiality of the coding. That the very same way of identifying these distances and directions is used in both perception and 'final' intention formation, where the latter connects immediately with the subject's capacity for basic spatial action, according to Peacocke's essential second link, illustrates that this coding is common. Personal level vision and action share a single type of distal, truly spatial content.

Thus, it is extremely tempting to make a superimposition of this personal level, folk psychological picture upon the subpersonal account outlined above: to identify the mental realm of intentional explanation with the central common coding system as this is realized in the brain.[8] As Jennifer Hornsby points out (Hornsby, 1986, pp. 95–6), two further considerations combine to make this kind of superimposition almost unavoidable. Firstly, there is the idea that the physiological systems which realize subpersonal common coding, most notably the brain, are parts of a person 'whose proper functioning is a necessary condition of that person's having the effects on the world she desires to have' (p. 96): they are parts of an agent whose operation is essential to her successful agency. Secondly, there is the thought that the casual chain leading from a person's psychological states to her intended worldly effects is bound to pass through the very biomechanical bodily events which are programmed by the common coding system. On pain of unacceptable overdetermination, and particularly in combination with the first idea above, this thought leads immediately to a superimposition of personal level psychology onto subpersonal level common coding onto neurophysiology.

Assume that there is, at least in principle, a complete causal story at the subpersonal level, which fully explains how physically realized common codes parameterize biodynamic structures. This follows almost by definition from the working conception of the subpersonal level as a closed and complete level of causal explanation. Then the idea that mechanical movements of the body are caused both by suitable mental events of intending or willing and neurophysiological events realizing subpersonal common coding, forces a choice between superimposition and a view on which an agent's active movements would equally have occurred even in the absence of her intending or willing them. 'This we certainly ordinarily take to be false, and it is not clear why we should change the belief' (Peacocke, 1979, p. 135; see pp. 134–43 for the necessary amplification of this argument). So the collapse of personal level psychological explanation into the subpersonal level seems inevitable. To quote Hornsby again on precisely this collapse:

> The dependence of the person's functioning on the [common coding] functioning of her brain may make one think of the brain as a mechanism inside the person which is responsible for producing the effects in virtue of which she has her distinctive effects on the world. But then the common properties of the brain's states and of the person's mental states – states of each sort being seen as causal intermediaries – may make one think that in placing the brain inside the person one locates the propositional attitude states there. Many will therefore feel compelled to say that particular beliefs and desires [and intentions and acts of will] *are* the neurophysiological states of a person. (Hornsby, 1986, p. 96)

Very crudely, what this amounts to in our current context, is the thesis that the brain *is* the common coding system, *is* the agent's mind. More precisely, the upshot is some form of narrow token identity theory, on which the relevant mental antecedents of bodily action are identified, as event tokens, with the neurophysiological events which bring about biomechanical bodily movement.

In essence then, the result of superimposition is a view on which the subpersonal visual system delivers (normally conscious) distal, spatial perceptions to the subject, which then, in the context of relevant beliefs and desires, transfer their spatial content immediately to the appropriate intentions (and resultant acts of will), which in turn directly drive the Bernsteinian bodily machine (see Figure 13.1). The view shares an important feature with William James' theory of ideo–motor action (1890, pp. 522–8). Although it allows room for the very acts of will which James is explicitly concerned to deny here, it embraces the notion of a causal explanatory connection between personal level, psychological items such as ideas of movement, intentions or willings, and *quite independent*, subpersonal, physiological movements in bits of the subject's body. It is precisely this proposed *psychophysical* link between intention or will and bodily movement, the idea of relevant mental states as prior, distinct and independent, *mere causes* of appropriate movements of an agent's body, which rules both out as adequate accounts of visually guided agency. James' plausible intuition of immediacy, which is played out here in the conjunction of Prinz's common coding, Bernstein's

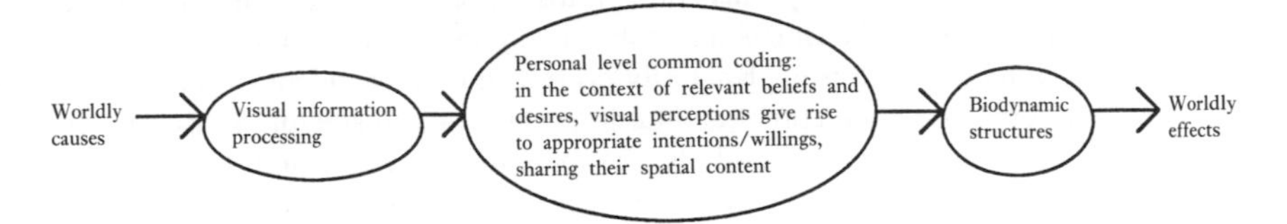

Figure 13.1 The result of personal level/subpersonal level superimposition for the integration of spatial vision and action.

autonomously self-regulating biomechanics and our current superimposition of personal level psychological explanation upon the underlying neural processing, can nevertheless be respected without appeal to this problematic conception of the relation between personal and subpersonal level explanation, as I shall try to show towards the end of this chapter.

2 *Objections*

First though, to bring out the cluster of related difficulties with the superimposition view. These are motivated by consideration of the picture of purported agency which Wittgenstein is characterizing in the following passage from the *Tractatus*.

> Even if all that we were to wish for were to happen, still this would only be a favour granted by faith, so to speak: for there is no *logical* connection between the will and the world, which would guarantee it, and the supposed physical connection itself is surely not something that we could will. (Wittgenstein, 1961, § 6.374)

Superimposition produces a similar situation. Visually guided spatial behaviour becomes more like a mentally induced reflex. For little sense can be made of the subject's being in control of what is going on, of its being something she is doing. She looks around, thinking about what would be desirable, useful, worthwhile and so on. She immediately perceives what should be done, intends (and wills) it to happen. Then things are in the lap of the gods. Either she strikes lucky, and her wish is granted, as if by magic as far as she is concerned; or she draws a blank and nothing happens at all. When she is successful, this just seems like a happy accident. So the idea of her as *agent* is completely out of place.

According to the current picture, a visually guided spatial action involves a mental event of intending or willing – which takes its content immediately from the subject's perceptual experience of the world around her, in the context of related beliefs and desires, and is identical with some complex neural event – causing a bodily movement event. Neither is essentially an *active* bodily movement of that kind: the agent's moving her body in that way. For the latter, biomechanical event might have occurred quite inactively, in the absence of the right mental antecedents; and, in a way more importantly, the former intending or willing might have been totally unsuccessful in bringing about any appropriate bodily effect.

What we have here then, is an analogue of a *common element* causal theory of perception (Grice, 1961; Pears, 1976; Peacocke, 1979). The crucial feature of such a theory is the idea that for a subject *S* to be perceiving an object *o*, is for *S* to be in some *e*-type psychological state (experientially thus and so), tokens of which are common between genuine perceptions of *o*-like things and subjectively indistinguishable illusions, and the characterization of which is explanatorily prior to and independent of the question whether its occurrence is perceptual or illusory, which is also suitably caused by the presence of *o* in his environment. Similarly, the consequence of superimposing personal level psychology upon subpersonal common coding is a view on which an agent *A*'s actively moving limb *l* is just her being in some *w*-type psychological state (willing thus and so), tokens of which are common between genuinely active movements of *l* and subjectively indistinguishable attempts (in which *l* is completely paralysed say), and the characterization of which is explanatorily prior to and independent of the question whether its occurrence is physically active or inert, which also happens suitably to cause an appropriate movement in *l*. An agent's psychological state is quite neutral on the question of her successfully moving her body or not. To adapt an expression of John McDowell's (1982, p. 457): for all *she does*, her limb might not move. Thus, it is difficult to make sense of the idea of her as the *agent* of her bodily movements.

Of course, activity requires both neural and bodily events in suitable causal relation. And a failed attempt may on occasion be subjectively indistinguishable from successful active bodily movement. Being subjectively certain one has tried or willed in the normal way does not guarantee success. Nevertheless, a successful willing attempt is essentially an active bodily movement, not a combination of two inactive events. Only given this logical connection between the will and the body in active bodily movement, does the idea of physical agency ultimately make sense. What we need here then, is something more like an analogue of the disjunctive account of perceptual experience. On such an account, cases of things experientially seeming thus and so to one fall into two *psychologically* quite distinct classes: those in which things' actually being thus and so is 'within the reach of one's subjective access to the external world', and those which are 'mere appearances' (McDowell, 1986, p. 150; see also Hinton, 1973; Snowdon, 1981; and McDowell, 1982). In contrast with common element views, these are intrinsically distinct, and not to be thought of simply as different worldly embeddings of the very same *e*-type mental state. The basic idea in our connection, would be that cases of *A*'s willing *l*-movement also fall into two psychologically quite distinct classes, and should not be thought of simply as different worldly embeddings of the very same *w*-type mental state. They are either cases of *A*'s actively moving, or merely her failed attempts at *l*-movement.[9]

The psychological transmission of spatial content from visual perception to intention and will is, when assimilated to the role of the central, subpersonal, common coding system, just one possible cause among many, of the body's biomechanical operation. And although it is not an unfamiliar view (apparently held in some form or other by Descartes, 1954; Locke, 1975; Hume, 1975, 1978; Davidson, 1980; and most modern functionalists),[10] the idea that such physical, or actively neutral, bodily movements as result qualify as actions just in case their cause is the right one – the appropriate intention or act of will – reduces successful agency to fortuitous wishful thinking of the kind Wittgenstein describes.

The same goes for the related view, on which the action itself is identified with the purely mental act of will, the common element between actively moving and failed trying, perhaps also providing this has the right physical effects (in his own way, Berkeley, 1965; cf. Taylor, 1985; Hornsby, 1980; Smith, 1988). Within the superimposition context, this mental art is presumably identified in turn with the central subpersonal act code formation, or its execution. Then the difficulty is that bodily action is only quite incidentally related with the appropriate movements of the agent's body. All we ever *do* is move our purely mental wills: any *bodily* movement is out of our control. The Wittgensteinian picture applies again; and again this distances the view from genuine agency.

On each account, agency is presented as something like reflective, mental manipulation of what happens to be one's body, as an object: the object which mysteriously obeys one's will. But in actively moving one's body the will flows into the movement itself. These are not two independent, causally related events. One does not move oneself, so to speak, but moves, when successful, just like that; or occasionally, one tries and fails. Yielding to the current temptation distances the subject from movement in her body in such a way as to threaten her status as its agent.

I shall briefly develop three further ramifications of these superimposition views, which compound their failure faithfully to represent visually guided spatial action.[11]

2.1 Agent's knowledge Firstly, as Charles Taylor notes (Taylor, 1985, pp. 80–4), there is a difficulty in giving an adequate account of the special features of an agent's knowledge of what he is actively doing. The thought here is that

> We are capable of grasping our own action in a way that we cannot come to know external objects and events. . . there is a knowledge we are capable of concerning our own action which we can attain as the doers of this action; and this is different from the knowledge we may gain of objects we observe or scrutinize. (Taylor, 1985, p. 80)

But the current suggestion, that our bodily actions simply involve movements of our bodies suitably caused by the right type of mental event, faces a dilemma here. If what is to be known really is to include the physical activity itself, then the knowledge is on a par with mere observation. For the body's movement is 'an external event like any other, only distinct in having a certain kind of cause' (Taylor, 1985, p. 81). On the other hand, if the knowledge is to be special in some way, then all that can be known is the purely inner, mental cause, available with introspective privilege. In this case, agent's knowledge amounts to nothing more than subjective awareness of some intention, wish or will, the common element between physical action and failed attempt, logically independent of any actual movement in one's body. Once again we see the threat of Wittgensteinian caricature. 'One does not watch it [one's hand] in astonishment or with interest while writing; does not think "what will it write now?" But not *because* one had a wish that it should write *that*. For that it writes what I want might very well throw me into astonishment.' (Wittgenstein, 1980, § 267)

One does not come to know what one is actively doing by observation (see O'Shaughnessy, 1980, vol. 2, ch. 8). Nor does mere introspective awareness of one's intention or will constitute agent's knowledge, on the current view. For the will is logically detached from the body. That the latter obeys the former is psychologically quite incidental, and thus unable to underwrite this special epistemological grasp of one's own action.

In any case, a distinction between privileged, introspective knowledge of one's own mental states and defeasible knowledge of public states of affairs is not the right kind of epistemic contrast. One would like to be able to say that we have special knowledge of our intrinsically directed bodily acts themselves in virtue of our directing them, where this is as corrigible as any other knowledge: we try as best we can to characterize *what we are doing*, as opposed simply to *what is going on*. But this kind of contrast does not seem to be available on either version of the current view. For agency is reduced to the occurrence of two independent, causally related events, neither essentially active: an inner, mental occurrence, which is regarded as an action in virtue of its distant bodily effects; or a mere bodily movement, which might equally have occurred completely inactively, but which is regarded as an action in virtue of its peculiar mental cause. Yet what we need to make the intuitive contrast is an intrinsically directed event, flowing from initiating will into body, known through its agent's imprinting this direction on it, which gives it its identity and activity. In his closely related context, Taylor gets this just right.

> There can be two kinds of knowledge. One kind is gained by making articulate what we are doing, the direction we are already imprinting on events in our action. As agents we will already have some sense, however dim, inarticulate or subliminal, of what we are doing; otherwise, we could not speak of directing at all. So agent's knowledge is a matter of bringing this sense to formulation, articulation or full consciousness. It is a matter of making articulate something we already have an inarticulate sense of.
>
> This evidently contrasts with knowledge of other objects, the things we observe and deal with in the world. Here we are learning about things external to our action, which we may indeed act with or on, but which stand over against action. (Taylor, 1985, p. 80)

2.2 Agent's ability A second difficulty concerns practical knowledge, in the sense of an agent's knowledge *how* to act. Superimposition reduces this to what we might call a *mere capacity*, mysteriously to get things done rather than to do them or control them as agent. All that we have so far, is the freedom, or impression of freedom, to intend or will, along with the reliable but fallible functioning of certain well designed bodily machinery, as an effect. But the will needs to be pointed in just the right direction, and to extend into the bodily movement as directed by the agent. The body's appropriate movement is itself a part of the agent's subjective extension into the world of material things. And again, this is far more than the incidental, lap-dog attendance of the right biomechanics to a neutral psychological common element between active success and failure.

This *genuine ability* for basic bodily action, to direct the will's extension into the body, is something I do not yet have in Wittgenstein's (1958, § 617) case of my fingers being crossed in a special way such that I am unable to move a particular finger when instructed to do so simply by your pointing at it, even though all the right machinery is ready and waiting, and I may be desperately willing it to move. More to the point, I am no better off here even if the right finger happens to move, magically as it seems to me, in response to my willing. The missing ingredient providing true practical knowledge, is the *control* I gain as soon as you touch the finger I am supposed to move (Wittgenstein, 1958, § 617). Here, this enables me to direct my activity into the finger itself. In fact I can retain the genuine ability even in the absence of any such bodily feeling, in slightly more basic situations (Kelso, 1977). Furthermore, a failed attempt to move a briefly paralysed limb might also manifest true practical knowledge (as an illusion has genuine, objective representational content), but only in virtue of its derivative relation with successful, active bodily movement.[12]

So neither actual mechanical success nor salient sensuous awareness of the relevant limb, as helps in Wittgenstein's case, seem either necessary or sufficient. Indeed, I would claim that their disjunction is not necessary, nor their conjunction sufficient. More than *simply* willing limb movement, as I might will my favourite team to win, or will the correct finger to go up when you point to it in the earlier case (this can be little more than a wish), more than this along with the felt presence of the limb, more than all of this along with the fortuitous obedience of the body machine – the mere capacity delivered by superimposition – what is required is *my moving my limb*, actually doing it, or if not, then derivatively failing to do so: merely trying to move.

Assimilating the personal level psychological story to the operation of the central, subpersonal, common coding system in selecting and parameterizing physiological biodynamic structures, leaves the subject without the appropriate control over her bodily movements to constitute her their agent. Vision may inform her how she is to move. But this falls short of telling her how to move like that: how to really try. She does not know how to direct her will, how to bring it into direct contact with her body in moving it in the required way. She is reduced to inner willing and hoping for the best.

2.3 Agent's experience This connects immediately with my third point about the discrepancy between the results of superimposition and genuine agency. Thomas Baldwin puts the contrast nicely as follows, drawing as I do here, very much on Brian O'Shaughnessy's (1980) work on bodily action. 'The experience of agency is not one of acts of will just regularly conjoined with bodily movements; it is one of acts of will that extend themselves to those parts of the body that are under the direct control of the agent' (Baldwin, 1990, p. 17).

On the current view, bodily action is movement of the body caused by an independent, purely inner, mental act, which wills it (Figure 13.1). The common element *w*-type mental state, itself quite neutral on any question of bodily involvement, suitably causes a physical bodily movement event, itself quite neutral on any question of activity. In truth, this movement must be integrated into the act of will in bodily action. The experience of agency is the experience of actively moving one's body, not that of intrinsically inert mental effort, which may or may not engage with the body-machine. Willing grows into and becomes active movement, controlled and directed by the agent, in successful cases. A person's psychology spreads into her body, and beyond, in actively moving it. She is not trapped in her head hoping it will mysteriously obey her commands. This single psychological whole, of the will coming to 'fruition' (O'Shaughnessy, 1980, vol. 2, p. 214) in active bodily movement, covers, or includes both the common coding system's adoption of a certain act code and the biomechanical effects this brings about (see Figure 13.2).

Conclusion

A central thesis of this final section then, is that in visually guided spatial action, the realm of personal level psychology extends well beyond that of the subpersonal common coding system, out into the world in the direction of the physiological mechanisms underlying bodily movement, and their external consequences. Indeed, I would also suggest it extends equally in the 'opposite direction', into the world via the early visual system and its outer objects.[13] So Bernstein's bodily machinery, Prinz's central

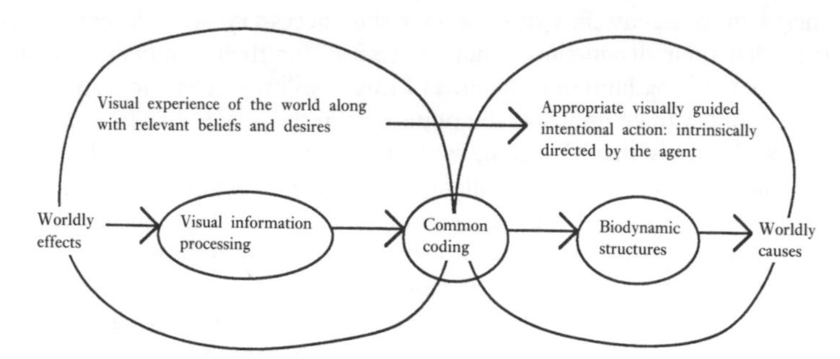

Figure 13.2 The relation between the personal and subpersonal levels in the integration of spatial vision and action.

common coding system which drives it and the visual information system which feeds this – and indeed, some of the external, worldly causes and effects of this whole subpersonal complex – are each important components of a larger, personal level, psychological picture integrating spatial vision and action (Figure 13.2). The events at this higher level are articulated out of the underlying story by their role in a relatively autonomous causal explanatory theory, by which organisms' behaviour is made intelligible as approximating more or less to what is rational, given their perspective on their needs and environmental context, or to what they ought to do given their place in the world (cf. Hornsby, 1985; McDowell, 1985).

On this account, the actions we have been concerned with are rightly regarded as intrinsically directed movings of the agent's body, under his immediate spatial control, as informed by visual perception of the world around him. James' intuitive visuo–motor immediacy emerges as the personal level reflection of subpersonal common coding, as this is supplied by the visual system and itself supplies the motor apparatus. These subpersonal components are the ingredients which they are, of the wider psychological picture, in virtue of their embedding in an environment in which they serve the needs and preferences of the organism of which they are a part.

To summarize, there are two basic elements in the overall picture I am proposing. The first, which has been my main concern so far in this section, is a disjunctive account of bodily willing. The second has been more in the background up to now, and can only really be touched on here. This is a parallel to Wiggins' (1980) conceptualist realism about substances, for events (an approach also adopted by Hornsby, 1985). I shall take these in turn.

The core of the disjunctive account of bodily willing, is the idea that in understanding a person's intentional spatial behaviour, her active bodily movement is fundamental. This is essentially active. It involves a whole package of subpersonal systems' setting the bodily biodynamics into operation in such and such ways in given circumstances. It is not to be identified with such a package, but some such events are important components. It is certainly *not* a fortunate causal combination of two intrinsically inactive events: a mental common element between active success and failure, and a bodily event also quite neutral between the subject's being active or passive. Failed attempts by an agent to move in various ways on the other hand, are to be understood only as derivative of this notion, not as psychologically identical with an independent physical consequence unfortunately missing. They fall short of genuine spatial action for all sorts of physical, physiological or neuropsychological reasons, but they make sense in the context of the appropriateness of certain genuine active bodily movements. This is the point of contact with successful physical agency: failing, rather than succeeding, in a similar context of rationalizing explanation. The connection is not a shared, inner mental happening. The two kinds of event are psychologically quite distinct, although a person need not be infallible at distinguishing between them. Willing bodily movement is *either* actively moving *or* trying and failing. I claim that an account along these lines is necessary to respect the epistemology of spatial action.

The second element is not strictly necessary for my overall argument as it stands, but it is clearly crucial to the general question of the relation between the personal and subpersonal levels of explanation and has been implicit in some of my comments above. It supplements the general argument, by simultaneously grounding the intuitions in favour of a disjunctive approach to bodily willing and providing the beginnings of an alternative to the superimposition of explanatory spaces which I reject. The leading idea might loosely be put as follows. Which events one finds in the world depends essentially upon how one is looking. There is not a single stock of neutral events, and causal relations between them, which are variously picked out, with variable explanatory illumination, by different kinds of descriptions. It is rather that the very identification and singling out of particular events as unitary items worthy of attention, is dependent on and controlled by the role of their types in the lawlike generalizations of some causal explanatory theory, which makes it intelligible to us why things of one kind follow those of another. Jennifer Hornsby put the point similarly as follows.

> Inasmuch as it is in the nature of continuants to persist, we expect individual continuants to be members of kinds whose instances have intelligible, individuation-sustaining persistence conditions; inasmuch as it is in the nature of events to cause and be caused, we expect individual events to be members of kinds that pull their weight in illuminating accounts of why one thing followed another. The items which are events, like the items which are continuants, need to be singled out not merely as occupiers of space and time, but by reference to a suitable ideology; and the suitable ideology for events is conditioned by the need to construct an *explanatory* causal nexus. (Hornsby, 1985, p. 454, my italics)

Of course this is a controversial position. And I can do little more than appeal to Hornsby's (1985) case in support of it here, as providing a way of recasting and developing Wiggins' (1980) arguments concerning substances in the domain of events. I want to end though, by sketching how conceptualist realism of this form adds to the current picture of visually guided spatial action, and by pointing to a thorny issue which remains.

Crudely, what this kind of conceptualist realist conception of events does, is attach psychological event identification and individuation firmly to folk psychological, rational explanation. This in turn is what ultimately grounds the externalist, disjunctive nature of bodily willing, and reinforces its distinction from the internal events of the subpersonal common coding system.

The crucial point is that the articulation and nature of willing events are responsible to the distinctive character of personal level, psychological action explanation. They only make sense as coherent individuals from the point of view of this kind of explanatory project. So there is no reason to believe that there will be a *subpersonal event* at all, at the place occupied by somebody's raising their arm at a given time, say, or their trying and failing to do so. The integrity and unity of these entities is essentially tied to their role in a folk psychological causal explanatory story, and is therefore likely to be obscure in the context of any radically different causal explanatory concern at a lower level. For this will naturally bring out only the relations between events of quite different types. It will bring into focus a different batch of explanatory units; an agent's raising her arm, or any other personal level event, may well made up, in some sense, out of such lower level events organized in some way or another, on any given occasion. But it is no mere fusion of these. For the principles of organization have their rationale only at the psychological level, and appear coincidental or arbitrary from any other sufficiently distant perspective.

Of course, this raises the serious question of what exactly event composition is supposed to be. If the story is correct so far, then this is weaker than token identity. But it is clearly also stronger than mere spatio–temporal coincidence (and not only because it is asymmetric). Supervenience of some kind is a standard starting point. This would be the idea that there can be no difference in the psychological type of composed macroscopic events, without some underlying difference in their composition out of microscopic subpersonal events. It is surely only a starting point though, because it fails to give any *intelligibility* to the systematic dependence of the nature of personal level psychology on its subpersonal components. It merely states the form of the dependence. Yet the composition relation must be what provides such intelligibility. For it is precisely to this proposed composition relation that we must look for any illumination we hope for about why, or how, the mental supervenes on the physical in the correct sense.What we need is some notion of the truth of the personal level story being *in virtue of* the truth of the subpersonal level story. Psychological facts obtain in virtue of the underlying neurophysiological and surrounding physical situation, even though the former provide a kind of structure and intelligibility quite hidden in the latter. But what precisely does this mean?[14]

The really thorny issue here can be brought into focus from a slightly different angle by asking what the connection is between causation at the personal level and causation at the subpersonal level. The distinct explanatory commitments of the theories conjoin with conceptualist realism about events to require a certain independence between the causal relations illuminated at each level. Yet the causal chains discerned by personal level psychological explanation must respect the contours of subpersonal level causation to a significant extent. For otherwise, visually guided spatial action becomes indistinguishable from paranormal phenomena like levitation. At present though, I have no completely satisfactory account of the correct course to be taken between the unacceptable extremes of autonomy and isomorphism.[15]

Following a proposal of Kathleen Lennon's (1990, p. 120), we can, I think, impose a substantial constraint on any such account of the relation between personal and subpersonal level causation: that it should guarantee the following supplement to the preferred form of supervenience. In suitably close possible worlds to the actual world, in which the personal level antecedents of a basic bodily act also obtain, then something sufficiently close to the actual subpersonal level causal story must also obtain, in which similar neurophysiological events give rise, or fail to give rise, to biomechanical events by which the same type of active bodily movement can be performed. For this minimal co-ordination between personal and subpersonal level explanations is a necessary condition on the very integrity of realistic psychological ascriptions: it is a requirement on the proper functioning of a person.

ACKNOWLEDGEMENTS

Many thanks to John Campbell, David Charles, Bill Child, Naomi Eilan, Roz McCarthy, Tony Marcel, Mike Martin, Julian Pears and Timothy Williamson for very helpful comments on earlier drafts. The work was supported by a grant from the Leverhulme Trust to King's College Cambridge Research Centre.

NOTES

1 He gives the following famous example of their successful integration on p. 522. 'Whilst talking I become conscious of a pin on the floor, or of some dust on my sleeve. Without interrupting the conversation I brush away the dust or pick up the pin. I make no express resolve, but the mere perception of the object and the fleeting notion of the act seem of themselves to bring the latter about.' The point here is that vision supplies the spatial parameters for a simple bodily movement in such a way as to enable its immediate motor execution. And the principle of ideo–motor action he endorses as an account of this immediacy is as follows: 'Think of the movement purely and simply, with all the brakes off [that is, in the absence of inhibiting thought about or preference for any conflicting course of action or inaction]; and presto! it takes place with no effort at all' (p. 527). His claim is that the uninhibited coming to mind of the idea of an act is sufficient to bring about the appropriate bodily movement.
2 The contrast between this traditional approach and his preferred 'common coding' framework goes back to Prinz (1987). In fact, approval for something very like common coding is already implicit in Turvey's work on the relations between perception and action (esp. Turvey, 1974 and 1977), in which he draws heavily on the Russian school of physiologists lead by Nicolai Bernstein. My own views on the nature of visuo–motor control owe a great deal to Prinz's work in this area in general and to his extended concentration on the presuppositions and consequences of the separate and common coding approaches in particular.
3 His general discussion of the connection between possession of structured visuo–motor abilities in this sense, and possession of psychological states with genuine spatial content (Peacocke, 1983, pp. 61ff), is very illuminating in this context. It has certainly shaped my own views in the area. The parallel Peacocke gives with the structure in linguistic understanding is also helpful here. For more on this see Davies (1981, ch. 4) and Evans (1981).
4 This idea of a connection between common coding and 'distal focussing' is also originally due to Prinz (1992). The point I want to emphasize though, is that the contents of common codes are genuinely spatial. This is the crucial feature of their distal focussing.
5 Part V of Kelso (1982), which contains this paper and two others by the same authors, constitutes an excellent introduction to the essential features of 'The Bernstein Perspective'.
6 For helpful reviews of this, and a lot more experimental work bearing on Bernstein's account, see Turvey (1977) and Bizzi and Mussa–Invaldi (1990).
7 More needs to be said about an independent rationale for this distal, truly spatial individuation of representations carried by the visual system, although its correctness follows from the common coding approach to visuo–motor integration along with our reasons for this way of collecting token bodily movements into spatial types. Perhaps the explanation has essentially to do with the role of perceptual experience in controlling and co-ordinating behaviour. But this is a difficult issue which falls outside the scope of this chapter.
8 This is in fact a particular instance of the general temptation which Jennifer Hornsby (1986) notes, to superimpose folk psychology upon neurophysiology. I am in agreement with much of her diagnosis of and suspicion about the temptation.

9 I develop this parallel between disjunctive accounts of experiential seeming and my preferred approach to bodily willing a little further in the Conclusion.

10 Hornsby (1986) gives a detailed account of various difficulties encountered by the standard functionalist version of this conception of behaviour.

11 In doing so, I move freely between the two versions, or remain neutral where possible, for ease of exposition, and to save space. As far as I can see, each problematic ramification affects both views significantly, although it may be true that the three considerations vary slightly in strength with respect to the two versions. The general idea, that action is either an outer physical movement, provided this has the right kind of inner cause, or an inner willing event, provided this has the right outer effect, fails to capture its nature as an integrated whole, extending from inner to outer as directed by the agent.

12 However it should be correctly spelled out, this dependence of the notion of a genuine failed attempt at bodily movement upon successful, basic physical action surely underlies the incoherence, or at least obscurity, of the idea of genuinely willing extra-bodily events. See O'Shaughnessy (1980, vol. 1, chs. 2–4) for an excellent discussion of this issue.

13 See Putnam (1975); Fodor (1980); Searle (1983, ch. 8); Woodfield (1982); and Pettit and McDowell (1986) for this more standard version of the internalism versus externalism debate.

14 See Lennon (1990), particularly ch. 5, for a preliminary characterization of such an 'in virtue of' relation.

15 Three possible intermediate positions here, increasing in strength, would be the following: (1) some subpersonal component of a psychological cause must cause some component of its psychological effect; (2) every subpersonal component of a psychological cause must cause some component of its psychological effect, and/or every subpersonal component of a psychological effect must be caused by a component of its psychological cause; (3) all the effects of every subpersonal component of a psychological cause must be components of its psychological effect, and/or all the causes of every subpersonal component of a psychological effect must be components of its psychological cause. It seems to me that the correct account will be found somewhere on the continuum joining them.

REFERENCES

Abbs, J. H. and Gracco, V. L. 1983: Sensorimotor actions in the control of multi-movement speech gestures. *Trends in Neurosciences*, 6, 391–5.

Baldwin, T. 1990: Objectivity, causality and agency. Unpublished paper read to the Spatial Representation Seminar on Perception and Action at King's College Research Centre, Cambridge.

Belen'kii, V. Yi., Gurfinkel, V. S. and Pal'tsev, Ye. I. 1967: Elements of control of voluntary movements. *Biophysics*, 12, 154–61.

Berkeley, G. 1965: Principles of Human Knowledge. In D. M. Armstrong (ed.) *Berkeley's Philosophical Writings*. London: Collier Macmillan.

Bernstein, N. 1967: *The Coordination and Regulation of Movements*. Oxford: Pergamon.

Bizzi, E. and Mussa-Invaldi, F. A. 1990: Muscle properties and the control of arm movement. In Daniel N. Osherson, Stephen M. Kosslyn and John M. Hollerbach (eds), *Visual Cognition and Action*. Cambridge, Mass: MIT, 213–42.

Davidson, D. 1980: Actions, reasons and causes. In his *Essays on Actions and Events*. Oxford: Clarendon Press, 3–19.

Davies, M. 1981: *Meaning, Quantification, Necessity*. London: Routledge and Kegan Paul.

Descartes, R. 1954: *Philosophical Writings*. London: Thomas Nelson.

Evans, G. 1981: Reply to Wright: Semantic theory and tacit knowledge. In S. H. Holtzman and

C. M. Leich (eds), *Wittgenstein: To Follow a Rule*. London: Routledge and Kegan Paul, 118–37.

Fel'dman, A. G. 1966: Functional tuning of the nervous system during control of movement or maintenance of a steady posture – III. Mechanographic analysis of the execution by a man of the simplest motor tasks. *Biophysics*, 11, 766–75.

Fodor, J. 1980: Methodological solipsism considered as a research strategy in cognitive psychology. *Behavioural and Brain Sciences*, 3, 63–73.

Grice, H. P. 1961: The causal theory of perception. *Proceedings of the Aristotelian Society Supplementary Volume*, 35, 121–52.

Hinton, J. M. 1973: *Experiences*. Oxford: Clarendon Press.

Hornsby, J. 1980: *Actions*. London: Routledge and Kegan Paul.

Hornsby, J. 1985: Physicalism, events and part–whole relations. In E. LePore and B. McLaughlin (eds), *Actions and Events*. Oxford: Clarendon Press, 444–58.

Hornsby, J. 1986: Physicalist thinking and conceptions of behaviour. In P. Pettit and J. McDowell (eds), *Subject, Thought and Context*. Oxford: Clarendon Press, 95–115.

Hume, D. 1975: *Enquiries Concerning Human Understanding and Concerning the Principles of Morals*. Oxford: Clarendon Press.

Hume, D. 1978: *A Treatise of Human Nature*. Oxford: Clarendon Press.

Jackson, F. and Pettit, P. 1988: Functionalism and broad content. *Mind*, 97, 381–400.

James, W. 1890: *The Principles of Psychology*, vol. 2. New York: Henry Holt.

Kelso, J. A. S. 1977: Motor control mechanisms underlying human movement reproduction. *Journal of Experimental Psychology: Human Perception and Performance*, 3, 529–43.

Kelso, J. A. S. 1982: *Human Motor Behaviour*. Hillsdale, NJ: Erlbaum.

Lennon, K. 1990: *Explaining Human Action*. London: Duckworth.

Locke, J. 1975: *An Essay Concerning Human Understanding*. Oxford: Clarendon Press.

McDowell, J. 1982: Criteria, defeasibility, and knowledge. *Proceedings of the British Academy*, 68, 455–79.

McDowell, J. 1985: Functionalism and Anomalous Monism. In E. LePore and B. McLaughlin (eds), *Actions and Events*. Oxford: Clarendon Press, 387–98.

McDowell, J. 1986: Singular thought and the extent of inner space. In P. Pettit and J. McDowell (eds), *Subject, Thought and Context*. Oxford: Clarendon Press, 137–68.

Marr, D. 1982: *Vision*. New York: W. H. Freeman.

O'Shaughnessy, B. 1980: *The Will*, 2 vols. Cambridge: Cambridge University Press.

Peacocke, C. 1979: *Holistic Explanation*. Oxford: Clarendon Press.

Peacocke, C. 1983: *Sense and Content*. Oxford: Clarendon Press.

Peacocke, C. 1992: Scenarios, concepts and perception. In T. Crane (ed.), *The Contents of Experience: Essays on perception*. Cambridge: Cambridge University Press, 105–35.

Pears, D. F. 1976: The causal conditions of perception. *Synthese*, 33, 41–74.

Pettit, P. and McDowell, J. (eds), 1986: *Subject, Thought and Context*. Oxford: Clarendon Press.

Pribram, K. H. 1971: *Languages of the Brain: Experimental Paradoxes and Principles in Neuropsychology*. Englewood Cliffs, NJ: Prentice Hall.

Prinz, W. 1987: Ideo–motor action. In H. Heuer and A. F. Sanders (eds), *Perspectives on Perception and Action*. Hillsdale, NJ: Erlbaum, 47–76.

Prinz, W. 1990: A common coding approach to perception and action. In O. Neumann and W. Prinz (eds), *Relationships between Perception and Action: Current Approaches*. Berlin: Springer, 167–201.

Prinz, W. 1992: Why don't we perceive our brain states? *European Journal of Cognitive Psychology*, 4, 1–20.

Putnam, H. 1975: The meaning of 'meaning'. In his *Mind, Language and Reality*. Cambridge: Cambridge University Press, 215–71.

Searle, J. 1983: *Intentionality*. Cambridge: Cambridge University Press.

Smith, A. D. 1988: Agency and the essence of actions. *Philosophical Quarterly*, 38, 401–21.

Snowdon, P. 1981: Perception, vision and causation. *Proceedings of the Aristotelian Society*, 81, 175–92.

Taylor, C. 1985: Hegel's philosophy of mind. In his *Human Agency and Language*. Cambridge: Cambridge University Press, 77–96.

Taylor, C. C. W. 1985: Action and Inaction in Berkeley. In J. Foster and H. Robinson (eds), *Essays on Berkeley*. Oxford: Clarendon Press, 211–25.

Tuller, B., Turvey, M. T. and Fitch, H. L. 1982: The concept of muscle linkage or coordinative structure. In J. A. Scott Kelso (ed.), *Human Motor Behaviour*. Hillsdale, NJ: Erlbaum, 253–70.

Turvey, M. T. 1974: A note on the relation between action and perception. In M. Wade and R. Martens (eds), *Psychology of Motor Behaviour and Sports*. Urbana, Ill: Human Kinematics.

Turvey, M. T. 1977: Preliminaries to a theory of action with reference to vision. In R. Shaw and J. Bransford (eds), *Perceiving, Acting and Knowing: Toward an Ecological Psychology*. Hillsdale, NJ: Erlbaum, 211–65.

Wiggins, D. 1980: *Sameness and Substance*. Oxford: Basil Blackwell.

Wittgenstein, L. 1958: *Philosophical Investigations*. Oxford: Basil Blackwell.

Wittgenstein, L. 1961: *Tractatus Logico–Philosophicus*. London: Routledge and Kegan Paul.

Wittgenstein, L. 1980: *Remarks on the Philosophy of Psychology*, vol. 2. Oxford: Basil Blackwell.

Woodfield, A. (ed.), 1982: *Thought and Object*. Oxford: Clarendon Press.

Part V

What and where

Introduction: What and where

Rosaleen McCarthy

The question of the interrelationship between spatial representations and our conception and perception of objects in space is a core concern of many of the chapters in this volume: those in Part V deal with the question of 'what and where' as it bears on some very specific problems. A unifying theme for each of these accounts is their common reference to the distinctions between what and where that have been proposed on the basis of neurobiological studies of the visual system. Work based on the recording of activity from single cells in the brain, together with studies of the behavioural effects of brain injuries, has challenged the conventional and intuitive view that the perception of objects is closely tied to the perception of space. Rather than being closely intertwined and mutually dependent, what and where are thought to be processed independently from the earliest levels of perception onwards. If this challenge to conventional and intuitive views is valid and useful at the level of cognition (as well as biology) then there are many profound implications.

Background

There are two main classes of neurobiological theory of concern to the authors in Part V. The first type of theory is based on the coding of spatial information across arrays of individual nerve cells or processing units. Electrophysiological studies of the nervous system have provided plentiful evidence for neural 'maps' consisting of units that are jointly triggered by particular classes of information arising in particular retinotopic locations. Whilst location and attribute information are jointly encoded by such arrays the question remains as to whether such codes make any direct contribution to vision. The second type of theory is based on material specific processing systems. It is well established that there are many parallel visual processing routes in the primate brain. Each of these routes appears to be specialized for different classes of information. Theorists concerned with what and where processing have focused on three of these pathways: a major subcortical route which is of primary importance in lower mammals and is probably involved in eye movements and primitive cross-modal spatial processes in man; there are also two principal cortical systems found in higher primates. It has been proposed that one way of classifying the types of content delivered by the evolutionarily sophisticated cortical systems is in terms of 'what' information and 'where' information (e.g. Ungerleider and Mishkin, 1982).

Braddick builds on the physiological constructs of neural maps and retinotopic units to ask how the relationships between objects might be coded in simple 'arrays' of feature detectors. He points out that whilst such arrays might support local differencing operations (sufficient for segregating texture boundaries) there is no 'obvious' way in which their local sign information can signal the *alignment* of two features. We could postulate multiple receptive fields with different resolutions and scales, but such a thesis clearly needs *a priori* bounds before it can be taken seriously. Ullman's theory of visual routines offers one possible means of explaining these capabilities. Ullman (1984) has proposed that there are a number of low-level cognitive procedures that may be carried out over array-like information to yield higher order information such as alignment and boundedness. Operations such as tracing lines or filling contours potentially can solve problems that are insoluble for a system made up of multiple arrays of detectors.

In addition to discussing constraints on the uptake of information, Braddick also considers the important computational problem of binding information across domains. Subjects are capable of making fast and accurate judgements of the orientation and position of stimuli which (on physiological grounds) seem to be processed separately by the cortical 'what' and 'where' systems. Braddick gives a summary review of some critical studies of this problem, taking as a starting point Anne Treisman's feature integration theory (Treisman, 1988). Braddick suggests that not only is there evidence for a degree of independence between the processes of stimulus identification and stimulus location, but there also may be multiple means of locating stimuli. There are at least two 'where' processes – termed coarse and fine 'where' – which may correspond to the subcortical orientation mechanisms and cortical 'where' processing pathways of the brain.

Atkinson returns to the issue of multiple wheres and multiple whats in her discussion of the development of visual sensory processing in infancy and early childhood. It is suggested that the distinction between 'cortical' and 'subcortical' processing systems may be critical in understanding some of the characteristics of visual processing in young babies: newborn babies may be reliant on the subcortical visual system. Electrophysiological recording techniques have made it possible to explore the uptake of different types of visual information from birth onwards. The method of Evoked Potentials depends on the use of electrical recordings from the scalp: changes in electrical activity in the brain can be detected by small surface electrodes whose signal can be amplified and analysed by specialized computer programmes. By presenting stimuli at known times and synchronizing the recording of changes in electrical activity that are evoked by stimulus input it is possible to find a measure of the brain's electrophysiological response to the stimulus. Evoked potentials are essentially harmless and have the advantage that they do not require the infant to make a specific motor response.

Changes in brain activity may be shown to some types of stimuli but not to others. The likelihood of response appears to depend both on the characteristics of the stimulus and on the age of the viewer. Atkinson reports that newborn babies are responsive to stripes changing (flickering) between black on white and white on black: phase alternation. Unlike older babies they do not respond to orientation alternation or changes in motion. Atkinson proposes that phase alternation might well be processed by subcortical visual processors. Changes in orientation and motion evoke electrophysiological responses in older babies and this change may reflect the switch in cortical information processing streams.

Although babies respond electrophysiologically to changes in orientation from the age of two months, they are unable to orient to forms defined by local changes in the orientation of texture features until over four months of age. Atkinson speculates that this discrepancy may reflect a more pervasive difficulty in integrating information between 'registration and response systems' or 'within and across cortical streams'. Her chapter therefore makes contact with Braddick's in its concern with the issue of the Binding Problem. Atkinson speculates that babies' failure to bind or integrate information may have some resemblance to neurological impairments of various sorts. In the older child, failure of integration may manifest as a difficulty with constructional tasks and planning.

McCarthy's chapter raises the issue of multiple wheres and multiple whats again. Building on unpublished work by Eilan, she considers specific types of task and different sub-types of neurological impairment as a means of analysing the very different ways in which the constructs of 'what' and 'where' have been used. Rather than reify the established partitioning of 'what' and 'where' along neuroanatomic lines it is suggested that the core distinction may actually be between assembled visuo–motor routines on the one hand an stored representations of knowledge on the other. The evidence from neurological patients is at least as consistent with the alternative framework as with the more established view.

The evidence from animal studies is also broadly in keeping with this formulation. Ungerleider and Mishkin reported that place-reversal learning was impaired in animals by damage to the 'where' system but landmark reversal learning was impaired by damage to the 'what' system. Whilst this *could* implicate the loss of specific cognitive codes for objects and spaces it could also arise as a consequence of damage to different types of procedure. Failure at the level of assembling visuo–motor routines might be primary. Difficulty in setting up spatial memories or in acting on the basis of memory might be a secondary consequence. Certainly, in neurological cases, failure on spatial learning tasks appears to arise for many different reasons (McCarthy and Warrington, 1990). Moreover, where the disorder is clearly one of lost memories (such as amnesia for familiar landmarks and routes) the hypothetical neuroanatomical 'what' system appears to be the critical site of damage. Within the domain of landmark knowledge it may even be possible to trace the disorder further – to a failure to retain specific classes of feature conjunction abstract as 'places'.

At this level, the issue of 'what and where' clearly makes contact with the questions raised by Campbell in chapter three. Just what is meant by 'what'? The concern with objects as material entities with causal coherence that is evidenced in chapter three, together with those in Parts I, II and IV has been largely avoided by those approaching the problem from the neurobiological standpoint. McCarthy introduces the notion of schemas as a possible means of linking the rather sparse neurobiological constructs of what and where with the richer cognitivist conceptualizations that are presented elsewhere in this volume.

Tye uses the distinction between what and where as a means of exploring a constructivist theory of visual imagery. Much debate on visual imagery has been focused on the issue of representational format – are images *analogue* (e.g. Kosslyn, 1987) pictures in the head or *propositional* descriptions (e.g. Pylyshyn, 1981). Tye avoids the pitfalls of this particular debate and suggests that images are really like drawings. They are constructed using whatever resources and information the subject has available. Sometimes the drawing is a sketch – partial and inaccurate like a cartoon or half-remembered scene. Sometimes it is wrong – as when children erroneously recall the water-level of a tilted beaker as being 180 degrees from the beaker's side. Tye points out that we do not need to postulate a 'blurred' picture to account for inaccuracies – neither do we have to resort to the notion that images are 'nothing but' error prone descriptions. He appeals to the dissociations between spatial processing and object recognition that have been observed in patients with neurological lesions as a potential source of support for his constructive/drawing theory of visual imagery. A hypothetical patient who cannot remember the appearances of objects might nevertheless be able to report on their location. Conversely, a patient who could not recall the locations of objects might be able to report on their appearances. In each case, the subject is constructing a visual image on the basis of the resources or information that they have available.

Tye considers Kosslyn's a 'bit-mapped' theory of the medium of visual imagery which seems to have problems in accounting for such putative profiles of impairment. According to Kosslyn, images correspond to the activation of elements in an array or visual buffer (possibly equivalent functionally to the 2.5D sketch in Marr's theory). if images correspond to the activation of cells in an array, then there ought to be a correspondence between the spatial activation of elements of an object and the spatial organization of multiple objects in the image 'buffer'. Imaging the handle of a cup at position M should be like in kind to imaging the position of the spout of a teapot at position M′. Since patients may lose spatial imagery but not object imagery – and vice versa (see also chapter 17) – then the simple 'bit-mapped' theory is challenged.

At an abstract level this imagery-array problem has much in common with the local sign/receptive field-array problem discussed by Braddick. It would also seem to run into the same problems, unless allowances were made for multiple spatial scales as a means of encoding multiple spatial interrelationships. Tye's solution (analogous to Ullman's solution) is to appeal to higher order operations on the information represented in an array. He suggests that failures in visual imagery might be attributable to impairments in the cognitive operations of inspection and categorization. As such, they might be separable from damage to the imagery medium itself.

This hypothesis and these contrasts have much in common with Bisiach and Vallar's (1988) discussion of imagery in the neurological syndrome of spatial neglect. Patients with neglect may fail to notice things on one side of space due to damage to the opposite side of the brain. For some cases this neglect may also be manifest in visual imagery. Bisiach, Luzzatti and Perani (1979) described patients who made errors in their imagery of scenes that consisted of omissions of items on the left side. By asking the subjects to shift perspective and re-visualize a single scene from two locations, Bisiach was able to show omissions were dependent on the imagined viewpoint. Subjects were asked to pretend that they were either facing towards the cathedral in the Piazza del Duomo of Milan, or standing on the cathedral steps and looking across the square. They simply had to tell the examiner what shops and buildings could be seen from each perspective. When imagining the scene subjects omitted details on the left of their imagined viewpoint on both occasions. They recalled details when they were 'imagined' on the right and left them out if they were imagined on the neglected left side. In order for the subjects to report the stimuli under any condition they must have had the information available in memory. However, the action of retrieving that information, of scanning the information, or perhaps even of holding and retaining the information in an active form, led to errors.

Implications

The relationship between information uptake and higher order planning and integrative processes is clearly a complex one and moreover a critical issue for any comprehensive theory of 'what' and 'where'. Both Braddick and Atkinson acknowledge that their neurobiological 'bottom up' approaches can only provide a partial solution to the wider problems of what and where processing. McCarthy also acknowledges the limitations of the traditional 'bottom up' approaches to analysing cognitive deficit and points to the importance of individual subjects' capacity for configuring and programming their own information processing systems.

This series of chapters, perhaps more than any other in this book, emphasizes a common 'hard-wired' approach to the analysis of the relationship between 'what' and 'where'. It both illustrates the utility of the framework as a means of bringing together diverse strands of research and also demonstrates the massive gaps in knowledge that exist once we get more than a few milliseconds of processing time into the human brain. Comparatively little contact has been made between the relatively simple automaton-like processes of 'what' and 'where' of the bottom up neurobiologists and the rich cognitive data base of those working at the levels of plans, schemas and expectancies. However, it is this level of theory and investigation which is a core concern of some of the philosophical contributions to other parts of this book (see especially chs. 3, 6 and 10).

My own view is that work at this level is potentially very important for the wider issues and in closing I shall attempt to propose and defend an evolutionary and computational perspective on its implications. The biologists, physiologists and psychologists have elegantly demonstrated that there do appear to be important differences between the ways in which different types of information are extracted from the sensory array. These differences are apparent across species (phylogenetically) and in the course of development (ontogenetically). At its simplest, we have evidence that more advanced, behaviourally flexible and widely distributed primates tend to have a larger number of cortical maps, visual streams, etc. They have multiple whats and multiple wheres. Animals living in more restricted ecological niches do not seem to require so many 'options'.

The availability of multiple means of computing and extracting information has presumably conferred a selective advantage. Perhaps a part of this advantage lies in being able to assign independent default values to a highly differentiated set of sensory inputs within higher order programs or schemas. The ability to weight different types of input independently may introduce a greater computational power to these higher order organizational programs. If the subject can value or weight their expectancy for objects, places or events independently, and assign these values independently of the weighting assigned to location or position, then their schemas will have far greater generality and power than those with fewer degrees of freedom (i.e. in which such weights are fixed or close-coupled). To be able to expect x to have some property p irrespective of variation in location l or to be able to orient to location l despite variance in the values of x or p (e.g. anomalies, occlusion or distortion of sensory input) would seem to introduce a useful degree of computational flexibility into the system. The availability of multiple options at the level of the cognitive program is one deliverance of the highly differentiated biological system. What about the wider implications? Do we need to close-couple certain sets of parameters in order to deliver a particular cognitive product? Whilst the system may be capable of assigning independent values to 'what' and 'where' it may be necessary to link or otherwise associate the values assigned to these informational types in order to carry out some types of cognitive operation. Independence at the level of sensory or mnemonic codes does not preclude interdependence at other levels. Whether such interdependence is a normative requirement or merely an option requires a very different type of theory to that discussed here. The ways in which the deliverances of the spatial/assembled or object/addressed informational systems are combined and integrated to yield information about self location and environmental continuity remains an issue for further speculation and research.

REFERENCES

Bisiach, E., Luzzatti, C. and Perani, D. 1979: Unilateral neglect, representational schema and consciousness. *Brain*, 102, 609–18.

Bisiach, E. and Vallar, G. 1988: Hemineglect in Humans. In F. Boller and G. Grafman (eds), *Handbook of Neuropsychology*, vol. I. Amsterdam: Elsevier

Kosslyn, S. 1987: Seeing and imagining in the cerebral hemispheres; A computational approach. *Psychological Review*, 94, 148–75.

McCarthy, R. and Warrington, E. 1990: *Cognitive Neuropsychology: A clinical introduction*. San Diego; Academic Press, Inc.

Pylyshyn, Z. 1981: Imagery and Artificial Intelligence. In N. Block (ed.), *Readings in Philosophy and Psychology*, vol. 2, Cambridge, Mass: Harvard, 174–6.

Treisman, A. 1988: Features and objects. *Quarterly Journal of Experimental Psychology*, 40A, 201–38.

Ullman, S. 1984: Visual Routines. *Cognition*, 18, 97–159.

Ungerleider, L. G. and Mishkin, M. 1982: Two cortical visual systems. In D. J. Ingle, M. A. Goodale and R. J. W. Mansfield (eds), *Analysis of Visual Behaviour*. Cambridge, Mass: MIT, 549–86.

14
A neurobiological approach to the development of 'where' and 'what' systems for spatial representation in human infants

Janette Atkinson

Introduction

Many different sub-systems have been considered under the heading of 'spatial representation'. For example, Kritchevsky (1988), using specific spatial dysfunction from patients with brain damage, has delineated five categories: spatial perception (object localization, line orientation discrimination and spatial synthesis), spatial memory, spatial attention, mental rotation and spatial construction. He argues that clinical cases can manifest isolated dysfunction in one of these areas, which can sometimes be related to damage in specific brain modules. By analogy, development of brain modules for representing space can be inferred from the changing abilities on spatial tasks shown by infants and young children during development.

In this chapter different theoretical models of spatial representation, using a neurobiological approach, are briefly described using examples from the infants' abilities on different tasks of spatial perception, attention and cognition. This approach is largely 'bottom-up' compared to those from cognitive psychology (e.g. chapter five). The biological approach largely considers an object as the integrated sum of its parts; each part or 'attribute' will depend on the operation of particular neurones in particular neural pathways. The infant's identification of an object may not always depend on integration, but merely on identification of a single salient part or attribute. For example, the infant may initially recognize the mother's face on the basis of a particular shape of high contrast hair outline around the face. Later this information may be integrated with that of the characteristic carrier frequencies of the voice, as well as with information concerning internal features of the face. The cognitive psychologists' approach is largely top-down and would consider that the infant might start the familiar face recognition process by recognizing a face as a unitary concept. The infant would initially understand the global holistic properties of the stimulus and react to these, possibly using detailed information from bottom-up processing to add to the richness of the percept. Both approaches are valid methods of study, although they use different vobacularies, signifying different levels of analysis.

The most basic task of spatial representation can be broken down into defining the spatial position ('where') and defining the nature ('what') of a single object, in relation either to oneself or to other objects in an array. Three biologically based theoretical approaches, attempting to define the mechanisms subserving 'where' and 'what' processing, have been put forward:

1 One cluster of theories, arising out of comparative studies of visual systems across different species, has delineated 'two visual systems' (Schneider, 1969). Here subcortical systems control orienting responses which define crudely 'where' an object is located and trigger foveation to this location, while cortical mechanisms are used to define 'what' is actually in the foveated area. Several models have been developed from this idea and applied to development of human vision, for example those of Bronson (1974), Maurer and Lewis (1979), Atkinson (1984) and Johnson (1990).

2 A 'where' and a 'what' system have been defined solely within the cortical pathways. The initial evidence for selective processing of visual information in primate extrastriate cortex was obtained by Zeki and his co-workers. They defined an area selective for the direction of stimulus motion which has since been called V5 or MT (Dubner and Zeki, 1971). In other studies (Zeki, 1973; 1977) they identified a colour specific area, V4. Ungerleider and Mishkin (1982) suggested that these two streams are associated with different visual capacities – a largely parietal module is involved with localizing objects within a spatial array, while those involving the temporal lobe contain mechanisms tuned to 'what' aspects such as form and colour. Clinical observations of patients with specific focal lesions have shown a dissociation between loss of position or movement perception and deficits of object recognition (e.g. Ratcliff and Davies–Jones, 1972; Damasio and Benton, 1979; Pearlman et al., 1979; Zihl et al., 1983).

3 A third approach is based on two types of anatomically and morphologically distinct ganglion cells (parvocellular and magnocellular). The two streams are separate at the subcortical levels with parvocellular and magnocellular neurones projecting to different regions of primary visual cortex, V1 (Livingstone and Hubel, 1988; Van Essen and Maunsell, 1983; Maunsell and Newsome, 1987). In V1 the superficial layers contain an array of dot-like patches of cytochrome oxidase-rich tissue called 'blobs' (Horton and Hubel, 1980; Humphrey and Hendrickson, 1980; Livingstone and Hubel, 1984) which have a largely parvocellular input. Blobs seem to be specific for colour while interblobs are important for orientational tuning. Within V2 there are regions of high and low cytochrome oxidase-rich activity. Those that are high are in thin and thick stripes with neurones in the thin stripes projecting to V4, while those in the thick stripes project to V5. A parvocellular-based system subserves detailed form, vision and colour while the magnocellular system subserves movement, perception and some aspects of stereoscopic vision. Comparisons have been made between psychophysical data on adults and the functioning of these two distinct pathways. Similar comparisons can also be made when looking at the development of vision in human infants (Atkinson and Braddick, 1989; Van Sluyters et al., 1990). As yet there is little anatomical or morphological data on the relative time course of development of these two subsystems in human infants (only Hickey, 1981) and so the model is based on behavioural and visual evoked potential measures to delineate the developmental time course of various cortical streams, each subserving different aspects of spatial vision.

Below, recent data from studies on the development of sensitivity of the infant to different visual attributes necessary for deciding 'where' and 'what' is discussed in relation to each of these three theoretical approaches.

1 Systems representing spatial form

The most basic spatial information concerning 'what' an object is must involve measures of acuity and contrast sensitivity. Although there is some debate as to the precise values of newborn acuity, there is general agreement that a very rapid improvement in analysis of spatial detail, reflected in changes in acuity and contrast sensitivity, takes place between birth and nine months of age. Previous reviews have considered the relative role of photoreceptors, retinal neural mechanisms and post-retinal mechanisms in limiting acuity (e.g. Banks and Bennett, 1989; Atkinson and Braddick, 1990). While retinal development plays a major role in determining the limits on spatial information transmitted by the infant's visual system, the ability to encode that information in the ways needed to perform any visual task (for example for shape recognition) must depend on central cortical processing. We know that in the human visual cortex there is a massive increase in connectivity in the first six months of life (Garey and de Courten, 1983) and presumably these new synapses are defining and refining stimuli to which cortical cells maximally respond. Using an approach where stimuli are 'designed' to elicit a response from cortical cells that have developed particular selective properties, we can track the time course of changing spatial abilities in the human infant. The responses are measured either as changes in behaviour or as visual evoked potentials.

Such an approach was used to put forward the theory that the visual cortex plays a relatively minor role in newborn spatial vision, but comes to achieve executive control over subcortical mechanisms between two and four months post-natally (Atkinson, 1984; Braddick and Atkinson, 1988). This is the first theoretical approach outlined above. Many visual attributes are important in object identification and localization. Analysis of many of these must be cortically based, as subcortical mechanisms are not capable of detecting changes in these attributes. Three such attributes discussed below are orientation or slant, relative directional motion and binocularity. Anatomical and physiological findings allow identification of populations of neurones specific for detection of each attribute. Information for different attributes within these cortical streams appears to be segregated at many different levels of the cortex (for reviews see Maunsell and Newsome, 1987; Zeki and Shipp, 1988; Felleman and Van Essen, 1991). Of course, there must be elaborate integrative processes following attribute analysis to enable the appropriate responses to be made. For example, the six-month-old infant is able to shape his or her hand appropriately in order to reach accurately in a specific location in space for a desirable toy, or to smile on recognizing a familiar face.

2 Orientation

There is now general agreement that responses can be obtained from infants in the first few weeks of life to indicate that they can discriminate between a static grating pattern oriented at 45 degrees and one at 135 degrees. Here infant control habituation procedures have been used (Slater et al., 1988; Atkinson et al., 1988). An alternative approach is to record evoked potentials (VEPs), minute electrical potentials, detected by means of electrodes placed on the infant's head, which are then amplified and averaged across many repetitions of the same visual event. A significant 'steady state' VEP can be identified by its consistent temporal relationship to the visual stimulus event. For example, the amplitude of the potential may increase and decrease at the same rate as the transitions between two stimulus patterns, with a consistent interval between the transition in the visual display and that of the brain waves (constant phase relationship). A steady state VEP (OR-VEP) can be elicited when a grating oriented at 45 degrees is replaced by one at 135 degrees in infants around six weeks post-natally (Braddick et al., 1986a). A statistically reliable response (Wattam–Bell, 1985) at the frequency of the orientation change (or at harmonics of this frequency) is evidence for orientation-selective mechanisms. These mechanisms appear to have different orientation tuning and temporal tuning curves at different ages. At a rapid reversal rate between the two orientations (e.g. eight reversals per second) a significant response is found only in the second post-natal month. With slower alternations (three reversals per second), the median age for the response is three weeks (Braddick et al., 1989). Interestingly, this rapid improvement in temporal sensitivity appears to be specific to the orientation response and not to non-orientational mechanisms. If the infant is shown a phase reversing (PR) grating, where the grating is in a constant orientation but the black and white lines are periodically interchanged at 3 rps or 8 rps, a significant PR-VEP can be recorded for infants from birth. This supports the view that the OR-VEP is generated by a distinctive rapidly developing mechanism which is different from that responsible for the conventional PR response. Different ranges of temporal and spatial sensitivity have been found in the cell properties of the magnocellular and parvocellular pathways (Derrington and Lennie, 1984). It is quite possible that the OR-VEP may reflect orientation selectivity in cortical cells forming part of the 'parvo stream', while the 'magno stream' is involved in transmitting non-oriented information. However, another plausible interpretation is that non-oriented information is carried by subcortical pathways and analysis and that the magnocellular pathways are involved in fast temporal analysis, while the parvocellular carry only slow temporal information.

There is ample evidence from animal studies that the development of cortical function is strongly influenced by the nature of the visual input. To ask how far the development of orientation selectivity in human infants is a determinate process of maturation and how far it is a response to visual stimulation, we have looked at the onset of a significant orientation-reversal VEP in a group of healthy pre-term infants who have had many weeks of extra stimulation from the visual environment because of their prematurity compared to term infants (Atkinson et al., 1990). To date, no systematic difference is apparent in the proportion of pre-term infants showing positive OR-VEP responses in comparison with control term infants (nor are the pre-term infants significantly delayed compared to the term infants). This suggests that the initiation of cortical orientation selectivity may reflect a fixed time course of neural maturation which cannot easily be accelerated by additional visual input. Of course, it is quite possible that although the onset of cortical functioning is in some sense preset, the fine tuning of orientationally tuned mechanisms and use of this information in more perceptuo–cognitive tasks may be sensitive to the environmental input.

3 Directional motion

It is well established that young infants prefer moving to static visual displays. However, this in itself cannot be taken as evidence of the operation of true motion mechanisms. Any moving stimulus also produces a temporal modulation and it is well known that infants show a preference for full field flicker, i.e. temporal modulation without coherent motion. In general, true motion detectors are in evidence if a differential response to different directions of motion can be demonstrated. To measure such mechanisms in infants we have used designer stimuli consisting of two-dimensional random dot displays, which can have a particular direction of motion without the confounding presence of any dominant orientation component. In a similar way to the OR-VEP technique, above, we can generate a VEP to a change in the direction of motion of a set of random dots (Wattam–Bell, 1988; 1991). The first significant motion VEP for a velocity of 5 deg./sec. appears at around two months of age, with onset for higher velocities occurring later (Wattam–Bell, 1991). Parallel behavioural studies show once again that the velocity is a critical determinant in obtaining discrimination of relative motion at different ages (Wattam–Bell, 1990).

Many models of directional motion involve the comparison of spatially separated locations across a temporal delay. The maximum velocity at which the infant sees coherent motion (Vmax) will be determined by the spatial range and the minimum time interval at which relative motion detectors can operate. A change in the spatial range has been found to be the limiting factor determining changes in Vmax with increasing age from 12 weeks upwards (Wattam–Bell, 1990).

Many behavioural studies have concentrated on the lower rather than upper velocity threshold for detecting whether an object is moving at all (e.g. Aslin et al., 1988). The lowest detectable velocity (Vmin) or the corresponding minimum discrete displacement (Dmin) increases between eight and 20 weeks of age. However, most of these tasks do not require relative directional information, and the limit may be set by the simple spatial and temporal resolution of the infant's visual system at a particular age.

In conclusion, it appears that both the upper and lower limits for discriminating direction motion increase with age, although true directional detectors have not as yet been demonstrated prior to eight weeks of age. The magnocellular stream running from ganglion cells to V1, V2, V3 and MT seems to be related to information about direction. In the previous section we saw that some discrimination of orientations was possible at birth and it seems strange that information about direction of movement should not also be a built-in capability of the newborn, to provide fundamental information about 'where?' Indeed, the newborn does have a crude directional system already operating, as is evidenced by the optokinetic system giving optokinetic nystagmus (OKN) – a stabilizing mechanism which is present in the visual system of virtually every species. In adult humans, control of OKN is both cortical and subcortical, whereas in newborn infants only the subcortical mechanisms are functioning (Atkinson, 1979; Atkinson and Braddick, 1981). Thus it seems that the infant has at least two potential 'where' systems. One system is subcortical – the OKN mechanism, which operates maximally when the whole visual field is moving in a uniform direction. In normal adult perception this subcortical mechanism is normally suppressed by using the pursuit cortical mechanism to fixate and smoothly track a single object in the field of view. This later cortical system presumably operates within the magnocellularly based system and in developmental terms becomes functional later than the subcortical system (i.e. a few months post-natally).

Interestingly, several perceptual discriminations dependent on relative motion detection can be demonstrated in relatively young infants. For example, four month-olds can distinguish between three dimensional forms which differ only in terms of 'structure from motion' cues in random dot displays (Arterberry and Yonas, 1988). Many of the experiments on infants' abilities to make avoidance responses to looming objects (e.g. Yonas, 1981) and to reach for nearby objects demonstrate how the infant integrates spatial and temporal information into a unified percept of an object in a given location. However, many of these abilities may depend on applying crude heuristic rules (e.g. getting bigger means nearer) rather than using specifically tuned orientation and relative direction detectors to calculate the precise shape and direction in which the object is moving. At present there seems to be a conflict between the results of detailed experiments on discrimination of orientation and direction and the apparently sophisticated discriminations demonstrated in the newborn's capacity to make facial imitations. In particular, some have claimed that newborns use the higher order variable of 'pure movement' to imitate facial movements. Bower (1989) has suggested that newborns not only imitate real faces but will also correctly imitate the facial expression represented only as a series of strategically placed light bulbs (on the lips and corner of the mouth). This type of capacity seems to involve at the very least relatively finely tuned spatial and temporal detectors, which have not so far been in evidence in many of the psychophysical studies on older infants cited above.

4 Binocularity

As signals from the two eyes first interact at the cortex, any response dependent on detection of binocular correlation or disparity must be dependent on the operation of cortical rather than subcortical mechanisms. An example of such a response was first demonstrated using a sucking habituation paradigm (Atkinson and Braddick, 1976). Such responses are first seen on average around three to four months (Braddick et al., 1980; Fox et al., 1980; Held et al., 1980; Petrig et al., 1981; Braddick and Atkinson, 1983; Braddick et al., 1983). Binocular correlation and disparity detection appear to have the same onset time in individual infants, with significant VEP responses usually being recorded a little before significant behavioural preferential looking responses (Smith et al., 1988). These studies indicate that the onset of functioning of cortical binocular mechanisms is post-natal and such mechanisms cannot be used in any depth or distance judgements involved in spatial localization made prior to three months of age. As the thick cytochrome oxidase staining stripes in V2, which receive input largely from the magnocellular system, are thought to be the predominant location of disparity selective neurons, it seems likely that development of functioning in this pathway takes place a few months post-natally in the case of human development.

5 Differential onset of function in parvocellular and magnocellular streams

Above, the findings on development of three types of cortical detectors necessary for spatial representation have been considered. Each type is specific for coding information concerning a particular spatial attribute: orientation (necessary for shape analysis), directional movement (for analysing an object's trajectory) and correlation/disparity detection (for judging the object's relative depth or distance from other objects). Development of sensitivity to these attributes takes place post-natally and must depend on cortical rather than subcortical analysis.

Infants show sensitivity to changes of orientation before they show differential responses to the direction of movement and before they detect disparity change. Although not discussed here, infants in the first few weeks after birth show relatively sophisticated ability to detect differences based on colour alone, using isoluminant displays. When we consider the ideas expressed under the third approach above, involving segregation of information in magnocellularly based and parvocellularly based streams, it would seem plausible to suggest that the parvo become functional slightly before the magno. This dissociation between colour/form information and motion/depth information in early infancy has been noted previously by several investigators. For example, Bower (1984) suggested that infants up to around five months of age identified an object either in its static form or when moving along a trajectory – a static and moving object was treated as two objects. Others (e.g. Baillargeon et al., 1985) have noted that analysis of movement and local depth cues are linked. The infant sees two parts of a rod as one if they move along a common path behind a barrier. Two-month olds tend to perceive the bar as in two parts if the display is static. Perhaps it is easier to integrate features if both are first rapidly analysed in specialized detectors. If one analysing stream is poorly operating (e.g. that for directional movement) then integration cannot take place across the two streams to perceive a unified object.

6 Additional localizing mechanisms for analysing 'where' objects are in space – attentional changes in development

In addition to the 'where' mechanisms of relative direction and distance discussed above, we now have evidence of developmental changes in a number of spatially localizing mechanisms which seem to be intrinsically linked to development of attentional mechanisms. Two such mechanisms are discussed below – the first is the saccadic localizing processes for foveation and focal attention, the second is segmentation processes useful for triggering the refixation mechanism.

Saccadic refixation: Localization based on differences between target and background

If a conspicuous large moving object is made to appear suddenly in the peripheral visual field, infants will often move their eyes to foveate the object centrally. Many studies indicate that this response is present but unreliable and inaccurate at birth, but becomes a robust response by three months of age. We have tested such refixations in one to three month olds, systematically varying the salience of the peripheral target, with an identical, initially fixated central target either constantly visible or extinguished at the moment the peripheral stimulus appears (Braddick and Atkinson, 1988; Atkinson et al. in press). Even when an attempt is made to equate the peripheral targets for visibility above contrast threshold, one-month olds tend to be a little slower in refixating, and often fail to refixate if a competing central target is visible.

Segmentation: Localization based on texture detection

Julesz (1981) has proposed a theory of pre-attentive processing, defining classes of local features called 'textons' (such as edge segments and edge terminators) which enable rapid discrimination of texture patterns necessary for segmenting one object from another. Textons embody local phase relationships, i.e. the relative phase of the spatial frequency components making up the pattern elements. Relative phase information, together with amplitude or intensity, enables a complete specification of any pattern. Certain phase relationships will be 'special' in that they correspond to physical features of the outside world, e.g. the peaks–subtract phase relationship specifies a square-wave edge. A number of physiological models of relative phase detection in the brain have been proposed (e.g. Movshon et al., 1978; Pollen and Ronner, 1981). In general, all these models propose perception of configuration to be dependent on comparator detectors across spatial frequency channels. Such phase selectivity might, for example, be embodied in channels having even- and odd-symmetric receptive fields, and such mechanisms exist in the cortex rather than subcortex.

Applying these models, we have used tests of relative phase and texton discrimination to infer cortical functioning in young infants. It appears that one-month olds are insensitive to changes in relative phase, when viewing complex gratings made up of several superimposed spatial frequencies (Braddick et al., 1986b). Nor do they discriminate between texture patterns, differing in the type of texton detectors proposed for adults (Atkinson et al., 1986). Three-month olds have no difficulty with such discriminations.

One might imagine that an even simpler form of texture segmentation would be based on merely recognizing areas containing identical single features, such as an area composed of similarly oriented short line segments. This segmentation process would take place after primary visual mechanisms have registered the simple stimulus properties of orientation, and would be used to define distinct surfaces, objects and events in the visual world. In adults, the presence of orientation differences between line or edge segments making up two textures is very effective in determining visual segmentation (Beck, 1966; Olson and Attneave, 1970; Nothdurft, 1990). Several models of cortical mechanisms responsible for these pre-attentive processes have been proposed (e.g. Sagi and Julesz, 1985; Nothdurft, 1985). Their principal component is local differencing operations, acting on the output of neurons which are visual filters for specific properties. The differencing function may depend on connections within the cortex providing inhibitory information between pools of neurons specifying differing orientations (for more detailed discussion see Atkinson and Braddick, 1992).

We have started to look for evidence of these cortical inhibitory mechanisms in infants by measuring their ability to orient towards a rectangular area in the visual field, the boundary being defined by a change in orientation of the line segments making up the display. The rectangle is presented to either the left or right of a central fixation point. Using the forced choice preferential looking technique, we measure the infant's ability to detect the change of orientation within the rectangular patch compared to the background. What we have found is that this appears to be an impossible task for infants up to four months of age, and even at this age the response appears relatively weak (Braddick and Atkinson, 1991; Atkinson and Braddick, 1992). This relatively late onset of ability to segment on the basis of orientation suggests that other mechanisms for segmenting and localizing objects in the visual world, such as contrast cues and differential motion, are more robust and useful in the first few months. Of course, it may be that even in adult vision the primary segmenting mechanisms use crude intensity differences to parse the visual scene and that these decisions are then confirmed by additional information provided by elaborate texture and phase comparators. It may also be that the orienting mechanism necessary for rapid detection of objects in the peripheral field does not use information processed in the cortex very effectively, but relies instead on crude information provided by the superior colliculus. Because of these possibilities we are now using a new paradigm, testing orientation segmentation in central vision, without the need for the peripheral orienting response. If infants younger than four months are indeed able to use local orientation differences to define the boundary of an object, although this discrimination is not used in orienting, then it would seem that the deficit in young infants is one of integration of information between registration and response systems, rather than a failure to register *per se*. An analogy might be made here to a similar lack of integration – infants of a few months of age register certain 'impossible' perceptual events (such as one object passing through another solid object) but do not necessarily make the appropriate manual responses until a much later age (e.g. Baillargeon et al., 1985).

7 Some similarities between infant behaviour and deficits of spatial representation within certain neurological syndromes

Blind-sight

A comparison can be made between the visual behaviour of newborn infants, adults with neurological deficits resulting from cortical damage and primates with V1 lesions. Monkeys with V1 lesions (Humphrey, 1974) have difficulty seeing and reaching for a small object if a competing contour surrounds the object. A similar 'externality effect' is seen in infants under six weeks of age in discriminating shapes if surrounded by an identical outer contour and in recognizing faces if the outer hair lines of both faces are made identical using swimming caps (Milewski, 1976; Bushnell, 1982). The difficulty has been interpreted as one of selective attention, with the most salient contour (often the outer one) capturing visual attention so that other objects in the field are ignored. Evidence for this idea has been found in studies where infants fail to process the outer contour, their attention being captured by dynamic inner contours (Bushnell, 1982). This 'reverse' externality effect supports the idea that the deficit is one of limited selective attention, rather than an attentional salience *per se* of large external contours.

'Blind-sight' patients (reviewed in Weiskrantz, 1986; Cowey and Stoerig, 1991) can detect and discriminate flicker, orientation, wavelength and simple shapes in their blind half field, similar to the abilities of newborns. However, they have difficulty in matching complex patterns or identifying letter shapes and in discriminating between different directions of motion. There are several plausible interpretations of these deficits. One is that processing is intact but the patient lacks conscious access to the outcome of these discriminatory processes. Another possibility is that these difficulties result not only from deficits of retrieval but also from deficits of integration – cortical neurons underlying relative phase and motion discriminations must pool their outputs across different positions in space and time. If such integration does not take place, it would seem that relative discriminations might be impossible, while simple discriminations (e.g. moving versus stationary) might be possible.

Balint's Syndrome

A second neurological condition, called 'Balint's Syndrome', is characterized by what has been called 'sticky fixation'. Patients have difficulty disengaging from a centrally fixated target and shifting their eyes to a new target of interest in the periphery. The problem is not one of 'neglect' or optic apraxia but rather concerns the co-ordination of mechanisms controlling saccadic eye movements and selective attention to provide accurate spatial localizing. Balint's Syndrome involves bilateral lesions in parieto–occipital areas but may also involve the circuitry between the superior colliculus and parietal lobes for controlling shifts of attention (review by de Renzi, 1988). Very similar behaviour to Balint's patients has been seen in primates with bilateral parietal damage (Mountcastle, 1978) although Schiller has reported similar deficits with damage to the superior colliculus and frontal eye fields (Schiller, 1985). Again, the evidence for development of cortical streams would lend support to the idea that crude localizing of single targets can be carried out by subcortical collicular mechanisms, while more elaborate selective processes, to shift attention from one object to another, require cortical executive control from the striate and extrastriate cortex.

In summary, young infants would seem to manifest deficits which could be interpreted as a lack of integration of information across neurons both within a cortical stream and across different cortical streams. In some circumstances the lack of integration between the perceptual, cognitive and motor parts of the different systems representing spatial information is also apparent in the behaviour of older children carrying out spatial tasks of categorization. Below, a brief description of some preliminary research on spatial constructions is briefly discussed. Here developmental work can be compared with studies on adult neurological patients with spatial problems, namely constructional apraxia.

8 Spatial grouping activity – block construction tasks

Results from many studies have shown disorders of spatial analytic functioning in patients with focal lesions. Differences have been suggested for left and right lesions (e.g. McFie and Zangwill, 1960; Warrington et al., 1966; Arena and Gainotti, 1978). Patients with left damage tend to oversimplify and omit detail, while right posterior lesions cause difficulties with the overall configuration. Similar differences have been found in children with focal lesions (Stiles-Davis, 1988; Stiles-Davis et al., 1985). It appears that in spontaneous construction, young children normally progress through a series of stages. Simple spatial stacking strategies characterize the designs of 18 month olds, whereas by three years elaborate multiple constructions are generated, showing understanding of a variety of spatial relationships ('on', 'inside', 'next to', 'parallel to', 'joined to'). Children with right focal lesions do not progress beyond simple stacking strategies no matter what their age. In collaboration with Stiles we have been designing constructional tasks involving copying block designs, each design containing different spatial relationships appropriate to the abilities of different ages. Our aim was to extend the tests down to infants of 12 months and to incorporate the task into a whole set of functional vision tests (Atkinson et al., 1989).

The findings to date are interesting but puzzling. Firstly, it is almost impossible to persuade a one year old to construct anything but isolated stacks, placing one brick on top of another, no matter what design is presented to them to copy. Even taking into account their somewhat limited motoric skills and their love of demolition, this result suggests that they attend focally to one point in space at a time, without attending to the more global properties of a display. Infants around two years of age tend to build multiple towers, conjoined towers and lines of bricks end to end, and constructions containing arrays in at least two directions from a focal point. At a later stage (around three years) elaborate multifocal constructions involving enclosures, bridges and lateral extensions are both spontaneously made and copied. There are of course a number of different interpretations of these changes in behaviour. It could be that the limitation in one year olds is not in perceiving different spatial relations but in planning a strategy for construction to reproduce these relationships. Alternatively, the salience of particular relationships may change with age, although it is difficult to rationalize why multiple stacks and lines with a space between them should be more salient than the conjoined version. To attempt to dissociate these possibilities we are concurrently running a choice discrimination task in which we are attempting to elicit three different modes of response from each infant: visual perceptual discriminatory responses between the different constructions (without construction being involved), spontaneous construction and construction in a copying task. Preliminary results to date show that while 18 month olds readily discriminate between constructions containing either conjoined or separated stacks or lines of bricks, they usually only construct from a single locus at one time, building isolated stacks and lines.

By three years of age children appear to both discriminate and construct elaborate edifices, with multiple loci and relations. Again, this seems to demonstrate a lack of integration in younger children between information pertaining to perceptual discrimination and information necessary for planning and initiating motor-appropriate actions. It seems that construction of 'next-to' relations may require more analysis of the spatial array than a spatially well defined construction involving either 'on' or 'in' relations. In stacking, elements in the construction provide support for new items, whereas 'next-to' relations are not defined within a single locus – new items can be placed in any position around other items. As suggested above, infants by six months of age seem to be able to apply segmenting principles to separate one object from another in the orientation displays. However, representing the global layout of these objects spatially over a relatively large area of visual field requires not only parallel segmentation of multiple objects, but also continuous updating of this map to take into account self movement and object movement, plus the ability to switch between global and focal attention in order to plan and execute the appropriate actions. Although young infants seem to possess functioning basic mechanisms necessary for spatial representation, they seem to lack functioning integrative processes. Maturation of integration must develop largely in the period beyond infancy, so that the mature system can unconsciously represent the visual world, smoothly and continuously, perpetually updating the spatial analysis.

Conclusions

In summary it appears that infants are born with operating discriminatory mechanisms linked to appropriate orienting responses which enable them to set out a crude spatial map of the world around them. These mechanisms for 'where' and 'what' decisions are initially largely under subcortical control. In the first few weeks post-natally many of the specialized detectors in the cortex start to operate to carry out discriminations of various visual attributes such as orientation, colour and size. Separate cortical streams based on parvocellular and magnocellular distinctions have been hypothesized to underlie changes in visual capacity over the first six months of post-natal life. Parvocellular spatial representation systems for colour and form appear to be operational somewhat earlier post-natally than magnocellular-based systems for relative position, motion and depth. Integration of information within channels and across channels subserving different attributes takes at least four or five months to develop post-natally, and some of these processes do not seem to be mapped onto appropriate response systems until late in infancy. Even the spatial representations of two and three year olds seem to show some immaturities compared to the adult, although we are not yet confident that we know where the immaturities lie. The puzzle appears to be why the fine discriminatory systems such as those for discerning slant and disparity develop so rapidly post-natally, whereas other very necessary systems for manipulating objects and organizing the spatial layout seem to take years to become mature. Hopefully, close liaison between developmental psychology, neuroscience and cognitive philosophy will give us new insights in the future.

REFERENCES

Arena, R. and Gainotti, G. 1978: Constructional apraxia and visuoperceptive disabilities in relation to laterality of cerebral lesions. *Cortex*, 14, 463–73.

Arterberry, M. E. and Yonas, A. 1988: Infants' sensitivity to kinetic information for three-dimensional object shape. *Perception and Psychophysics*, 44, 1–6.

Aslin, R. N., Shea, S. L. and Gallipeau, J. M. 1988: Motion thresholds in 3-month-old infants. *Investigative Ophthalmology and Visual Science*, (suppl.), 29, 26.

Atkinson, J. 1979: Development of optokinetic nystagmus in the human infant and monkey infant: An analogue to development in kittens. In R. D. Freeman (ed.), *NATO Advanced Study Institute Series*. New York: Plenum Press.

Atkinson, J. 1984: Human visual development over the first six months of life. A review and a hypothesis. *Human Neurobiology*, 3, 61–74.

Atkinson, J. and Braddick, O. J. 1976: Stereoscopic discrimination in infants. *Perception*, 5, 29–38.

Atkinson, J. and Braddick, O. J. 1981: Development of optokinetic nystagmus in infants: An indicator of cortical binocularity? In D. F. Fisher, R. A. Monty and J. W. Senders (eds), *Eye movements: Cognition and visual perception*. Hillsdale NJ: Lawrence Erlbaum Associates.

Atkinson, J. and Braddick, O. J. 1989: Development of basic visual functions. In A. Slater and G. Bremner (eds), *Infant Development*. London: Lawrence Erlbaum.

Atkinson, J. and Braddick, O. J. 1990: The developmental course of cortical processing streams in the human infant. In C. Blakemore (ed.), *Vision: Coding and efficiency*. Cambridge: Cambridge University Press.

Atkinson, J. and Braddick, O. J. (1992): Visual segmentation of oriented textures by infants. *Behavioural Brain Research*, 49, 123–31.

Atkinson, J., Hood, B., Wattam-Bell, J., Anker, S. and Tricklebank, J. 1988: Development of orientation discrimination in infancy. *Perception*, 17, 587–95.

Atkinson, J., Braddick, O. J., Anker, S., Hood, B., Wattam-Bell, J., Weeks, F., Rennie, J. and Coughtrey, H. 1990: Visual development in the VLBW infant. *Transactions of the IVth European Conference on Developmental Psychology*, University of Stirling, 193.

Atkinson, J., Gardner, N., Tricklebank, J. and Anker, S. 1989: Development for examining functional vision (ABCDEFV). *Ophthalmic and Physiological Optics*, vol. 9, October.

Atkinson, J., Hood, B., Wattam-Bell, J. and Braddick, O. J. (in press): Changes in infants' ability to switch visual attention in the first three months of life. *Perception*.

Atkinson, J., Wattam-Bell, J. and Braddick, O. J. 1986: Infants' development of sensitivity to pattern 'textons'. *Investigative Ophthalmology and Visual Science*, (suppl.), 27, 265.

Baillargeon, R., Spelke, E. S. and Wasserman, S. 1985: Object permanence in 5-month-old infants. *Cognition*, 20, 191–208.
Banks, M. S. and Bennett, T. J. 1989: Optical and photoreceptor immaturities limit the spatial and chromatic vision of human neonates. *Journal of the Optical Society of America*, A, vol. 5 (12), 2059–79.
Beck, J. 1966: Effect of orientation and shape similarity on perceptual grouping. *Perception and Psychophysics*, 1, 300–2.
Bower, T. G. 1984: *Development in Infancy*. San Francisco: Freeman.
Bower, T. G. 1989: The perceptual world of the new-born child. In A. Slater and G. Bremner (eds), *Infant Development*. London: Lawrence Erlbaum Associates.
Braddick, O. J. and Atkinson, J. 1983: Some recent findings on the development of human binocularity: A review. *Behavioural Brain Research*, 10, 141–50.
Braddick, O. J. and Atkinson, J. 1988: Sensory selectivity attentional control, and cross-channel integration in early visual development. In A. Yonas (ed.), *20th Minnesota Symposium on Child Psychology*. Hillsdale NJ: Lawrence Erlbaum Associates.
Braddick, O. J. and Atkinson, J. 1991: Infants' and adults' segmentation of oriented textures. *Investigative Ophthalmology and Visual Science*, (suppl.), 32 (4), 1045.
Braddick, O. J., Atkinson, J., Julesz, B., Kropfl, W., Bodis-Wollner, I. and Raab, E. 1980: Cortical binocularity in infants. *Nature*, 288, 353–65.
Braddick, O. J., Atkinson, J., Wattam-Bell, J. and Hood, B. 1989: Characteristics of orientation-selective mechanisms in early infancy. *Investigative Ophthalmology and Visual Science*, (suppl.), 30: 313.
Braddick, O. J., Wattam-Bell, J. and Atkinson, J. 1986a: Orientation-specific cortical responses develop in early infancy. *Nature*, 320 (6063), 617–19.
Braddick, O. J., Wattam-Bell, J. and Atkinson, J. 1986b: Development of the discrimination of spatial phase in infancy. *Vision Research*, 26 (8), 1223–39.
Braddick, O. J., Wattam-Bell., J., Day, J. and Atkinson, J. 1983: The onset of binocular function in human infants. *Human Neurobiology*, 2, 65–9.
Bronson, G. W. 1974: The postnatal growth of visual capacity. *Child Development*, 45, 873–90.
Bushnell, I. W. R. 1982: Discrimination of faces by young infants. *Journal of Experimental Psychology*, 33, 298–30.
Cowey, A. and Stoerig, P. 1991: The neurobiology of blindsight. *Trends in Neurosciences*, 14 (4), 140–5.
Damasio, A. R. and Benton, A. L. 1979: Impairments of hand movements under visual guidance. *Neurology*, 29, 170–8
de Renzi, E. 1988: Oculomotor disturbances in hemispheric disease. In C. W. Johnston and F. J. Pirozzolo (eds), *Neuropsychology of Eye Movements*. Hillsdale NJ: Lawrence Erlbaum Associates.
Derrington, A. M. and Lennie, P. 1984: Spatial and temporal contrast sensitivities of neurons in lateral geniculate nucleus of macaque. *Journal of Physiology*, 357, 219–40.
De Yoe, E. A. and Van Essen, D. C. 1988: Concurrent processing streams in monkey visual cortex. *Trends in Neurosciences*, 11, 219–26.
Dubner, R. and Zeki, S. M. 1971: Response properties and receptive fields of cells in an anatomically defined region of the superior temporal sulcus. *Brain Research*, 35, 528–32.
Felleman, D. J. and Van Essen, D. C. 1991: Distributed hierarchical processing in the primate cerebral cortex. *Cerebral Cortex*, 1, 1–47.
Fox, R., Aslin, R. N., Shea, S. L. and Dumais, S. T. 1980: Stereopsis in human infants. *Science*, 207, 232–324.
Garey, L. and de Courten, C. 1983: Structural development of the lateral geniculate nucleus and visual cortex in monkey and man. *Behavioural Brain Research*, 10, 3–15.
Held, R., Birch, E. E. and Gwiazda, J. 1980: Stereoacuity of human infants. *Proceedings of the National Academy of Sciences USA*, 77, 5572–4.
Hickey, T. L. 1981: The developing visual system. *Trends in Neurosciences*, 4, 41–4.

Horton, J. C. and Hubel, D. H. 1980: Cytochrome oxidase stain preferentially labels intersections of ocular dominance and vertical orientation columns in macaque striate cortex. *Society of Neuroscience*, Abstr., 6, 315.

Humphrey, A. L. and Hendrickson, A. E. 1980: Radial zones of high metabolic activity in squirrel monkey striate cortex. *Society of Neuroscience*, Abstr., 6, 315.

Humphrey, K. 1974: Vision in a monkey without striate cortex: a case study. *Perception*, 3, 241–56.

Johnson, N. H. 1990: Cortical maturation and the development of visual attention in early infancy. *Journal of Cognitive Neuroscience*, 2, 81–95.

Julesz, B. 1981: Textons, the elements of texture perception and their interactions. *Nature*, 290, 92–7.

Kritchevsky, M. 1988: The elementary spatial functions of the brain. In J. Stiles-Davis, M. Kritchevsky and U. Bellugi (eds), *Spatial Cognition, Brain Bases and Development*. Hillsdale NJ: Lawrence Erlbaum Associates.

Livingstone, M. S. and Hubel, D. H. 1984: Anatomy and physiology of a color system in the primate visual cortex. *Journal of Neuroscience*, 4, 309–56.

Livingstone, M. S. and Hubel, D. H. 1988: Segregation of form, color, movement, and depth: anatomy, physiology, and perception. *Science*, 240, 740–49.

Maunsell, J. H. R. and Newsome, W. T. (1987): Visual processing in monkey extrastriate cortex. *Annual Review of Neuroscience*, 10, 3416–68.

McFie, J. and Zangwill, O. L. 1960: Visual–constructive disabilities associated with lesions of the left cerebral hemisphere. *Brain*, 83: 243–60.

Maurer, D. and Lewis, T. L. 1979: A physiological explanation of infants' early visual development. *Canadian Journal of Psychology*, 33, 232–52.

Milewski, A. E. 1976: Infants' discrimination of internal and external pattern elements. *Journal of Experimental Child Psychology*, 22, 229–46.

Mountcastle, V. B. 1978: Brain mechanisms for directed attention. *Journal of the Royal Society of Medicine*, 71, 14–28.

Movshon, J. A., Thompson, I. D. and Tolhurst, D. J. 1978: Spatial summation in the receptive fields of simple cells in the cat's striate cortex. *Journal of Physiology*, 283, 53–77.

Nothdurft, H. C. 1985: Sensitivity for structure gradient in texture discrimination. *Vision Research*, 25, 1957–68.

Nothdurft, H. C. 1990: Texton segregation by associated differences in global and local luminance distribution. *Proceedings of the Royal Society of London*, B, 239, 295–320.

Olson, R. and Attneave, F. 1970: What variables produce similarity grouping? *American Journal of Psychology*, 83, 1–21.

Pearlman, A. L., Birch, J. and Meadows, J. C. 1979: Cerebral color blindness: An acquired defect in hue discrimination. *Annals of Neurology*, 5, 253–61.

Petrig, B., Julesz, B., Kropfl, W., Baumgartner, G. and Anliker, M. 1981: Development of stereopsis and cortical binocularity in human infants: Electrophysiological evidence. *Science*, 213, 1402–5.

Pollen, D. and Ronner, S. F. 1981: Phase relationship between adjacent simple cells in the visual cortex. *Science*, 212, 1409–11.

Ratcliff, G. and Davies-Jones, G. A. B. 1972: Defective visual localization in focal brain wounds. *Brain*, 95, 49–60.

Sagi, D. and Julesz, B. 1985: 'Where' and 'what' in vision. *Science*, 228, 1217–19.

Schiller, P. H. 1985: The superior colliculus and visual function. In I. Darian-Smith (ed.), *Handbook of Physiology – The Nervous System III. Sensory Processes*, Part I. Bethesda MD: American Physiological Society.

Schneider, G. E. 1969: Two visual systems. *Science*, 163, 895–902.

Slater, A., Morison, V. and Somers, M. 1988: Orientation discrimination and cortical function in the human newborn. *Perception*, 17, 597–602.

Smith, J., Atkinson, J., Braddick, O. J. and Wattam-Bell, J. 1988: Development of sensitivity to binocular correlation and disparity in infancy. *Perception*, 17, 395–6.

Stiles-Davis, J. 1988: Spatial dysfunctions in young children with right cerebral hemisphere injury. In J. Stiles-Davis, M. Kritchevsky and U. Bellugi (eds), *Spatial Cognition, Brain Bases and Development*. Hillsdale NJ: Lawrence Erlbaum Associates.

Stiles-Davis, J., Sugarman, S. and Nass, R. 1985: The development of spatial and class relations in four young children with right cerebral hemisphere damage. Evidence for early spatial constructive deficit. *Brain and Cognition*, 4, 388–412.

Ungerleider, L. G. and Mishkin, M. 1982: Two cortical visual systems. In D. G. Ingle, M. A. Goodsale and R. J. Q. Mansfield (eds), *Analysis of Visual Behavior*, 549–86. Cambridge Mass: MIT.

Van Essen, D. C. and Maunsell, J. H. R. 1983: Hierarchical organization and functional streams in the visual cortex. *Trends in Neurosciences*, 6, 370–5.

Van Sluyters, R. C., Atkinson, J., Banks, M. S., Held, R. M., Hoffmann, K.-P. and Shatz, C. J. 1990: The development of vision and visual perception. In R. C. Van Sluyters (ed.), *Visual Perception: The neurophysiological foundations*. San Diego: Academic Press.

Warrington, E. K., James, M. and Kinsbourne, M. 1966: Drawing disability in relation to laterality of cerebral lesion. *Brain*, 89, 53–82.

Wattam-Bell, J. 1985: Analysis of infant visual evoked potentials (VEPs) by a phase-sensitive statistic. *Perception*, 14, A33.

Wattam-Bell, J. 1988: The development of motion-specific cortical responses in infants. *Investigative Ophthalmology and Visual Science*, (suppl.), 29, 24.

Wattam-Bell, J. 1990: The development of maximum velocity limits for direction discrimination in infancy. *Perception*, 19 (3), 369.

Wattam-Bell, J. 1991: Displacement limits for the discrimination of motion direction in infancy. *Investigative Ophthalmology and Visual Science*, (suppl.), 32 (4), 964.

Weiskrantz, L. 1986: *Blindsight. A case study and implications*. Oxford: Clarendon Press.

Yonas, A. (1981): Infants' responses to optical information for collision. In R. N. Aslin, J. R. Alberts and M. R. Petersen (eds), *The Development of Perception: Psychobiological perspectives*, New York: Academic Press.

Zeki, S. M. 1973: Colour coding in rhesus monkey prestriate cortex. *Brain Research*, 53, 422–7.

Zeki, S. M. 1977: Colour coding in the superior temporal sulcus of rhesus monkey visual cortex. *Proceedings of the Royal Society of London*, Series B 197, 195–223.

Zeki, S. M. and Shipp, S. 1988: The functional logic of cortical connections. *Nature*, 335, 311–17.

Zihl, J., von Cramon, D. and Mai, N. 1983: Selective disturbance of movement vision after bilateral brain damage. *Brain*, 106, 313–40.

15
Computing 'where' and 'what' in the visual system

Oliver Braddick

The ability to extract spatial relations seems to be a skill that is deeply embedded in our visual system: we need little in the way of deliberation or special strategy to judge, say, that points A B C lie in a straight line; that D is not vertically above E but slightly to the right; that X is above the centre of circle O, and so on. This chapter looks at some evidence that may help us to understand how such tasks are performed – or at least to pose the question better. It outlines the possible roles of two kinds of neural encoding: first in terms of receptive fields responding to specific spatial configurations, and second in terms of the 'local sign' of neurons responding to features at particular locations. Broadly these might be thought of as signalling 'what' and 'where' respectively, although the encoding of spatial relations, by either mechanism, is potentially relevant both for identifying an object and for locating it. Insofar as location and identity are distinguishable kinds of information, we can ask whether they are necessarily associated, so that a feature and its location are always registered together in perception. In visual search experiments we can find examples of dissociations of 'where' and 'what'. In particular, it is sometimes possible to perceive the location of a target without reliable information about what is located there. However, the relation between 'what' and 'where' at various scales can only be understood in terms of perception as a flexible and controllable process, which can use multiple forms of spatial information to meet the demands of different visual tasks. The particular neural systems discussed in the first part of this chapter may provide the elements that are used in this process, but the control must depend on other, less 'bottom-up' mechanisms.

1 Parallel arrays and 'neural images'

At the levels where we think we understand the functional architecture of the visual system, it consists of richly parallel and topographical arrays. That is, we can identify a population of nerve cells, for example the alpha class of retinal ganglion cells, which all have a common form of receptive field, and these fields cover the visual field in a complete and orderly array. (In this particular case the coverage can be elegantly visualized anatomically; Wässle, Peichl and Boycott, 1983). Other types of cell have different receptive field properties, and although the different types may be intermingled anatomically, each appears to have its own coverage of the field. They can be thought of as forming multiple arrays. At the level of the visual cortex, the number of such distinct arrays is large, if each different value of preferred orientation and spatial frequency is taken as defining a distinct array. The two-dimensional pattern of information carried by a specific array is sometimes referred to as a 'neural image' (Robson, 1980).

Speaking this way makes some strong assumptions about the organization of vision. It has anatomical implications: such an array would only be a functional entity if the connections between cells making it up were systematically structured, in a way different from the connections between cells that were members of different arrays. This systematic structuring should somehow reflect the spatial relationship between different receptive fields. Cells with similar receptive field properties do not necessarily form a separate, anatomically coherent structure, and the evidence that they have systematic patterns of interconnection is in fact slight, although there are some clear instances (Wiesel and Gilbert, 1983). Nevertheless, the idea of parallel arrays which carry information about neural images is basic to our current understanding of spatial vision. It is embodied in the idea of neural maps (see, for example, Cowey, 1979), although an orderly topographic layout on a neural surface is not actually an essential criterion for such an array.

2 A special status for 'location' as a feature

If we wanted to avoid the assumptions implied by speaking of these arrays, we could simply consider each cell as showing a combination of response selectivities for a number of different dimensions, i.e. as best activated by a particular size of spot, a particular colour, a particular direction of movement, a particular location and so on. This would not require the relationship between two cells, responding to the same size spot at different locations, to be treated any differently from the relationship between two cells responding to different sized spots at the same location. However, location is not usually treated simply as another item in the list of dimensions showing specificity, but rather as the organizing principle for maps or images describing the other dimensions. This reflects a special status for spatial location as against other visual properties. Perhaps the reason for this status is that spatial coherence is the primary defining property of a visual object. Kubovy (1981) draws an interesting contrast with auditory objects. In vision, two colours at one location lead to perception of one object, but one colour at two locations gives two objects. In audition, space is not primary in the same way, so that the same frequency or modulation from two spatially separate sources can be fused into a single auditory object, while two frequencies or different modulation patterns from the same source can be segmented as two auditory objects.

The special role of space in vision is also intrinsic to Treisman's feature-integration theory of attention (Treisman and Gelade, 1980; Treisman, 1988). Her theory rests largely on evidence that, at least in some circumstances, conjunctions of visual features require focused attention to bind them together. However, this clearly does not apply to the conjunction of a feature with its location, as illustrated by Figure 15.1. Spatial location, then, does not act simply in a way analogous to other features, but rather as a metric or medium within which other features or properties can be represented. The

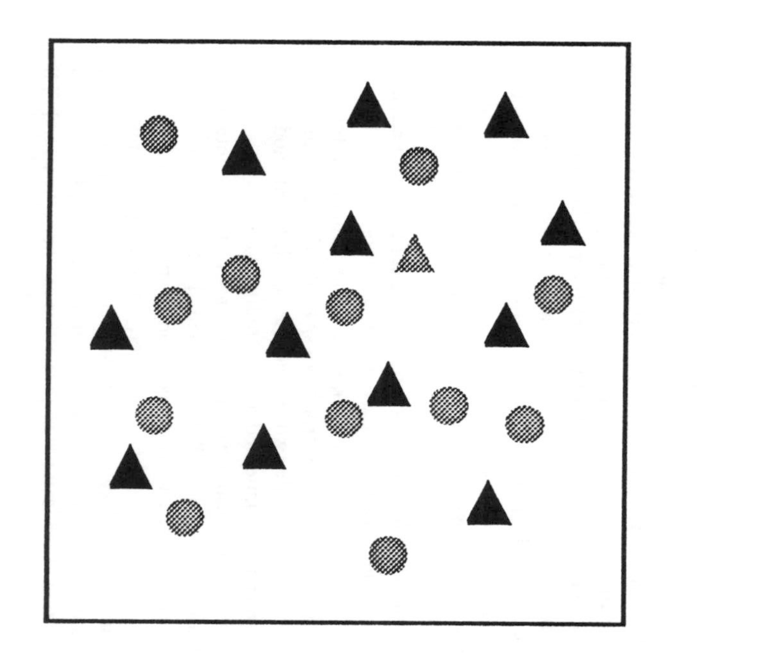

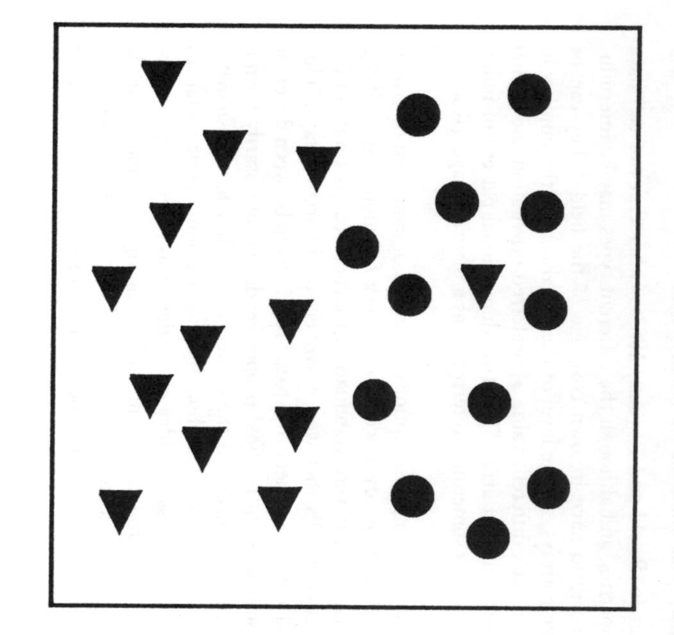

Figure 15.1 Targets for visual search defined by a conjunction of properties. (a) The target is the only shaded triangle in the array. Some of the other objects are shaded and some are triangles, i.e. the target cannot be distinguished by any single feature. Detection of this 'conjunction target' is slow and appears to depend on focal attention (Treisman & Gelade, 1980). (b) The target is the only triangle in the right half of the array. Some of the other objects are triangles and some are in the right half. However, when defined by this conjunction of a feature with spatial location, the target 'pops out' very readily.

relation of focal attention to the perceptual *use* of location information will be discussed further below.

3 Two limitations

Most work relating spatial judgements to the framework of neural spatial arrays has accepted, usually implicitly, two limitations. First, although in real life spatial judgements are commonly three-dimensional, the majority of psychophysical experiments have been concerned with spatial relations in a fronto-parallel plane. Most visual neurophysiology, similarly, has mapped receptive fields two-dimensionally and has related cortical layout to the two-dimensional layout of the visual field. Some cortical neurons do respond to stereoscopic disparity (Poggio and Poggio, 1984) and so can be thought of as having three-dimensional receptive fields. Early suggestions that arrays of these neurons might represent three-dimensional topography in an orderly way (Blakemore, 1970) have not been pursued, although much theorizing about stereopsis rests on proposals for systematic connections within such arrays (Blake and Wilson, 1991). In any case, stereo disparity is only one out of many sources of visual depth information, and our general ability to appreciate three-dimensional relationships must rest on a much more integrative system than disparity-selective neurons alone. At present, we have essentially no idea what the neural details of such a system might look like.

Second, the neural basis of spatial judgements has usually been discussed without much consideration of different possible frames of reference. Many judgements are concerned with relationships between locations, e.g. linear alignment or the equality of spatial intervals, that can be specified independently of the frame of reference. In any case, most of the experimental work, both psychophysical and physiological, has been carried out in ways which would not distinguish whether visual operations were performed in a retinotopic, body centred or externally defined frame. Receptive fields and topographic mapping in visual cortical areas are generally believed to be defined retinotopically, but explicit tests of this are rare (Wurtz, 1969). A parietal brain area has been studied (Andersen, Essick and Siegel, 1985) where eye position information is integrated with retinal information, but cells here appear to have retinotopic receptive fields modulated by gaze direction, rather than receptive fields and arrays organized directly in a head-centred frame. The brain mechanisms of head-body or object-centred representation are therefore still unknown.

The two-dimensional, retinotopic framework of the discussion in this chapter results from the simplifications of experimentation, rather than from any assertion that this is necessarily the nature of the process. Our perceptual world is certainly three-dimensional and in many respects invariant for changes in observer viewpoint. However, this should not lead us to dismiss the possibility that many operations encoding visual spatial relations may indeed be carried out on two-dimensional retinotopic information, with the outcome being transformed or transplanted into other frames of reference.

4 Receptive field structure and local sign

If we accept the framework of spatial arrays, then a visual neuron belonging to one of these arrays potentially carries spatial information in two distinct ways. First, its receptive field contains a particular layout of excitatory and inhibitory regions which determines how it will respond to any spatial distribution of light and dark within the field.[1] This *receptive field structure* means that the neuron's response will reflect certain spatial relationships in the stimulating pattern. For example, the presence of a vernier break in a vertical line will lead to activation of some neurons, with receptive fields including the region of the break, whose preferred orientations are slightly off vertical (Sullivan, Oatley and Sutherland, 1972; Watt, Morgan and Ward, 1983). More precisely, neurons whose preferred orientations are slightly off vertical one way, say clockwise, will be more strongly activated than those off vertical in the opposite direction. This balance of activation carries information about the presence and direction of the vernier break. Second, the neuron's receptive field is in a particular location in the array and so its activation can serve to indicate stimulation at the corresponding particular location in the visual field. To use a term going back to Lotze in the nineteenth century, each neuron in the array has a *local sign*. The local signs of the set of active neurons then contain information about spatial relationships between the visual features that have activated them.

On the account in terms of receptive field structure, all these spatial judgements are essentially forms of shape judgement, and presumably our ability to make them reflects the sensitivity of mechanisms whose primary purpose is to describe what kind of object is in front of us. On this view *what* and *where* questions are two ways of reflecting the same processes. The local sign approach, in contrast, posits a second kind of information which is primarily concerned with *where* something is (although it leaves open the possibility that such *where* information may be combined from different locations to yield information about *what* shape is displayed).

5 Joint representation in space and spatial frequency

To understand current thinking and experiments on how spatial relationships are encoded, it is necessary to describe patterns and receptive field properties in spatial frequency terms. A wealth of psychophysical and physiological experiments have shown that spatial contrast information is transmitted by visual mechanisms ('channels' or 'filters') which each respond to a restricted band of spatial frequencies (Braddick, Atkinson and Campbell, 1978; De Valois and De Valois, 1988). For instance, a pattern consisting of three parallel, equally spaced lines can be analysed as containing a wide range of spatial frequencies, or in other words information at different spatial scales. Channels responding to the lower, coarse-scale frequencies will not clearly resolve the lines but could signal the broad symmetry or otherwise of the light distribution; higher spatial frequencies provide fine-grain information defining the individual lines and the gaps between them. If the middle line is shifted to a slightly asymmetrical, off-centre position between the other two, this will in fact produce changes over a wide range of scales. Observers can detect extraordinarily small deviations from the symmetry in this display, and Klein and Levi (1985) have analysed the possible basis of this spatial judgement in some detail. They argue that the signals used by the visual system are best described in terms of what they dub a 'viewprint': a joint distribution of activity across channels responding to different spatial scales and across different locations. ('Viewprint' is an analogy to 'voiceprint' – the graphic representation of speech in terms of acoustic frequencies varying over time.) Their computations show that the critical differences between patterns with the middle line centred or off-centre will occur in quite specific regions of the viewprint, i.e. at particular combinations of location and channel scale.

6 Spatial judgements of Gabor patches

One way of analysing the issue experimentally is to test spatial judgements with stimuli that cannot activate a wide range of spatial frequency channels, such as 'Gabor patches' of the kind illustrated in Figure 15.2. The receptive fields of cortical neurons are quite well modelled by functions of this kind (Daugman, 1980), and the rather narrow band of spatial frequencies that they typically respond to fit well with psychophysical findings of spatial-frequency-selective channels. A pattern of the kind shown in Figure 15.2b would be good at activating neurons or channels which responded to the high frequency of the stripes within each patch. However, the relationship between the patches would have very little effect on the level of high-spatial-frequency activity; it would be reflected rather in its distribution over space, that is in the local signs of the activated high-spatial-frequency neurons. To reflect the spatial relationship between the patches directly in terms of receptive-field structure, a large receptive field would be required, and such receptive fields would typically respond only to low spatial frequencies. Geisler and Davila (1987) and Burbeck (1987) have shown that, for a wide range of separations, performance in judging the alignment and separation of Gabor patches is good even though the frequencies contained in the patch are very far from those which could be detected by a receptive field spanning both patches. That is, precise spatial judgements can be made that must be based on local-sign-type information. Only when very small separations were involved, leading to the ultra-high sensitivity to positional relationships that is sometimes known as hyperacuity (Westheimer, 1981; Braddick, 1984), did the results suggest that information specific to the receptive field structure was required for optimal performance.

A related but slightly different approach has been taken by Kooi, De Valois and Switkes (1987). They asked their subjects to make judgements of the alignment of Gabor patches which could either be the same or different in terms of their spatial frequency, their orientation or the axis of colour space in which their modulation was defined. Thus, if the upper patch was a horizontal grating, modulated in luminance, of centre frequency 2 cycles/degree, the lower might be identical, but it might instead be vertical, or modulated between red and green at constant luminance, or of centre frequency 6 cycles/degree. In the latter cases, all our knowledge of visual channel properties implies that the two patches would not be registered within a common array. Nevertheless, Kooi et al. found very little degradation in performance on the alignment judgement when the two patches differed in any of these ways. The conclusion must be that location information can be obtained from one array, and compared with location from other arrays, without any marked loss of precision. Again, spatial judgements do not seem to rest on the activation of particular types of receptive field, but can use local sign information in a rather flexible way. Even greater flexibility is suggested by the work of De Valois et al. (1989). They tested a wide variety of displays, in all of which subjects had to judge the proportions in which a spatial interval was

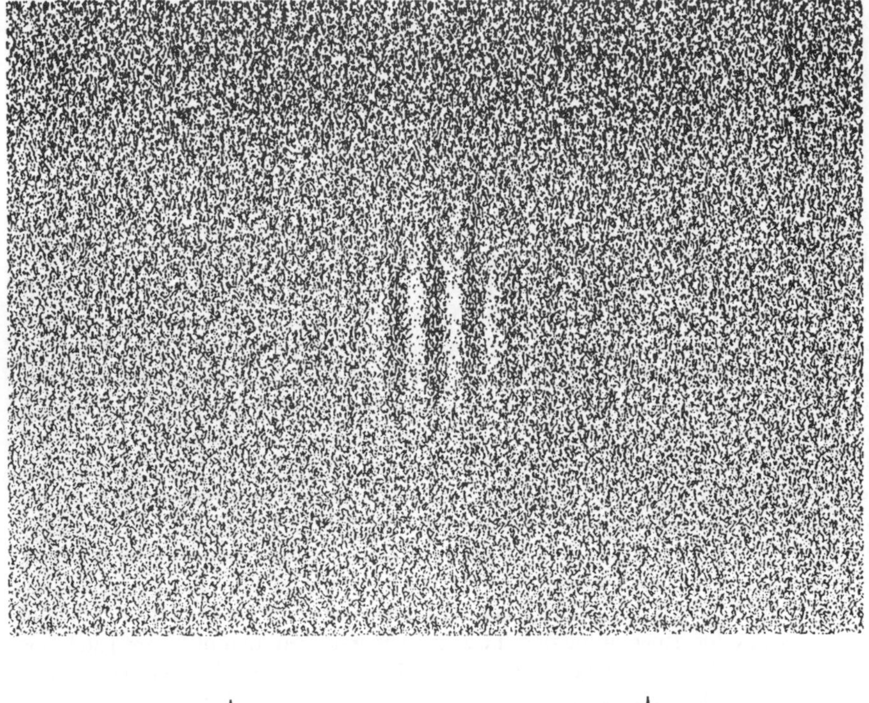

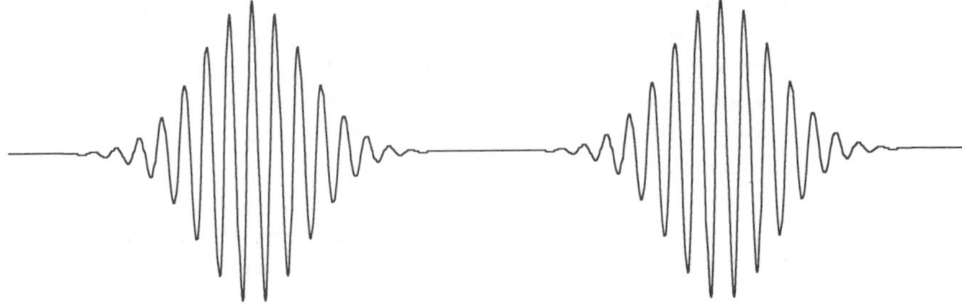

Figure 15.2 (a) A two-dimensional 'Gabor patch'. The pattern consists of a sine-wave grating modulated in contrast horizontally and vertically by Gaussian functions. It has a well-defined location but also contains a limited band of spatial frequencies around the centre frequency of the sine wave. (b) The profile of luminance across a display containing two separated Gabor patches. In experiments by Geisler and Davila (1987) and Burbeck (1987), subjects had to make judgements of the separation of the patches.

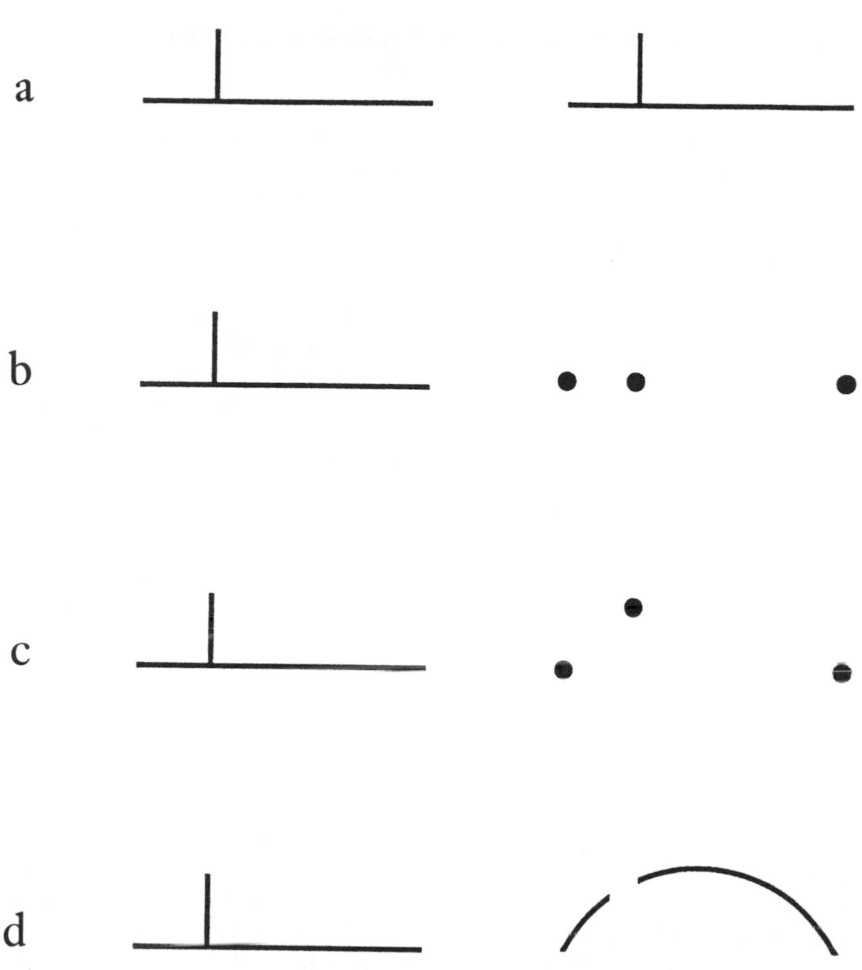

Figure 15.3 Displays used by De Valois et al. (1989). In each case, subjects had to judge whether the line on the right is divided by the marker in the same proportions as the line on the left. The precision of judgement when the two displays were identical (a) was hardly impaired when the line was implicitly defined by its end-points (b); when the marker was not horizontally aligned with the line (c); or when the whole line was transformed into a curved arc (d).

divided by some kind of mark. Figure 15.3 illustrates some examples. They found that proportions could be compared across the very different kinds of display illustrated, with minimal effects on the precision of the comparison. Clearly, the spatial relationships which the subjects were judging do not have to be represented within a common array of similar receptive fields for the comparison to be made; whether the judgement is regarded as one of relative position or one of shape, it must be based on a very generalized kind of spatial information.

7 Use of local sign information: Higher-level receptive fields v. visual routines

Invoking the concept of local sign, however, cannot by itself give a satisfactory account of how we compute spatial relations. If two neurons with appropriate local signs are activated by two spots vertically above one another, we still need to know what operations on the neural array can determine that this is an instance of vertical alignment, as distinct from some other pairings of activation which would indicate misalignment. One possibility is that the operations are in fact performed by higher-level neurons receiving a pattern of inputs from the array we are considering, a pattern which is systematically related to the spatial relationship we are interested in. That is to say, the relationship is signalled by the selective activity of higher-level receptive fields. Certainly, at the levels where we are familiar with receptive field structure, that structure is determined largely by the local signs of direct or indirect inputs to the neuron concerned. So, for example, chains of connections originating from photoreceptor cells in appropriate locations establish the concentric centre and surround of retinal ganglion cell receptive fields, and the elongated form of simple cortical receptive fields must depend at least in part on the locations of the retinal ganglion receptive fields which (via the l.g.n.) feed the cortical cells. Thus, it might be argued, local sign is only a stepping stone in articulating receptive-field-based theories of how spatial relationships are represented.

This view, though, seems hard to maintain. The experiments we have discussed emphasize that the process of making precise spatial judgements is highly flexible. Local sign information can be drawn from an array of elements having common spatial frequency, orientation and colour properties, or it can be combined almost equally efficiently from arrays responding to very different stimuli. It can be integrated over distances that are very large compared to the structures from which the local sign information is derived. It can be used to make very varied kinds of judgement, of alignment, symmetry, equality of intervals that may be in different positions and directions and so on. To explain all this in terms of the structures of higher-level receptive fields would require an extraordinary variety of such receptive fields, spanning pairs of locations on every scale up to a large fraction of the field of view, and drawing on every possible combination of types of input information. It must be conceded that this is a plausibility argument only, and that it might be possible to propose some general but economical receptive field organization, which could extract the information for all the different possible kinds of spatial judgement. However, parallel patterns of hard-wired connections are unlikely to be the only means of information processing available to the visual system, and 'visual routines', as proposed by Ullman (1984), may offer a promising alternative. Ullman discusses judgements such as continuity, and interior v. exterior of a closed curve (Figure 15.4), which human beings can perform very readily, but for which receptive-field-based mechanisms, if they were feasible at all, would suffer an even worse combinatorial problem than for the metrical spatial judgements we have considered so far. He outlines how these tasks could be approached by 'routines' composed of a few basic operations (such as 'indexing' or marking a location, stepping incrementally along a contour, etc.), with these operations being combined in different, controlled sequences for different tasks. There is some experimental evidence (e.g. Jolicoeur, Ullman and Mackay, 1986)

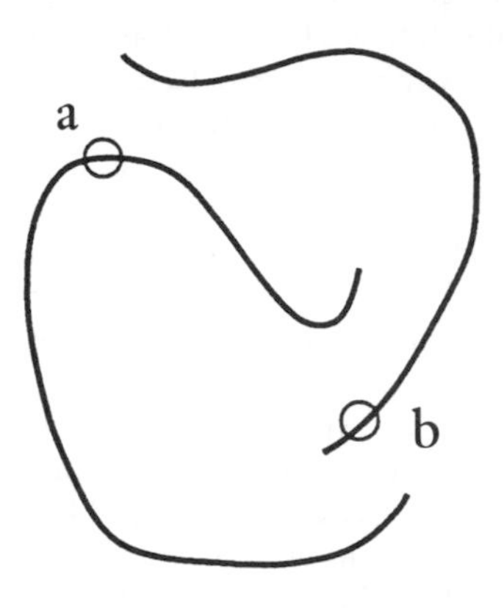

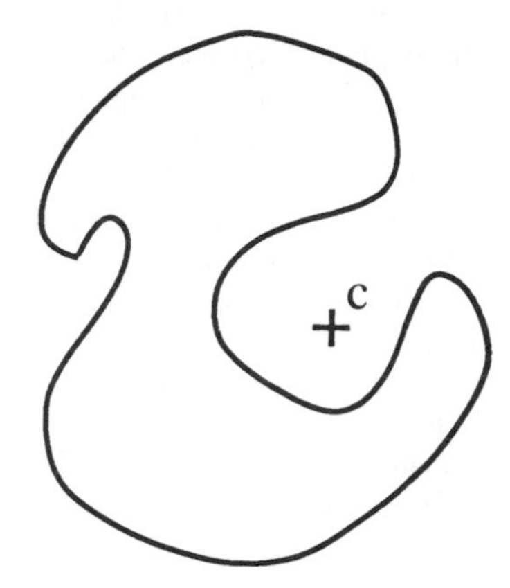

Figure 15.4 Do points A and B lie on the same continuous contour? Does point C lie inside or outside the closed curve? These are examples of spatial judgements which people find quite easy, but which are difficult to account for in terms of uniform processing by parallel arrays of spatial filters. Ullman (1984) suggests that they are performed by 'visual routines'.

that the time required for such judgements is consistent with this kind of serial sequence of operations.

If we are to develop a 'visual routines' approach also to judgements of metrical properties such as alignment, we need proposals for the 'instruction set' of elementary operations from which they are made up, and the nature of the information on which they operate. It seems reasonable to suppose that the latter would include both the local sign of indexed features (although in what format we do not know) and local configurational properties derived from receptive field structures.

8 Coupling of local sign and feature identity

In the neural image, or its elaborated version in the 'viewprint', local sign is a parameter within an array, and each array is characterized by a particular spatial filtering operation or receptive field type. Which of the multiple parallel arrays are most strongly activated would depend on the feature that was being located. In such a system, then, one might expect that location information within the array ('where') would be closely coupled to information about the identity of the located feature ('what') derived from which array was functioning. However, the evidence from Kooi et al.'s experiment implies that location information can be used in a way more or less independent of the filtering operation from which it was derived. Does this imply that 'where' information can be partly or wholly detached from information about 'what' is 'there'?

9 Access to 'where' and 'what' information in visual search

This question of the separation of 'where' and 'what' information has been closely linked to debates about the distinction between pre-attentive and post-attentive visual processing. In particular, Anne Treisman's feature-integration theory (Treisman and Gelade, 1980; Treisman, 1988) proposes that each of the different features of an object, such as its colour and shape properties, is represented within a separate feature map. The idea of a map suggests that location information is implicitly associated with every feature, and this might be expected to provide a basis for linking the features that share a common location. However, according to Treisman's theory, such linkage is not automatic, but requires focal attention to access the common location in the different maps and so to bind the features into an object. Visual search tasks, where subjects have to find a target object in an array, have provided the main experimental basis for this account. The theory implies that when a target has been identified by a combination of features, its location has been the subject of explicit attention and so should be available for accurate report. There is experimental data to support this (Treisman and Gelade, 1980: experiment VIII). However, when a target object can be identified by a single feature (e.g. the only orange letter among a mixture of pink and blue letters), the theory does not require any access to location information. In fact, in the same experiment, Treisman and Gelade found that subjects made large errors in reporting the location of a single-feature target on many trials when they correctly reported its *identity* (i.e. whether it was distinctive as the only orange letter, or as the only H).

Visual search experiments (Treisman and Gelade's experiment I, and many since) indicate that the detection of a single-feature target is achieved by a process which acts in parallel over the whole array, yielding what has been graphically described as 'pop-out' of the target. Treisman and Gelade's finding of errors in reporting location would imply that this process can extract 'what' information about the nature of the target from the feature maps without accompanying 'where' information. Perhaps local sign, although intrinsic to the organization of a feature map, can only become explicit by means of a serial, spatial indexing operation?

Treisman and Gelade's experiment has recently been the subject of some detailed technical criticisms by Johnston and Pashler (1990). They argue that the form of Treisman and Gelade's arrays was not optimal for subjects' reporting of location, and also that subjects may be able to infer the identity of an undetected target simply because they find that shape-defined targets, say, are harder to detect than colour-defined targets. In experiments that were basically similar to Treisman and Gelade's, but designed to minimize these problems, they found a much closer statistical association between correct reports of target location and identity. In fact, on the relatively small number of occasions on which these did not go together, target location was usually reported correctly with an error in target identification. That is, subjects could sometimes access 'where' information without 'what', the opposite of Treisman and Gelade's result.

'Where' without 'what' also summarizes the findings of Sagi and Julesz (1985a, b). In this case, only one stimulus dimension was involved. Both the background and the target elements were oriented line segments: all the background elements were parallel obliques, and the targets were either vertical or horizontal. The briefly flashed arrays were followed by a masking pattern. Subjects could perform a task requiring target location with a stimulus–mask interval so short that they could not reliably perform a task requiring identification of the target as horizontal or vertical.

The ability to report the location of objects we cannot identify is not, of itself, surprising or particularly interesting. For instance, if a small letter is presented in isolation in peripheral vision, it could be readily localized as a dark spot even if the details needed to identify it were below the limit of resolution in that part of the field. In visual search experiments, however, there is an apparent paradox in 'where' without 'what', because the target to be located is only defined by virtue of its distinctive features. So to know that the target is present at all, the features that identify it must have been processed, and yet they are apparently not available for overt identification.

However, in displays like those of Sagi and Julesz, target detection does not depend on extracting absolute orientation. Since the background elements are homogeneous, the presence of local differences in orientation will locate the target. Sagi and Julesz take their results to indicate that the location information is associated with the detection of such differences or 'feature gradients'. It is plausible that local differencing operations should play an important part in pre-attentive processing, and many cortical cells show wide-field inhibitory interactions of a kind which could provide a physiological basis for such operations (Allman, Miezin and McGuinness, 1985). Local differencing could serve to define texture boundaries for segmentation of surfaces and objects, and a simple signal of a local discrepancy could be a crude but effective guide for a fast orienting response – the 'pop-out' of Treisman's parallel process. For these purposes the difference signal would have to be associated with explicit location information. Absolute identification would then require further processing (which according to Sagi and Julesz requires focal attention, although this is a separate issue).

10 Coarse and fine localization

We re-examined this question experimentally for two reasons (Atkinson and Braddick, 1989). First, Sagi and Julesz tested what they called 'localization' and 'identification' by rather complex and indirect tasks involving multiple targets. Secondly, it seems extremely likely that localization is not a process of fixed resolution, and that a broad localization of a target as being in one general region of the field of view can be made under circumstances in which more precise judgements of location are impossible. We therefore devised two versions of the 'where' task; for 'coarse where' judgements the subjects had to report only whether the target was in the upper or lower half of the 6 × 6 array, while 'fine where' required judgements accurate to a single array position, as shown in Figure 15.5. The difficulty of each task could be increased by shortening the time interval between the array and a following masking pattern. Thus, the threshold stimulus-to-mask interval for each of these judgements could be compared with that for a 'what' judgement of whether the target was horizontal or vertical.

In this experiment, the 'coarse where' judgement was possible at intervals for which the mask made either 'fine where' or 'what' judgements impossible. The latter two had thresholds that were not reliably different. Thus, our subjects could report location information for targets whose orientation they could not identify, but the localization was relatively crude. Finer localization required at least as long an interval as identification.

The results imply that at least coarse location information can be dissociated from feature identity, a result in line with those of Sagi and Julesz. However, we did not find the same relation between identification and fine localization. This dependence on the precision of localization could be interpreted in at least two ways:

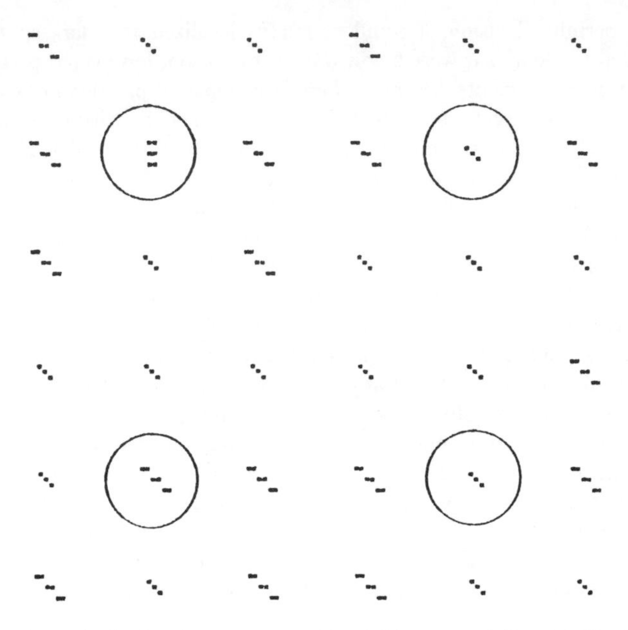

Figure 15.5 A typical stimulus array used in Atkinson and Braddick's (1989) visual search experiment. The circles are shown to mark the four 'inside corner' positions and were not present in the displays seen by the subjects. For the 'fine where' task, subjects had to distinguish whether the target (here a vertical line segment) was (a) in any of these four 'inside corner' positions, or (b) in any other position in the array; this requires localization accurate to one array position. In the 'coarse where' condition they had to distinguish targets in the upper half from those in the lower half of the display. In the 'what' condition they had to distinguish horizontal from vertical targets, regardless of location.

1 Any process of making locality information explicit is independent of the processes that enable feature identification. It must require something like a local differencing of orientation, but not its absolute value. However, location judgements on a finer scale require more time to process the stimulus information. (Watt (1987), and others, have independent experimental evidence on this point.) The variation of spatial precision might be a graded continuum, but it must then be the case that at some level of precision (approximately that required for our 'fine where' judgements) the judgement requires as long an interval as is also needed for correct identification, even though the localization and identification processes are not directly related.
2 'Coarse where' and 'fine where' judgements depend on qualitatively distinct processes. The additional processing required for the fine over the coarse judgement is associated with the processing needed for absolute identification of orientation (as an example only, both might depend on focal attention). Thus, 'coarse where' can be faster than 'what', but 'fine where' and 'what' judgements require similar stimulus timing.

11 Multiple mechanisms of spatial information processing

On either view localization judgements, at least at the coarse scale, have a different neural basis from the identification of shape. It has been proposed that the perception of space and the identification of objects are served by two broad and distinct divisions of the visual nervous system, with the former involving a pathway to the parietal lobe and the latter to temporal regions of the cortex (Mishkin, Ungerleider and Macko, 1983). A third broad subdivision of the visual system is the subcortical route to the superior colliculus. This latter system must be concerned with localization, since it is closely involved in the control of fixation eye movements (Wurtz and Albano, 1980) and probably more generally in visual orienting responses (Schneider, 1969), but has little or no capacity for form recognition. It is possible that the level of spatial precision required by the 'coarse where' task (but not the fine) could be achieved by a subcortical mechanism whose basic function was a fast, crude alerting to potentially significant visual events. Of course, the subcortical, parietal and temporal pathways have numerous interconnections and must act in a closely integrated way in normal visual processing.

Although the evidence for separate neural sub-systems for space and object perception is quite strong, the issues discussed earlier in this chapter make it clear that location and shape information cannot necessarily be neatly partitioned along this divide. The relative activation of receptive-field types is most obviously a means of transmitting shape information, but we have seen that it is at least potentially a source of information about the relations between locations. The 'where' tasks in all the experiments discussed above have required subjects to report a target's position within a well-defined spatial matrix: a matrix whose boundaries and structure have been generally laid out conspicuously in the masking pattern as well as in the transitory stimulus array. Arrays with various possible target locations, as illustrated in Figure 15.6, differ in their symmetry properties and proportions, and can be quite plausibly thought of as shape differences on an appropriate scale. If receptive field structures on this scale are available, they could provide a suitable means of encoding target location (although these 'receptive fields' would need to have inputs of a moderately high level of abstraction, to encode the position of a target defined by the discrepancy of its orientation or other features). Alternatively, as suggested above, the relationship between target and array might be extracted by some kind of visual routine, possibly a routine which can achieve finer resolution by iterating its operations for a longer time. Our understanding of the parietal space perception pathway and the temporal object recognition pathway has not yet reached the stage where we know whether these systems are distinguished at all by the degree to which they use receptive-field-based processing compared with visual routines.

The principal conclusion from our experiment is that it is misleading to look for a unitary and homogeneous process underlying visual localization. Rather, localization is a function or functions which can operate at a variety of different scales, and probably using a variety of different neural mechanisms. It is also clear that experiments which differ in detail in the tasks required of the subject (such as Treisman and Gelade's, Sagi and Julesz's, and ours) may yield apparently quite different results on the relationship between localization and identification. These differences emphasize the fact that it is no longer adequate to think of vision as purely a bottom-up or data-driven system, transforming and encoding image information by a series of hard-wired stages. Rather,

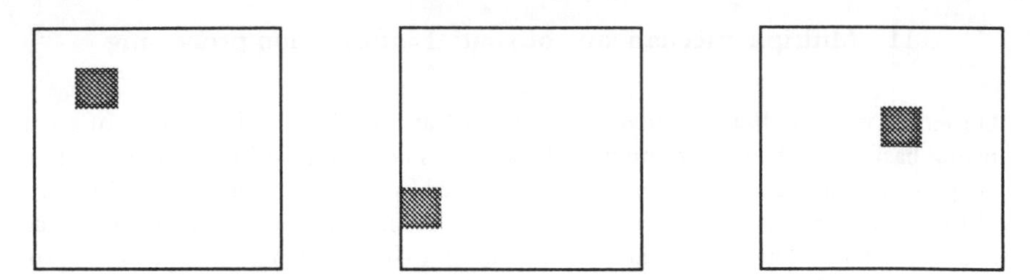

Figure 15.6 Different possible target locations, in arrays like that in Figure 15.5, can be thought of as giving structures of different overall shape.

the visual machinery must be subject to active control processes, specifying the information that the perceiver currently requires and setting up necessary but temporary routes of information flow. The routes set up to handle different tasks may integrate, or separate, 'where' and 'what' information in different ways. The two modes of representing spatial information that we have discussed, by means of receptive field specificity or by local sign, are best seen as two kinds of building block available within an actively controlled system. These and other components can be used in the flexible assembly of processes to provide the spatial information that will meet our continuously changing visual needs.

NOTE

1 Strictly, many neurons have selective responses to spatial patterns which cannot be derived from their layout of excitatory and inhibitory regions, since they combine inputs in a more complicated way than simple linear summation. However, receptive field structure provides the most intuitive way of thinking about a neuron's spatial properties, and including the complications of non-linearity would not alter the way issues of spatial representation are treated in this chapter.

REFERENCES

Allmann, J., Miezin, F. and McGuinness, E. 1985: Stimulus-specific responses from beyond the classical receptive field: Neurophysiological mechanisms for local–global comparisons in visual neurons. *Annual Review of Neuroscience*, 8, 407–30.

Andersen, R. A., Essick, G. K. and Siegel, R. M. 1985: Encoding of spatial location by posterior parietal neurons. *Science*, 230, 456–8.

Atkinson, J. and Braddick, O. J. 1989: 'Where' and 'what' in visual search. *Perception*, 18, 181–90.

Blake, R. and Wilson, H. R. 1991: Neural models of stereoscopic vision. *Trends in Neurosciences*, 14, 445–52.

Blakemore, C. 1970: The representation of three-dimensional space in the cat's striate cortex. *Journal of Physiology*, 205, 471–97.

Braddick, O. J. 1984: Visual hyperacuity. *Nature*, 308, 228–9.

Braddick, O. J., Atkinson, J. and Campbell, F. W. 1978: Channels in vision: Basic aspects. In R. Held, H. Leibowitz and H. L. Teuber (eds), *Handbook of Sensory Physiology, vol. VIII, Perception*. Heidelberg: Springer–Verlag.

Burbeck, C. 1987: Position and spatial frequency in large-scale localization judgements. *Vision Research*, 27, 417–27.

Cowey, A. 1979: Cortical maps and visual perception. *Quarterly Journal of Experimental Psychology*, 31, 1–17.

Daugman, J. G. 1980: Two-dimensional spectral analysis of cortical receptive field profiles. *Vision Research*, 20, 847–56.

De Valois, K. K., Lakshminarayanan, V., Schlussel, S. and Nygaard, R. 1989: Perceptual space as an elastic manifold. *Investigative Ophthalmology and Visual Science* (suppl.), 30, 486.

De Valois, R. L. and De Valois, K. K. 1988: *Spatial Vision*. Oxford: Oxford University Press.

Geisler, W. S. and Davila, K. D. 1987: Can receptive-field properties explain hyperacuity thresholds? *Investigative Ophthalmology and Visual Science* (suppl.), 28, 137.

Johnston, J. C. and Pashler, H. 1990: Close binding of identity and location in visual feature perception. *Journal of Experimental Psychology: Human Perception and Performance*, 16, 843–56.

Jolicoeur, P., Ullman, S. and Mackay, M. 1986: Curve tracing: A possible basic operation in the perception of spatial relations. *Memory and Cognition*, 14, 129–40.

Klein, S. A. and Levi, D. 1985: Hyperacuity thresholds of 1 sec arc: Theoretical predictions and empirical validation. *Journal of the Optical Society of America A*, 2, 1170–90.

Kooi, F. L., De Valois, R. L. and Switkes, E. 1987: Vernier acuity with Gabor patches. *Investigative Ophthalmology and Visual Science* (suppl.), 28, 360.

Kubovy, M. 1981: Concurrent-pitch segregation and the theory of indispensible attributes. In M. Kubovy and J. R. Pomerantz (eds), *Perceptual Organization*. Hillsdale, NJ: Erlbaum.

Mishkin, M., Ungerleider, L. and Macko, K. A. 1983: Object vision and spatial vision: Two critical pathways. *Trends in Neurosciences*, 6, 414–17.

Poggio, G. F. and Poggio, T. 1984: The analysis of stereopsis. *Annual Review of Neuroscience*, 7, 379–412.

Robson, J. G. 1980: Neural images: The physiological basis of spatial vision. In C. S. Harris (ed.), *Visual Coding and Adaptability*. Hillsdale, NJ: Erlbaum.

Sagi, D. and Julesz, B. 1985a: 'Where' and 'what' in vision. *Science*, 228, 1217–19.

Sagi, D. and Julesz, B. 1985b: Detection vs discrimination of visual orientation. *Perception*, 14, 619–28.

Schneider, G. E. 1969: Two visual systems. *Science*, 163, 895–902.

Sullivan, G. D., Oatley, K. and Sutherland, N. S. 1972: Vernier acuity as affected by target length and separation. *Perception and Psychophysics*, 12, 438–44.

Treisman, A. 1988: Features and objects. *Quarterly Journal of Experimental Psychology*, 40A, 201–38.

Treisman, A. and Gelade, G. 1980: A feature-integration theory of attention. *Cognitive Psychology*, 12, 97–136.

Ullman, S. 1984: Visual routines. *Cognition*, 18, 97–159.

Wässle, H., Peichl, L. and Boycott, B. B. 1983: A spatial analysis of on- and off-ganglion cells in the cat retina. *Vision Research*, 23, 1151–60.

Watt, R. 1987: Scanning from coarse to fine spatial scales in the human visual system after the onset of a stimulus. *Journal of the Optical Society of America A*, 4, 2006–21.

Watt, R., Morgan, M. J. and Ward, R. 1983: The use of different cues in vernier acuity. *Vision Research*, 23, 991–5.

Westheimer, G. 1981: Visual hyperacuity. *Progress in Sensory Physiology*, 1, 1–30.

Wiesel, T. N. and Gilbert, C. D. 1983: Morphological basis of visual cortical function. *Quarterly Journal of Experimental Physiology*, 68, 525–43

Wurtz, R. H. 1969: Comparison of the effects of eye movement and stimulus movement on striate cortex neurons of the monkey. *Journal of Neurophysiology*, 23, 987–94.

Wurtz, R. H. and Albano, J. E. 1980: Visual motor function of the primate superior colliculus. *Annual Review of Neuroscience*, 3, 189–226.

16

Image indeterminacy: The picture theory of images and the bifurcation of 'what' and 'where' information in higher-level vision

Michael Tye

Introduction

A long and impressive philosophical history attaches to the view that visual images represent the world in something like the manner of pictures. This century, however, the pictorial view of images has come under heavy philosophical attack. Three basic sorts of objections have been raised. First, there have been challenges to the sense of the view: mental images are not seen with real eyes; they cannot be hung on real walls; they have no objective weight or colour. What, then, can it mean to say that images are pictorial? Secondly, there have been arguments that purport to show that the view is false. Perhaps the best known of these is founded on the charge that the pictorial conception of images cannot satisfactorily explain the indeterminacy of many mental images. Finally, there have been attacks on the evidential underpinnings of the view. Historically, the philosophical claim that images are picture-like rested primarily on an appeal to introspection. And today introspection is taken to reveal a good deal less about the mind than was traditionally supposed. This attitude towards introspection has manifested itself in the case of imagery in the view that what introspection really shows about visual images is not that they are pictorial but only that what goes on in imagery is experientially much like what goes on in seeing.

So the pictorial view of visual images has been variously dismissed by many philosophers this century as nonsensical, or false or without support. Perhaps the most influential alternative has been a view that is now commonly known as descriptionalism (Dennett, 1981; Shorter, 1952; Pylyshyn, 1981 and Tye, 1991). This view has some similarity with the claim made by the behaviourist, J. B. Watson (1928, pp. 75–6), that imaging is talking to oneself beneath one's breath; for the basic thesis of descriptionalism is that mental images represent in the manner of linguistic descriptions. This thesis, however, should not be taken to imply that during imagery there must be present inner tokens of the imager's spoken language either in any movements of the imager's larynx or in the imager's brain. Rather, the thought is that there is a neural code within which the relevant descriptional representations are constructed.

Descriptionalism remains popular in philosophy today, and it also has significant support in contemporary psychology. Nevertheless, the tide has begun to turn again back towards the pictorial view, and I think it is fair to say that the pictorial conception of images is now quite widely regarded as intelligible, perhaps true, and certainly not a theory in search of facts to explain. What has been responsible for this change in attitude more than anything else is the work of some cognitive psychologists, notably Stephen Kosslyn. In response to a large body of experimental data on imagery, Kosslyn and his co-workers have developed an empirical version of the pictorial view that seems much more promising than any of its philosophical predecessors.

I have discussed in detail elsewhere (Tye, 1988 and 1991) the question of whether it is really possible to formulate a coherent account of imagistic representation in pictorial or quasi-pictorial terms. My primary interest in the present chapter is not in this question (although I shall make some comments on behalf of pictorialism here) nor in an evaluation of the experiments that allegedly support the pictorial approach to images. Instead, I want to focus for the most part on the issue of whether the pictorial view can accommodate image indeterminacy. I shall pursue this issue in the first instance with reference to some well-known philosophical objections. What I shall argue is that these objections are successful against one possible pictorial model of imagery but there remain two other models which are not touched by them.

I shall then turn, in section II, to certain pieces of psychological data which have been held to create difficulties for the pictorial approach. I shall try to show that objections based on these pieces of empirical data are only partially successful, since they carry no weight against one of the two pictorial models of imagery that emerge unscathed from the previous section. The objections I shall address here are not themselves directly concerned with image indeterminacy – in so far as indeterminacy enters into the discussion it does so in some, but not all, of the responses that are available to adherents to the pictorial view.

In section III, I shall argue that the one surviving pictorial model can handle satisfactorily the indeterminacies involved in certain neuropsychological cases of impaired imagery. It will be seen in the course of my examination that my answer to the question of how these cases are to be handled draws on the hypothesis that there are two higher level visual systems in primates, one concerned with the identification of *what* is present and the other with *where* it is positioned.

Finally, in section IV, I shall comment on the relationship between my preferred pictorial model of imagery and the view of Stephen Kosslyn; I shall also make some further remarks on image indeterminacy and 'what' and 'where' information in connection with Kosslyn's theory.

I

Most philosophers are prepared to grant that mental images are frequently indeterminate. There have been some notable exceptions, however. Bishop Berkeley, for example, seems to have believed that images can never be abstract or sketchy in any way. Exactly why Berkeley takes this view is not entirely clear. It appears that he thinks of mental images as we would think of clear photographs. It also appears that, as far as Berkeley is concerned, introspection dictates such a position. It is interesting to note that most philosophers take precisely the opposite view. Indeed, the standard philosophical argument against a pictorial or photographic conception of images (and percepts) assumes, on the basis of what is 'given' in introspection, that images (and percepts), unlike pictures or photographs, are sometimes less than fully determinate. For example, Daniel Dennett says the following:

> Consider the Tiger and his Stripes. I can dream, imagine or see a striped tiger, but must the tiger I experience have a particular number of stripes? If seeing or imagining is having a mental image, then the image of the tiger *must* – obeying the rules of images in general – reveal a definite number of stripes showing, and one should be able to pin this down with such questions as 'more than ten?', 'less than twenty?' (Dennett, 1981, p. 55)

When Dennett speaks of the rules of images in general here, he is referring to the rules of pictorial representation. Since he holds that a mental image of a tiger can be indeterminate with respect to the issue of stripes, the conclusion he draws is that mental images are not pictorial.

This argument evidently has considerable force, if visual images are conceived on the model of clear photographs. But it ignores two other possible pictorial models. The first of these has it that images are like photographs generally, blurred in some cases and sharp in others. Since a blurred photograph of a tiger need not have a definite number of picture stripes on it, nor relatedly need it represent exactly how many stripes are on the pictured tiger, the fact that a visual image of a tiger may fail to reveal a definite number of stripes does not show that the pictorial view of images is mistaken (Fodor, 1981).

The second alternative model for images is provided by drawn pictures.[1] Here again indeterminacy with respect to the tiger's stripes is no obvious problem: think, for example, of an outline drawing of a tiger, or a sketch done in the style of an impressionist painting. In the former case, the absence of stripes on the picture does not mean that it represents an anomalous stripeless tiger. Rather, the question of how many stripes the tiger has is not addressed. In the latter, the indeterminacy runs exactly parallel to that for the blurred photograph.

Dennett (1981) has also offered a more sophisticated version of the standard philosophical objection from indeterminacy. Dennett now explicitly acknowledges that pictures can sometimes be fuzzy. However, he holds that for any visual property *P*, you can look at what is pictured by any given picture and determine whether or not *P* is present, *unless* the relevant portion of the picture is vague or unclear. Dennett maintains that this is not so for mental images. Suppose, for example, I tell you to imagine a woman wearing a red dress. If I then ask you such questions as 'Does your imaged woman have shoes with laces?', 'What colour are her shoes?' and 'Does her dress extend below her knees?' you may not be able to say. The problem is not that the relevant portions of your image are obscured. Rather, in Dennett's view, your mental image simply doesn't go into some of the details.

The general point of Dennett's objection, then, can be put this way. Imagining, unlike picturing, is subject to a fourfold distinction: it's one thing to imagine an object with visual property *P*; it's another to imagine that object without *P*; it's a third thing to imagine the object with the relevant portion obscured so that one cannot tell if *P* is present; and it's a fourth to imagine the object without one's image going into the presence or absence of *P*. By contrast, in the case of picturing, there are only the first three distinctions – the fourth is inapplicable.

This argument is an improvement over the earlier one. However, it is still badly flawed. The most obvious problem concerns the assumption that pictures cannot be

Figure 16.1

Figure 16.2

representationally indeterminate with respect to the presence or absence of any visual property *P*, unless either they are partially blurred or indistinct (or the objects they represent are represented as seen from a viewpoint relative to which *P* is not visible). Consider, for example, a clear black and white photograph of an apple. It neither represents that the apple is red nor that it is not red. Similarly, the cartoon picture of a man about to bowl a ball surely does not represent him as lacking ears, nostrils, eyebrows and lips (Figure 16.1). But equally it conveys no positive information about these facial features and their aspects (for example, shape, size). It simply does not go into such matters. In this respect, it is like the following description: 'the tall thin man with a large nose who is stooped over and about to bowl a large ball'.

In like fashion, an outline drawing of a crocodile (Figure 16.2) is representationally indeterminate. For example, there is no representation of the crocodile's colour, whether

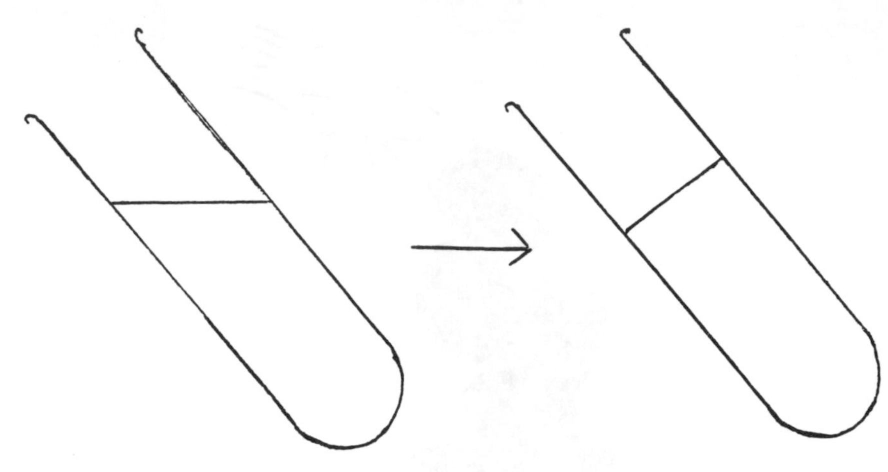

Figure 16.3

it has scales on its back or legs, and whether there is any difference in the texture of its skin on its upper and lower parts.

In none of these cases is there any blurring or obstruction due to the viewpoint relative to which the depicted items are represented. So Dennett's second argument is also unsuccessful.

I know of no other philosophical arguments from indeterminacy that carry any weight against the picture theory. I want next to look at some empirical data cited by Zenon Pylyshyn (1981).

II

Let me begin with Pylyshyn's appeal to the inclined beaker experiment. When four-year-old children are shown an inclined beaker containing a coloured fluid, and are later asked to draw what they saw, they usually draw the fluid level perpendicular to the sides of the beaker (Figure 16.3).

Pylyshyn maintains that the pictorial view of images cannot explain this fact. If images are like inner pictures that simply reproduce 'snapshots' taken in perception then there will be no differences between the children's images in the above case and those of adults. So the errors the children produce will be inexplicable. By contrast, according to Pylyshyn, if visual images are inner descriptions, the children's errors are easily accounted for. In imaging, what the children do is to generate a description out of concepts they possess. Since they lack the concept 'geocentric level', or so Pylyshyn supposes, they cannot construct the same description as an adult.[2] So they bring into play a related concept, usually 'perpendicular', and they produce a description that does not quite match the observed situation. So an error results in their drawings.

Is there really a problem here for the pictorial view? Let us accept, for the sake of argument, Pylyshyn's supposition that, on a pictorialist approach, the *percepts* the children form, as they see the inclined beaker, are like clear photographs. The question I want to address is whether a pictorial view of the *mental images* the children subsequently generate in order to draw what they saw is threatened by the results of the experiment.

To answer this question, three pictorial possibilities need to be distinguished: the children's images are also like clear photographs, image generation here being simply a matter of retrieving a stored percept of the beaker and fluid; alternatively, their images are like blurred photographs, blurring being due to the decay or fading of the percept during storage or else due to a defect in the retrieval system; finally, their images are like drawn pictures, constructed with the assistance of processes that examine the stored percept. What I shall now argue is that only the third of these possibilities is defensible.

The suggestion that the children's images are like blurred photographs straightforwardly explains why the children do not draw the fluid level correctly, since an appropriately blurred photograph could easily fail to specify precise information about the fluid level orientation. Of course, an explanation is also needed as to why the fluid level is drawn *perpendicular* to the sides. But this can be provided by supposing that the children, lacking any precise inner representation of the fluid level, decide to draw it in a way that comports with typical fluid levels (i.e. ones in untilted containers). Still, there remains a serious problem for the view that the children's images are like blurred photographs: no plausible account is available for why the *fluid level* rather than some other aspect or part of the inclined beaker is blurred.

The claim that the children's images are like clear photographs provides no immediate explanation of the fact that the children draw the fluid level perpendicular to the sides. Moreover, as Ned Block (1983) has noted, experiments on children who are eidetic imagers are incompatible with the claim. For example, when eidetic children are shown a picture of an elephant projected onto a surface, they can later image it very clearly on the same surface. But when they are asked to trace around their images, they produce stereotyped drawings of an elephant similar to those produced by other children of their age who have been asked to draw an elephant from a presented picture. By contrast, when they are asked to trace out the elephant outline in the original picture using tracing paper, they do so without any significant distortions.[3] This makes no sense, if images are simply retrieved percepts akin to clear photographs. For, on that hypothesis, both perceptual and image tracing should produce the same result.

A further objection is that the image generation process cannot simply be a matter of retrieving a stored photographic percept, blurred or clear. This is shown by a number of experiments conducted by Kosslyn (1980; also Kosslyn et al., 1982). For example, it has been found that subjects require more time to form images of more detailed line drawings of animals. Similarly, when subjects are shown pictures of parts of a single animal on separate sheets of paper and are then asked to put them together mentally, thereby memorizing the appearance of the composite animal, the time it takes to form an image of the *whole* animal picture, upon later request, increases linearly with the number of pages earlier used to present it. These results demonstrate that images are generated from separate components rather than being simply retrieved all of a piece.

So mental images cannot be either mechanically produced blurred copies of pictorial percepts (as perhaps John Locke supposed) or mechanically produced clear copies of such percepts (as, for example, Berkeley held). But there remains the possibility that images are like drawn pictures constructed with the aid of concept-driven processes that inspect, decompose and then reconstitute the original percept. This proposal can

Figure 16.4

explain the drawings of Pylyshyn's children as follows: when the children view the inclined beaker, they pay little attention to the orientation of the fluid level, and so they fail to store any specific information about it. Instead, they store information about what the beaker itself looks like, together with *some* information about the fluid, e.g. its colour, and the general information, couched in a quasi-linguistic or non-pictorial format, that the beaker contains the fluid.[4] They call upon this information when later they draw inner pictures. Since they lack any precise stored representation of the fluid level, they represent it, in their mental pictures, in a way that matches standard fluid levels with which they are familiar in pots, pans and other ordinary containers.[5] So they construct inner drawings of the beaker with the fluid level perpendicular to its sides. Inspection of these drawings leads to the production of public drawings similar to the one on the right side of Figure 16.3. So, *contra* Pylyshyn, the results of the inclined beaker experiment can be accounted for without adopting the view that images are inner descriptions.

The hypothesis that images are like drawn pictures also accommodates the data above from Kosslyn on image generation. Moreover, it finds strong support in the results of another experiment performed by Kosslyn. Here subjects were first told to observe grids in which there were block letters (Figure 16.4). They were, then, presented with a grid that was empty (except for containing one or two X marks in its cells) together with a lowercase cue, and they were asked whether the corresponding uppercase block letter *would* occupy the cells with one or two X marks. This is shown in Figure 16.5.

The positions of the X marks in the empty grid were varied for each of the letters. Now if images are constructed a part at a time (as they will be, if generating an image is like drawing a picture) then the response times should change with the positions of the X marks (Kosslyn, 1988, p. 1622). This was found to be the case. Furthermore, the times of response varied in just the ways that would be expected on the hypothesis that the image parts are constructed in the order in which the letters are typically drawn. For example, it was found that when a separate group of subjects were asked to *copy* the letters into empty grids, there was a highly consistent order to the placement of the letter segments. This order was mirrored in the image evaluation task: more time was required to respond when the X mark (or one of the two X marks) occupied a cell on which a letter segment, that was typically drawn later, would fall.[6]

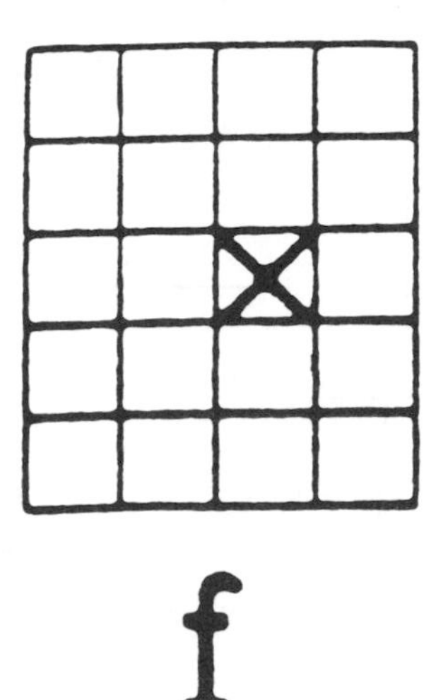

Figure 16.5

Before I turn to the second experiment cited by Pylyshyn as evidence for a descriptional approach to images and against a pictorial one, there is one further experiment performed by Kosslyn (Kosslyn et al., 1982) that I want to mention. Subjects were shown three by six arrays of letters (Xs in one case and Os in another). In each case the array was removed, and the subjects were told to think of the presented array as three rows of six or as six columns of three. Finally, at the sound of a tone, subjects were told to form an image of the array, and to push a button when the image was complete. It was found that more time was required for image generation, if the array had been conceptualized in terms of six columns rather than three rows.

This is again just what one would expect if images are like drawn pictures. For normally it takes a little longer to draw six columns of three Xs (or Os) than three rows of six (there are typically five hand movements from column to column in the former case and two hand movements from row to row in the latter). The conclusion Kosslyn himself urges on the basis of this experiment is that the way one conceptualizes what one sees affects how one stores and later regenerates the visual percept as an image. It is evident that this conclusion is compatible with the drawn picture proposal.

I come now to the second experiment described by Pylyshyn. This experiment required small children to view a beaker being placed beside a jug and then, a little later, to duplicate the action they had just seen. What the children actually did was to place the beaker *inside* the jug (Figure 16.6).

Pylyshyn believes that pictorial approaches to imagery cannot explain this phenomenon, whereas descriptional approaches can. However, it seems to me that there is no real problem for the pictorialist here. Assuming that the children use visual *images* at all in responding to the instructions, the pictorialist can maintain that the images they form are again like drawn pictures constructed, by processes that utilize available concepts, from distinct parcels of stored information. One possible pictorialist account, then, of the children's actions is as follows: when the children view the beaker being

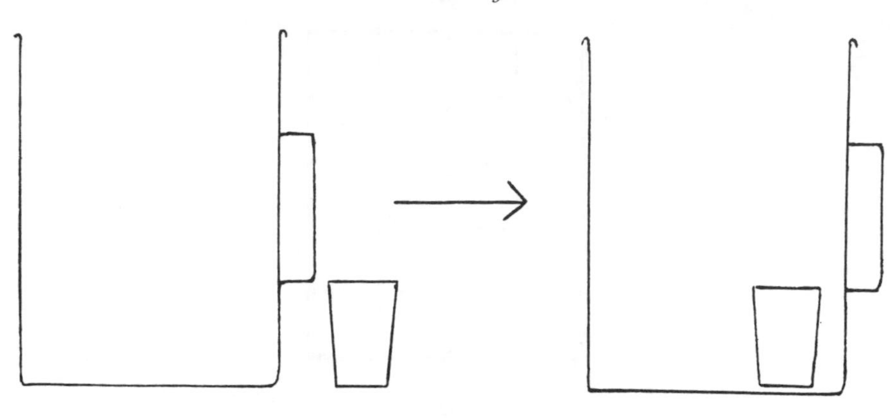

Figure 16.6

placed beside the jug, they fail to cognitively extract this specific information from their percepts, recognizing only that the beaker stands in some proximate relation to the jug. So, they store information about what the beaker looks like, information about what the jug looks like and non-pictorial information to the effect that the beaker is close to the jug. A little later, in responding to the instructions, they draw an inner mental picture of the beaker and the jug that represents the two as being close. On the hypothesis that this picture is drawn so as to represent also that the beaker is *inside* the jug (because, it may be supposed, the children think of the jug as a container, and they think of containers as standardly containing things that are close), the action they perform, upon picture inspection, is the one which would be expected (Kosslyn, 1980, ch. 9; Tye, 1991).[7]

The last piece of evidence Pylyshyn cites that I wish to discuss concerns the vastly superior memory chess masters have over duffers for real board positions (after only a very brief exposure). This difference between the two groups vanishes in the case of random arrangements of chess pieces. It seems to me that this result presents no serious difficulty for the pictorialist who, like Kosslyn, thinks that the way in which one conceptualizes what one sees affects the way in which one decomposes one's percept and stores it. Since the chess masters have a hugely superior knowledge for standard board positions, they are able to construct a much richer description of what they see upon being shown a configuration of pieces from a real game. Having such descriptions at their disposal enables them to carve apart mentally each presented board position into a number of chunks, as it were, and to store information both about the appearance of each of these chunks and their connections (both geometrical and tactical) with one another. When they later generate mental images of the board positions, they construct their images piece by piece from this stored information rather as if they are drawing pictures. The reason, then, that the chess masters do better than the mediocre players only in the case of real board positions is that it is only in this case that there is a significant difference in the stored information from which the mental pictures are constructed, a difference that is traceable to their superior knowledge of chess and their richer conceptualizations of actual positions.

I conclude that Pylyshyn's data is powerless against the sort of pictorial view of

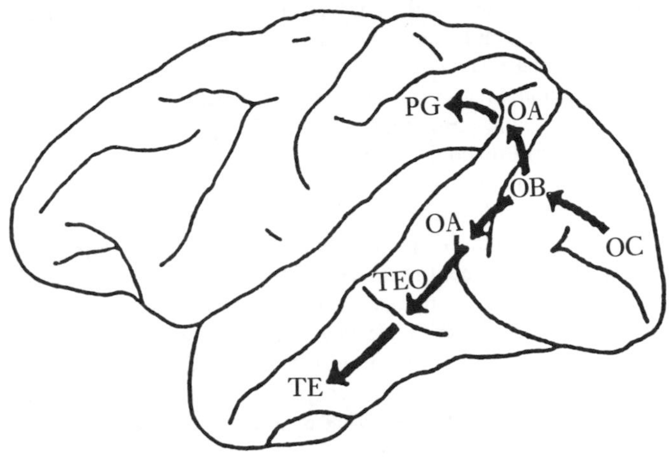

Figure 16.7

images I have presented above. So long as images are like drawn pictures, constructed from a combination of stored pieces of information, there is no problem whatsoever in explaining the results of the above experiments.

III

I want next to take up some neuropsychological data pertaining to image indeterminacy. Ungerleider and Mishkin (1982) have hypothesized that there are two higher-level visual systems in primates. One of these – the ventral system – runs from area OC (primary visual cortex) through area TEO to the inferior temporal lobe. This system is concerned with the identification of *what* is present. The other system – the dorsal system – analyzes the location of the seen object, *where* it is positioned. It runs from circumstriate area OB to OA and then on to PG (in the parietal lobe). See Figure 16.7.

The two systems together enable us to recognize objects when they appear in different positions in the visual field. The existence of these systems is shown, according to Ungerleider and Mishkin, by neuroanatomical investigations documenting two separate pathways, by neurophysiological investigations of the functions of cells in monkey brains, and by behavioural data obtained from animals with one or other of the relevant neural regions ablated.

Now, there is evidence that patients who have impairments in the visual recognition of objects on the basis of their appearances but not in the identification of the objects' relative positions, often have corresponding deficits and abilities in imagery (Levine et al., 1985).[8] Patients whose imagery is impaired in this way do badly at describing and drawing objects from memory, but they correctly specify the spatial relationships of objects to one another in previously witnessed scenes. It appears, then, that their images are indeterminate in certain respects but not in others. How is this indeterminacy to be explained on the proposed pictorial view of images?

The answer is straightforward enough. If there are two higher-level visual systems of the sort distinguished above, then information about the visual appearances of objects and information about their spatial relations are stored separately in memory (Kosslyn, 1987).[9] Mental images are like drawn pictures, constructed from combinations of such information, or so we are presently supposing. Thus, a patient, who lacks detailed information in long-term memory about the appearances of objects, but who has adequate spatial information, will draw inner pictures that are very sketchy or inaccurate with respect to the visual appearances of objects in previously witnessed scenes, but are accurate with respect to the spatial layout of those scenes.

It is also interesting to note that patients who have the reverse impairment in vision often have the reverse impairment in imagery. In these cases, the patients do poorly at drawing and describing spatial arrangements of objects in scenes that are no longer present, but they do much better at drawing or describing the objects' visual appearances.[10] Here, then, there appears to be image indeterminacy at the level of spatial relations between objects. This indeterminacy, like the one above, is easy to explain on the drawn picture approach to images. What is now lacking is a set of instructions for combining drawings of the individual objects into a single inner picture of the whole scene. Consider, for example, a person who has the given impairment, and who is asked to describe the contents of his sitting room at home. He accurately describes the chair, the sofa, the lamp, the clock. But he cannot say how they are spatially related. It seems to me that one plausible explanation of this behaviour is that he draws four different inner pictures for the purposes of inspection, one for each of the objects. *Four* pictures are produced here rather than a single integrated one, since there is no information available in memory about the spatial arrangement of the objects. These pictures may be drawn and mentally erased one after the other, or there may be a time at which they are all present together in something like the manner of four separate sketches all done on a single sheet of paper. In neither case is there any pictorial representation of the spatial relations that actually obtain between the objects.

On one understanding of the proposal I have just made, some mental images are modelled on *groups* of pictures. So, for example, an image of a square and a triangle may be composed of two unrelated parts, one a picture of a square and the other a picture of a triangle. This claim may perhaps seem to face a problem that I have not fully put to rest. For if I draw a square and I also independently draw a triangle on the same sheet of paper, my two drawings must be directionally related: my picture square must be left, right, up or down relative to my picture triangle. So, it may be urged, contrary to what I have claimed, an image of a square and a triangle must represent egocentric spatial information.

This line of reasoning presupposes that if one pictorial representation p bears a certain spatial relation R to another pictorial representation p′, then what p represents is thereby represented as bearing R to what p′ represents. This is plainly false, however. Suppose, for example, that a museum has a number of different sketches of body parts by Leonardo da Vinci, say, and it arranges them on a wall so that the sketch of a foot is above the sketch of an arm. Obviously, Leonardo himself did not thereby represent the foot as being above the arm. In parallel fashion, if I construct two independent drawings of a square and a triangle for which I happen to use the same sheet of paper, I have not thereby represented a square as bearing some definite viewer-centered spatial relation to a triangle. On the contrary, I have not represented anything about the spatial relationship of the two objects.

There is another intuitively more natural way of taking the present proposal: instead of holding that a person who images a square and a triangle, say, and who cannot identify their relative location, has a single mental image consisting of two separate pictures, we might suppose that he is actually subject to two different simultaneous or temporally overlapping images, one representing a square and the other a triangle, each of which is like a single drawn picture. This version of the proposal seems compatible with our ordinary conception of images – we talk as if it is possible for a person to have two red after-images at the same time, for example, or for a person to form an image of a dog, say, and then to generate a second image of a cat without allowing the first to fade. Moreover, it permits a person to image two objects without representing their relative direction for exactly the same underlying reasons as the previous account: in each such case, it is as if there are two drawn pictures.

I should perhaps add here that, so far as I can see, the difference between the two accounts just offered is really terminological: which one we prefer will depend only upon how broadly we wish to apply the term 'image'. I might also add that the general approach I have taken is, I believe, compatible with other related pieces of neuropsychological data. For example, Deleval, De Mol and Noterman (1983) cite a patient whose power to form images was impaired by left hemisphere damage and who described himself thus: 'When I try to imagine a plant, an animal, an object, I can recall but one part, my inner vision is fleeting, fragmented; if I'm asked to imagine the head of a cow, I know that it has ears and horns, but I can't revisualize their respective places.' Here, we may suppose, the patient is subject to incomplete, fragmented inner drawings that separately depict certain object or animal parts; for example, the ears and horns of a cow.

The version of the picture theory of images I have been defending is similar to, and much influenced by, the view of Stephen Kosslyn. There are certain features of Kosslyn's view, however, that are no part of the above proposal and that preclude Kosslyn from handling some of the cases of impaired imagery in the ways I have suggested. I want now to take a quick look at Kosslyn's position and how it differs from the proposal I have been defending; I want also to consider whether there is some alternative way of accommodating these cases that is compatible with Kosslyn's theory.

IV

According to Kosslyn (1980 and 1983), mental images are to be conceived of on the model of displays on a cathode ray tube screen attached to a computer. Such displays are generated on the screen by the computer from information that is stored in the computer's memory. Since there are obvious differences between mental images and screen displays, a question arises as to just what the respects are in which the former are supposed to be like the latter.

Kosslyn suggests that before we answer this question, we reflect upon how a picture is formed on a monitor screen and what makes *it* pictorial. We may think of the screen itself as being covered by a matrix in which there are a large number of tiny squares or cells. The pattern formed by placing dots in these cells is pictorial, Kosslyn (1983) asserts, at least in part because it has spatial features which correspond to spatial features of the represented object. In particular, dots in the matrix represent points on the surface of the object and relative distance and geometrical relations among dots match the same relations among objects points. Thus, if dots *A*, *B* and *C* in the matrix stand respectively for points P_1, P_2 and P_3 on the object surface, then if P_1 is below P_2 and to P_3's left (as the object is seen from a particular point of view) then likewise A is below B and to C's left (as the screen is seen from a corresponding point of view). Similarly, if P_1 is further from P_2 than from P_3 then A is further from B than from C.[11]

Kosslyn's reasoning now becomes more opaque. The main strand of thought which is to be found in Kosslyn's writings seems to be that although mental images lack the above spatial characteristics, they none the less function *as if* they had those characteristics. Thus, in Kosslyn's view, it is not literally true that mental images are pictures. Rather, the truth in the picture theory is that mental images are *functional* pictures.

This claim itself is in dire need of clarification, of course. But before I try to clarify its meaning, I want briefly to sketch certain other aspects of Kosslyn's position. Consider again a cathode ray tube screen on which a picture is displayed. The screen may be thought of as the medium in which the picture is presented. This medium is spatial and it is made up of a large number of basic units or cells, some of which are illuminated to form a picture. Analogously, according to Kosslyn, there is a functional spatial medium for imagery made up of a number of basic units or cells. Mental images, on Kosslyn's view, are functional pictures in this medium.

Kosslyn hypothesizes that the imagery medium, which he calls 'the visual buffer', is shared with visual perception. In veridical perception, any given unit in the medium, by being active, represents the presence of a just-noticeable object part at a particular two-dimensional spatial location within the field of view. In imagery, the same unit, by being active, represents the very same thing. Thus, imaged object parts are represented within an image as having certain viewpoint-relative locations they do not in fact occupy, namely those locations they would have occupied in the field of view had the same object parts produced the same active units during normal vision. Kosslyn hypothesizes further that the visual buffer is roughly circular in shape. What he means by this hypothesis is not that the buffer is *literally* circular, but rather that if all its component cells were active, the object represented would be circular (or at least would appear to be circular).

We are now in a position to see what Kosslyn means by the thesis that mental images are functional pictures. The basic idea is that a mental image of an object O (and nothing else) functions like a picture of O in two respects: (1) Every part of the image that represents anything (i.e. any active cell in the buffer) represents a part of O; (2) Two-dimensional relative distance relations among parts of O are represented in the image via distance relations among corresponding image parts (active cells). Distance in the imagery medium is not a matter of actual physical distance, however. Instead, it is to be explicated in terms of the number of intervening cells – the greater the distance represented, the greater the number of active cells representing adjacent portions of the object. Thus, Kosslyn's proposal does not require that cells representing adjacent object parts themselves be physically adjacent (as in a real picture). Rather, like the cells in an array in a computer, they may be widely scattered. What matters to the representation of adjacency is that the processes operating on such cells treat them *as if* they were adjacent.

Much more needs to be said about the thesis that images are functional pictures.[12] Still, the above comments should at least suffice to distinguish Kosslyn's pictorial view from that of philosophers and psychologists who take images to be akin to descriptions. For a description of object O will have parts (e.g. predicates) that represent items other than parts of *O* (e.g. properties of *O*); moreover, a description will fail to represent relative distance relations among parts of *O* in the required manner.

The remaining elements of Kosslyn's theory concern the processes that operate on images in the visual buffer. Kosslyn postulates that there are three sets of processes, namely those that 'generate', 'inspect' and 'transform' the images. As we saw earlier, the generation process itself is decomposable, on Kosslyn's view, into further processes. The inspection process is also really a number of different processes that examine patterns of activated cells in the buffer, thereby enabling us to recognize shapes, spatial configurations and other characteristics of the imaged objects. For example, if I form an image of a racehorse, it is the inspection process that allows me to decide if the tip of its tail extends below its rear knees. Similarly, if I image two equilateral triangles of the same size, one upright and one inverted with its tip touching the middle of the base of the upright one, the inspection process is what enables me to recognize the diamond shaped parallelogram in the middle. Finally, there are transformation processes. These processes 'rotate', 'scale in size' or 'translate' the patterns of activated cells in the buffer.

That, then, in crude outline is Kosslyn's theory. As I noted in section II, the proposal I have made on behalf of the pictorialist is similar to Kosslyn's in its account of image generation: generating an image, on my proposal, is, in the ways I specified earlier, like drawing a picture. Moreover, the central component of Kosslyn's view, namely that mental images are functional pictures in a visual buffer composed of cells, can be put to work to flesh out my proposal further, so long as it is supplemented by some account of how appropriate overall interpretations are supplied for the contents of the functional pictures. However, it may appear that if cases of impaired imagery with respect to spatial location are to be handled in the manner suggested in the last section, then one element of Kosslyn's theory must go, namely the claim that the cells in the visual buffer are each dedicated to representing, when active, the presence of a tiny patch of object surface at one and only one (two-dimensional) viewer-centered location (so that each cell always has the same corresponding location). For if this is so then an image of a single object cannot fail to represent, at least implicitly, its egocentric location; and an image of two objects, for example a lamp and a clock, cannot fail to contain information about their relative direction. Thus, an image of a previously witnessed scene containing several objects will now apparently have to represent spatial relations between the objects, contrary to my earlier claim.

Altering Kosslyn's theory in the manner mentioned above would be a relatively minor matter, since nothing in the rest of his theory requires that each cell in the visual buffer has a corresponding fixed egocentric location. Another response which Kosslyn himself might give is to retain the original conception of the representational role of the active cells in the visual buffer and to maintain that the above imagery deficit derives from an inability to extract, and cognitively process, information that is implicitly contained in the buffer. Let me explain.

The cells in the visual buffer are individually concerned solely with the representation of tiny object parts. There is no segmentation at this level of what is represented into distinct objects and shapes. Segmentation occurs via the inspection processes that examine patterns of activated cells in the buffer. Thus, consciously imagining that a lamp bears a certain directional relation to a clock, say, requires the subject to inspect the contents of the buffer and to categorize them appropriately. One reasonable hypothesis is that impaired imagers with dorsal defects can only attend to, and extract information from, isolated parts of the buffer, that is, parts that have been categorized as representing individual objects. This being so, it will be possible for such a person to form a single image of two objects without being able to identify their relative direction.

It appears, then, that Kosslyn's view is not directly threatened by the cases I have adumbrated. The general upshot of my discussion is that the pictorial conception of images cannot easily be discredited by appeals to indeterminacy. Not only can the pictorial view accommodate quite extreme cases of indeterminacy but it also has the resources to do so in more than one way. There remain problems for the pictorial approach to imagery, but they lie, I believe, elsewhere – specifically, in the phenomenal aspects of imagery and in the representation of the third dimension (Tye, 1991 and 1992).

ACKNOWLEDGEMENTS

I would like to thank Naomi Eilan, Andrew Meltzoff and Tony Marcel for helpful comments.

NOTES

1 This model is suggested by Ned Block (1983). In several places in this chapter I am indebted to Block's discussion.
2 There is a potential problem for Pylyshyn here, since it is not obvious that the children do not have the concept 'parallel to the ground'. So it is not clear that they cannot generate the same description as an adult.
3 The experiment described here was performed by Neisser and Dirks and is reported by Neisser (1979).
4 This claim is in keeping with the hypothesis that information about the visual appearances of things and information about their spatial relations are stored separately in memory. See here section 3.
5 It seems plausible to suppose that for four-year-old children the relevant comparison class of containers here does not include the tilted feeding bottles they often saw when they were very young.
6 The effect of X mark position did not occur in a perception control task conducted by Kosslyn (1988), in which subjects were asked to decide if the X mark actually did fall on a perceptually presented letter.
7 I want to stress here that I am not claiming that the children definitely do use visual images in the above experiment. My claim is rather that if they do then there is no problem for the pictorialist. An alternative possibility is that the children, instead of generating images, simply retrieve the stored descriptional information that the beaker is close to the jug. They then act in a way that is guided by this information, together with their preference to place things that are close to containers, such as the jug, inside the containers. This possibility, of course, does not itself present any threat to pictorialism. For it is not being claimed that the above experiment provides any evidence *for* the pictorial view. Rather, my aim is to demonstrate that the experiment does not count *against* that view, since both the hypothesis that the children use pictorial visual images and the hypothesis that they do not use images at all can explain the data. As far as the other experiments in this section are concerned, it is, I think, implausible to deny that visual images are generated and used. For a general discussion of when visual images are typically used in cognitive tasks which supports this claim, see Kosslyn (1988, ch. 9) and Tye (1991).

8 I should perhaps add that it is really not clear from the evidence Levine et al. cite just how common the link is between the impairments in vision cited in the text and the corresponding impairments in imagery. The same is true for the reverse impairments discussed a little later. Part of the problem here is that some of the case studies Levine et al. review from the literature on imagery are subject to more than one interpretation, due to lack of accuracy in some of the questions put to the patients on their imagery by the original researchers.

9 Actually this may be a little oversimplified. For Kosslyn has recently argued that the dorsal system – the one concerned with the analysis of location – may be used to represent not only locations of objects in a scene but also, for separate encodings of object parts (formed during shifts of attention), relative locations of parts within a single object. I might add that it is not easy to see how Kosslyn's proposal is consistent with the data in this section on impaired imagery, unless the dorsal system is taken to contain two distinct sub-systems, one concerned with the representation of locations of objects and the other with the representation of locations of parts within objects.

10 However, if the objects have parts that are viewed as separate objects in their own right (e.g. a bicycle) then spatial relations among these parts are not depicted or described accurately although the individual part shapes are.

11 It is perhaps worth noting that this claim needs qualification, if it is to be generally applicable. For P_1 may be further from P_2 than from P_3 on the object surface and yet appear to be closer to P_1 within the context of the relevant point of view. It would be more accurate to say, then, that if *P appears* further from P_2 than from P_3 (relative to the relevant point of view) then *A* is further from *B* than from *C*. But even this claim is too strong. Distortions of one sort or another can be accommodated by weakening the stated requirement so that it obtains only in a sufficient number of cases and only when P_1, P_2 and P_3 do not appear to be at different distances away. Similar qualifications are also needed, I might add, in Kosslyn's claim concerning geometrical relations.

12 For a detailed discussion of Kosslyn's pictorialism and the concept of distance in the imagery medium, see Tye (1988 and 1991).

REFERENCES

Berkeley, G.: *The Principles of Human Knowledge*. In D. M. Armstrong (ed.), *Berkeley's Philosophical Writings*. London: Collier Macmillan.

Block, N. 1983: The photographic fallacy in the debate about mental imagery. *Nous*, 654–64.

Deleval, J., De Mol, J. and Noterman, M. 1983: La perte des images souvenirs. *Acta Neurologia Belgique*, 83, 61–79.

Dennett, D. C. 1981: The nature of images and the introspective trap. In N. Block (ed.), *Imagery*. Cambridge, Mass: MIT, 51–61.

Fodor, J. 1981: Imagistic representation. In N. Block (ed.), *Imagery*. Cambridge, Mass: The MIT, 63–86.

Kosslyn, S. 1980: *Image and Mind*. Cambridge, Mass: Harvard University Press.

Kosslyn, S. 1983: *Ghosts in the Mind's Machine*. New York: W. W. Norton.

Kosslyn, S. 1987: Seeing and imagining in the cerebral hemispheres: A computational approach. *Psychological Review*, 94, 148–75.

Kosslyn, S. 1988: Aspects of a cognitive neuroscience of mental imagery. *Science*, 240, 1621–6.

Kosslyn, S., Pinker, S., Smith, G. and Shwartz, S. 1982: On the demystification of mental imagery. In N. Block (ed.), *Imagery*. Cambridge, Mass: MIT, 131–50.

Levine, D. N., Warach, J. and Farah, M. J. 1985: Two visual systems in mental imagery: dissociation of 'what' and 'where' in imagery disorders due to bilateral posterior cerebral lesions. *Neurology*, 35, 1010–18.

Neisser, U. 1979: Tracing mental imagery. *Behavioral and Brain Sciences*, 2, 4.

Pylyshyn, Z. 1981: Imagery and artificial intelligence. In N. Block (ed.), *Readings in the Philosophy of Psychology*, vol. 2. Cambridge, Mass: Harvard, 174–6.
Shorter, J. M. 1952: Imagination. *Mind*, 61, 528–42.
Tye, M. 1988: The picture theory of mental images. *The Philosophical Review*, 97, 497–520.
Tye, M. 1991: *The Imagery Debate*. Cambridge, Mass: MIT/Bradford Books.
Tye, M. 1992: Visual qualia and visual content. In T. Crane (ed.), *The Contents of Experience*. Cambridge: Cambridge University Press, 158–76.
Ungerleider, L. G. and Mishkin, M. 1982: Two cortical visual systems. In D. J. Ingle, M. A. Goodale and R. J. W. Mansfield (eds), *Analysis of Visual Behavior*. Cambridge, Mass: MIT, 549–86.
Watson, J. B. 1928: *The Ways of Behaviorism*. New York: Harper, 75–6.

17
Assembling routines and addressing representations: An alternative conceptualization of 'what' and 'where' in the human brain

Rosaleen McCarthy

Introduction

At an intuitive level it seems obvious that there is a close relationship between 'what' and 'where' in visual perception: in order to perceive an object it is necessary to segregate out a portion of visual space as an object; objects stand out from their backgrounds by virtue of their spatial properties; they are even differentiated from each other because they have different spatial organizations. The intuitive notion of a very particular type of dependence between 'what' and 'where' has been brought into question by recent psychological and neuropsychological evidence. An extreme but widely supported view is that the processes of spatial analysis and object perception depend on specific, independent computational operations. These cognitive distinctions are mirrored at the anatomical level by separate neural pathways (see Figure 17.1). The integration of 'what' and 'where' information is construed as a computationally complex process rather than a basic property of information extraction.

One of the reasons for continuing debate concerning the status of 'what' and 'where' is definitional:[1] these terms are used to encompass a huge range of different operations. Table 1 gives a brief summary of some major relevant subtypes. Given the range of processing types identified even in this short table it would be surprising if there were any simple answer to the interdependence or separability of 'what' and 'where' information. The answer may depend on which what and which where we are working with. The seemingly clear-cut distinction does have value, but as a useful heuristic device and means of focusing empirical questions, rather than as a fact to be explained.

A general model of skilled information processing

In this chapter I shall be presenting some neuropsychological evidence from tasks that seem broadly to correspond to the different levels outlined in Table 1. Figure 17.2 presents a very general information-processing framework of the type that has become conventional in cognitive neuropsychology.[2] It may be read as a labelled graph in which

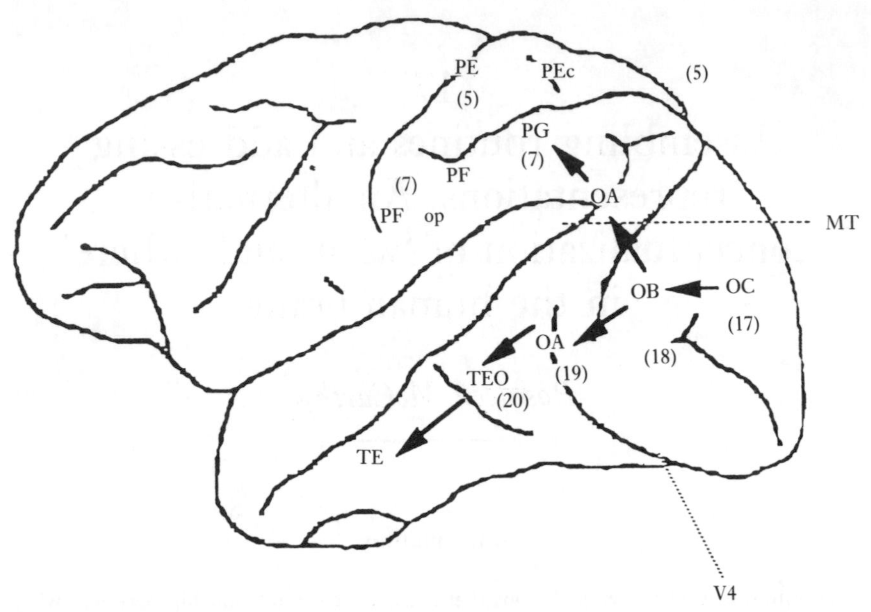

Figure 17.1 Brain map showing lateral view of rhesus monkey brain with the separate 'what' and 'where' pathways superimposed. Areas V4 and MT do not appear on the surface of the brain. Their approximate geographical location is indicated by dashed lines. Many critical regions are referred to by parallel nomenclatures: these are indicated by letter and number designations on this map.

the boxes show three major processing subdivisions corresponding to levels A–C in Table 1. Figure 17.2 incorporates at least two means of proceeding from sensory input to output. One route (A–B) is based on computations carried out 'on line' and contemporaneously via *assembled routines*. The other processing route (A–C) is dependent on reference to stored knowledge or *addressed representations*. The evidence from neuropsychology is as consistent with these subdivisions of assembled and addressed representations as with the more conventional 'what' and 'where'. I shall be using this alternative framework to structure this chapter, since it avoids any premature commitment to specific processing types and may avoid confusion between different levels of explanation.

Sensory processing: Detecting, integrating and grouping

The current framework for analysing this level of visual information processing is usually pitched at a level of explanation that is a mixture of engineering, psychophysics and anatomy. In essence it is proposed that there are a number of sensory analysers and filters in the visual regions of the brain that are tuned to different 'trigger' features such as colour, motion, luminance contrast, etc. Alone or in combination these finely tuned detectors may serve to provide a preliminary structuring or parsing of sensory input

Table 17.1 Basic subtypes of 'what' and 'where'

What	Where
1 Sensory processing: Detection and discrimination	
Levels of analysis	*Levels of analysis*
Sensory attributes/ trigger stimuli	Receptive field/ retinocentric location
2 Assembled routines: On-line action control	
Levels of analysis	*Levels of analysis*
Target, missile obstacle	Visuo–motor integration
3 Addressed representations: Recognition/modulation	
Levels of analysis	*Levels of analysis*
Type recognition	Object orientation?
Form (structure)?	Object axes?
Meaning (semantics)	
Episodic indexing	Topographic knowledge
Orientation cues	Routes
Autobiographic cues	Places

into edge-like regions (cf. Marr, 1982; Frisby, 1979; Watt, 1991). Further processes may be required to (a) synthesize information across different sensory filters and (b) to make explicit grouping or linkages between parsed edge-like regions.

Detection and discrimination

Early visual analysers deliver evidence that is necessary for the higher order processes of structured perception and recognition. Damage within the sensory processing regions of the brain may impair the individual's ability to discriminate and detect some kinds of sensory information whilst sparing others. There are multiple reciprocal clinical dissociations between different sensory processing domains: some patients lose the ability to discriminate colour and retain the ability to perceive subtle luminance (light–dark) contrasts. Other cases retain their colour discrimination but are impaired in the luminance domain. These fractionations go beyond colour and brightness. One patient has lost the ability to perceive motion cues but can detect displacement, whereas other cases cannot perceive displacement but can discriminate motion (Zihl et al., 1991) (see Table 17.2). These reciprocal dissociations imply that information about different sensory properties may be extracted by multiple specialized processors that operate in parallel on the visual array.

Many sensory visual disorders implicate one side or even one quadrant of the retinotopic visual field (e.g. Albert, Reches and Silverberg, 1975; Zeki, 1990; Warrington, 1985).[3] Loss of colour discrimination can be 'one sided' as can loss of motion perception.

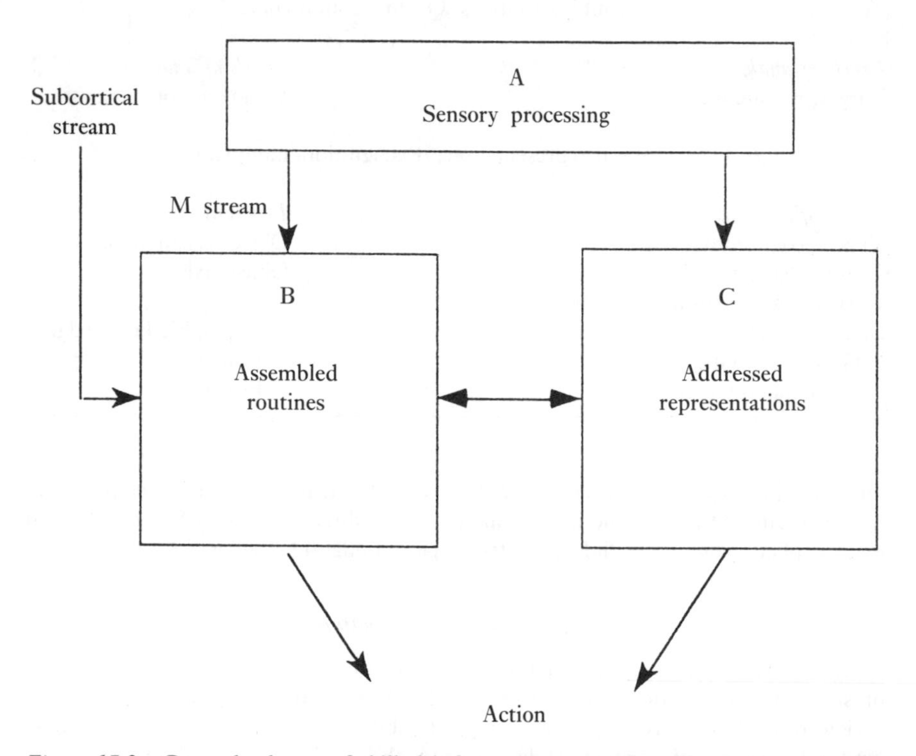

Figure 17.2 General scheme of skilled information processing. Two separate inputs are shown to the 'assembled routine' component corresponding to cortical (M) and subcortical channels.

The side of vision that is affected is determined by the laterality of the patient's brain damage; damage in the sensory areas of the left hemisphere results in impairment within the right visual field. Complete loss of colour or motion perception occurs when the patient has a bilateral and symmetrical brain injury resulting in a double deficit.

In summary, the clinical evidence indicates that stimulus detection and initial parsing of the visual array is carried out by systems which are tuned to particular physical properties *and* tuned to a finite set of positions in retinotopic space.

Table 17.2 The reciprocal dissociation of early visual sensory attributes

	Impaired				
Normal	Hue	Grey scale	Motion	Form	Acuity
Hue	/	+	+	+	+
Grey scale	+	/	+	?	?
Motion	+	+	/	+	+
Form	+	?	+	/	+
Acuity	+	?	+	+	/

Summary of dissociations within early sensory processing domains. (+) indicates a profile in which one component has been documented as being impaired, sparing the other. (?) refers to dissociations which have not been unequivocally demonstrated.

Physiological substrata

Maps: The clinical evidence converges extremely well with anatomical and physiological studies of visual sensory processing in infrahuman species. Anatomically defined areas of visual cortex have (statistical) preferences for certain trigger features (e.g. Cowey, 1982). These regions are jointly sensitive to sensory attributes and retinotopic location and so are sometimes dubbed 'maps'. Figure 17.3 shows some of the specialist regions within the visual cortex according to their anatomical subdivisions. Area V1 is the initial cortical receiving area, information is passed from here via complex reciprocally interconnected pathways to areas V2–V5 and on to areas such as MT. The retinotopic 'gaps' in the vision of neurological patients are almost the direct inverse of the electrophysiologists' maps, in terms of their functional characteristics, their spatial sensitivity and their anatomical correlates (Zeki, 1990).

Streams: Maps do not stand alone, rather they are interlinked into anatomically and physiologically distinct parallel processing streams summarized in Figure 17.3.[4] The major subdivisions are:

The cortical M (magnocellular) pathway comprised of large, rapidly triggered cells that are particularly sensitive to luminance contrasts, to motion, orientation and stereoscopic depth. It has poor colour sensitivity (biased towards the low and medium wavelengths).

The cortical P pathway (or pathways) consisting of cells that are slow to respond and are particularly sensitive to colour and/or small stimuli (end stopped cells).

In addition to the systems shown in Figure 17.3 there is also a subcortical (SC) system made up of retinotopic regions that are relatively rapidly triggered by changes in luminance and may be dynamically co-ordinated with input from other modalities (e.g. audition) at the level of a structure termed the superior colliculus.

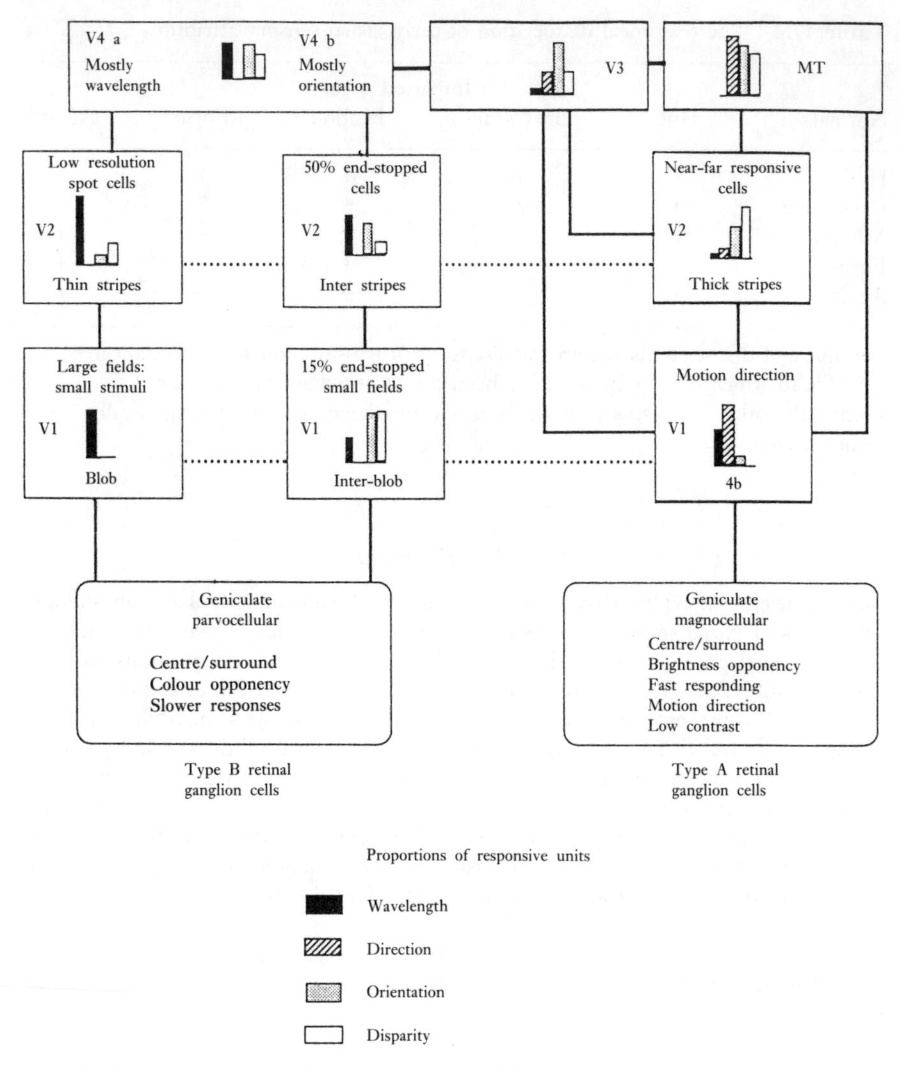

Figure 17.3

Processing domains

Parsing and beyond: It has been suggested that those same regions that are specialized for detection may *also* be capable of yielding a preliminary structuring of visual input. So cells with partially overlapping receptive fields may produce a particularly strong or powerful response if they are triggered in synchrony by continuous edges in retinotopic space. Relatively simple computational algorithms have shown that such 'overlap' models provide a powerful means of parsing visual sensory information (see, for example, Hildreth and Ullman, 1989, for discussion). Relative position as well as edge information might be encoded by similar retinotopic mechanisms operating at different spatial scales (Watt, 1991).

Integration: Does the fact that information about attributes and retinocentric location are conjointly encoded have implications for processing beyond the level of preliminary parsing of the visual input? Retinotopic spatial sensitivity may be an engineering solution to a complex wiring problem or the consequence of epigenetic or ontogenetic constraints. It is not necessarily a computational code. Treisman and her colleagues have proposed that the common spatial organization of early visual processing systems simplifies the problem of binding disparate visual attributes across multi-item arrays (e.g. integrating a colour map, a luminance map). Conjunction processing requires an active attentional process operating on a hypothetical spatial *master-map* of locations (e.g. Treisman and Gelade, 1980; Treisman, 1988). Whilst there may be more than a germ of truth in Treisman's theory, the boundary conditions under which it applies have been difficult to establish unambiguously (see chapter 15 of this volume). Developments of conjunction theory will probably need to take into account the physiological fact that the various sensory processing sub-systems differ in more than their tuning to physical attributes (e.g. Humphreys and Bruce, 1989). They have vastly different temporal and spatial resolutions. Moreover, there may be preferential patterns of feature integration arising from the specific interconnections and inter-linking of processing maps into sub-systems or processing streams.

Structures from edges? Some cues to form are based on the spatial organization of the constituent regions or features of input; this type of parsing appears to require more than simple edge detecting algorithms. The distinction between a square and a rectangle, the parsing of overlapping contours into separate forms, the perception of cues to occlusion and overlap and the derivation of contours from good continuation and cotermination all appear to require a close interaction between 'what' and 'where'. These 'spatial' cues to the segregation of figure and ground were given prominence as theoretically challenging phenomena by the Gestalt psychologists. Experimental work with normal subjects has demonstrated both the importance and the relative automaticity and speed with which such operations are carried out (e.g. Donnelly et al., 1991). Livingstone and Hubel (1988) have suggested that those cues that the viewer uses in order to connect parsed edge-like regions into spatial 'objects' are extracted by the (spatial) M system.

Normal subjects are very handicapped in extracting these spatial linking cues under conditions in which the M system is disadvantaged (as when viewing displays varying only in hue). Under isoluminant conditions with displays varying in wavelength, edges look jazzy and unstable – but are still edges. They do not link together as immediately and directly into structures. These findings suggest that the derivation of spatial linking features is not merely a change in scale from feature (at the lower end) to objects (at a coarser resolution): spatial linking information and object features may be derived independently. There are also suggestions that this level of processing may be mediated by retinotopic systems (Weiskrantz, 1980).

The neuropsychological evidence is broadly consistent with the hypotheses put forward by Livingstone and Hubel. Kartsounis and Warrington have recently described a patient who was unable to use the cue of texture density to segregate objects from a noisy background, was unable to parse overlapping geometric shapes into their

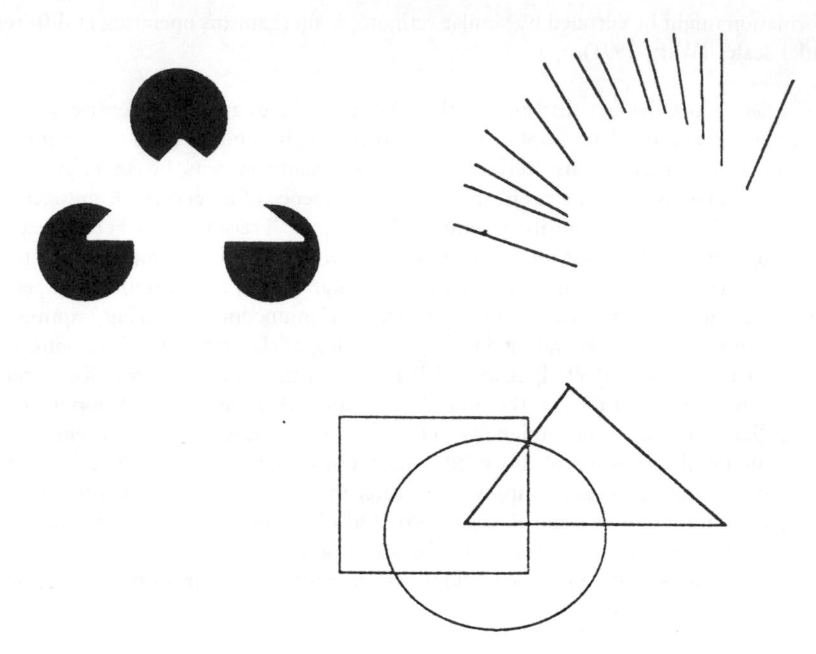

Figure 17.4 Examples of implicit contours and overlapping figures. See text for details.

elements, and failed to perceive 'cognitive contours' such as the Kanisza forms (shown in Figure 17.4). The patient was not totally 'form blind' since she could discriminate between squares and rectangles matched for total area. Following the analysis offered by Livingstone and Hubel, it was suggested that this patient had an impairment of the spatial M system with relative sparing of the P system (Kartsounis and Warrington, 1991). A further case has since been documented who showed a complimentary dissociation. He had a normal texture density parsing and normal detection of cognitive contours, but impaired discrimination of squares and rectangles (Davidoff and Warrington, 1992). These complimentary dissociations would be consistent with there being at least two sub-processes involved in constructing parsed structures from edges.

Synthesis

There appears to be a close coupling between 'what' and 'where' in early visual processing – from the level of information extraction through to parsing and structuring of the visual input. A picture has emerged in which arrays of retinocentric filters and detectors extract sensory information and may provide a preliminary parse into edges and regions of potential interest. The *independent* extraction of information within multiple domains may require the use of attentional integration. These filters are further interconnected into systems that extract spatial linking features of possible objects and sensory attributes of objects, in parallel.

Assembled procedures: On line visuo–motor control

In acting and interacting with the environment objects have to be negotiated as well as detected. In reaching for an article from one's desk-top, or in taking a path through a room, one needs to know how to move so as to economize effort and avoid obstacles. Processing at this level does not necessarily require the recognition of objects as members of a functional class – except in so far as such recognition can compliment action (e.g. Goodale and Milner, 1992). For example, it may be helpful to know that an object can serve as a step, as a support or as a prop in the course of action, although it is largely irrelevant that the object is a stool rather than a chair or a coffee-table. Similarly, if an object is likely to hit the observer, or if the observer is on a collision course with the object, the precise species and genera of the missile or obstruction are less relevant than are taking the appropriate avoiding action in time. Our interactions with material objects require knowledge of their likely displacement with respect to ourselves. Moreover, we must be able to make very simple default assumptions regarding their permanence or transience and their material properties (see chapters five and six of this volume).

Three main classes of human neurological disorder implicate this assembled level of information processing and will be the focus of this section: dysmorphopsia/blindsight; visual disorientation and optic ataxia. The principal characteristics of each of these syndromes are given in Table 17.3.

Dysmorphopsia

Patients with globally impaired 2D shape discrimination (or dysmorphopsia) are unable to discriminate or identify simple forms on conventional matching and discrimination tests. They fail on tasks such as judging whether a stimulus is a square or a rectangle, a circle or an ellipse. They are also typically unable to parse forms from the background on the basis of information such as texture density. At the same time they may be able to navigate in new environments and locate single stimuli. These cases are sometimes considered as providing unambiguous evidence for 'where without what' (e.g. Humphreys and Riddoch, 1987; Farah, 1990). However, this seemingly clear distinction has been blurred by recent evidence. At least some dysmorphopsic patients may be able to show accurate form discrimination and recognition if assessed by visuo–motor tasks rather than perceptual matching or verbal identification.

Efron (1968) described one patient, Mr S, who could not discriminate (or identify) shape on the basis of sight alone. For example, he was unable to state whether a form was a square or a rectangle unless the rectangle was a long narrow line. Mr S attempted to use boundaries of luminance and colour contrast as a means of parsing objects from the visual array but this strategy was of very limited functional use. Remarkably, however, he was able to find his way around an unfamiliar hospital using a wheelchair and without bumping into objects. He could also visually 'track' moving objects along complex paths and could count scattered stimuli – provided that they were of sufficient contrast with the background. Milner et al. (1991) have described a further very similar case and have taken the analysis of preserved and impaired function one stage further. On conventional matching tests the patient is impaired in discriminating orientations and matching forms. She confused orientations differing by as much as 45 degrees.

Table 17.3 Clinical disorders of assembled routines

Deficit	Impaired	Preserved	Neurological correlates
Dysmorphopsia	1 Discrimination of square/rectangle 2 Detection of figure on basis of texture density or colinearity	Stimulus position Colour and motion Reaching	Often due to carbon monoxide poisoning Possibly implicates M pathway
Blindsight	Spontaneous detection of presence/absence of stimuli	Position Motion Retinal colour Reaching	Damage to V1
Visual disorientation	Stimulus position Relative size Distance Reaching	Form identification and recognition Motion Colour	Damage to posterior parietal regions including area 7
Optic ataxia	Reaching	Position Motion Colour Form identification and recognition	Damage to superior parietal regions including area 5

Despite these grave difficulties on discrimination and matching tests the patient could reach towards a narrow slot placed at different orientations with high levels of accuracy – and could orient her hand appropriately. She is also reported as being able to produce anticipatory pre-shaping movements of her hand which accurately reflected the orientation and size of rectangular block-targets.[5] Milner et al. have suggested that their patient could produce actions on the basis of *tacit* knowledge of the stimuli despite failure on the explicit task of visual discrimination.

Blindsight

Blindsight cases may be unable to detect the presence or absence of stimuli in the affected sector of their visual field on conventional testing. However, if they are tested using 'forced choice' guesses by making eye-movements towards the source of unseen stimulation, or by asking them to reach for objects presented in the blind field, then surprisingly competent visuo–motor control may be evident (Perenin and Jeannerod, 1975; Weiskrantz et al., 1974; Weiskrantz, 1986). Marcel (personal communication) has found that such patients may be able to reach accurately and they may show evidence of form discrimination in the anticipatory pre-shaping of their hands in reaching tasks.

Blindsight even appears to be sufficient for much navigation in space. However, despite extensive practice, this residual vision cannot be brought under verbal conscious control (Weiskrantz, 1986).

The evidence for relatively intact anticipatory reaching in blindsight and dysmorphopsia makes a simplistic partitioning of function along 'what' and 'where' lines highly implausible. Form and location appear to be encoded *tacitly* in both disorders. Such encoding may be insufficient or inappropriate for operations such as comparison and fine discrimination but it is sufficient and appropriate for the visuo–motor control of reaching and grasping.

Disorientation

A direct contrast with the preceding disorders is seen in visual disorientation.[6] Holmes (1918) and his colleagues provided some classical descriptions of the impairments in visual localization experienced by patients with missile wounds affecting the junction of the occipital and parietal lobes. Case WF described by Holmes and Horax (1919) was a soldier who had sustained a penetrating missile wound (probably from a machine-gun bullet) which pierced through the back of his skull. He was bright and alert and by no means demented. However, he had very profound impairments in localization. Holmes and Horax's own descriptions of the case are compelling and will be quoted directly:

> He moved his eyes to order readily and accurately in every direction, but from the very first day he came under observation he had much difficulty in fixing objects which he could see. If spoken to suddenly he first stared in a wrong direction and then moved his eyes about until they fell, as if by chance, on the observer's face; he was extremely slow and inaccurate in his attempts to fix or bring into central vision any object when its image fell in the periphery of his retinae, especially if it were small. When, however, requested to look at his own finger or to any point of his body which was touched he did so promptly and accurately, moving as a rule both his head and eyes in the normal manner. (Holmes and Horax, 1919, p. 388)

WF is described as having some preserved central vision and once he had managed to find an object in visual space then he could sustain fixation upon it. However, his vision in depth was affected profoundly and his eyes did not converge to a stimulus brought slowly towards him. If he used his own finger as a fixation mark then convergence and occulo–motor movements were entirely normal. 'Another noteworthy symptom was the complete absence of the blink reflex; if the observer's hand or any large object, such as a book, was unexpectedly swung towards his eyes he never blinked, withdrew his head, or reacted as the normal person does, unless the movement produced a current of air' (p. 388). He was profoundly disoriented in walking-space:

> On one occasion. . . he was led a few yards from his bed and then told to return to it; after searching with his eyes for a few moments he identified the bed, but immediately started off in a wrong direction. . . some of the most interesting observations were made on a large flat roof that was divided by a wooden paling about four feet high with an open gate in it through which it was possible to pass from one into the other portion; there was a step about one foot high at this gate. Day after day he was brought to this roof and told to walk from one spot in its larger division. . . to a point at the balustrade around the smaller

portion. He always remembered that he had to pass through the gate and that there was a step there, but even after he had been repeatedly over this route he frequently set out in a wrong direction. . . his course was often at right angles to the correct one, but as a rule he walked into the paling some distance from the gate and then guided himself to it by hand. . . . When he had succeeded in passing through the gate he often took a wrong direction and reached the target destination only by groping with his hands.' (p. 395)

Holmes and Horax initially considered that the problem in case WF might be one of a limitation of his visual attention to single items, rather than a failure in representing spatial information. Subsequent testing revealed a very profound impairment with *single* objects, making a simple attentional account unsatisfactory. Moreover, the patient *could* process multiple objects if the physical organization of the input was structured. Thus, WF was *unable* to judge his distance from single objects, the absolute length of line and the absolute size of single objects. If presented with scattered objects which were randomly distributed on a desk, or asked to make relative position judgements of stimuli which were 'unpredictable', he was gravely impaired. However, when coins were placed closely in a horizontal or vertical row he counted them promptly and without difficulty. By contrast with dysmorphopsic patients, visual testing also showed relative preservation of WF's ability to discriminate squares and rectangles, and to identify simple forms and discriminate the direction and extent of motion. Visually disoriented patients such as WF may be able to recognize the semantic significance of an object, identify specific faces and even parse individual objects from overlapping stimuli – provided these are located at the point of fixation.

Visual disorientation seems to be a paradigm case of 'what without where'. However, it is a very particular type of 'what' that is spared. The ability to discriminate and recognize objects, even objects as complex and special as the individual human face, appears to be intact. However, the ability to track objects, to judge their size and interact with them, has been disrupted. It is also a very particular type of 'where' that is damaged: such cases can make discriminations between forms (e.g. between squares and rectangles) detect displacements and use spatial cues in counting scattered stimuli. However, their ability to interact with objects is gravely impaired.

Optic ataxia

The clinical syndrome of 'optic ataxia' also affects the patient's ability to *reach* for stimuli under visual guidance. The patient may have normal visual point localization and be capable of fixating stimuli spontaneously; failure occurs when reaching towards a visual stimulus. Optic ataxic patients appear clumsy: their reaching errors are not simply failures in localization, their anticipatory hand orientation and pre-shaping are also affected. The disorder is not simply one of motor control since if subjects are reaching to an auditory target or if they view their limb before moving then they may be normally accurate. The disorder may be both retinotopic and somatotopic: it primarily affects the arm contralateral to the lesion when reaching into the visual field contralateral to the lesion (e.g. Perenin and Vighetto, 1983; 1988).

Processing substrata

Psychobiology: The disorders of dysmorphopsia, blindsight, optic ataxia and disorientation implicate a set of visuo–motor control systems. At the level of psychological architecture this system corresponds to the node labelled 'assembled routines' in Figure 17.2 (see Goodale and Milner, 1992, for a broadly similar account). Reaching appears to be mediated by a system that computes visuo–motor co-ordinates in head and body-centred space. Under normal circumstances the assembled procedures of visuo–motor control are mediated via a cortical system that receives two major types of input: one is cortical and based on the M pathway; the second is subcortical and based on the SC collicular processing stream.

The evidence from pathology, single cell work and computational modelling is largely consistent with a very simple three-part analysis. Visuo–motor location and form may be processed by a system that does not *require* modulation by explicit processing but which is normally 'open' to reflection and introspection by virtue of information from, or modulation by the M stream. The precise characteristics, extent and limitations of the information delivered by this system have yet to be fully defined.

1 Input from the M stream to the cortical visuo–motor system is disrupted in blindsight and dysmorphopsia. Patients may be reliant on subcortical input to the visuo–motor system (see, for example, Weiskrantz, 1986; Cowey and Stoerig, 1991; Milner et al., 1991).
2 Patients with visual disorientation may have damage *within* the cortical visuo–motor processing system itself. More specifically, they may have damaged a sub-component that is normally required for 'subtracting' the effects of actual or intended eye and head movements from retinal displacement (probably mediated via systems within area seven in the superior parietal lobe). Loss of this sub-component effectively destroys the patient's ability to compute *somatocentric vectors* for locating visual target stimuli in head or body-centred space.
3 Optic ataxic patients may also have damage within the cortical visuo–motor system but at the level of computing *reaching vectors*. Their impairment is at the 'output' end of the assembled procedure sub-system outlined in Figure 17.2. Single cell analysis of the region damaged in optic ataxia suggests that the location of targets in somatic hand or shoulder-centred co-ordinates is computed here via coarse coding procedures (e.g. Georgopoulos, 1987). These reaching vectors seem to be computed in advance of an action and thereby provide a reference copy against which actual actions can be compared.

Cognitive substrata: As emphasized by Evans, visuo–motor space appears to be a cross-modal space. At the single cell level there is evidence for plentiful cross-modal interaction right from the earliest input levels. Assembled visuo–motor representations are mediated via common perceptuo–motor codes in networks that can subtract out local (or transient) input-channel information. These networks compute target, action and location information in a variety of cross-modal subject-centred reference frames (depending on the circumstances and the action: see chapter 13 of this volume). These reference frames are action frames.

Cognitive psychologists of the Ecological or Gibsonian tradition have analysed this level of visuo–motor information processing in considerable detail (e.g. Gibson, 1966; 1979). It is clear that information available directly from vision provides many cues sufficient for *on line* contemporaneous control of action and navigation. A considerable amount of information processing can proceed on the basis of deriving 'invariances', basic, highly consistent parameters from the visual input (e.g. Lee, 1980). For example, visual cues from motion parallax or from optic flow patterns provide information about depth, distance, direction and egocentric displacement (see also chapter one of this volume). This information does not have to be analysed for meaning or significance in order to contribute to visuo–motor control. Cues sufficient for reaching, touching, grasping and dodging may be assembled as and when they are required.

Gibson's metaphor was that the system 'resonated' when triggered by appropriate stimuli. Contemporary accounts of visuo–motor control rely on similar physical metaphors. They are often cast in terms of networks and dynamical systems that can describe visuo–motor co-ordination in precise engineering terms. In both types of analogy, the direct assembly of information allows for very rapid initiation of adaptive behaviour in the face of novel input. It eliminates the need for much complex learning and decision making.

Overview: What about what and where? The foregoing analysis suggests that specific local visuo–motor frames of reference may be implicated in certain neurological disorders. It does not offer much support to the notion that there is a single core or unified 'egocentric' map at the basis of form perception – or to the other extreme that objects are separable from the space which they occupy. There may be local computations in hand, shoulder and head-centred frames allowing the precise control of anticipatory reaching and grasping. Cognitive studies have shown that hand shaping and reaching direction may be independently 'programmed' components of visuo–motor action (e.g. Jeannerod, 1988). Both reach and grasp components may be modulated (quasi-independently) in the light of specific expectancies. Reaching direction may be modulated if displacement is anticipated, and grasp shape and orientation may be varied in accordance with expectations about the physical properties of the target such as weight, friction, etc. (Athènes and Wing, 1989). It seems entirely possible that these higher-order anticipatory and control aspects of visuo–motor action are mediated by different neurological sub-systems.

Addressed representations: Objects and maps

In this section I want to turn away from the many problems involved in seeing and moving through the environment and broach the more complex problem of knowing and recognizing objects and places. Familiar objects have multiple types of significance to the individual: they may be analysed as structures, as tokens of larger object classes and as landmarks. These three aspects of object recognition are selectively affected by neurological disorder (see Table 17.4).

Object recognition: Object structure

If an object is known or familiar, then the perceptual processing system allows the individual to go beyond the information given and 'clean up' partial, degraded or distorted visual information. The individual perceiver may not have to know in advance that the subject of perception belongs to a known category; rather, noisy input is referred via perceptual processing routines to a range of possible candidate objects. Such operations are thought to depend on stored knowledge of objects. If these processes

Table 17.4 Disorders of addressed representations

Deficit	Impaired	Preserved	Neurological correlates
Structural agnosia (Apperceptive agnosia)	Discrimination, recognition or identification of 'noisy' or 'degraded' stimuli	Recognition of stimuli presented in a clear format	Right hemisphere Lateral Parietal lobe
Associative agnosia	Identification and recognition of common objects Categorization Utilization	Structural percreption Gesture, definition and naming from auditory or tactile input	Left hemisphere Occipito–temporal junction
Topographical amnesia for places	Place recognition Buildings and landmarks	Common object recognition across modalities Spatial learning	Right hemisphere Occipito–temporal regions
Topographical amnesia for routes	Spatial recognition and spatial learning	Common object recognition across modalities Place and landmark recognition	Right hemisphere Occipito–temporal regions

do not result in a solution to the perceptual problem, if the structure is 'unknown', then information-extraction may be restricted to stimulus-specific codes or referred to the 'next likeliest' candidate structure by inference and analogy (Rock, 1973).

Clinical characteristics: Patients with damage to the posterior sectors of the right hemisphere may have particular problems in recognizing familiar objects if they are presented in a distorted or visually 'noisy' format. Examples of such tasks include unusual-view photographs, fragmented line drawings and overlapping letters. Such puzzle pictures may also present difficulties for normal subjects and very often require a 'second look' in order to interpret them. Analyses of this type of perceptual task have either emphasized the role of directly activated stored knowledge (e.g. Hinton, 1981; Warrington and James, 1988) or else have considered the role of 'normalizing' transformations on the visual input (e.g. Humphreys and Riddoch, 1987). The neurological correlates of 'unusual view' and 'fragmented picture' perception appear to implicate the lateral parietal regions of the right hemisphere (areas 39 and 40 according to the Brodmann anatomical map). These regions may be unique to the human brain.

Processing substrata: The 'direct access' and 'normalizing transform' interpretations of unusual view and fragmented stimulus processing actually epitomise the extreme independence and interactionist positions with regard to 'what' and 'where' information in object recognition. On the direct activation interpretation, features that are available in the input array directly trigger stored structural descriptions of objects. These structural descriptions are composed of object features together with information about their relative position. (It has been demonstrated computationally that input consisting of partial featural information may be sufficient to access a stored structural description). The triggering of stored knowledge can be cross-checked interactively with other information – such as contextual cues. By contrast, on the normalizing transform interpretation, the stimulus is considered *a priori* as a member of a particular class of canonical volumetric structures – such as a generalized cone (e.g. Marr, 1982). Transforming operations (such as mental rotation) are carried out on the visual input which can then be matched against stored representations in which objects are represented in an object-centred spatial framework (e.g. Humphreys and Riddoch, 1987).

These hypotheses should, in principal, be straightforward to distinguish, but the evidence thus far is equivocal. Indeed, it has been suggested that both options may be available (e.g. Humphreys and Riddoch, 1984). On the transforming operations hypothesis, failure in deriving object structure when the target is rotated into the picture plane arises because of a basic spatial deficit. Consistent with this position, Layman and Greene (1988) found that matching of nonsense stick figures and matching of real objects across rotations were *both* adversely affected by right hemisphere strokes, and Humphreys and Riddoch (1984) found that rotation of object photographs in the picture plane hindered object recognition in affected patients. This evidence must somehow be reconciled with findings of a dissociation between spatial operations and the perception of degraded and fragmented material (e.g. Paterson and Zangwill, 1944). Furthermore, there is no evidence that affected patients have particular difficulties with foreshortening rotations of objects as compared with those that maintain the principal axis of an object but occlude distinctive features (Warrington and James, 1988).

The role of spatial codes and spatial operations in deriving object structure remains uncertain. Indeed, the evidence for the existence of a stored vocabulary of object structures or stored structural descriptions is lacking except in the very special case of face stimuli.

Object recognition: Object meaning

The derivation of object meaning appears to be dependent on neural systems which are quite separate from those required for computing object structure. Patients with 'visual associative agnosia' are usually diagnosed because they have problems in naming visually presented real objects or object pictures. Their problem seems to go beyond that of labelling and implicates object recognition itself. Not only do such patients fail to name objects, they also have problems in demonstrating their use and in classifying them as members of a particular category (e.g. items of cutlery v. items of stationery). At the same time, the patient may have no difficulty with the task of tactile object identification, and may be able to carry out any of the tasks they have previously failed if the stimuli are presented via touch. Two cases whom I studied in collaboration with Elizabeth Warrington showed problems in day-to-day life on tasks such as finding the appropriate objects to set the table and in selecting the appropriate clothing to wear. These tasks could be performed but were slow and laborious, presumably because the patients had to cross check their visual information with tactile knowledge (McCarthy and Warrington, 1986; 1990).

The disorder may also be material specific – affecting small man-made objects far more than knowledge of the built environment. Patients recognize where they are and do not get lost. Face recognition may also be quite normal. In its pure or selective form this loss of object recognition is associated with damage to the left hemisphere (or the hemisphere dominant for language). Within the left hemisphere the critical site seems to lie within the regions identified as part of the 'what' processing system (see Table 17.4).

Processing substrata: This semantic level of object recognition appears to be separate from (and probably organized in parallel with) structural analysis. In the absence of visual recognition the integrity of structural processing is often difficult to establish quantitatively – the patient cannot name, cannot pantomime or match objects and the number of tests that are appropriate for evaluating perception is therefore very limited. In the two cases we examined it was possible to demonstrate good performance on tests requiring subjects to parse known objects from noisy and degraded arrays, even to the level of tracing the outline of fragmented stimuli and detecting 'real' objects from amongst perceptually confusing distracters. The inference of independence between structural and semantic systems is further supported by evidence for a double dissociation in group studies (De Renzi et al., 1969).

Topographic orientation

Navigation in the environment and finding one's way around locomotor space seems to impose very specific demands and requires very particular types of representational knowledge. This knowledge can be lost very selectively. In topographical amnesia, patients may get lost in once familiar places such as their own homes or within their own immediate neighbourhood. Topographical amnesia is differentiated from general or global memory disorders. Patients are not simply without memory, rather they have specifically lost the ability to find their way around and typically the disorder affects both old memories and new learning. The clinical evidence suggests there is a major distinction between the recognition of *places* and the ability to navigate. The first class of deficit is often considered as a highly specific agnosia (or object recognition disorder). The latter problem in navigation is viewed as a deficit in 'spatial' learning.

Places: Pallis (1955) described a patient who had particular problems in recognizing places. The profound nature of his disorder is shown by the following quote: 'My reason tells me that I must be in a certain place yet I don't recognise it. It all has to be worked out each time. It's not only the places that I knew before all this happened that I can't remember. Take me to a new place now, and tomorrow I couldn't get there myself.' He was less certain of his surroundings than a blind acquaintance and also commented that when travelling on a bus the blind man had offered to tell him when he had reached his destination! Pallis' patient had intact recollection of how to proceed from one place to another along a map; it is reported that he could outline on a plan of Cardiff the appropriate route for getting from the hospital to the railway station or football stadium. His knowledge of routes was spared but failure to recognize places meant that performance of the actual task was impaired.

Can failure in the recognition of places shed any light on the fundamental questions of how 'landmarks' are coded? Very recent evidence from a patient I have studied in collaboration with John Hodges and Jon Evans suggests that non-verbal aspects of place recognition may require memory for feature juxtapositions and conjunctions. SE is a case with gravely impaired place recognition but relative sparing of spatial memory.[7] He has difficulty in learning new places, and in recognizing and navigating around environments that were very familiar to him before his illness. For example, he reported that he became lost and did not recognize where he was in his home town unless he was able to get to a single well-known street of shops; once he had found this place he could navigate home. Our tests demonstrated that SE failed to recognize famous buildings and landmarks. For example, he confused the Leaning Tower of Pisa with the Empire State building and thought that the Statue of Liberty might be associated with Rome or Greece. Despite this failure in recognition, he was able to learn a 16-choice stylus maze at a normal speed and also learned spatial sequences normally. He could recall the direction and number of turns on a route and indicate their approximate distance on a plan. Moreover, he was able to draw a map of the streets around his home and could recall the spatial layout of his local town centre and the names of the main streets.

Detailed probing of SE's knowledge of his home revealed excellent recall of the floor plan and the positioning of furniture. However, he was quite unable to recall any information about the colour or pattern of walls, floor covering or curtaining, even though he could and did volunteer information about where the decoration was damaged and in need of replacement. Tests of pattern learning revealed a very marked impairment. Our tentative interpretation is that SE's failure of 'place' recognition is attributable to a more fundamental problem in learning and recognizing those critical conjunctions of sensory attributes such as colour and texture that are necessary so that one place or one building can be differentiated from others. This core deficit in non-verbal learning may well account for a substantial part of SE's difficulty in topographic orientation.[8]

Fortunately, SE is able to make use of verbal cues in compensating for his disorder. The names of shops, roads and unique and distinctive easily labelled objects provide him with some alternative means of locating himself in his environment. This compensatory strategy does have its limitations as the following anecdote illustrates. On one occasion SE became very confused when recalling a test route around the hospital. On later questioning he admitted that the problem arose because he had coded a particular turn by the presence of somebody sitting in a chair. Both chair and person had moved by the time SE was tested and as a consequence he did not know where to turn!

Buildings and places are 'special' and are 'recognized' in a number of particular and distinctive ways. The subject has to

1 Recognize the object as an examplar of a particular class (e.g. a church, a pub, a college).
2 Recognize the building itself as a specific instance (e.g. St Paul's).
3 Make use of buildings or other landmark tokens to index the current location (e.g. outside St Paul's, i.e. therefore in the city of London; cf. O'Keefe and Nadel, 1978).

4 Retain an episodic record of places so that
   (a) one has a record of approach to one's goal (e.g. St Paul's followed by Fleet Street = westerly direction in London).
   (b) one can construct other complimentary autobiographical indices based on 'where' information (e.g. I met John outside St Paul's and Tim at University College).

SE's disorder appears to arise at level one or two and has effects right through to level four. In terms of the complexity of information processing involved in place recognition and the close links between such recognition and behavioural guidance, perhaps only the human face makes similar information processing demands. Indeed, deficits in face recognition and place recognition are frequently observed in the same patient (as in Pallis' case and in SE's). The two deficits arise from damage to adjacent cerebral regions suggesting that they may require similar (neural) computational substrates.

Navigation: A rather different type of disorder giving rise to topographic disorientation was shown by the patient MA studied by De Renzi et al. (1977). She was able to recognize places and landmarks – but could not recall how to find her way around. During her time in hospital 'she was not self confident and looked hesitantly to left and right in search of some familiar landmark. . . If she went downstairs to the ground floor, on her return to the first floor she was uncertain whether she had to turn to the right or the left in order to come back to the female ward.' The patient's problem was in knowing which way to turn or navigate on the basis of recognized places. MA was also very slow in learning a spatial maze and so De Renzi and his colleagues plausibly interpreted her deficit in terms of a failure of long term spatial memory. However, failure of spatial recall for locomotor routes does fractionate from the ability to learn and recall a maze (Habib and Sirgu, 1987). Habib and Sirgu describe a patient with problems in spatial navigation. They report that he recovered his ability to navigate by sight in a light aircraft and could learn paper-and-pencil mazes *prior* to being able to locate himself in his town. They suggested that in his case there was a specific disorder of locomotor space. Cases such as this make it likely that spatial navigation is based on representations at multiple scales. However, the exact bases of these deficits requires further elucidation.

Physiological substrata: The site of damage implicated in the class of disorders that we call topographical amnesia may come as something of a surprise to those who would like to draw a clear distinction between the parietal processing of space and the temporal lobe processing of objects. Both of the well documented topographic amnesias (i.e. for places and routes) are in fact caused by damage centred on the major area considered to be critical for 'object' recognition – the inferior occipito–temporal junction. Although the hippocampus, the region highlighted by John O'Keefe and Lyn Nadel (1978) as being the substrate of a cognitive map, has been normal or uninjured in a number of topographic amnesic patients (e.g. Landis, et al., 1986; Habib and Sirgu, 1987), the major cortical input to the hippocampus has been disrupted in all. Furthermore, evidence from patients such as SE suggests that the damage to the hippocampus itself may be critical in some cases.

The cognitive representation of navigable locomotor space may have at least as much in common with the representation of object structure and the processing of object meaning as it has with representing 'space'. The acquisition of a cognitive representation of a zone or region may in some respects be comparable to acquiring a structural description of an object. Memory for a path may require spatial, configural information. However, memory for a map may require establishing those types of invariance and higher-order constraint that are also characteristic of the structural descriptions of objects (e.g. O'Keefe and Nadel, 1978; Presson et al., 1989; Campbell, chapter three of this volume).

Summary: Of the different levels of information processing considered in this section only the semantic stage of object recognition seems to 'fit' the neat packaging of separate 'what' and 'where' processing. In the critical case of topographical memory the neuroanatomical evidence challenges the hypothesis put forward by writers such as Ungerleider and Mishkin (1982): disorders of topographical orientation whether for places or for routes are associated with damage in the '*what*' system. This raises the possibility that failures of spatial learning observed following parietal damage are really attributable to major disorders of assembled processing rather than to something as abstract as 'spatial' knowledge.

What and where or addressed and assembled?

This chapter has reviewed evidence for some sub-types of representation that are required for processing objects and space. The evidence which has been surveyed seems to fit just as conveniently within a partitioning of cerebral systems into 'assembled routines' and 'addressed representations' as it does into a 'what' system and a 'where' system. Indeed, the evidence from the topographical amnesias seems to offer a direct challenge to the classical view, but is consistent with the information processing analysis presented in Figure 17.2.

I have suggested that many of these assembled routines or procedures are mediated via a distinct information processing sub-system. As emphasized by Gibson, there is no need to invoke access to stored representations based on past learning and experience in order to give an account of one's basic visuo–motor interactions with objects. Neither do we need to invoke a 'sketch' that provides a cognitive buffer between the observer and the ecosphere in which he or she operates (as proposed by Marr, 1982). Temporary memory representations may have a role – but in monitoring the progress of actions rather than as an interface with the environment. Visuo–motor interactions can be supported by the extraction of invariant properties of the input display (such as the rate of change in a texture gradient) and their direct co-ordination with somatic and vestibular/gravitational cues. As anticipated by Gibson, and suggested by most recent neurobiology, this type of computation appears to be 'hard wired' into the neural processing apparatus of the brain. There is abundant independent evidence for the critical role of temporal lobe structures in setting up memories: addressed representations. If we are to partition the brain into anatomically defined sub-processors, it seems at least as plausible to make the division according to whether the information required is stored in memory (places, routes and objects) or whether it has to be computed on line (visuo–motor routines and procedures). Adoption of the alternative convention also has a theoretical payoff in that it precludes a premature commitment to the type of *cognitive* information mediated by a neurobiological system.

Postscript: Cognitive control, interlacing what and where

The philosophical analysis of object perception has given at least equal weight to object permanence as to the problems of image categorization/recognition. The analysis of those attributes that are constitutive of material objects is at least as important as the fancy image processing procedures which form the central concern of cognitive neuroscientists (see Part II; chapters three and ten). Is there a means of bringing these approaches together? The most relevant linkages have been offered by cognitive psychologists who take conceptual structure and problem solving as their starting point for inquiry – rather than those whose primary concern is with the identification (fractionation and dissociation) of processing sub-systems and broad processing domains.

Such cognitive accounts are pitched at the level of informational content or 'virtual' structures rather than informational systems. At the level of content it is possible to analyse the range of possible expectancies computed by an organism when confronted by a physical display or dynamic system. So we can ask whether a subject expects a stimulus to be displaced to point x given initial motion y; which constraints are in effect in determining whether a subject expects a particular type of displacement to yield a particular transformation in the image; whether occlusion is equivalent to obliteration; whether the subject's expectancies are in accord with general physical laws such as gravity, etc. (see chapters four and five). The organization and programming of expectancies can be considered in terms of virtual systems: cognitive structures and systems which are themselves the products of interaction between computational elements or sub-systems.

The evidence for intention and attention interacting with processing at the *single cell* level is now strong – at least for systems subsequent to the primary visual cortex (V2 onwards). For example, neurons in area seven and five (see sections on disorientation and optic ataxia, above) show very pronounced and significant modulation effects. If stimuli have motivational significance then responses to the target are enhanced in cells which are 'triggered' by stimuli in the target location (see, for example, Mesulam, 1981; Anderson, et al., 1985). The phenomena of enhancement have now been documented over wide regions of cerebral cortex – and subcortex – and they are by no means unique to posterior parietal neurons. They have even been documented in V4 which is supposedly part of the 'non-spatial' system. If the activity of individual cells can be modulated by expectancy and by motivation then the relationship between different neural processing systems may also be modifiable in real time.

The performance of many (if not all) every-day tasks requires the synchronized ordered coupling of multiple processing systems. Some neural processing configurations may have privileged connectivity or synchronization (as in the case of the M and P streams of early vision). However, the human cognitive system has a tremendous capacity for re-scheduling and reconfiguring itself according to requirements. Our expectancies guide the way in which information is integrated and prioritized (e.g. Allport, 1989). There are some clues as to how these higher-order operations are modulated, but the evidence is tenuous and preliminary (e.g. Shallice, 1988). For the adult there appears to be a degree of flexibility in the programming of expectancies about position and identity. For example, we do not *have* to resolve spatial information prior to resolving target identity. Target discrimination and identification are speeded both by cues to the location of a stimulus *and* by the contingent probability of finding a particular target in a specific location. Having detected a particular object we might also have some better idea of where we are looking or where we are located (e.g. Lambert and Hockey, 1986; Kingstone and Klein, 1991). It therefore seems that under some circumstances 'where' may be dependent on 'what', but under others 'what' may be dependent on 'where'.

Much of learning entails a 'ceasing to be surprized' and an abstracting of expected regularities over longer spells of time and over larger spatial windows. Even the very young human cognizer has expectancies about the way the world is, and this affects the way the world is perceived and structured. These expectancies become more and more sophisticated over time until by the age of around seven, children have established expectations for the specific objects in particular contexts and the spatial organization of specific objects in complex environments (e.g. Mandler, 1979). Expectancy plays a crucial role in daily perception, indeed cognisance and perception would be extremely impoverished were it not for the complimentary constraints of expectancies about the way objects are and how the world is.

Under normal conditions there are multiple ways in which the individual's expectancies can be organized or configured. We are flexible and can change our expectations in accordance with past experience and in the light of immediate demands and requirements. Current theorizing suggests that this level of information processing is mediated via meta-programmes (sometimes termed *schemas*) that are made up of subroutines 'calling up' specific processing systems as and when required (see, for example Mandler, 1979).[9] To take a concrete 'everyday' example; the schema of 'going to a restaurant' will have routines at the upper levels of the hierarchy that are as abstract as 'booking a table', 'driving to town' and 'dining'. Instructions to call up addressed representations for discriminating a soup spoon or dessert spoon and very specific assembled commands for reaching in order to retrieve that object from its expected location will occupy lower levels in the control hierarchy. Schema analyses are also applicable to the content of task-specific anticipations, such as those concerning position and identity. For example, programmes can take account of task-specific *a priori* probabilities such as the likelihood that position is a cue to identity – or that identity is a cue to position (Kingstone and Klein, 1991).

Breakdown in information processing may arise because of loss of the information required to fill the slot in one sub-routine of the schema. Naive interpretations of neuropsychological deficit are often predicated on the assumption that all deficit falls into this category – so, for example, we can think of the agnosic or dysmorphopsic patient as having 'lost' the information necessary to recognize an object. Schema theory highlights the other major cause of failure, failures that arise because the *means* of programming a specific routine have become disordered. So the blindsight patients may not have lost some specific 'conscious ability to process form'. They have lost the ability to encode that information within a schema for explicit verbal identification. They may have retained the ability to encode that same information within a schema for visuo–motor interaction. Such failures of encoding may of course arise at the level of data, access to data or in the organization of programming resources.

These considerations probably underpin the very common observation that patients' disorders are often specific to particular contexts or settings or to particular sets of task demand. So a patient may seem to fail to recognize a toothbrush or demonstrate its use when in the examiner's office, but proceed to the bathroom and brush their teeth normally when getting ready for bed (e.g. Heilman and Rothi, 1985). In the case cited above the basic action system and recognition system associated with 'toothbrushes' appears intact, but the patient has limited or restricted means by which this information may be addressed. Whilst the representation may have been 'lost' to the schema that was active in the examiner's office or under testing conditions, it became available when embodied in a more familiar routine and well established schema. The knowledge was not lost, merely unavailable. Instances of contextual relativity are not confined to object recognition. In one of the earliest discussions of the impact of spatial impairments on 'everyday life' it was shown that the subject was able to cope with the demands of a responsible job which seemed to require a considerable amount of spatial knowledge despite terrible failures on clinical tests (e.g. inability to discriminate orientation or even to count blocks in a three-dimensional construction). The subject continued in practice as a surgeon (McFie, Piercy and Zangwill, 1950). Such observations should stand as a warning to those who have tended to assume that reducing degrees of freedom and operating under controlled conditions necessarily results in a transparent relation between the virtual system as configured by the schema and the computational sub-systems embodied in neural architecture.

Conclusion

This chapter has given a very superficial survey of the cognitive neuroscience of 'what' and 'where' processing. I have only scratched the surface of the problem by considering the dissociations, and neuronal systems which are located 'upstream' in the parietal and temporal lobes. These systems seem to provide a range of processed perceptual attributes and information about objects. The information can be combined within virtual systems programmed by schemata and recombined according to the needs of the subject – and the demands of the moment and the setting of the action. However, the higher order control processes of expectancy, attention, integration, modulation and prioritizing of information require the operation and integrity of further systems which have only barely been explored (e.g. Goldman-Rakic, 1988).

Neuroscientific models of the relationship between spatial processing and object processing have sometimes seemed content to assume that having identified the deficit and located the system one has explained the process. As we have seen, current evidence only gets us part of the way towards understanding image processing and needs to be complimented by an approach that makes more of the content of expectancies and subject-driven configuration of their own informational systems. Flexibility in organization is possible and may have real practical significance. Analysing the dynamics and content of such organization may provide us with a key to unlock those complex philosophical issues at the core of our perception and interaction with the material world.

ACKNOWLEDGEMENTS

This chapter was prepared with the aid of a grant from the Leverhulme Trust to King's College, Cambridge and an MRC grant to myself. I am indebted to Bill Brewer, John Campbell, Naomi Eilan, John Hodges, Luke Kartsounis, Tony Marcel, John O'Keefe and Trevor Robbins who all took time to read and comment on earlier versions of this material.

NOTES

1 I am grateful to Naomi Eilan for aid in clarifying these conceptual ideas.
2 The same basic tripartite division has been successfully applied by many writers to a disparate range of cognitive skills (such as reading, writing, language production and face recognition; see McCarthy and Warrington, 1990).
3 The cortical 'retinotopic' field is defined in terms of the visual field of stimulation, *not* according to the individual retina receiving this stimulation. The optic nerve from each eye is like a multi-cored communication fibre with fibres from adjacent retinal ganglion cells being located next to each other. The optic nerve divides at the level of the optic chiasm and the fibres cross over to map two different hemi-retinas onto one side of the brain. At a general level we can say that the nasal side of the retina of the right eye and the temporal (lateral) side of the retina of the left eye both 'feed' visual processing areas in the left hemisphere. The nasal retina of the left eye and the temporal retina of the right eye send information to the right hemisphere, therefore the brain codes sensory input in retinal co-ordinates but *occular* co-ordinates are lost *very* early in processing (i.e. in the first few thousandths of a second).

   Clinical testing for visual field defects has typically involved the subject being asked to detect the presence or absence of stimuli with different physical properties whilst oriented vertically with eyes, head and trunk aligned. Whether more complex procedures with head or body rotation would reveal high-order interactions (such as an egocentric 'rotation' of the field) remains unestablished for attributes such as colour, form and luminance contrast. Egocentricity *has* been demonstrated in certain types of position judgement and detection task in patients whose disorders have been labelled as 'neglect'. (I am grateful to Tony Marcel for highlighting these definitional problems.)
4 These sub-systems are graded phylogenetically. The SC system is the oldest and has homologies in fish and reptiles as well as in most mammals. The M system is common to all primates and the P system is the most recent and is only found in advanced primates.
5 In normal subjects reaching towards a gap or reaching for an object is a co-ordinated motor act which involves the anticipatory orientation and positioning of the hand. Jeannerod and his colleagues have shown that such pre-shaping occurs very early in the overall reaching 'programme' in normal subjects (e.g. Jeannerod, 1988).
6 This perspective on visual disorientation owes much to discussions with John Campbell and to views expressed in his unpublished seminars and papers presented to the Spatial Representation Group in 1990–1.
7 SE's difficulties arose as a consequence of an encephalitic illness that destroyed the right hippocampal formation and amygdala.
8 At the time of writing we are planning further investigations to establish whether SE's remaining topographic knowledge is based on maps or routes (O'Keefe and Nadel, 1978).
9 Mandler (1979) offers the following definition of a schema: 'a spatially and/or temporally organised structure in which the parts are connected on the basis of contiguities that have been experienced in space or time. A schema is formed on the basis of past experience with objects, scenes or events and consists of a set of (usually conscious) expectations about what things look like and/or the order in which they occur. The parts of units of a schema consist of a set of variables or slots, which can be filled, or instantiated, in any given instance by values that have greater or lesser degrees of probability of occurrence attached to them. Schemata vary greatly in their degree of generality – the more general the schema, the less specified or the less predictable, are the values that may satisfy them' (p. 263).

REFERENCES

Albert, M. L., Reches, A. and Siverberg, R. 1975: Hemianopic colour blindness. *Journal of Neurology, Neurosurgery and Psychiatry*, 38, 546–9.

Allport, A. 1989: Visual attention. In M. I. Posner (ed.), *Foundations of Cognitive Science*. Cambridge, Mass: MIT, 631–82.

Anderson, R. A., Essik, G. K. and Siegel, R. M. 1985: Encoding of spatial locations by posterior parietal neurons. *Science*, 230, 456–8.

Athènes, S. and Wing, A. M. 1989: Knowledge directed co-ordination in reaching for objects in the environment. In S. A. Wallace (ed.), *Perspectives on the co-ordination of movement*. North-Holland: Elsevier Science Publishers, 285–301.

Cowey, A. 1982: Sensory and non-sensory visual disorders in man and monkey. *Philosophical Transactions of the Royal Society*, B 298, 3–13.

Cowey, A. and Stoerig, P. 1991: The neurobiology of blindsight. *Trends in Neurological Sciences*, 14, 140–5.

Davidoff, J. and Warrington, E. K. 1992: A dissociation of shape discrimination and figure ground perception in a patient with normal acuity. *Neuropsychologia*, in press.

De Renzi, E. 1983: *Disorders of Space Exploration and Cognition.* Chichester: Wiley.

De Renzi, E., Faglioni, P., Villa, P. 1977: Topographical amnesia. *Journal of Neurology, Neurosurgery and Psychiatry*, 40, 498–505.

De Renzi, E., Scotti, E. and Spinnler, H. 1969: Perceptual and associative disorders of visual recognition: Relationship to the site of lesion. *Neurology*, 19, 634–42.

Donnelly, N., Humphreys, G. W. and Riddoch, M. J. 1991: Parallel computation of primitive shape descriptions. *Journal of Experimental Psychology, Human Perception and Performance*, 17, 561–70.

Efron, R. 1968: What is perception? *Boston Studies in the Philosophy of Science*. New York: Humanities Press Inc., 137–73.

Evans, G. 1982: *The Varieties of Reference*. J. McDowell (ed.), Oxford: Clarendon Press. New York: University Press.

Farah, M. 1990: *Visual Agnosia*. Cambridge Mass: MIT.

Frisby, J. 1979: *Seeing: Illusion, Brain and Mind.* Oxford: Oxford University Press.

Georgopoulos, A. P. 1987: Cortical mechanisms subserving reaching. In *Motor areas of the cerebral cortex*: Ciba foundation symposium 132. Chichester: Wiley-Interscience, 125–31.

Gibson, J. J. 1966: *The Senses Considered as Perceptual Systems.* Boston: Houghton Mifflin.

Gibson, J. J. 1979: *The Ecological Approach to Visual Perception.* Boston: Houghton Mifflin.

Goldman-Rakic, P. S. 1988: Topography of cognition: Parallel distributed networks in primate association cortex. *Annual Review of Neuroscience*, 11, 137–56.

Goodale, M. A. and Milner, A. D. 1992: Separate visual pathways for perception and action. *Trends in Neurosciences*, 15, 20–5.

Habib, M. and Sirgu, A. 1987: Pure topographic disorientation. A definition and anatomical basis. *Cortex*, 23, 73–85.

Heilman, K. M. and Rothi, L. J. 1985: Apraxia. In K. M. Heilman and E. Valenstein (eds), *Clinical Neuropsychology*. New York: Oxford University Press.

Hildreth, E. and Ullman, S. 1989: The computational study of vision. In M. I. Posner (ed.), *Foundations of Cognitive Science*. Cambridge Mass: MIT, 631–82.

Hinton, G. E. 1981: A parallel computation that assigns canonical object-based frames of reference. *Proceedings of the International Joint Conference on Artificial Intelligence*, Vancouver, Canada.

Holmes, G. 1918: Disturbances of visual orientation. *British Journal of Opthalmology*, 2, 449–68.

Holmes, G. and Horax, G. 1919: Disturbances of spatial orientation and visual attention with loss of stereoscopic vision. *Archives of Neurology and Psychiatry*, 1, 385–407.

Humphreys, G. W. and Riddoch, J. 1987: *To See But Not to See: A case study of visual agnosia*. London: Lawrence Erlbaum.

Humphreys, G. W. and Bruce, V. 1989: *Visual Cognition: Computational, experimental and neuropsychological perspectives*. Hillsdale, NJ: Lawrence Erlbaum Associates.

Humphreys, G. W. and Riddoch, M. J. 1984: Routes to object constancy: Implications from neurological impairments of object constancy. *Quarterly Journal of Experimental Psychology*, 26A, 385–415.

Jeannerod, M. 1988: *The Neural and Behavioural Organisation of Goal Directed Movements*. Oxford: Clarendon Press.

Jeannerod, M. 1987: The posterior parietal area as a spatial generator. In M. Jeannerod (ed.), *Neurophysiological and Neuropsychological Aspects of Spatial Neglect*. North Holland: Elsevier Science Publishers.

Kartsounis, L. D. and Warrington, E. K. 1991: Failure of object recognition due to a breakdown of figure ground discrimination in a patient with a normal acuity. *Neuropsychologia*, 29, 969–80.

Kartsounis, L. and Warrington, E. K. 1992: *Neuropsychologia*. In press.

Kingstone, A. and Klein, R. 1991: Combining shape and position expectancies: Hierarchical processing and selective inhibition. *Journal of Experimental Psychology, Human Perception and Performance*, 17, 512–19.

Lambert, A. and Hockey, G. R. J. 1986: Attention and performance within a multi-dimensional visual display. *Journal of Experimental Psychology, Human Perception and Performance*, 12, 484–95.

Landis, T., Cummings, J. L., Benson, F. and Palmer, P. 1986: Loss of topographic familiarity: An environmental agnosia. *Archives of Neurology*, 43, 132–6.

Layman, S. and Greene, E. 1988: The effect of stroke on object recognition. *Brain and Cognition*, 7, 87–114.

Lee, D. N. 1980: Visuo–motor coordination in space–time. In G. E. Stelmach and J. Requin (eds), *Tutorials in Motor Behaviour*. Amsterdam: North Holland Publishing Company.

Livingstone, M. and Hubel, D. 1988: Segregation of form, color, movement and depth: Anatomy, physiology and perception. *Science*, 240, 740–9.

Mandler, J. 1979: Categorical and schematic organisation in memory. In C. R. Puff (ed.), *Memory Organisation and Structure*. New York: Academic Press.

McCarthy, R. A. and Warrington, E. K. 1986: Visual associative agnosia: A clinico-pathological study of a single case. *Journal of Neurology, Neurosurgery and Psychiatry*, 49, 1233–40.

McCarthy, R. A. and Warrington, E. K. 1990: *Cognitive Neuropsychology: A Clinical Introduction*. San Diego: Academic Press.

McFie, J., Piercy, M. F. and Zangwill, O. L. 1950: Visual spatial agnosia associated with lesions of the right cerebral hemisphere. *Brain*, 73, 167–90.

Marr, D. 1982: *Vision*. San Francisco: Greeman.

Mesulam, M. M. 1981: A cortical network for directed attention and unilateral neglect. *Annals of Neurology*, 10, 309–25.

Milner, A. D., Perret, D. I., Johnston, R. S., Benson, P. J., Jordan, T. R., Heeley, D. W., Bettucci, D., Mortara, F., Mutani, R., Terazzi, E. and Davidson, D. L. W. 1991: Perception and action in visual form agnosia. *Brain*, 114, 405–28.

O'Keefe, J. and Nadel, L. 1978: *The Hippocampus as a Cognitive Map*. Oxford: Oxford University Press.

Pallis, C. A. 1955: Impaired identification of locus and places with agnosia for colours. *Journal of Neurology, Neurosurgery and Psychiatry*, 18, 218–24.

Patterson, A. and Zangwill, O. L. 1944: Disorders of visual space perception associated with lesions of the right cerebral hemisphere. *Brain*, 67, 331–58.

Perenin, M. T. and Jeannerod, M. 1975: Residual vision in cortically blind hemifields. *Neuropsychologia*, 13, 1–7.

Perenin, M. T. and Vighetto, A. 1983: Optic ataxia: A specific disorder in visuomotor coordination. In A. Hein and M. Jeannerod (eds), *Spatially Oriented Behaviour*. New York: Springer, 305–26.

Perenin, M. T. and Vighetto, A. 1988: Optic ataxia: A specific disruption in visuomotor mechanisms. *Brain*, lll, 643–74.

Presson, C. C., De Lange, N. and Hazelrigg, M. D. 1989: Orientation specificity in spatial memory: What makes a path different from a map of the path. *Journal of Experimental Psychology: Learning, Memory and Cognition*, 15, 887–97.

Rock, I. 1973: *Orientation and Form*. New York: Academic Press.

Shallice, T. 1988: *From Neuropsychology to Mental Structure*. Cambridge: Cambridge University Press.

Treisman, A. M. 1988: Features and objects: The fourteenth Bartlett Memorial Lecture. *Quarterly Journal of Experimental Psychology*, 40A, 210–37.

Treisman, A. M. and Gelade, G. 1980: A feature integration theory of attention. *Cognitive Psychology*, 12, 97–136.

Ungerleider and Mishkin, M. 1982: Two cortical visual systems. In D. J. Ingle, M. A. Goodale and R. J. W. Mansfield (eds), *Analysis of Visual Behaviour*. Cambridge, Mass: MIT, 549–86.

Warrington, E. K. 1985: Visual deficits associated with occipital lobe lesions in man. Vatican symposium 'Pattern Recognition Mechanisms'. In C. Chagas, R. Gattas and C. Gross (eds), *Pontificae Academiae Scientiarum Scripta Varia*, 54, 247–61.

Warrington, E. K. and James, M. 1988: Visual apperceptive agnosia: A clinico–anatomical study of three cases. *Cortex*, 24, 13–32.

Watt, R. J. 1991: *Understanding Vision*. London: Academic Press.

Weiskrantz, L. 1980: Varieties of residual experience. *Quarterly Journal of Experimental Psychology*, 32, 365–86.

Weiskrantz, L. 1986: *Blindsight: A case study and its implications*. Oxford: Oxford University Press.

Weiskrantz, L., Warrington, E. K., Sanders, M. D. and Marshall, J. 1974: Visual capacity in the hemianopic field following a restricted occipital ablation. *Brain*, 97, 709–28.

Zeki, S. A. 1990: A century of achromatopsia. *Brain*, 113, 1721–78.

Zihl, J., Cramon, D., von, Mai, N. and Schmid, C. 1991: Disturbance of movement vision after bilateral posterior brain damage, further evidence and follow-up observations. *Brain*, 114, 2235–52.

Workshop participants

Janette Atkinson
Edoardo Bisiach
Oliver Braddick
Bill Brewer
Michele Brouchon
Peter Bryant
George Butterworth
John Campbell
Bill Child
Lynn A. Cooper
Adrian Cussins
Anthony Dickinson
John Duncan
Naomi Eilan
Richard Evans
Usha Goswami
Bruce Hood
Julie Jack
Roberta Klatzky
Adam Lowe
Roz McCarthy
Theresa McCormack
Nick Mackintosh
Ian McLaren
Tony Marcel
Michael Martin
Hugh Mellor
Andrew N. Meltzoff
Ruth Millikan
John O'Keefe
Christopher Peacocke
Julian Pears
Herbert Pick
Jim Russell
Mark Sainsbury
Gabriel Segal
Roger N. Shepard
Elizabeth Spelke
Michael Tye
Elizabeth Warrington

Index